Mathematical Models and Immune Cell Biology

Carmen Molina-París • Grant Lythe
Editors

Mathematical Models and Immune Cell Biology

Springer

Editors
Carmen Molina-París
Department of Applied Mathematics
School of Mathematics
University of Leeds
Woodhouse Lane
LS2 9JT Leeds, UK
carmen@maths.leeds.ac.uk

Grant Lythe
Department of Applied Mathematics
School of Mathematics
University of Leeds
Woodhouse Lane
LS2 9JT Leeds, UK
grant@maths.leeds.ac.uk

ISBN 978-1-4419-7724-3 e-ISBN 978-1-4419-7725-0
DOI 10.1007/978-1-4419-7725-0
Springer New York Dordrecht Heidelberg London

Library of Congress Control Number: 2011926883

Printed on acid-free paper

Springer is part of Springer Science+Business Media (www.springer.com)

a nuestros hijos, Federico y Sofía

Preface

The new experimental and imaging techniques of the past few decades allow us to observe, study and ponder the world of cells and molecules inside our bodies, stimulating the development of new mathematics and putting a new perspective on our view of our own world.

Whole new areas of immunological research are emerging from the analysis of experimental data, going beyond statistics and parameter estimation into what an applied mathematician would recognise as modelling of dynamical systems. Stochastic methods are increasingly important, because stochastic models are closer to the Brownian reality of the cellular and sub-cellular world.

This volume contains chapters on mathematical modelling, on immunology, and on mathematical modelling in immunology. Although there is a bias towards the adaptive immune system, and towards T cells in particular, the reader will find Chapters on dendritic cells, B cells and germinal centres.

We hope the list of abbreviations will help to indicate the type of research that is being carried out at the interface of mathematics and immunology.

This book would not have been possible without the passion and perseverance of Joseph Burns. We are extremely grateful for the support, assistance and patience of Andrea Macaluso, Jeffrey Ciprioni and Melanie Wilichinsky.

That the two of us are doing research in mathematical immunology at all is due to the influence of David Rand, Nigel Burroughs and Hugo van de Berg.

2011

Carmen Molina-París
Grant Lythe

Contents

Contributors

Asma Ahmed Department of Biochemistry, Indian Institute of Science, Bangalore 560012, India, asma.127@gmail.com

Graham Anderson Medical Research Council Centre for Immune Regulation, Institute for Biomedical Research, Medical School, University of Birmingham, Birmingham B15 2TT, UK, g.anderson@bham.ac.uk

Belen de Andrés Centro Nacional de Microbiología, ISCIII, Majadahonda, 28220 Madrid, Spain, bdandres@isciii.es

Becca Asquith Department of Immunology, Wright-Fleming Institute, Imperial College London, Norfolk Place, London W2 1PG, UK, b.asquith@imperial.ac.uk

José A.M. Borghans Department of Immunology, University Medical Center Utrecht, Lundlaan 6, Utrecht, The Netherlands, j.borghans@umcutrecht.nl

Rowann Bowcutt Faculty of Life Sciences, AV Hill Building, University of Manchester, Oxford Road, Manchester M13 9PL, UK, Rowann.Bowcutt@postgrad.manchester.ac.uk

Kevin Burrage The Oxford Computing Laboratory and Oxford Centre for Integrative Systems Biology, University of Oxford, Oxford OX1 2JD, UK, and
The Institute for Molecular Biosciences, The University of Queensland, Queensland, Brisbane, QLD 4072, Australia, kevin.burrage@comlab.ox.ac.uk

Jorge Carneiro Instituto Gulbenkian de Ciência, Oeiras, Portugal, nunosep@igc.gulbenkian.pt

Mario Castro Grupo Interdisciplinar de Sistemas Complejos (GISC), Grupo de Dinámica No Lineal, Universidad Pontificia Comillas, Madrid E-28015, Spain, mariocastro73@gmail.com

Daniel Coombs Department of Mathematics and Institute of Applied Mathematics, University of British Columbia, Vancouver, BC, Canada V6T 1Z2, coombs@math.ubc.ca

Isabel Cortegano Centro de Biología Molecular, CSIC-UAM, Campus, de Cantoblanco, 28049 Madrid, Spain, icortegano@isciii.es

Sheena Cruickshank Faculty of Life Sciences, AV Hill Building, University of Manchester, Oxford Road, Manchester M13 9PL, UK, sheena.cruickshank@manchester.ac.uk

Mark Day Department of Applied Mathematics, University of Leeds, Leeds LS29JT, UK, jhs5msd@leeds.ac.uk

Omer Dushek Centre for Mathematical Biology, University of Oxford, Oxford OX1 3LB, UK
and
Sir William Dunn School of Pathology, University of Oxford, Oxford OX1 3RE, UK, omer.dushek@path.ox.ac.uk

Jose Faro Edificio de Ciencias Experimentáis, Universidade de Vigo, Campus As Lagoas-Marcosende, 36310 Vigo, Spain,
and
Estudos Avançados de Oeiras, Instituto Gulbenkian de Ciência, Apartado 14, 2781-901 Oeiras, Portugal, jfaro@uvigo.es; jfaro@igc.gulbenkian.pt

Marc Thilo Figge Applied Systems Biology, Friedrich Schiller University Jena, Leibniz Institute for Natural Product Research and Infection Biology – Hans Knöll Institute, Beutenbergstrasse 11a, 07745 Jena, Germany, thilo.figge@hki-jena.de

Helen Fryer The Institute for Emerging Infections, The Oxford Martin School, Department of Zoology, University of Oxford, Oxford OX13PS, UK, helen.fryer@zoo.ox.ac.uk

Karina García-Martínez Center of Molecular Immunology, Havana, Cuba, karina@cim.sld.cu

María-Luisa Gaspar Centro Nacional de Microbiología, ISCIII, Majadahonda, 28220 Madrid, Spain, mlgaspar@isciii.es

Edwin D. Hawkins The Walter and Eliza Hall Institute of Medical Research, 1G Royal Parade, Parkville, Victoria 3052, Australia
and
Department of Medical Biology, The University of Melbourne, Parkville, Victoria 3010, Australia, Edwin.Hawkins@petermac.org

Stephen Hickling Peter Medawar Building for Pathogen Research, University of Oxford, South Parks Road, Oxford OX1 3SY, UK, stephen.hickling@lincoln.ox.ac.uk

Phillip D. Hodgkin The Walter and Eliza Hall Institute of Medical Research, 1G Royal Parade, Parkville, Victoria 3052, Australia
and
Department of Medical Biology, The University of Melbourne, Parkville, Victoria 3010, Australia, wellard@wehi.edu.au

Sharmili Jagtap Centro de Biología Molecular, CSIC-UAM, Campus de Cantoblanco, 28049 Madrid, Spain, Sharmili.Jagtap@dartmouth.edu

Eric Jenkinson Medical Research Council Centre for Immune Regulation, Institute for Biomedical Research, Medical School, University of Birmingham, Birmingham B15 2TT, UK, e.j.jenkinson@bham.ac.uk

William Jenkinson Medical Research Council Centre for Immune Regulation, Institute for Biomedical Research, Medical School, University of Birmingham, Birmingham B15 2TT, UK, w.e.jenkinson@bham.ac.uk

Kalet León Center of Molecular Immunology, Havana, Cuba, kalet@cim.sld.cu

Grant Lythe Department of Applied Mathematics, University of Leeds, LS29JT, UK, grant@maths.leeds.ac.uk

Shev MacNamara The Oxford Computing Laboratory and Oxford Centre for Integrative Systems Biology, The University of Oxford, Oxford OX1 2JD, UK
and
The Institute for Molecular Biosciences, The University of Queensland, Brisbane, QLD 4072, Australia, shev.macnamara@maths.ox.ac.uk

Miguel-Angel Rodríguez Marcos Centro de Biología Molecular, CSIC-UAM, Campus de Cantoblanco, 28049 Madrid, Spain

John F. Markham The Walter and Eliza Hall Institute of Medical Research, 1G Royal Parade, Parkville, Victoria 3052, Australia
and
Department of Medical Biology, The University of Melbourne, Parkville, Victoria 3010, Australia, jmarkham@wehi.edu.au

Angela McLean The Institute for Emerging Infections, The James Martin 21st Century School, Department of Zoology, University of Oxford, Oxford OX13PS, UK, angela.mclean@zoo.ox.ac.uk

Michael Meyer-Hermann Department of Systems Immunology, Helmholtz Centre for Infection Research (HZI), Inhoffenstr. 7, D-38124 Braunschweig, Germany, michael.meyer-hermann@helmholtz-hzi.de

Carmen Molina-París Department of Applied Mathematics, University of Leeds, Leeds LS29JT, UK, carmen@maths.leeds.ac.uk

Dipankar Nandi Department of Biochemistry, Indian Institute of Science, Bangalore 560012, India, nandi@biochem.iisc.ernet.in

Michal Or-Guil Systems Immunology Research Group, Institute for Theoretical Biology, Humboldt University, 10115 Berlin, Germany
and
Research Center ImmunoSciences, Charité, 10115 Berlin, Germany, m.orguil@biologie.hu-berlin.de

Rodney Phillips Peter Medawar Building for Pathogen Research, University of Oxford, South Parks Road, Oxford OX1 3SY, UK, rodney.phillips@ndm.ox.ac.uk

Katie Quinn Department of Applied Mathematics, University of Leeds, Leeds LS29JT, UK

Ana R. Sánchez-Archidona Centro Nacional de Microbiología, ISCIII, Majadahonda, 28220 Madrid, Spain, asarchidona@sanfordburnham.org

Nuno Sepúlveda Instituto Gulbenkian de Ciência, Oeiras, Portugal
and
London School of Hygiene and Tropical Medicine, Keppel Street, London WC1E 7HT, UK, nuno.sepulveda@lshtm.ac.uk

Natalia Serrano Centro Nacional de Microbiología, ISCIII, Majadahonda, 28220 Madrid, Spain

Andrew K. Sewell Department of Infection, Immunity and Biochemistry, Cardiff University School of Medicine, Henry Wellcome Building, Heath Park, Cardiff CF14 4XN, UK, sewellak@cardiff.ac.uk

Emily Stirk Department of Applied Mathematics, University of Leeds, Leeds LS29JT, UK, emily@maths.leeds.ac.uk

Hugo A. van den Berg Warwick Systems Biology Centre, University of Warwick, Coventry CV47AL, UK, hugo@maths.warwick.ac.uk

P. Anton van der Merwe Sir William Dunn School of Pathology, University of British Columbia, Oxford OX1 3RE, UK, anton.vandermerwe@path.ox.ac.uk

Cameron Wellard The Walter and Eliza Hall Institute of Medical Research, 1G Royal Parade, Parkville, Victoria 3052, Australia
and
Department of Medical Biology, The University of Melbourne, Parkville, Victoria 3010, Australia, wellard@wehi.edu.au

List of Abbreviations

Ab	antibody
ACE	abundance coverage estimator
AID	activation-induced deaminase
APC	antigen-presenting cell
APP	antigen presentation profiles
BCR	B cell antigen receptor
BM	bone marrow
BrdU	5-bromo-2'-deoxyuridine
CD	cluster of determination cell surface molecules
CDR	complementarity-determining region of the TCR
CFSE	carboxyfluorescein diacetate succinimidyl ester
CP	colonic patch
CpG	C–phosphate–G region of DNA
CSR	class switch recombination
cTEC	cortical thymic epithelial cells
CTL	cytotoxic T lymphocyte
DC	dendritic cell
DN	double negative
DP	double positive
EBV	Epstein-Barr virus
ER	endoplasmic reticulum
FDC	follicular dendritic cell
FSP	finite state projection
GC	germinal centre
GIT	gastrointestinal tract
HCMV	human cytomegalovirus
HIV	human immunodeficiency virus
HLA	human leukocyte antigens
HSC	haematopoietic stem cells
IEC	intestinal epithelial cell
IFN	interferon
Ig	immunoglobulin
ILF	isolated lymphoid follicle

IL-2	interleukin-2
IL-7	interleukin 7
IR	immune response
IS	immune system
ITAM	immunoreceptor tyrosine-based activation motif
ITIM	immunoreceptor tyrosine-based inhibition motif
MAP	methionine aminopeptidas
mTEC	medullary thymic nepithelial cells
LCD	limiting conditional probability distribution
ME	master equation
MHC	major histocompatibility complex
MHCI	class I MHC
MHCII	class II MHC
MLN	mesenteric lymph node
MZ	marginal zone
NK	natural killer cell
NP	nitrophenyl
ODE	ordinary differential equation
PCR	polymerase chain reaction
PDE	partial differential equation
PDF	probability density function
PGF	probability generating function
PLC	peptide loading complex
pMHC	peptide-MHC
PRR	pattern recognition receptor
QSD	quasi-stationary distribution
RAG	recombinase activating gene
SHM	somatic hypermutation
SHIV	simian-human immunodeficiency virus
SIV	simian immunodeficiency virus
SP	single positive
SSA	stochastic simulation algorithm
TAP	transporter associated with antigen presentation
TCR	T cell antigen receptor
TDL	thoracic duct lymphocyte
TNF	tumour necrosis factor
TREC	T cell receptor excision circle
V	immunoglobulin variable region

Chapter 1
Thymocyte Development

William Jenkinson, Eric Jenkinson, and Graham Anderson

Abstract T cell development within the thymus involves the dynamic interaction of thymocytes with a unique heterogeneous microenvironment formed predominantly by a three-dimensional network of specialized thymic epithelium. Multiple developmental checkpoints have been identified during thymocyte maturation, including selective recruitment to the thymus, $\alpha\beta$ versus $\gamma\delta$ T cell fate decisions, positive and negative selection and finally regulated egress of mature T cells from the thymus to the periphery. The controlled migration of thymocytes within the thymus ensures that developing T cells undergo a series of tightly-regulated interactions with stromal cells of the thymus, ensuring that, firstly, only cells bearing a functional TCR are selected for survival and onward differentiation via positive selection and, secondly, that autoreactive T cells capable of responding to self-antigens and causing autoimmune disease are deleted via negative selection. The stringent selection mechanisms enforced within the thymus are demonstrated by the fact that only 2–5% of thymocytes generated within the thymus mature to form naïve T cells capable of forming a functional, self-tolerant component of the peripheral adaptive immune system. Whilst many of the developmental processes occurring during thymocyte development have become elucidated over the past several years, systems directed towards modelling the dynamic migratory patterns and real-time cellular interactions will radically advance our understanding of how such pivotal cells of the immune system are generated.

Introduction

During T cell development, thymocytes undergo a programme of sequential interactions with stromal cells of the thymus, predominantly of an epithelial nature, providing essential developmental cues to ensure a carefully orchestrated

W. Jenkinson (✉)
Medical Research Council Centre for Immune Regulation, Institute for Biomedical Research, Medical School, University of Birmingham, Birmingham B15 2TT, UK
e-mail: w.e.jenkinson@bham.ac.uk

C. Molina-París and G. Lythe (eds.), *Mathematical Models and Immune Cell Biology*, DOI 10.1007/978-1-4419-7725-0_1, © Springer Science+Business Media, LLC 2011 maturational programme lasting 3–4 weeks [1]. Similar to B cells, thymocytes bear a cell surface receptor capable of recognizing antigen, inducing cellular activation and driving adaptive immune responses, although unlike B cells, T cells require presentation of antigen by MHC class I and II. The huge diversity of potential antigen that a single individual may encounter over the course of a lifetime requires a highly efficient and specialized mechanism to ensure the production of sufficient T cells, each bearing a receptor of a single specificity, to confer the ability to recognize and respond to such diversity of antigen. This variety of T cell receptors (TCR) is generated during thymocyte development via the ability of individual cells to rearrange multiple gene segments encoding the α and β-chains of the TCR in a random fashion. Whilst the capacity to generate TCR diversity via random gene recombination provides the capacity to produce a huge range of TCR specificities, estimated at 10^{15}, an intrinsic drawback of such a system is the potential to generate multiple non-functional TCR arrangements and also TCRs capable of recognizing the host's own tissue and driving potential autoimmune responses. It is therefore essential that a strict quality-control mechanism operates within organisms capable of generating lymphocytes bearing randomly generated receptors. In most vertebrates the thymus represents the site of selection, sorting the good from the bad and ensuring that only useful T cells progress to populate the periphery and contribute to immunological defence. As a result of this, while approximately 50 million thymocytes are generated within the thymus every day, only around 1 in 20 thymocytes survive the stringent selection events required to become a mature T cell.

The thymus until relatively recently, remained an organ of mystery. The essential function of the thymus in the generation of lymphocytes was established in the 1960s through pioneering research by Jacques Miller, demonstrating that neonatal thymectomy resulted in immunodeficiency and furthermore indicating that the thymus was a site where lymphocytes were selected and tolerance induced. Extensive research into thymocyte development and thymus biology have begun to uncover the nature of the mechanisms regulating T cell development and the cellular interactions involved in this process. In this chapter, an overview of thymocyte development will be provided summarizing the relationship between thymocyte and thymic microenvironment resulting in the generation of functional, self-tolerant T cells.

Thymic Microenvironments

The thymus represents a bilobed organ, located in the superior mediastinum of the thorax anterior to the outflow tract of the heart. The mature thymus shows a clear degree of compartmentalization, displaying both anatomically and functionally distinct regions of cortex and medulla (Fig. 1.1). A fibrous capsule surrounds the mature thymus, with a rich vascular network penetrating throughout the organ. The parenchyma of the thymus consists predominantly of a unique three-dimensional arrangement of epithelial cells providing a mesh-like network, within which developing T cells are able to migrate and receive essential signalling interactions.

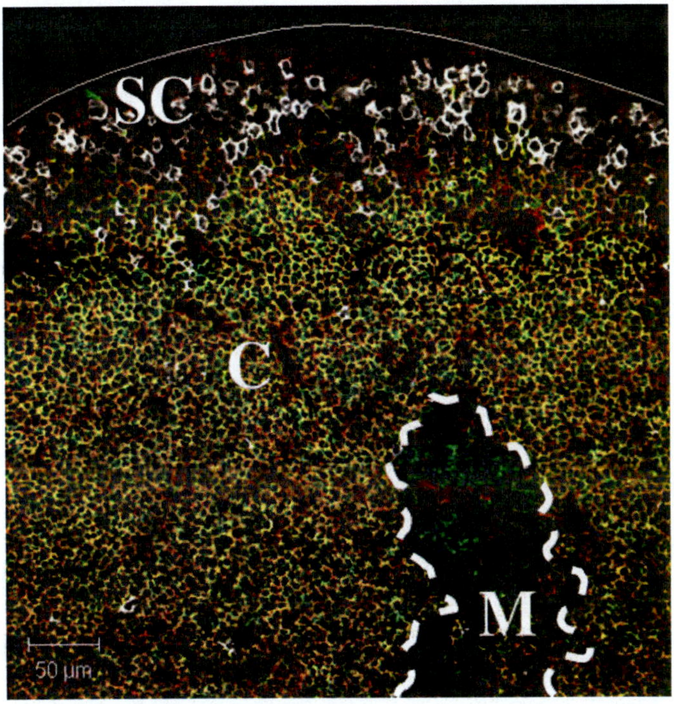

Fig. 1.1 The thymus displays a high degree of organization. Confocal analysis of mouse thymus for CD4 (*green*), CD8 (*red*) and CD25 (*white*) reveals the distinct thymic compartments of outer subcapsular region (SC), cortex (C) and inner medulla (M). Thymocytes at different stages of development differentially express the cell surface markers CD4, CD8 and CD25 and specifically localize to defined thymic regions. The subcapsular region is predominantly populated by immature $CD4^{-}8^{-}25^{+}$ thymocytes (*white*), forming a discrete rim to the thymus. The cortex is visualized as a densely packed region predominantly populated by $CD4^{-}8^{-}$ double positive thymocytes (appearing *yellow*), which undergo interactions with a network of interspersing cortical thymic epithelial cells. The inner thymic medulla consists of a more sparsely populated region formed by thymocytes at the single positive $CD4^{+}8^{-}$ (*green*) and $CD4^{-}8^{+}$ (*red*) undergoing maturational interactions with a combination of medullary thymic epithelium and dendritic cells

In addition to thymic epithelium and thymocytes, a heterogeneous mixture of additional cell types reside within the thymus, including stromal fibroblasts, thymic macrophages and dendritic cells, all of which play an important role in thymus biology.

During embryogenesis, the thymus develops as a bilateral outgrowth of endoderm, which buds off from a region of the foregut termed the third pharyngeal pouch. A series of reciprocal signalling interactions with neural crest-derived mesenchyme ensheathing the thymic rudiment leads to a cascade of transcription factor activity within the endoderm-derived thymus primordium, resulting in both outgrowth and differentiation of the thymus. Recent experimental data indicate that endoderm-derived thymic epithelium producing cortex and medulla is not only generated from a single embryonic origin, but that both cortical and medullary thymic epithelium derive from a bipotent progenitor population. Whether such a common thymic epithelial progenitor population persists within the adult thymus and whether such a population may represent a potential stem cell population with clinical manipulation implications remains an open area of research.

T cell development is not a cell autonomous process. The thymic microenvironment nurtures the development of thymocytes, providing a wide range of signalling interactions at defined developmental checkpoints in discrete anatomical locations. Signals provided by thymic stromal cells regulate commitment to the T cell lineage, regulation of proliferation, survival and importantly, selection of functional T cells that are self-tolerant on the basis of thymocyte T cell receptor capability and specificity. Due to the strict compartmentalization of thymic epithelial cells providing defined signals required at specific stages of thymocyte maturation, developing T cells display a highly ordered pattern of migration within the thymus regulating stepwise interactions and sequential maturation. Amongst the important signals provided by thymic epithelium to developing thymocytes, Delta-like 4 (Dl4) a ligand for the Notch signalling pathway is expressed by thymic epithelium. Signalling through the Notch receptor, expressed by thymus colonizing thymocyte precursors, is critical for commitment to the T cell lineage. Mice demonstrating defects in the Notch signalling pathway lack the capacity to generate thymic T cells, instead B cells are found to populate the thymus indicating that Notch plays an essential role in the commitment of lymphoid precursors to the T cell lineage [2]. In addition, thymic epithelium provides essential survival factors to developing thymocytes, including IL-7 and stem cell factor [3]. Interestingly, the requirement for IL-7 appears to be less important for thymocyte development within the fetal thymus compared to the adult thymus, highlighting subtle differences in the developmental requirements of T cells at defined temporal stages [4]. Of critical importance, thymic epithelium also expresses high levels of both MHC class I and MHC class II. MHC expression within the thymus plays an essential role in the selection of both CD4 and CD8 T cells through the capacity to test whether thymocytes with randomly generated T cell receptors are able to recognise MHC presenting self-peptides with sufficient affinity to ensure the TCR is functional but also ensure that those cells recognizing MHC–self-peptide complexes with too high affinity are deleted to avoid potentially autoreactive immune responses. Lack of MHC molecules on thymic epithelium results in a complete block in thymocyte development at the $CD4^{+}8^{+}$ double positive (DP) stage, as developing thymocytes are unable to test their randomly generated receptors and fail to receive TCR transmitted survival signals.

It is clear that thymic epithelium provides essential signals to developing thymocytes in a highly ordered stepwise manner. However, such signalling interactions are not simply a one-way process. Analysis of the thymi of mice demonstrating blocks in thymocyte development at defined stages demonstrates a clear corresponding defect in differentiation, organization and maintenance of cortical and medullary epithelium. Such findings highlight a complex reciprocal signalling mechanism operating within the thymus whereby thymocytes provide important signals, such as through the Lymphotoxin signalling pathway, that act to regulate the microenvironments that regulate their own survival and maturation [5].

The thymus is not an organ that is maintained at a constant size throughout life. A programmed decrease in the proportion of functional thymic tissue ensues generally from the onset of puberty, resulting in a consequent decrease in T cell output from the thymus of older individuals. The decrease in functional thymic tissue is termed thymic involution or atrophy. The precise mechanics of thymus involution remains unclear although several lines of evidence point to a pivotal role for sex steroids in this process [6]. During thymocyte development, approximately 95% of thymocytes are deleted due to production of either non-functional or auto-reactive T cell receptors. Such huge wastage may provide an explanation as to the reason for thymic involution, the necessity to maintain such a wasteful organ following the establishment of a peripheral T cell pool of sufficient size and diversity as to maintain immune protection may no longer be required. In support of this, thymectomy of young individuals does not lead to significant immunodeficiency, indicating that maintenance of a fully functional thymus with increasing age is not strictly necessary. However, improving healthcare and associated increases in lifespan of individuals may present problems in the future as elderly individuals may demonstrate a reduction in the T cell repertoire as a result of reduced thymic output. In addition, the advent of clinical therapies resulting in depletion of lymphoid compartments followed by bone marrow transplant presents a problem in terms of the reduced presence of functional thymic microenvironments capable of supporting rapid reconstitution of T cell compartments. Indeed, post-bone marrow transplant patients demonstrate a high susceptibility to opportunistic infection due to a severely reduced capacity to produce T cells. Identification of potential thymic epithelial stem cell populations and the precise signalling pathways involved in regulating thymic epithelial development and expansion currently present intense areas of research aimed at developing methods of regenerating functional thymic tissue and T cell reconstitution.

Colonization and Export from the Thymus

The thymus lacks a resident population of self-renewing haematopoietic stem cells (HSC). Unlike B cell development, where HSC exist within the same localized microenvironment as their B cell progeny, the remote location of the thymus therefore requires the selected recruitment of T cell progenitors throughout life to ensure a continuous supply of mature T cells. Recruitment of T cell progenitors to the thymus, whilst continuous throughout life, does not occur in a steady stream but rather entry is regulated in a periodically-gated manner [7]. The precise mechanisms regulating such stop-start thymic entry remains unclear, although evidence suggests that this may at least in part be regulated by the availability of space within intrathymic niches capable of nurturing developing thymocytes [8]. Following entry to the thymus, T cell precursors undergo a series of maturational events including bursts of proliferation resulting in a single thymus colonizing cell generating up to one million progeny [9]. The recruitment and specific entry of lymphoid progenitor cells into the thymus therefore forms the first hurdle that potential thymocytes must overcome in their journey to becoming a mature T cell.

Colonization of the thymus within the mouse appears to occur via two different mechanisms operating within temporally defined windows. At approximately day 11.5 of gestation, recruitment of circulating fetal liver and AGM-region haematopoietic cells to the embryonic thymus initiates and continues for a period of approximately 2 days. During the initial window of thymus colonization, the thymus remains an avascular structure, with entry of blood vessels into the thymus proper occurring after day 14 of gestation. As a result, specific recruitment of haematopoietic precursors occurs via the induction of haematopoietic precursor extravasation from capillaries in proximity to the thymus followed by migration through perithymic mesenchyme and finally entry to the thymus through the presumptive capsule. Attraction of haematopoietic cells to the early fetal thymus appears to be a highly specific process regulated by the action of a restricted set of chemokines comprising CCL25, CCL21 and CXLC12. Absence of expression of CCR7 and CCR9, the receptors for CCL21 and CCL25 respectively, on haematopoietic cells results in a severe reduction in pre-vascularized thymus colonization and an associated absence of T cell subsets normally generated from these cells [10]. The importance of CCL25 in mediating colonization of the fetal thymus is further highlighted in Nude mice. Nude mice display a mutation in the transcription factor Foxn1. Lack of functional Foxn1 expression results in a cell-autonomous defect in thymic epithelial differentiation after day 11.5 of gestation and a corresponding absence of thymic epithelial CCL25 expression. Whilst CCL25 expression is absent in Nude thymic epithelium, CCL21 is expressed normally in a Foxn1-independent manner in cells of the adjacent parathyroid. Importantly, whilst haematopoietic cells are attracted to the proximity of the Nude thymus, no cells enter into the thymus itself. Such findings suggest a differential two-step role for CCL21 and CCL25, whereby CCL21 attracts haematopoietic cells into proximity of the thymus, whereupon CCL25 mediates entry of cells into the epithelial microenvironment of the thymus itself. Whilst the diversity of chemokine expression within the embryonic thymus appears relatively restricted, analysis of the chemokine receptors expressed by thymus colonizing cells demonstrates a diverse heterogeneous mixture of cells, whether such cells demonstrate differential developmental potential remains unclear [11].

Within the mature adult thymus a rich network of blood vessels penetrate into the thymic tissue. Colonization of the thymus in vascularized adult stages occurs via extravasation of thymus colonizing cells at post-capillary venules located predominantly at the junction between cortex and medulla. The role of chemokines in the control of adult thymus colonization remains less clear than for the fetal thymus. Recent evidence however, suggests that CCL25-CCR9 signalling may also be involved in recruitment of cells to the adult thymus. In addition to chemokine action, the protein P-selectin expressed on the endothelium of thymic post-capillary venules binds the carbohydrate P-selectin ligand-1 (PSGL1) expressed by thymus colonizing cells, absence of either receptor or ligand results in a significantly reduced entry of cells into the mature thymus highly implicating this pairing in thymus colonization. Importantly, expression of P-selectin has been shown to be regulated by the availability of thymic stromal niches identifying a potential mechanism regulating the gated entry of haematopoietic precursors into the thymus [12].

Until recently, the mechanisms regulating T cell exit from the thymus have remained ill-defined. Again, chemokines have been implicated in the regulation of this process. CCL19, an additional ligand for the chemokine receptor CCR7 expressed by mature thymocytes, has been identified on blood vessels present within thymic medulla. Whilst such chemokine expression is able to attract mature thymocytes to exit thymic parenchyma into perivascular spaces within the thymus, additional signals are required to facilitate full thymic egress. The cell surface receptor sphingosine 1-phosphate type 1 (S1P1) appears to play an essential role in exit from the thymus in response to a gradient of expression of the ligand S1P between thymus and blood. As such, mice lacking the capacity to signal through the S1P1 receptor expressed on haematopoietic cells demonstrate a traffic jam of mature thymocytes within the thymus and perivascular spaces [13] as cells are unable to receive the correct signals to exit the thymus. Together, such data imply that thymic exit of mature thymocytes occurs in a two step process driven firstly via initial chemokine attraction to blood vessels and into perivascular spaces followed secondarily by S1P1 receptor mediated exit into the bloodstream. Previous reports have also suggested that exit from the thymus occurs in a lucky-dip haphazard manner whereby some thymocytes leave early in maturation and some thymocytes leave late [14]. However, recent data now suggest that thymic emigration is a strictly ordered process ensuring that only the most mature thymocytes, having completed the full developmental program, have a preferential ability to leave the thymus. Such mechanisms are thought to operate at least partly based on differential thymocyte expression of the S1P1 receptor [15].

$\alpha\beta$ Versus $\gamma\delta$ T Cell Development

Two main lineages of T cells are produced within the thymus, a prevailing $\alpha\beta$ T cell lineage, and a minor $\gamma\delta$ T cells lineage. Whilst multiple aspects of $\alpha\beta$ T cell development and function have been defined, those relating to the $\gamma\delta$ T cell lineage remain vague. Both $\alpha\beta$ and $\gamma\delta$ T cells develop from a common haematopoietic progenitor population colonizing the thymus. Branching of the $\gamma\delta$ T cell lineage from $\alpha\beta$ T cell lineage is though to occur within the window of the late DN2 to DN3 stage [16]. During DN thymocyte development, Recombinase activating gene (RAG) complex activity results in TCR gene rearrangement, in addition to β-chain rearrangements, DN thymocytes also undergo rearrangements of both γ- and δ-TCR chains. The fact that β-chain rearrangements may be found within mature $\gamma\delta$ T cells and conversely, $\gamma\delta$-chain rearrangements within $\alpha\beta$ T cells strongly supports the notion of a common precursor origin for both T cell lineages [17, 18].

The precise developmental requirements of the $\gamma\delta$ T cell lineage within the thymus remain uncertain. However, recent studies have indicated that a subset of $\gamma\delta$ T cells bearing a canonical TCR of restricted diversity, generated exclusively during fetal development, may undergo positive selection in response to interactions with thymic epithelium. In contrast, the interactions of $\gamma\delta$ T cells, bearing diverse TCR specificities, with thymic stromal elements are unclear. Within this chapter, we will therefore concentrate on describing development of the $\alpha\beta$ T cell lineage, which form the majority of thymocytes developing within the thymus and also mature T cells within the periphery.

Developmental Changes During Thymocyte Development

During thymocyte development, the linear progression of maturation can be characterised by the differential expression of numerous cell surface molecules. Initially, as thymic settling haematopoietic precursors colonize the thymus, the cells lack expression of both the T cell co-receptor molecules CD4 and CD8, as such these cells are termed double negative $CD4^{-}8^{-}$ (DN) and represent the most immature thymocyte subset. The precise identity of thymus settling cells remains contentious. While cells have been found within both bone marrow and blood with a lymphoid bias, thymus-settling cells are capable of generating multiple different lineages, with current data suggesting that multiple populations of haematopoietic cells colonize the thymus, with each subset displaying different degrees of T cell potential. Upon entry into the thymus, thymic settling progenitors undergo a series of interactions with thymic stromal cells that regulate the proliferation, survival and progressive differentiation of the haematopoietic cells. DN $CD4^{-}8^{-}$ thymocytes subsequently upregulate both CD4 and CD8 to become double positive $CD4^{+}8^{+}$ (DP) thymocytes, and finally mature to become single positive $CD4^{+}$ or $CD8^{+}$ (SP) cells following stringent selection events. Development of thymocytes is strictly regulated by sequential reciprocal signalling interactions with stromal cells of the thymic microenvironment. Consequently, it is essential that developing thymocytes are located in the right place at the right time to receive the right signals to drive efficient maturational events.

Double negative thymocytes can be further subdivided into subpopulations on the basis of CD25 and CD44 expression. The most immature thymocytes express CD44 but lack CD25 expression being termed $CD44^{+}25^{-}$ DN1 cells. After approximately 10 days, thymocytes next upregulate CD25 to become $CD44^{+}25^{+}$ DN2 cells. The DN2 stage lasts 2 days, during which thymocytes begin the process of gene rearrangement within the TCRβ chain locus. Downregulation of CD44 follows, marking the transition to the $CD44^{-}25^{+}$ DN3 stage, accompanied by ongoing V(D)J TCRβ gene rearrangement mediated by RAG1 and RAG2. Successful completion of TCRβ chain gene rearrangement leads to pairing of the TCRβ chain with the invariant surrogate pre-TCRα chain (pTα) at the cell surface in conjunction with CD3 molecules. Due to the imprecise nature of random TCRβ gene rearrangement, a high proportion of DN thymocytes are unable to generate a functional TCRβ chain due to failed gene rearrangement. The creation of the preTCR provides an early screening mechanism to eliminate thymocytes bearing non-functional TCRs through the process termed

β-selection. Transient expression of the preTCR at the cell surface in conjunction with CD3 elements results in the transmission of signal through the assembled receptor in what is currently thought to be a ligand-independent process. Successful preTCR signals result in several important outcomes:

(a) Prevention of apoptosis
(b) Initiation of cell division
(c) Inhibition of TCR β-chain rearrangement and
(d) Progression of differentiation through downregulation of CD25 to briefly become $CD44^{-}25^{-}$ DN4 thymocytes

The DN3 stage of development again lasts for approximately 2 days.

Following β-selection and multiple rounds of cell replication, DN4 thymocytes upregulate both CD4 and CD8 becoming double positive (DP) $CD4^{+}8^{+}$ thymocytes. During transit to the DP stage, cellular replication ends and rearrangement of TCRα genes is initiated through re-expression of the RAG complexes responsible for driving TCR gene rearrangement. During TCR α-chain rearrangement, thymocytes are provided with multiple opportunities to express a functional TCRα chain. Only thymocytes expressing a functional TCR of correct specificity receive survival signals allowing escape from programmed cell death. During this period, thymocytes will continue to rearrange α-chain genes until a functional α-chain paired with the β-chain allows for positive selection, or alternatively thymocytes run out of time and are subsequently deleted as a result of failure to generate a functional TCR capable of interacting correctly with thymic stromal elements. The precise mechanisms regulating DP thymocyte lifespan remain unclear. However, one potential mechanism regulating the lifespan of DP thymocytes is thought to operate via the action of the orphan nuclear receptor RORγ regulating expression of the anti-apoptotic factor Bcl-XL. Under steady-state conditions, it is thought that the pre-selection DP stage of thymocyte development may last in the region of 2–3 days. Regulation of DP thymocyte lifespan therefore plays an essential role in determining the diversity of the T cell repertoire through the linked interplay with the amount of time that thymocytes receive to continue TCR α-chain gene rearrangement and potential for creating a functional TCR [19]. Once thymocytes are generated bearing a functional TCR consisting of rearranged α- and β-chain subunits expressed at the cell surface, positive selection of thymocytes ensues assuming appropriate recognition of self-peptide:MHC complexes, resulting in downregulation of RAG and cessation of further TCRα gene rearrangement.

Up until the DP stage of development, thymocytes are confined to and mature within the cortex. Post-positive selection DP thymocytes upregulate CD69 and subsequently down regulate either CD4 or CD8 becoming single positive (SP) $CD4^{+}8^{-}$ or $CD4\text{-}8^{+}$. In addition, SP CD4 and CD8 thymocytes demonstrate a redistribution of compartmentalization, being primarily located within medullary areas. Within the medulla , thymocytes undergo further maturational events thought to last up to 12–14 days, although recent estimates suggest this may last as little as 4–5 days [15]. Within the medullary residency period, newly formed SP thymocytes demonstrate a $CD69^{HI}CD62L^{LOW}$ phenotype and exhibit relatively functionally immature properties, following a period of maturation involving as yet poorly defined events, SP thymocytes lose CD69 expression and increase CD62L expression at which stage thymocytes become functionally mature and exhibit a heightened responsiveness to thymic export signals.

Intrathymic Migration

After entry to the mature thymus at the cortico-medullary junction, thymocytes demonstrate a clear pattern of migration. Such migration helps to establish compartmentalization of cells at different stages of maturation. This process acts to ensure that cells are in the correct microenvironment at the right stage of development to ensure access to the signals provided by stromal cells that dictate whether individual thymocytes survive and differentiate or alternatively meet an early death through induction of apoptosis (Fig. 1.2). The most immature thymocytes being $CD4^{-}8^{-}$ DN exhibit an outward pattern of migration moving from the cortico-medullary junction towards a sub-capsular location. As such, DN1 thymocytes are located in relation to the cortico-medullary junction, extending into the deep cortex, DN2 thymocytes are present within mid-cortex and DN3 cells are present within the outer cortical region [20]. Within the DN3 population, β-selection induces proliferation and accumulation of pre-DP thymocytes within the subcapsular region. Following transition to the subcapsular region, progression to the $CD4^{+}8^{+}$ DP stage is initiated resulting in TCR α-chain rearrangement and cessation of proliferation. Subcapsular microenvironments may play a role in regulating the progression of DN thymocytes to the DP stage through production of TGFβ, which has been shown to negatively regulate pre-DP thymocyte proliferation [21]. The outward migration of thymocytes across the cortex is mediated by the action of several chemokine receptors inducing directed cellular migration. Mice lacking the chemokine receptors CXCR4 or CCR7 both display an accumulation of early DN thymocytes near the cortico-medullary junction and aberrant thymocyte development [22, 23]. In addition, mice lacking CCR9, whilst demonstrating normal DN2 and DN3 localization within mid- and outer-cortical regions display an inefficient localization of pre-DP thymocytes at subcapsular sites [24]. However, CCR9-deficient mice do not appear to demonstrate any gross defects in early thymocyte development suggesting that sub-capsular migration is not essential for thymocyte development.

In addition to directional cues provided by chemokine signalling gradients, thymic stromal cells provide a defined matrix over which immature thymocytes are able to crawl. Thymocytes express adhesion molecules differentially during their maturational program, with studies demonstrating that expression of the adhesion molecule VCAM1 by cortical stromal cells is necessary for normal thymocyte precursor migration [25].

As thymocytes enter the DP stage, they begin to rearrange the TCR α-chain and test the function of newly generated TCR$\alpha\beta$ receptors to recognize peptide:MHC complexes. DP thymocytes accumulate near the cortico-medullary junction but are

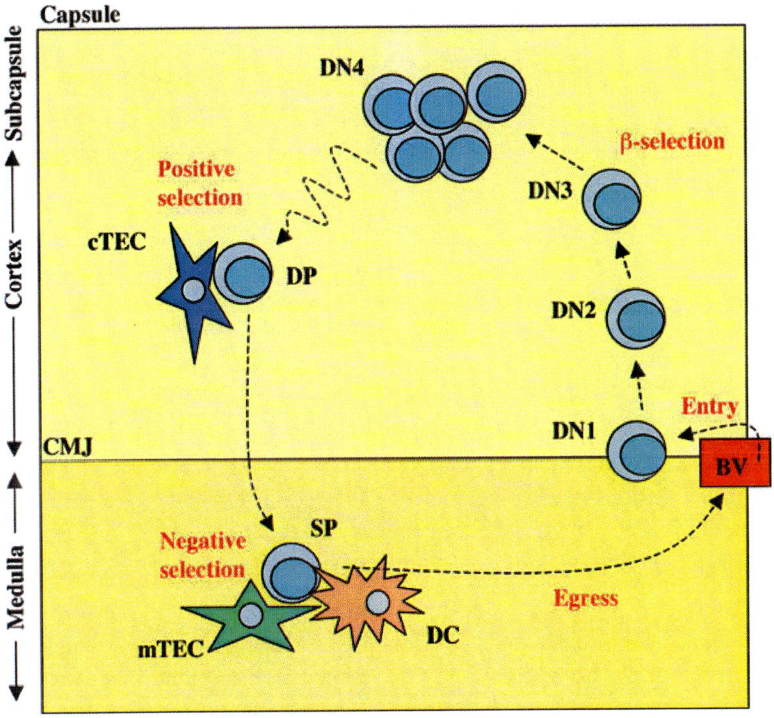

Fig. 1.2 Thymus settling cells enter the thymus through post-capillary venules located at the cortico-medullary junction (CMJ) as DN1 ($CD4^-CD8^-CD44^+CD25^-$) cells. Differentiation of thymocytes into DN2 ($CD4^-CD8^-CD44^+CD25^+$) and DN3 ($CD4^-CD8^-CD44^-CD25^+$) subsets is accompanied by outwards-directed migration towards the subcapsular region. DN3 thymocytes passing the β-selection checkpoint undergo both differentiation and extensive proliferation whilst passing through the DN4 ($CD4^-CD8^-CD44^-CD25^-$) stage. Double positive $CD4^+CD8^+$ thymocytes subsequently undergo inwards-directed migration back through the cortex, whilst undergoing a series of cognate interactions with cortical thymic epithelial cells in search of positively selecting ligands. Positive selection of thymocytes results in rapid relocation to thymic medulla where SP $CD4^+8^-$ and $CD4^-8^+$ thymocytes pass through further developmental checkpoints prior to export from the thymus. Presentation of self-antigens by medullary thymic epithelial cells (mTEC) and dendritic cells (DC) ensure that potentially auto-reactive thymocytes are deleted via the mechanism of negative selection. Functional, self-tolerant SP T cells subsequently exit the thymus through blood and lymphatic vessels via a combination of chemokine and S1P mediated signals

unable to migrate into medullary areas. Thymocytes receiving positive signals confirming the generation of a functional TCR forming low affinity interactions with self-peptide:MHC exhibit upregulation of the chemokine receptor CCR7 [26] and begin rapid migration towards thymic medullary areas preferentially expressing the CCR7 associated ligands CCL19 and CCL21. In support of this, mice lacking either CCR7 or both ligands CCL19 and CCL21 exhibit a clear block in cortex to medulla migration as a result of inhibited chemotaxis towards medullary regions. However, SP thymocytes are still capable of entering medullary areas albeit in a seemingly inefficient random manner occurring as a result of an absence of directed chemotaxis to medullary CCR7-ligands. Recent studies have further suggested that while CCR7 is required for SP thymocyte chemotaxis towards medullary areas, a second undefined G-protein coupled receptor mediated signal is required to enter thymic medulla and migrate on medullary substrates possibly via activation of specific adhesion molecules on SP thymocytes.

Positive Selection

The random nature of T cell receptor generation is accompanied by the intrinsic drawback of the potential for T cells to be generated bearing receptors that either are

(a) Incapable of recognizing peptide presented by MHC, therefore being non-functional or
(b) Capable of recognizing self-antigens to such an extent that they pose the dangerous potential to generate autoimmune actions against the bodies own tissues

The quality control process operating within the thymus provides essential mechanisms acting to eliminate both non-functional and auto-reactive thymocytes yet retain and facilitate the maturation of functional, self-tolerant T cells. Following rearrangement and pairing of TCRα and β-chains, three developmental outcomes are open to DP thymocytes. Thymocytes generating TCR pairings incapable of identifying self-peptide:MHC complexes with sufficient strength fail to generate TCR transduced signals of sufficient level to cross the threshold required to induce cell survival. In such cells, further TCR α-chain rearrangement provides a second-chance for DP thymocytes via continued generation of alternative new TCR$\alpha\beta$ pairings in order to try and generate a functional TCR. Alternatively thymocytes may fail to win a reprieve and undergo deletion via neglect due to a lack of survival signal provision. On the flip-side to lack of sufficient TCR signalling inducing death by neglect, an overly strong interaction between thymocytes bearing TCR and self-peptide:MHC complexes presented by antigen presenting cells again results in the induction of cell death in a process termed negative selection or clonal deletion, thought to be an important factor in preventing the generation of potentially autoreactive T cells. Interestingly, the threshold for TCR stimulation within DP thymocytes undergoing selection events in the thymus appears to be lower than that required by mature T cells responding to peptide in the periphery.

Sitting in a happy medium between these two negative outcomes of thymocyte deletion lies the third potential fate of DP thymocytes, positive selection. Positive selection occurs following engagement of DP thymocyte-expressed TCR with self-peptide:MHC complexes with low affinity recognition remaining below the signal strength threshold responsible for inducing negative selection. The three potential developmental outcomes open to developing thymocytes during selection has been described as following the Goldilocks hypothesis. The Goldilocks conditions command that positive selection and onwards development of thymocytes within the thymus may only proceed when the selecting signal is just right, as opposed to too little or too much signal leading to death by neglect and negative selection respectively [27]. Following positive selection, DP thymocytes undergo a host of developmental changes including termination of TCR α-chain rearrangement, reduction in the expression of the RAG complexes, downregulation of either CD4 or CD8 and induction of chemotaxis from the cortex towards the thymic medulla. In addition, positive selection ensures the viability of functional thymocytes through upregulation of the survival-related factors Bcl-2 and IL-7Rα. Interestingly, it is thought that following positive selection by self-peptide:MHC low affinity interactions, mature T cells subsequently generated reduce their sensitivity to such ligands, whilst retaining their capacity to respond to high affinity peptide:MHC antigens [28, 29]. Such alterations in the differential threshold levels of TCR sensitivity may reflect mechanisms operating to ensure that mature T cells do not retain a capacity to respond to self-antigen consequently reducing the risk of autoimmune reactions, yet ensure that mature T cells are capable of responding to foreign insult.

The precise nature and kinetics of the selecting signals involved within positive selection remain unclear. However, recent studies utilising two-photon real-time imaging techniques have begun to address the interactions between thymocytes and thymic epithelium to investigate the dynamics involved in thymocyte selection events. Within a thymic environment lacking a positively selecting capacity, thymocyte migration appeared random, presumably as thymocytes sample numerous surrounding thymic epithelial cells in search of a selecting MHC-self-peptide complex. Conversely, within a positively selecting environment, thymocyte migration appears restricted with sustained interactions occurring with thymic epithelial cells demonstrating both stable, persistent cell–cell contacts and shorter, dynamic interactions. The precise reasons for the mixture of interaction types and durations between thymocytes and epithelium remains unclear, although it has been proposed that such differences may result from the provision of signal variants and sub-stages of positive selection occurring [30, 31].

Further studies utilising two-photon microscopy have additionally demonstrated that within a wildtype thymus, the majority of cortical thymocytes display relatively slow and random patterns of migration. Within such wildtype mice, a small proportion of thymocytes demonstrated an increase in speed of movement and specific direction of movement towards thymic medullary areas, suggesting that positive selection induces a capacity for instructed directional migration from cortex to medulla as cells pass an important hurdle in development. Complementing this theory, analysis of mice bearing transgenic T cell receptors and a corresponding positively selecting thymic microenvironment demonstrated an increased proportion of rapidly migrating thymocytes demonstrating apparent guided chemotaxis [32]. Such experiments demonstrate the dynamic nature of thymocyte development during positive selection and highlight the complexity of getting the right cells together at the right time to ensure the provision of the right signals to drive thymocyte development.

An open question that remains however, is how does a rare thymocyte with a single receptor specificity sample enough self-peptide:MHC bearing cortical thymic epithelial cells in order to identify rare complexes of the corresponding type to receive positive selection and consequently survival signals before the estimated 2–3 day window prior to apoptosis induction. Recent studies suggest that during positive selection, thymocytes scan many thymic epithelial cells receiving multiple transient encounters with numerous cortical thymic epithelial cells in a manner independent of the formation of an immunological synapse [33] which sustains thymocyte survival. The requirement for re-exposure to multiple positively selecting complexes may explain the short, dynamic interactions between thymocytes and thymic epithelial cells observed in the two-photon studies by Bousso et al. [30]. In addition, the requirement for thymocytes to make numerous continued interactions with different thymic epithelial cells may help to prevent the selection of thymocytes bearing TCR with a restriction to a rare peptide:MHC specificity thus enforcing the onward development of only T cells tolerant to self-antigen. The huge potential array of peptides that the randomly generated TCR repertoire can recognize is easily likely to outnumber the capacity for a single antigen-presenting cell to provide to a developing thymocytes. As such, it would make sense that developing thymocytes would need to generate interactions with numerous thymic cells presenting a combined wide array of antigen in order to screen sufficient ligand combinations to ensure tolerance induction. The ability of such selective events would likely be far less efficient in the case of sustained interactions where thymocytes bound tightly to APC presenting a limited array of peptide–MHC combinations rather than in a system relying on transient, promiscuous interactions. The regulation of positive selection and receptor diversity appears to be controlled by competition amongst developing thymocytes for niches providing appropriate positively selecting ligands. This hypothesis is supported by studies utilizing transgenic mouse models, whereby competition for selecting ligands stimulates TCR editing [34]. Such findings indicate that positive selection is limited by the abundance of selecting ligands, highlighting the advantages of a dynamic system of numerous short-lived thymocyte:APC interactions.

Specificity of cTEC for Positive Selection

Positive selection of thymocytes requires specific cognate interactions with radioresistant epithelial cells making up the thymic cortex. Importantly, the unique three-dimensional mesh-like organization of cortical thymic epithelium (cTEC) appears to play an important role in the capacity to mediate positive selection as culture of cortical thymic epithelium in two-dimensional adherence cultures abrogates the capacity of cTEC to mediate positive selection most likely as a result of the downregulation of specific cell surface molecule expression. Cortical thymic epithelium appears to be unique in its capacity to mediate positive selection efficiently. In addition to expression of MHC class I and II molecules, it would appear that cTEC provide additional specialized signals to mediate positive selection that other cell types are unable to provide. The precise nature of such additional accessory signals remains unclear, although it is likely that any such signals will be cell surface associated due to the specific requirement for cell–cell cognate interactions.

In addition to accessory signalling molecules at the cell surface, the nature of mechanisms used by cTEC to allow association of MHC class II and self-peptides appears distinctive, in that cTEC utilize the protease Cathepsin L as opposed to the differential usage of protease Cathepsin S in the periphery. Recent studies have also identified a unique proteasome subunit β5t specifically within cTEC [35]. The function of β5t has been shown to play a highly important role in the processing of peptides for presentation via MHC class I and consequently CD8 T cell positive selection. Such differential processing of self-peptide–MHC complexes in the thymus may play a role in determining the differential signalling thresholds involved in positive selection in the thymus versus activation of mature T cells in the periphery to foreign-antigen:MHC complexes.

CD4 Versus CD8 Cell Fate Decisions

Within the periphery, mature CD4 and CD8 T cells generate specific immune responses through $\alpha\beta$TCR recognition of antigen presented by either MHC class II and MHC class I respectively. In addition to TCR mediated recognition of antigen, the TCR co-receptor glycoproteins CD4 and CD8 facilitate TCR peptide:MHC interactions through specific recognition and binding of either MHC class II or MHC class I molecules respectively. As such, CD4 T cells demonstrate an exclusive restriction to MHC class II presented antigens whereas CD8 T cells display a restriction to MHC class I presented antigens. During positive selection, DP CD4$^+$8$^+$ thymocytes undergo interactions with thymic epithelial cells selecting functional thymocytes through provision of survival and differentiation signals resulting in the generation of CD4 and CD8 $\alpha\beta$TCR thymocytes. A major maturational event during the stage of positive selection is the decision of selected thymocytes to become restricted to either the CD4 SP or CD8 SP sub-lineage. Studies analyzing mice with transgenic TCRs specific for peptide associated with MHC class I, demonstrate that all SP thymocytes within the thymus and SP T cells in the periphery are restricted to the CD8 branch, conversely, mice with TCRs specific for MHC class II associated peptides generate only CD4 T cells. It would appear therefore that whether a T cell expresses either CD4 or CD8 is determined by the aptitude of a $\alpha\beta$TCR pairing on a single T cell to recognize peptide provided in the context of either MHC class II or I respectively.

The management of CD4 versus CD8 lineage at the transcriptional level has recently begun to be unravelled. The zinc finger transcription factor cKrox (alternatively termed ThPOK) demonstrates specific upregulation during the selection of CD4 SP thymocytes but not CD8 SP thymocytes. In addition, experiments performed whereby imposed expression of cKrox within TCR transgenic thymocytes normally restricted to the CD8 lineage, have demonstrated that the cKrox transcription factor is capable of diverting such thymocytes to a CD4-expressing T cell fate [36]. In relation to enforcement of CD8 SP cell fate, the transcription factors TOX and Runx3 have been implicated [37, 38]. Whilst the identity of specific transcription involved in CD4:CD8 cell fate have begun to be revealed, further investigation is required to determine how signals transmitted through a single TCR activate specific transcription factor cascades and subsequently divert thymocytes to either a CD4 or CD8 fate. However, clues to this question are provided by studies suggesting that the strength of TCR transmitted signals may contribute to the decision of a thymocyte to assume either a CD4 or CD8 cell identity, thereby helping to co-ordinate TCR specificity for MHC class I or II antigens and CD4 or CD8 expression. Transient, low-level TCR signals have been suggested to induce a CD8 specification, whereas stronger, prolonged TCR signals are thought to result in CD4 lineage commitment [39]. In addition to TCR mediated signals, signalling through the Notch receptor on developing thymocytes may also play a role in supplementing CD4 versus CD8 cell fate decisions by active signalling promoting CD8 lineage commitment, although this view has been challenged [39, 40].

Negative Selection

Positive selection within the thymus operates via the induced survival of developing thymocytes capable of recognizing self-peptide:MHC complexes with the correct degree of affinity. Such low affinity ligands however, in general lack the capacity to stimulate and activate mature peripheral T cells, although continuous sub-optimal TCR mediated stimulation via low affinity interactions with self-peptide:MHC complexes are required for continued peripheral T cell survival [41]. Activation of mature T cells operates via recognition of foreign peptide:MHC complexes, thought to bear a structural relation to positively selecting ligands presented within the thymus, with a high level of affinity such that TCR stimulation crosses the threshold of T cell activation. The corresponding stimulation of thymocytes within the thymus by self-antigen:MHC complexes with high affinity however, leads to a very different outcome, inducing programmed cell death of the corresponding thymocyte, thereby ensuring that potentially auto-reactive thymocytes with a high level of specificity for self-antigen are deleted thereby enforcing central tolerance.

As indicated, positive and negative selection both operate via recognition of self-antigen:MHC complexes. However, these two events lead to two very different cell fates, on the one-hand survival and differentiation and conversely on the other hand cellular deletion. How these differential outcomes are induced via the same type of ligand has yet to be fully elucidated. Evidence suggests however, that the induction of negative versus positive selection may relate to the strength of signal propagated through the TCR in response to self-peptide:MHC. In response to TCR stimulation via ligation of self-peptide:MHC, it is thought that intracellular signalling cascades utilising MAP kinase activation is differentially induced. In this setting, a prolonged activation of the protein ERK at a low level is induced in positive selection , whereas negative selection is associated with ERK activated for a short period but at a high level. In addition, during negative selection, ERK activation is thought to synergize with the action of the additional protein kinases JNK and p38 [42, 43]. The utilization of different potential signalling pathways during either positive or negative selection may again relate to differences in the dynamics of TCR stimulation as previously described, in that positively selecting thymocytes endure short-lived interactions with thymic epithelium whereas negatively selecting thymocytes undergo sustained interactions. How the dynamics of TCR binding change the intracellular signalling pathways utilized during positive versus negative selection remains an open area of research.

Central tolerance imprinted within the thymus ensures that T cells with the potential to generate autoimmune responses through recognition of self-antigens and activation against the bodies own tissues are deleted. However, one potential barrier to this system is how are thymocytes within the restricted environment of the thymus exposed to the plethora of self-antigens associated with spatially diverse peripheral tissues. Mechanisms operating within the thymus provide a route around this problem through the action of the transcriptional regulator Aire. Within the thymic medulla, multiple populations of medullary epithelial cells express proteins normally associated with peripheral tissues, such that the thymus provides a miniature reflection of the bodies own tissues thereby providing a microenvironment in which autoreactive thymocytes can be deleted (Fig. 1.3). Absence of Aire expression results in multi-organ cellular infiltrates and autoimmune disease in both

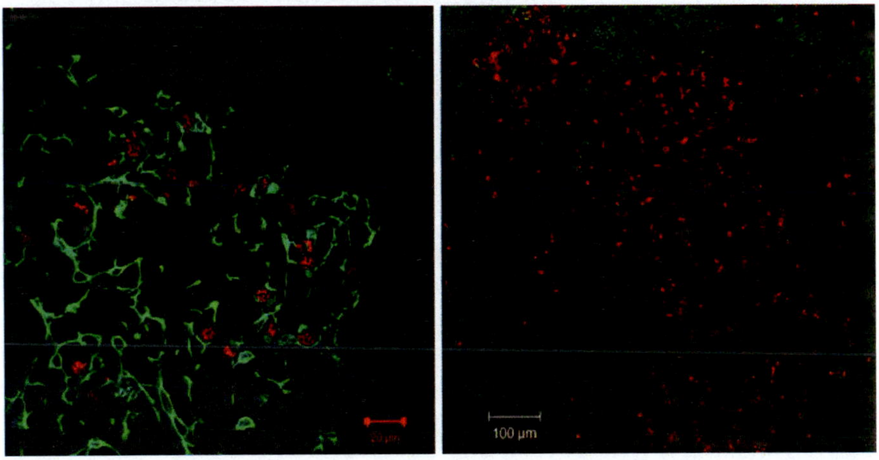

Fig. 1.3 The thymic medulla imposes central tolerance. (*Left panel*) The transcriptional regulator Aire (*red*), responsible for the control of ectopic peripheral tissue antigen expression within the thymus, is specifically expressed within medullary thymic epithelium, identified by reactivity for cytokeratin 5 (*green*). (*Right panel*) Dendritic cells, identified by expression of the cell surface marker CD11c (*red*) predominantly cluster within medullary thymic areas. Staining for CD4 (*green*) discriminates densely populated cortical regions and more sparsely populated medullary areas. Thymocytes are screened for auto-reactive specificities via their ability to recognize self-antigen. T cell recognition of self-antigen, provided by the combinatory action of mTEC and dendritic cells, with sufficiently high affinity results in deletion of auto-reactive T cells

mice and humans where Aire deficiency results in the autoimmune disorder autoimmune polyendocrinopathy-candidiasis-ectodermal dystrophy (APECED) [44]. In addition, the loss of a single Aire-regulated eye-related antigen within the thymus specifically leads to ocular-associated autoimmunity, highlighting the importance of peripheral tissue antigen expression within the thymus for the maintenance of self-tolerance [45]. Importantly, not all peripheral tissue antigen expression within the thymus appears to be regulated by Aire, although additional regulators of negative selection have yet to be identified. Development of Aire expressing medullary thymic epithelial cells is dependent on cellular interactions with lymphoid tissue inducer cells (LTi), previously demonstrated to be essential for secondary lymphoid tissue formation. Such interactions regulate $Aire^{+}$ mTEC generation via provision of RANK-ligand binding the RANK receptor on epithelial cells. In addition to LTi cells regulating Aire expression, mature $CD4^{+}$ SP thymocytes may also play a role in maintaining Aire expression in mTEC through provision of CD40-ligand to CD40 expressing epithelium [46].

As well as cell fate decisions generating CD4 and CD8 $\alpha\beta$TCR SP thymocytes during selection events, a distinct subset of thymocytes develop bearing a $CD4^{+}CD25^{+}$ phenotype, being termed regulatory cells (Treg). $CD4^{+}CD25^{+}$ T cells generated within the thymus display a regulatory function in the periphery, acting to suppress autoimmune T cells that have managed to bypass and evade negative selection mechanisms within the thymus. Regulatory T cells therefore act to provide a safety net of peripheral tolerance maintenance, acting to supplement negative selection and central tolerance induction imposed within the thymus. In addition to the cell surface phenotype of CD4 and CD25 expression, regulatory T cells may also be identified by the transcription factor Foxp3. Foxp3 acts as the master regulator of Treg development and also confers the capacity for a regulatory function. The exact mechanisms operating in the generation and commitment of Treg within the thymus remains unclear at present. It is thought that such regulatory cells are generated in response to high-affinity interactions with self-peptide:MHC complexes [47]. The basis for how some developing thymocytes generating high affinity interactions with self-peptide:MHC escape negative selection to become Treg remains unknown although again this may be related to narrow differences in the threshold of TCR stimulation.

The kinetics and progression of positive and negative selection whilst unclear do not seem to follow a linear path. Studies utilizing mice transgenic for specific TCR β-chain demonstrate that negative selection of MHC-peptide specific thymocytes within a polyclonal compartment occurring in response to the presence of a deleting peptide occurs throughout thymic development suggesting that passage through positive selection is not a prerequisite for negative selection [48]. As previously mentioned, expression of peripheral tissue antigens is thought to be restricted to medullary epithelium partially under the control of Aire. In addition to compartmentalization of peripheral tissue antigens, thymocytes display clear organization, such that DP thymocytes occur within the cortex and SP thymocytes within the medulla. However, if negative selection of thymocytes can occur throughout thymocyte development and even prior to positive selection, it follows that negative selection must also occur within the cortex. Multiple studies have demonstrated that the thymic medulla provides a unique site for negative selection, emphasized by studies demonstrating autoimmunity in mice lacking formation of normal medulla or inability of thymocytes to enter medullary areas [26,49]. Whilst medullary thymic epithelial cells play an essential role in the production and expression of peripheral tissue antigens in an Aire dependent manner, medullary epithelial cells themselves appear to be poor mediators of negative selection. Medullary thymic epithelial cells are capable of negatively selecting CD8 thymocytes, although they are unable to delete CD4 thymocytes. Bone marrow-derived dendritic cells represent efficient antigen presenting cells (APC), not only in the periphery but also within the thymus efficiently negatively selecting both CD4 and CD8 SP thymocytes. Efficient negative selection of thymocytes by dendritic cells may in part be related to the provision of co-stimulatory molecules such as B7-1/2 and CD40 providing an essential secondary signal supplementing that provided through TCR peptide:MHC interactions [50]. Such negative selection mediated by dendritic cells is thought to rely on production of peripheral tissue antigens by mTEC, transfer to and capture of antigen by dendritic cells and subsequent presentation to and screening of developing thymocytes [51]. Such cross-presentation of antigens from mTEC to dendritic cells potentially increases the number of cells presenting rare peripheral tissue antigens improving screening efficiencies and tolerance induction. In addition, migration of dendritic cells into the thymus from the periphery also raises the possibility that self-antigens may be brought into the thymus from anatomically distant sites. Whilst such mechanisms may aid in the deletion of autoreactive cells, it remains to be determined whether importation of antigen into the thymus by migrating dendritic cells influences selection of developing thymocytes during infection of the periphery.

Whilst it is clear that the thymic medulla provides an essential microenvironment for the negative selection of thymocytes through provision of peripheral tissue antigens, it has also been suggested that many thymocytes may be reactive to ubiquitous self-antigens expressed within cortical regions. Model systems have demonstrated that clonal deletion of thymocytes may occur within the cortex in response to ubiquitous self-antigens without any medullary involvement [52]. However, while cortical thymic epithelium is able to trigger signalling through the TCR, again it is only when dendritic cells are present that full negative selection of responding thymocytes is completed, resulting in the deletion of such reactive cells.

During thymocyte development, random patterns of movement are observed in pre-selection cells as thymocytes screen multiple cortical epithelial cells in a series of dynamic interactions before exhibiting directed migration to medulla following positive selection [30,32]. In relation to negative selection, recent studies have begun to model thymocyte dendritic cell interactions via two-photon microscopy real-time imaging techniques highlighting important interactions between these two cell types within the thymic cortex [53]. Thymocytes receiving positive selection signals demonstrate increased association with dendritic cells located near to cortical vasculature mediated by chemokine signalling via CCR7 expression on thymocytes. The purpose of such interactions between thymocytes and cortical dendritic cells may involve negative selection to ubiquitously expressed self-antigens found in the cortex as previously mentioned and may also involve negative selection of thymocytes responsive to peripheral tissue antigens provided to dendritic cells derived not only from medullary epithelial cells but also from peripheral tissues themselves. Together, such mechanisms may explain how negative selection is able to occur throughout the development of thymocytes and also overcome the anatomical barriers of compartmentalization of both thymocytes at defined stages of development and their corresponding negatively selecting ligands.

It would appear clear that bone marrow-derived dendritic cells play a key role in negative selection. The efficiency of dendritic cells to mediate negative selection of developing thymocytes has been demonstrated in studies utilizing reaggregate

Table 1.1 Models to study thymocyte development

Model type	Advantages	Disadvantages
In vitro		
Fetal thymus organ culture (FTOC)	Allows study of T cell development in 3-dimensional thymic environment in the absence of external influences. Easily manipulated via addition of growth factors, viral delivery systems etc. Can be depleted of thymocytes via addition of dGuO. Easy to image	Limited to fetal thymus. Loss of normal thymus organization. Relatively expensive requirement for timed pregnant mice
Reaggregate thymus organ culture (RTOC)	Allows isolation and reaggregation of defined thymic cellular elements allowing study of specific cellular interactions and requirements. Easy to image	Limited to fetal thymus. Loss of normal thymus organization
OP9-DL1 bone marrow culture system	Provides simple, uniform stromal environment. Allows easy analysis of single cell clonal assays of T cell development. Easy to image	Lacks 3-dimensional organization of thymus. Utilizes DL-1 ligand, whilst physiologically relevant Notch ligand expressed by thymic epithelium is DL4. Failure to support normal selection events by absence of MHC class II and Aire expression. Absence of normal thymocyte migration patterns
In vivo		
Bone marrow chimeras	Physiologically relevant. Allows study of adult thymus and T cell development. Provides capacity to study thymus colonization and T cell egress. Allows study of competitive T cell development in mixed bone marrow chimeras	Wide diversity of cell types present, external influences and waves of thymus colonization hamper study of isolated cellular interactions. Sub-lethal irradiation sometimes performed prior to bone marrow transplant may alter thymus biology due to creation of an empty thymus. Very difficult to image

thymic organ cultures (RTOC) [54]. In RTOC, fetal thymi were digested and populations of thymic epithelium and pre-selection DP thymocytes were reaggregated to form a composite thymus organ culture in the presence of carefully titrated proportions of dendritic cells. Analysis of dendritic cell numbers on negative selection remarkably revealed that maximal negative selection was achievable with dendritic cells representing just 1% of total cell numbers, highlighting the efficiency of dendritic cells in mediation of negative selection. In addition, within the same studies, the impact of peptide diversity on negative selection indicated that in a setting of reduced peptide diversity, negative selection was reduced when compared to negative selection in the presence of a diverse peptide repertoire. Such findings highlight the fact that a large degree of positively selected cells within the thymus display a potentially autoreactive repertoire and emphasize the importance of dendritic cell mediated negative selection in the presence of a diverse peptide repertoire in order to negate potential autoimmune disease.

References

1. Petrie H, Zúñiga-Pflücker J (2007) Zoned out: functional mapping of stromal signaling microenvironments in the thymus. Annu Rev Immunol 25:649–679
2. Feyerabend T, Terszowski G, Tietz A, Blum C, Luche H, Gossler A, Gale N, Radtke F, Fehling H, Rodewald H (2009) Deletion of Notch1 converts pro-T cells to dendritic cells and promotes thymic B cells by cell-extrinsic and cell-intrinsic mechanisms. Immunity 30: 67–79
3. Di Santo J, Rodewald H (1998) In vivo roles of receptor tyrosine kinases and cytokine receptors in early thymocyte development. Curr Opin Immunol 10:196–207
4. Crompton T, Outram S, Buckland J, Owen M (1998) Distinct roles of the interleukin-7 receptor chain in fetal and adult thymocyte development revealed by analysis of interleukin-7 receptor-deficient mice. Eur J Immunol 28:1859–1866
5. Boehm T, Scheu S, Pfeffer K, Bleul C (2003) Thymic medullary epithelial cell differentiation, thymocyte emigration, and the control of autoimmunity require lympho-epithelial cross talk via LT {beta} R. J Exp Med 198:757
6. Sutherland J, Goldberg G, Hammett M, Uldrich A, Berzins S, Heng T, Blazar B, Millar J, Malin M, Chidgey A (2005) Activation of thymic regeneration in mice and humans following androgen blockade. J Immunol 175:2741
7. Goldschneider I (2006) Cyclical mobilization and gated importation of thymocyte progenitors in the adult mouse: evidence for a thymus-bone marrow feedback loop. Immunol Rev 209:58–75
8. Prockop S, Petrie H (2004) Regulation of thymus size by competition for stromal niches among early T cell progenitors. J Immunol 173:1604
9. Shortman K, Egerton M, Spangrude G, Scollay R (1990) The generation and fate of thymocytes. Semin Immunol 2:3
10. Liu C, Saito F, Liu Z, Lei Y, Uehara S, Love P, Lipp M, Kondo S, Manley N, Takahama Y (2006) Coordination between CCR7-and CCR9-mediated chemokine signals in prevascular fetal thymus colonization. Blood 108:2531
11. Jenkinson W, Rossi S, Parnell S, Agace W, Takahama Y, Jenkinson E, Anderson G (2007) Chemokine receptor expression defines heterogeneity in the earliest thymic migrants. Eur J Immunol 37:2090–2096

12. Rossi F, Corbel S, Merzaban J, Carlow D, Gossens K, Duenas J, So L, Yi L, Ziltener H (2005) Recruitment of adult thymic progenitors is regulated by P-selectin and its ligand PSGL-1. Nat Immunol 6:626–634
13. Mori K, Itoi M, Tsukamoto N, Kubo H, Amagai T (2007) The perivascular space as a path of hematopoietic progenitor cells and mature T cells between the blood circulation and the thymic parenchyma. Int Immunol 19:745
14. Kelly K, Scollay R (1990) Analysis of recent thymic emigrants with subset-and maturity-related markers. Int Immunol 2:419
15. McCaughtry T, Wilken M, Hogquist K (2007) Thymic emigration revisited. J Exp Med 204:2513–2520
16. Ciofani M, Knowles G, Wiest D, von Boehmer H, Zúñiga-Pflücker J (2006) Stage-specific and differential notch dependency at the $\alpha\beta$ and $\gamma\delta$ T lineage bifurcation. Immunity 25:105–116
17. Livak F, Tourigny M, Schatz D, Petrie H (1999) Characterization of TCR gene rearrangements during adult murine T cell development. J Immunol 162:2575
18. Capone M, Hockett R, Zlotnik A (1998) Kinetics of T cell receptor β, γ, and δ rearrangements during adult thymic development: T cell receptor rearrangements are present in $CD44^+$ $CD25^+$ Pro-T thymocytes. Proc Natl Acad Sci USA 95:12522
19. Guo J, Hawwari A, Li H, Sun Z, Mahanta S, Littman D, Krangel M, He Y (2002) Regulation of the tcrα repertoire by the survival window of $CD4^+$ $CD8^+$ thymocytes. Nat Immunol 3: 469–476
20. Lind E, Prockop S, Porritt H, Petrie H (2001) Mapping precursor movement through the postnatal thymus reveals specific microenvironments supporting defined stages of early lymphoid development. J Exp Med 194:127
21. Takahama Y, Letterio J, Suzuki H, Farr A, Singer A (1994) Early progression of thymocytes along the CD4/CD8 developmental pathway is regulated by a subset of thymic epithelial cells expressing transforming growth factor beta. J Exp Med 179:1495
22. Plotkin J, Prockop S, Lepique A, Petrie H (2003) Critical role for CXCR4 signaling in progenitor localization and T cell differentiation in the postnatal thymus. J Immunol 171:4521
23. Misslitz A, Pabst O, Hintzen G, Ohl L, Kremmer E, Petrie H, Forster R (2004) Thymic T cell development and progenitor localization depend on CCR7. J Exp Med 200:481
24. Benz C, Heinzel K, Bleul C (2004) Homing of immature thymocytes to the subcapsular microenvironment within the thymus is not an absolute requirement for T cell development. Eur J Immunol 34:3652–3663
25. Prockop S, Palencia S, Ryan C, Gordon K, Gray D, Petrie H (2002) Stromal cells provide the matrix for migration of early lymphoid progenitors through the thymic cortex. J Immunol 169:4354
26. Ueno T, Saito F, Gray D, Kuse S, Hieshima K, Nakano H, Kakiuchi T, Lipp M, Boyd R, Takahama Y (2004) CCR7 signals are essential for cortex-medulla migration of developing thymocytes. J Exp Med 200:493
27. Yun T, Bevan M (2001) The Goldilocks conditions applied to T cell development. Nat Immunol 2:13
28. Davey G, Schober S, Endrizzi B, Dutcher A, Jameson S, Hogquist K (1998) Preselection thymocytes are more sensitive to T cell receptor stimulation than mature T cells. J Exp Med 188:1867
29. Lucas B, Stefanova I, Yasutomo K, Dautigny N, Germain R (1999) Divergent changes in the sensitivity of maturing T cells to structurally related ligands underlies formation of a useful T cell repertoire. Immunity 10:367–376
30. Bousso P, Bhakta N, Lewis R, Robey E (2002) Dynamics of thymocyte-stromal cell interactions visualized by two-photon microscopy. Sci STKE 296:1876
31. Bhakta N, Oh D, Lewis R (2005) Calcium oscillations regulate thymocyte motility during positive selection in the three-dimensional thymic environment. Nat Immunol 6:143
32. Witt C, Raychaudhuri S, Schaefer B, Chakraborty A, Robey E (2005) Directed migration of positively selected thymocytes visualized in real time. PLoS Biol 3:1062
33. Ebert P, Ehrlich L, Davis M (2008) Low ligand requirement for deletion and lack of synapses in positive selection enforce the gauntlet of thymic T cell maturation. Immunity 29:734–745

34. Wong P, Goldrath A, Rudensky A (2000) Competition for specific intrathymic ligands limits positive selection in a TCR transgenic model of CD4$^+$ T cell development. J Immunol 164:6252
35. Murata S, Sasaki K, Kishimoto T, Niwa S, Hayashi H, Takahama Y, Tanaka K (2007) Regulation of CD8$^+$ T cell development by thymus-specific proteasomes. Sci STKE 316:1349
36. Sun G, Liu X, Mercado P, Jenkinson S, Kypriotou M, Feigenbaum L, Galéra P, Bosselut R (2005) The zinc finger protein cKrox directs CD4 lineage differentiation during intrathymic T cell positive selection. Nat Immunol 6:373–381
37. Wilkinson B, Jeff Y, Han P, Rufner K, Goularte O, Kaye J (2002) TOX: an HMG box protein implicated in the regulation of thymocyte selection. Nat Immunol 3:272–280
38. Taniuchi I, Osato M, Egawa T, Sunshine M, Bae S, Komori T, Ito Y, Littman D (2002) Differential requirements for Runx proteins in CD4 repression and epigenetic silencing during T lymphocyte development. Cell 111:621–633
39. Germain R (2002) T-cell development and the CD4–CD8 lineage decision. Nat Rev Immunol 2:309–322
40. Robey E, Chang D, Itano A, Cado D, Alexander H, Lans D, Weinmaster G, Salmon P (1996) An activated form of Notch influences the choice between CD4 and CD8 T cell lineages. Cell 87:483–492
41. Kirberg J, Berns A, Boehmer H (1997) Peripheral T cell survival requires continual ligation of the T cell receptor to major histocompatibility complex-encoded molecules. J Exp Med 186:1269
42. Alberola-Lla J, Forbush K, Seger R, Krebs E, Perlmutter R (1995) Selective requirement for MAP kinase activation in thymocyte differentiation. Nature 373:620–623
43. Alberola-Ila J, Hogquist K, Swan K, Bevan M, Perlmutter R (1996) Positive and negative selection invoke distinct signaling pathways. J Exp Med 184:9
44. Anderson M, Venanzi E, Klein L, Chen Z, Berzins S, Turley S, von Boehmer H, Bronson R, Dierich A, Benoist C (2002) Projection of an immunological self shadow within the thymus by the aire protein. Science 298:1395
45. DeVoss J, Hou Y, Johannes K, Lu W, Liou G, Rinn J, Chang H, Caspi R, Fong L, Anderson M (2006) Spontaneous autoimmunity prevented by thymic expression of a single self-antigen. J Exp Med 203:2727
46. White A, Withers D, Parnell S, Scott H, Finke D, Lane P, Jenkinson E, Anderson G (2008) Sequential phases in the development of Aire-expressing medullary thymic epithelial cells involve distinct cellular input. Eur J Immunol 38:942–947
47. Jordan M, Boesteanu A, Reed A, Petrone A, Holenbeck A, Lerman M, Naji A, Caton A (2001) Thymic selection of CD4$^+$CD25$^+$ regulatory T cells induced by an agonist self-peptide. Nat Immunol 2:301–306
48. Baldwin K, Trenchak B, Altman J, Davis M (1999) Negative selection of T cells occurs throughout thymic development. J Immunol 163:689
49. Akiyama T, Maeda S, Yamane S, Ogino K, Kasai M, Kajiura F, Matsumoto M, Inoue J (2005) Dependence of self-tolerance on TRAF6-directed development of thymic stroma. Sci STKE 308:248
50. Amsen D, Kruisbeek A (1998) Thymocyte selection: not by TCR alone. Immunol Rev 165:209–230
51. Gallegos A, Bevan M (2004) Central tolerance to tissue-specific antigens mediated by direct and indirect antigen presentation. J Exp Med 200:1039
52. McCaughtry T, Baldwin T, Wilken M, Hogquist K (2008) Clonal deletion of thymocytes can occur in the cortex with no involvement of the medulla. J Exp Med 205(11): 2575–2584
53. Ladi E, Schwickert T, Chtanova T, Chen Y, Herzmark P, Yin X, Aaron H, Chan S, Lipp M, Roysam B (2008) Thymocyte-dendritic cell interactions near sources of CCR7 ligands in the thymic cortex. J Immunol 181:7014
54. Anderson G, Partington K, Jenkinson E (1998) Differential effects of peptide diversity and stromal cell type in positive and negative selection in the thymus. J Immunol 161:6599

Chapter 2
A Review of Mathematical Models for T Cell Receptor Triggering and Antigen Discrimination

Daniel Coombs, Omer Dushek, and P. Anton van der Merwe

Abstract Theoretical studies of T cell receptor signalling and T cell activation have become a well-known part of immunology and the models described in this chapter provide a good basis for future studies. Nonetheless it is crucial that, over the next few years, modelers seek to expand the scope of their efforts and provide a more comprehensive, predictive and multifaceted approach to T cell receptor signalling. Currently available models usually provide qualitative results and cannot be confidently parameterized. To obtain more precise and predictive models will be difficult but is plausible given improvements in quantitative experimental techniques and their quick adoption by experimentalists.

Introduction

The central event in the generation of adaptive immune responses is the binding of T cell receptors (TCR) to peptide-major-histocompatibility-complex (pMHC) molecules at the T cell-antigen-presenting-cell (APC) interface. It is important to understand how features of the molecular interaction determine the T cell response (potentially leading to a large-scale immune response in the body). This molecular recognition event is remarkably sensitive (as few as 5–10 antigenic pMHC can cause a robust cellular response) but also specific (a single amino-acid change in the presented peptide can dramatically alter the cellular response). Furthermore, the sensitivity to very small quantities of antigenic pMHC occurs in the context of a vast number of chemically similar but functionally irrelevant pMHC. These pMHC are derived from self proteins and are thought to weakly interact with TCR at the T cell-APC contact interface. Understanding the specificity and sensitivity of pMHC recognition in the presence of many self pMHC by T cells is further complicated by

D. Coombs (✉)
Department of Mathematics and Institute of Applied Mathematics, University of British Columbia, Vancouver, BC, Canada V6T 1Z2
e-mail: coombs@math.ubc.ca

C. Molina-París and G. Lythe (eds.), *Mathematical Models and Immune Cell Biology*, DOI 10.1007/978-1-4419-7725-0_2,

the fact that we do not understand how pMHC binding to TCR transmits a signal across the plasma membrane, a process termed TCR triggering. In this chapter we will review models that attempt to explain the interplay between sensitivity, specificity and TCR triggering.

In particular, we will focus on the organization of TCR proximal signalling events, that underlie three major classes of TCR triggering models (conformational change, aggregation, and segregation) and how the details of the TCR–pMHC interaction can affect signalling. We will begin the chapter with a very brief summary of the key experimental results that motivated these models.

Sensitivity and Specificity

T cells are challenged to detect the molecular signatures of infection, in the form of antigenic pMHC, from the background of noise, in the form of endogenous (self) pMHC. Antigen-presenting-cells express 10^5–10^6 diverse pMHC on their cell surfaces, of which relatively few are antigenic, capable of activating a T cell via its TCR. The key experiments on sensitivity, showing that very low numbers of antigenic pMHC can stimulate T cells were by Sykulev et al. (1996) [1] and Irvine et al. (2002) [2]. In the 2002 experiments, pMHC were individually labelled with a fluorescent marker, showing that CD4 and CD8 T cells will transiently respond (flux calcium) in response to a single antigenic pMHC. The level of calcium response correlated with the number of presented pMHC up to about 10. After this many pMHC are detected, a more complete response occurs. A further finding of this study was that cytotoxic T cells can kill antigen presenting cells after recognizing as few as three antigenic pMHC. This hair-trigger level of sensitivity is amazing and perhaps frightening given our knowledge of autoimmune disease. The specificity of the T cell response has been known for some time by the observation that a single amino acid substitution in the presented peptide can substantially alter the T cell response [3, 4]. Taken together, these experiments showed that T cells are able to respond to a few specific pMHC amidst a background of chemically similar endogenous pMHC, which when presented in the absence of specific pMHC do not elicit a T cell response.

Parametric Descriptions of TCR–pMHC Binding

It is natural to ask what quantitatively measurable features of the TCR–pMHC interaction determine the cellular response. The simplest model of the TCR–pMHC binding interaction treats it as a simple chemical binding with binding rate k_{on} and unbinding rate k_{off}. It is then possible to define the half-life $t_{1/2} = (\ln 2)/k_{\mathrm{off}}$ and the dissociation constant $K_{\mathrm{D}} = k_{\mathrm{off}}/k_{\mathrm{on}}$. It is important to note that k_{on} is a two-dimensional rate (units of $\mu\mathrm{m}^2\mathrm{s}^{-1}$), reflecting the fact that the pMHC and TCR are restricted to their respective cell membranes.

Commonly used methods for measuring biochemical rates, such as those using surface plasmon resonance [5, 6], yield three-dimensional rates which may or may not correlate well with the actual rates of binding and unbinding within the tight region of contact between a T cell and an APC. For example, it is possible that the bond in a physiological situation is stressed and therefore that k_{off} would be higher than a 3-d measurement would suggest [7]. On the other hand, experiments that directly measured the two-dimensional rates of CD2–CD58 and CD16–IgG Fc interactions found a 2-d off-rate that is 100-fold smaller than the 3-d measurement [8]. This observation may be due to rapid rebinding of the same ligand-receptor pair in the 2-d environment which is a consequence of diffusion-limited reactions [9, 10].

In any case, 3-d measurements must be converted to two dimensions for use in mathematical models. The most commonly implemented method is to keep k_{off} the same, but modify k_{on} by dividing by an appropriate length scale (typically 5–10 nm) [11]. An additional effect was proposed by Qi et al. [12] based on the work of Krogsgaard et al. [13], who used thermodynamic methods and a cartoon model of the TCR to obtain a formula for converting experimentally measured 3-d off-rates to 2-d parameters. This method requires additional measurements of the heat capacity of the bond to be made. Both of these methods still require experimental validation or refutation.

Importance of Stable TCR–pMHC Binding

With the caveat that the parameters still need to be measured in a physiological 2-d situation, the majority of authors have focused on the stability of the TCR–pMHC interaction (presumed to be governed by k_{off} from 3-d measurements) as the key determinant of T cell activation. Experimentally, this is supported by a range of studies showing T cell activation, as measured by cytokine production, is well correlated to k_{off} (reviewed in [14, 15]) but these results were challenged by others [16, 17] who did not find such correlations. In terms of signalling, the kinetic proofreading model (discussed in detail below) gives a rationale for superior signalling by long-lived complexes in terms of a series of essential signalling steps.

In order to explain experimentally-observed deviations from this rule, more complex models have been devised that take into account additional aspects of the interaction. For instance, it appears that the TCR coreceptor CD8 stabilizes the TCR–pMHC complex by binding in a peptide-independent fashion to MHC and therefore supports TCR signalling on cytotoxic T cells [18]. An alternative approach was taken by Krogsgaard et al. [13], who measured the heat capacity of the TCR–pMHC bond and found that this could be used, along with k_{off}, to provide an improved fit of T cell activation data. This is consistent with the importance of effects of molecular reorientation during bond formation [12].

We now move on to describe models for TCR signalling following pMHC binding. In turn, we look at the following:

– Kinetic proofreading models based on a linear sequence of signalling events proximal to the TCR and dependent on pMHC binding

- Detailed biochemical models for TCR signalling, involving more complex reaction networks and feedbacks
- Conformational change of the TCR upon binding
- Models that take special account of TCR aggregation into multimeric complexes
- Segregation-based models

These models should not be viewed as contradictory. TCR signalling is a multi-faceted process and each class of models takes into account only aspects of the biological situation. A full model that reconciles all experimental findings is a challenging task that will build on existing models and will only develop in tandem with appropriate experimental data.

Kinetic Proofreading Models for Antigen Discrimination

Kinetic proofreading models for TCR signalling are, at their heart, phenomenological models based on a cartoon version of biochemical signalling initiated upon TCR/pMHC binding. During signal transduction after receptor ligation, a series of biochemical events occur at the cytoplasmic tail of the receptor. These events build a signalling structure of modified components that eventually gives a complete signal. However if the ligand dissociates from the receptor, the chemical reactions are aborted and do not go to completion. The essence of the kinetic proofreading model is therefore to provide a reasonable mechanism for a time lag separating ligand binding from receptor signalling and hence allowing a receptor to discriminate between ligands with small differences in off-rate.

As a simple quantitative example, consider a receptor that signals after $\tau = 1\,\mathrm{s}$ of engagement, and two potential ligands with $k_{\mathrm{off}}^{A} = 1\,\mathrm{s}^{-1}$ and $k_{\mathrm{off}}^{B} = 5\,\mathrm{s}^{-1}$. The probability of signalling following a single engagement by ligand A is $\exp(-k_{\mathrm{off}}^{A}\tau) = e^{-1}$ while for ligand B it is $\exp(-k_{\mathrm{off}}^{B}\tau) = e^{-5}$. Ligand A is thus $\exp(4) \simeq 55$ times as likely to signal as ligand B, based on one binding event. Observe that this mechanism allows discrimination based on half-life (or equivalently, k_{off}) and that small differences in the half-life can be translated into large differences in signalling. A time-lag model of receptor signalling was considered in a more general context in [19].

Despite their simple nature, kinetic proofreading models have been extensively used in a variety of contexts. McKeithan introduced the kinetic proofreading model for TCR in his seminal paper [20], following on from earlier work in a different context [21, 22]. McKeithan's original scheme is illustrated in Fig. 2.1a. In this model, the TCR begins in an inactive state. Upon binding, it undergoes N sequential modifications, each representing an intermediary on the path to signal transduction. Upon completing N modifications, a signal is assumed to have been completely transduced. Furthermore, McKeithan supposed that if the pMHC unbinds from the TCR prior to the Nth step being reached, the TCR immediately reverts back to the inactive state. Mathematically speaking, this model essentially generates a sigmoidal response to pMHC based on their mean lifetime of binding to the TCR ($1/k_{\mathrm{off}}$). We may therefore say that the kinetic proofreading model allows

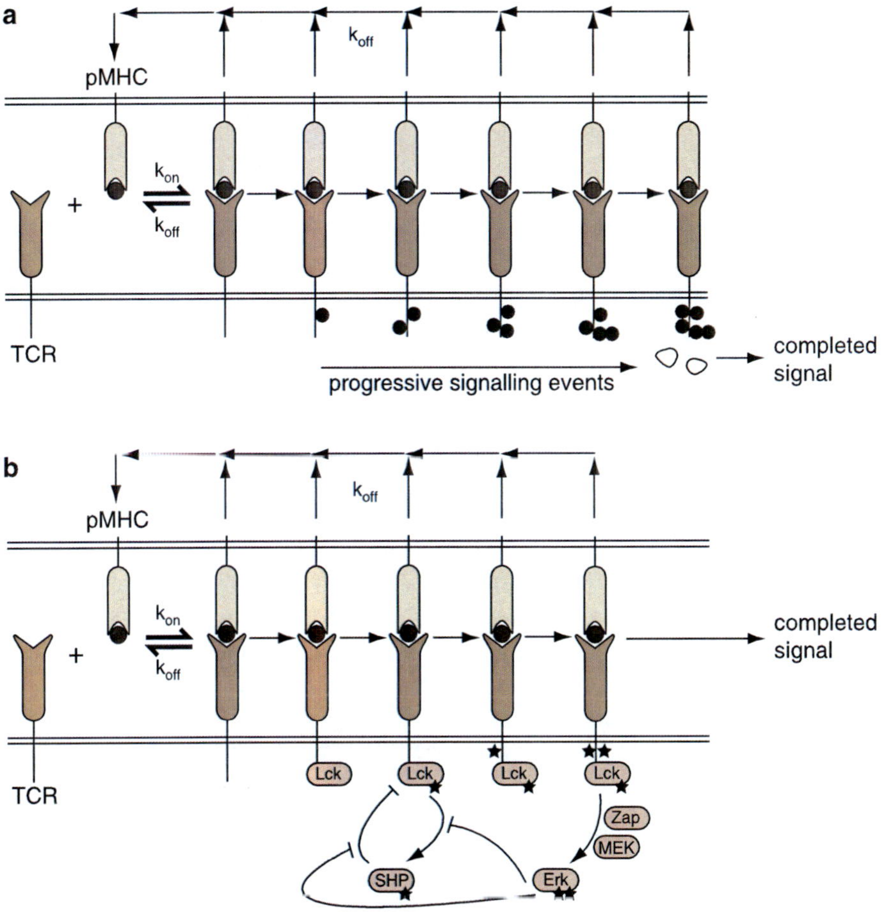

Fig. 2.1 Schematic descriptions of kinetic proofreading and feedback models. (**a**) We show the original kinetic proofreading model [20]. ***Black dots*** indicate unspecified signalling steps occuring at the cytosolic tail of the receptor. A full signal is transduced at the end of the signalling sequence. (**b**) A positive/negative feedback model of TCR signalling [36–38]. Schematic shown most closely resembles the feedback model by Lipniacki et al. [38]

the T cell to discriminate pMHC based on their chemical off-rate for the TCR. This model has been used by many researchers as a simple model of TCR signalling (e.g. [9, 23–27]).

One of McKeithan's observations about his model was that the specificity of discrimination grows with N, but at the expense of sensitivity (i.e. as N increases, the number of signalling events due to pMHC with high off-rates decreases, but the total number of signalling events from any pMHC also decreases). This point was examined in more detail in critical reviews of the kinetic proofreading model [25, 28], and shown to be an unavoidable feature of the basic model. It was proposed that to resolve this issue, fully activated TCR could have a decreased off-rate for the pMHC – requiring that the signalling machinery be able somehow lock the pMHC in place at the TCR [20]. Although we note that there is no experimental evidence for this effect at present, we can speculate as to how this might be achieved. Possible non-exclusive mechanisms could include

- A direct mechanical modification of the TCR upon signalling (possibly related to the piston action or receptor deformation described below). We note that, since single amino-acid substitutions on the presented peptide can change the cellular response, it is not inconceivable that small changes in the geometry of the TCR could have an effect.
- Introduction of additional nonspecific MHC-binding coreceptors CD4 and CD8 to the signalling complex, increasing the effective affinity of the signalling complex for pMHC and fixing the pMHC in place.
- Recruitment of additional TCR to the signalling region. For this idea to make sense, nearby TCR would need to be able to act effectively to share, or integrate signals. The pMHC would then serially ligate different TCR within the signalling cluster, but the kinetic proofreading steps would take place as if via a single receptor. Experimentally, TCR are known to form clusters in response to cognate presented pMHC [29, 30].
- By modulating the local geometry of the cell membranes at the signalling receptor to optimize the intermembrane separation from the point of view of the TCR–pMHC bond, and reduce the unbinding rate. We note that TCR signalling is known to be linked to the cortical actin cytoskeleton of the T cell, that could provide a feasible mechanism for some local control. An alternative or complementary means of control would be via the removal of cell-surface molecules with large extracellular domains from the local area.

We note that mechanisms for holding pMHC fixed and bound to a single TCR contradict the serial engagement hypothesis [31] which argues that the signalling capacity of rare agonist pMHC multiplies when each pMHC binds to multiple TCR during the cellular interaction. However, serial engagement of TCR by pMHC could occur within a TCR microcluster or other localized signalling region [32].

It is also important to note that within the basic model, ligands with lower probabilities of generating a signal can compensate for this weakness by being present in larger quantities. This is critical for the TCR given the large numbers of irrelevant peptides present on the APC. This problem is partially overcome by the strong nonlinearity in response achieved if there are many proofreading steps (albeit at the expense of sensitivity) but remains a point to be addressed by more complex models.

The biological basis for the kinetic proofreading model in the case of the TCR remains unclear. We may, however, draw some inspiration from the signalling pathway of the high affinity IgE receptor FcϵRI on mast cells. This receptor, closely related to the TCR, binds IgE with high affinity and the IgE on different receptors can then be crosslinked by an appropriate ligand. The subsequent signalling cascade has been studied in great detail, allowing a detailed mathematical model to be built [33, 34]. Experimental and modelling results reflect many aspects of kinetic proofreading, but certainly do not allow us to fit the parameters of McKeithan's formulation of the model, or obtain a clear biological interpretation of the parameters (reviewed in [35]).

To summarize our discussion of kinetic proofreading, we observe that a literal interpretation of McKeithan's kinetic proofreading model is certainly not correct. The basic model is unable to reconcile specificity and sensitivity, and the intermediate steps have not been found. However, detailed modelling of the FcϵRI receptor shows that the basic idea of kinetic proofreading in terms of a set of molecular events that must happen proximal to the receptor should remain a useful paradigm for describing TCR signalling. New models can be described in terms of their deviations from the kinetic proofreading model. Additionally, the kinetic proofreading model is a useful and easily implemented component that can be incorporated into larger or more complex models of TCR signalling. A good example of this is the model of Wedagedera and Burroughs (2006) which examines the whole process of T cell activation from a queuing theory perspective [27]. The kinetic proofreading component is a simple and natural choice to capture the essence of the signalling cascade without getting bogged down in the details.

Extensions of Kinetic Proofreading

TCR–pMHC Rebinding

The reactions between many cytosolic proteins are thought to be reaction-limited because diffusion coefficients in the cytosol are relatively large. In this regime the dynamics can be accurately captured by simple ordinary-differential-equations (ODEs) which rely on the so called well-mixed assumption. In contrast, the reactions between many membrane confined proteins are thought to be diffusion-limited, whereby molecular collisions occur at much lower frequencies but often lead to a reaction owing to the reaction on-rate being larger than an appropriate measure of diffusion, see for example Lauffenburger and Linderman [39]. A consequence of diffusion-limited reactions is that a receptor-ligand complex will break and reform several times before each binding partner diffuses apart. As discussed above, the 2D reaction rates between TCR and pMHC are presently unknown but it is reasonable that their binding rate, like the binding rate for other membrane confined proteins, is diffusion-limited. Therefore it is expected that a TCR–pMHC pair will unbind and rebind many times before the TCR and pMHC move apart. Incorporating this effect into the canonical kinetic proofreading model proposed by McKeithan [20] it was found that TCR–pMHC rebinding has only a small effect on productive signalling, Fig. 2.2a. However, by supposing that the TCR signal is not lost during the brief interval between rebinding events, pMHC rebinding can have a large effect on productive signalling, Fig. 2.2b. In particular, it was found that in addition to k_{off}, the on-rate can also have a critical role in determining signalling. Therefore this simple extension to the kinetic proofreading model showed that by explicitly modelling rebinding and signal persistence at the TCR, T cells are able to discriminate pMHC based on both k_{off} and k_{on}. This theoretical work is summarized in Dushek et al. [9]. In support of this model, the potency of pMHC can be well-predicted by an effective off-rate that accounts for rebinding but not by the off-rate alone [10].

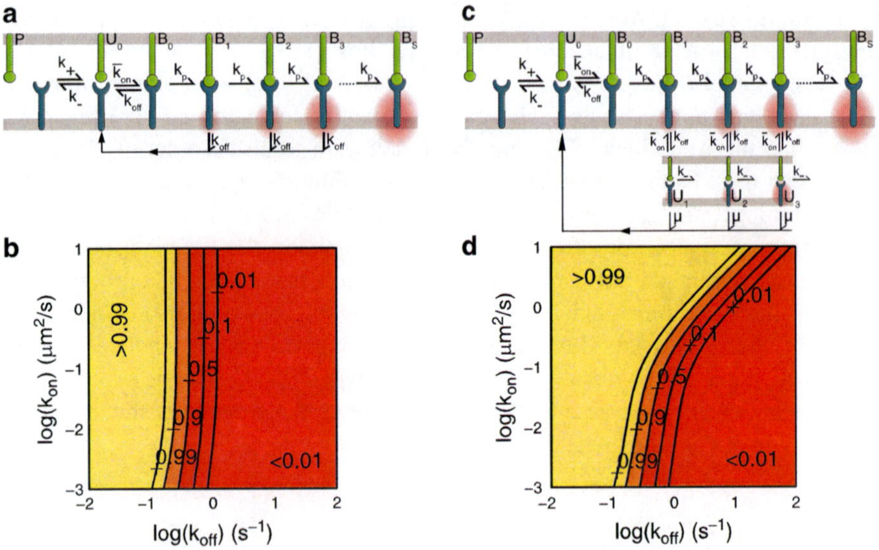

Fig. 2.2 Incorporating rebinding and signal persistence in kinetic proofreading models allows for pMHC discrimination based on both k_{off} and k_{on}. (**a**) Schematic of the canonical kinetic proofreading model modified to include the possibility of TCR–pMHC rebinding. In this model pMHC dissociation from TCR results in state U_0 which allows the two molecules to rebind or diffuse away at rate k_-. (**b**) Contours of the probability that at least 1 of 10 presented pMHC will transduce a productive signal (final step in the proofreading scheme) after $t = 30$ s as a function of k_{off} and k_{on} on a log–log plot. As expected, k_{off} is critical in determining productive signalling but increasing rebinding events, by increasing k_{on}, has little effect on productive signalling. (**c**) The scheme shown in panel (**a**) is modified to include the possibility that TCR signals persist during the brief intervals between rebinding events (states U_i). This modification is plausible given the finite time required for phosphatases to revert the phosphosphorylation state of the TCR to basal levels. (**d**) In this modified scheme (shown in panel (**c**)) it is found that k_{on} may also have a critical role in determining productive signals. For further details see Dushek et al. [9]

Ligand Antagonism

Certain pMHC are known to desensitize T cells to further stimulation. These pMHC are called antagonists and we should seek to understand how they work in forming a complete theory of TCR signal transduction. A 1996 study by Rabinowitz et al. [40] used a modified kinetic proofreading model to allow for partial and incomplete signals leading to cellular desensitization. They modelled TCR as existing in three states – inactive, partially modified, and fully modified. A receptor must be fully modified to generate a positive signal. Partially modified receptors are taken to give negative signals. The existence of the intermediate state can be motivated by the observation that ZAP-70 is not activated by antagonist pMHC, indicating that pMHC discrimination occurs upstream of ZAP-70 activation and so intermediate state(s) must exist. Further experimental work shows that the membrane phosphatase SHP-1 is recruited to TCR during pMHC binding, and that this leads to the inactivation of Lck kinase, but that as binding continues, this process is inhibited by recruitment of the MAP kinase ERK-1 [36] (discussed further below).

Theoretical modelling of ligand antagonism was also performed by van den Berg, Burroughs and Rand [23]. This work proposed a method whereby careful experimentation and data analysis could be used to distinguish between passive antagonism (where antagonist pMHC successfully compete with agonists in terms of binding TCR) and active antagonism (where the signalling ability of individual TCR is reduced after antagonist binding). The strength and mode of ligand antagonism was shown to depend on the density of presented pMHC, also suggesting that T cells may control their signalling capacity by modulating surface presentation of TCR.

Spatially Extended Models of TCR Activation

The basic ideas described by Rabinowitz et al. [40] were extended in a mathematical modelling paper by Cliburn Chan et al. [41]. In this model, a spatial Monte Carlo simulation of pMHC-induced TCR activation was used to examine spatial spread of activation and inhibition from a ligated TCR to its neighbours. Individual TCR in the model were supposed to exist in different states – empty, bound, partially activated, and fully activated. Furthermore, signals were supposed to spread to neighbouring TCR. These signals could be inhibitory (prohibiting further activation) or protective (protecting the TCR against inhibition). Biologically, these effects could occur via recruitment of SHP-1 phosphatase to the signalling region (inhibiting Lck activation) and ERK-1 activation (which protects Lck from SHP-1). Inhibitory signal spreading ("receptor crosstalk") improves the specificity of signalling based on k_{off} of the TCR–pMHC bond, although it decreases sensitivity to low numbers of presented pMHC. The loss of sensitivity in the model is restored in the full model by protective signal spreading. Importantly, this model addresses a crucial shortcoming of the basic kinetic proofreading model in that it shows how a low-density, long-lived (small k_{off}) ligand can be discriminated from a very high density, but short-lived (large k_{off}) ligand.

Detailed Biochemical Models and Feedback Control of the Signalling Cascade

A development of the kinetic proofreading model based on signalling feedbacks has been advanced by Germain and coworkers [15, 36, 37]. This model contains four TCR states – unbound, bound, partially activated and completely activated. Two feedback loops are proposed: a negative feedback from the partially activated state, reducing further activation steps, and a positive feedback from the fully activated state, enhancing further signalling. The model identifies the mediator of negative feedback as SHP-1, while the positive feedback is mediated by ERK-1. The model is described using ordinary differential equations, and broadly represents a spatially homogenized version of earlier models of Chan et al. [41, 42]. However, the focus here is on short-time (1–3 min) responses to ligand, rather than over 60 min as in Chan et al. The combination of positive and negative feedback responses in the model leads to a bistability in the T cell response. The power of this work is in the integration of modelling and detailed experimental work within a single laboratory. The model is shown to make a number of predictions, which are then tested experimentally, verifying the model. For example, the feedback model shows the response time decreases sharply as the number of pMHC is reduced. By measuring the ERK-1 response at different ligand densities, Altan-Bonnet and Germain were able to verify this behaviour [37]. Wylie et al. (2007) incorporated the same feedback mechanisms in another model for the role of CD4 coreceptors (constitutively associated with Lck) and nonagonist ligands in T cell activation [43]. This latter paper also emphasizes the importance of considering stochastic fluctuations when analyzing receptor signalling models.

The basic idea of combining kinetic proofreading with positive and negative feedbacks received further attention in a modelling paper by Lipniacki et al. [38]. In this model, TCR can exist in six states: (1) unbound; (2) bound; (3) associated with unphosphorylated Lck; (4) associated with phosphorylated Lck; (5) with phosphorylated Lck and TCR-ζ chain singly phosphorylated; (6) with phosphorylated Lck and TCR-ζ chain doubly phosphorylated, leading to cellular activation. The model also incorporates negative feedback via phosphorylated SHP from state (4) onto itself, and positive feedback from state (6) via a pathway from the receptor to ZAP-70, MEK and ultimately doubly phosphorylated ERK onto the negative feedback. The model, which explicitly contains SHP and ERK, is expressed as a set of ordinary differential equations which are solved deterministically and also in a stochastic framework. The results indicate that the T cell will respond, most of the time, to 5–10 agonist pMHC, but that the sensitivity is substantially reduced in the presence of antagonist pMHC. The modelling indicates a bistability in response (activation or nonactivation with no intermediate) but that the barrier between the basins of attraction is low enough that small fluctuations can change the response. It is shown that the deterministic solution of the system is therefore a poor descriptor of the actual behaviour in the presence of noise.

To summarize, models based on known biochemical events, and particularly the SHP-1 / ERK-1 feedback loops, allow experimentally testable predictions to be made (and tested), and surely represent the future of TCR signalling models. However, substantial challenges remain:

- The accurate parameterization of the models remains a thorny issue, especially given the multitude of interacting chemical species. This means that sensitivity analysis becomes extremely important (discussed in [43]). However, the fact that we can speak of measuring identifiable parameters at all is a major step forward.
- The models must handle the presence of two kinds of stochastic effects. First, the obviously stochastic nature of any biochemical reaction network with small numbers of players. Second, there is substantial variation in expression levels of signalling components between cells. This second point was examined in detail in a theoretical-computational study of Feinerman et al. [44].

– The role of spatial effects such as TCR clustering and segregation of signalling molecules must be carefully addressed. This requires the intelligent use of models defined by partial differential equations or (more likely) spatial Monte Carlo simulations. The development of efficient algorithms is an important ongoing concern [45–47].

The authors of competing feedback models are sometimes at pains to distinguish their work from kinetic proofreading. We find this, to some extent, to be a false dichotomy since in every model we have described, a core pathway can be distinguished, which is essentially determining a kinetic proofreading process. This scheme is modified by the presence of feedback loops, of course. As stated above, the paradigm of kinetic proofreading remains useful in describing the basic features of these models.

Models of TCR Triggering

The mathematical models described above have assumed that TCR proximal signalling is initiated upon pMHC binding to the TCR. For example, kinetic proofreading models assume that proofreading initiates when pMHC binds TCR and is terminated when pMHC unbinds. However, the exact mechanism by which the pMHC signal is communicated across the plasma membrane and initiates signalling is presently unknown. We now review three broad classes of TCR triggering models and discuss how kinetic proofreading is modified in these systems.

Models Relying on a Conformational Change

Conformational change models postulate that TCR binding to pMHC somehow results in a conformational change in the CD3 cytoplasmic portions. Early conformational change models postulated that a conformational change was transmitted allosterically through the TCR$\alpha\beta$ subunits. However this is implausible given the huge semi-random structural diversity of TCR/pMHC interfaces, and structural studies of the TCR/pMHC complexes have failed to identify any such long-range conformational change which is common to many TCR upon pMHC binding. More recent models have postulated that the TCR binding leads to a conformational change of two TCR/CD3 complexes with respect to each other, the TCR$\alpha\beta$ module with respect to the CD3 chains or a 'piston-like' change in the TCR/CD3 complex with respect to the plasma membrane. How could pMHC binding lead to such changes? We have noted that pMHC binding will automatically subject the TCR to a mechanical pulling force [7] and proposed that this pulling could be responsible for such conformational changes [48, 49]. A very similar 'receptor-deformation' model has been proposed more recently by others [50].

Support for conformational change models has come from the demonstration that the cytoplasmic portions of the CD3 chains undergo conformational change [51–53]. In one case the conformational change in a proline-rich motif of the CD3ϵ was shown to be induced by TCR binding to pMHC [54], although subsequent functional studies indicated that this motif is not required for TCR signalling but is instead involved in regulation of TCR/CD3 surface density in thymocytes [55]. The recent evidence that the CD3ϵ chain binds to the membrane, with the two ITAM tyrosine residues sequestered deep therein, suggests how conformational change might regulate tyrosine phosphorylation [53]. It is possible, however that phosphorylation of CD3ϵ regulates membrane association rather than vice versa, and it remains to be shown that TCR binding to pMHC can influence this CD3ϵ binding to the membrane.

In this model of TCR triggering, the basic kinetic proofreading scheme does not need any modification. The binding of pMHC to TCR transduces a conformational change at the TCR which initiates kinetic proofreading (signalling) while pMHC unbinding reverses the TCR conformational change which terminates kinetic proofreading (signalling). In this way, TCR triggering by a conformational change is the simplest kinetic proofreading mechanism.

Molecular Aggregation Models

A common mechanism of signal transduction across the plasma membrane is the dimerization (or oligomerization) of cell surface proteins [56]. For example, receptor tyrosine kinases (RTKs) are transmembrane receptors composed of an extracellular ligand binding site and an intracellular tyrosine residue which can become phosphorylated by a specific kinase domain also located on the RTK. Typically, a single RTK cannot phosphorylate itself because these intracellular domains are physically separated. Ligand binding induces RTK dimerization which brings these domains into close proximity and allows each receptor to phosphorylate the other, a process known as trans-autophosphorylation. Unlike RTKs, the TCR does not have intrinsic kinase domains that allow for autophosphorylation. However, tyrosine kinases of the Src family (SFKs) (e.g. Lck, Fyn) may associate with the TCR/CD3 complex even in the basal state such that subsequent dimerization of TCRs bring SFKs into close proximity of tyrosine residues on the other TCR. Therefore it is possible that signal transduction is initiated by TCR aggregation.

Several experimental studies have demonstrated that TCR aggregation is sufficient for TCR triggering. Stern and colleagues [57] have used *soluble* pMHC oligomers to demonstrate that homo-dimers, -trimers, and -tetramers are able to induce T cell activation (by crosslinking TCR) but monomeric pMHC cannot. However, in a physiological setting agonist pMHC are present at low concentrations making it improbable to find agonist pMHC homodimers. To address this, a subsequent study revealed that even agonist-endogenous pMHC heterodimers can drive

T cell activation [58]. These studies have convincingly demonstrated that TCR aggregation via pMHC oligomers can drive TCR triggering.

A key drawback to a model where pMHC oligomers drive TCR triggering by aggregation is that, to date, there is little evidence that pMHC form oligomers when presented to T cells on the APC membrane [49]. However, quantitative analysis of the dependence of TCR internalization rates on TCR surface density suggests that TCR internalization following exposure to pMHC pulsed APCs is preceded by TCR dimerization [59,60]. Since only triggered TCR are marked for internalization, these result imply that TCR triggering is accompanied by TCR dimerization. Although suggestive, it does not follow that TCR triggering is the result of and follows dimerization/aggregation. Moreover, it is possible that TCR dimerization/aggregation follows and is the result of TCR triggering.

Assuming that TCR aggregation via pMHC binding is required for TCR triggering, it is natural to ask what are the effects of TCR aggregation on pMHC discrimination. This question was investigated using a mathematical model by Salzmann and Bachmann [61] (reviewed by Bachmann and Ohashi [62]). The model assumes that two pMHC species, denoted with superscript plus and minus, undergo reversible reactions with TCR,

$$
\begin{aligned}
C^+ &= K^+ P^+ T \\
C^- &= K^- P^- T
\end{aligned}
$$

where T, P, and C represent the free TCR, free pMHC, and bound TCR–pMHC concentrations. Assuming that all TCR–pMHC complexes rapidly partition into dimers they approximate the dimer concentrations as

$$
\begin{aligned}
D_{+/+} &= \beta (C^+)^2 / C_{\text{tot}} \\
D_{+/-} &= 2\beta C^+ C^- / C_{\text{tot}} \\
D_{-/-} &= \beta (C^-)^2 / C_{\text{tot}}.
\end{aligned}
$$

where $C_{\text{tot}} = C^+ + C^-$, β is a dimensionless proportionality constant, and the subscripts indicate the dimer composition. The model next assumes that both TCR in the newly formed dimer undergo basic kinetic proofreading and that a productive signal is transduced only if both TCR remain bound to pMHC. They do not model kinetic proofreading explicitly, but instead assume a simple lag time (τ) between TCR/pMHC binding and full TCR activation. The probability that both TCR in an agonist homodimer remain bound after time t is simply $e^{-k_{\text{off}}^+ t} e^{-k_{\text{off}}^+ t}$. Signals generated by each pMHC can be computed as

$$
\begin{aligned}
A_{+/+} &= \beta \frac{(C^+)^2}{C_{\text{tot}}} \exp^{-2k_{\text{off}}^+ t} \\
A_{-/-} &= \beta \frac{(C^-)^2}{C_{\text{tot}}} \exp^{-2k_{\text{off}}^- t}.
\end{aligned}
$$

Extending the analysis for an oligomer consisting of n TCR–pMHC complexes the relative signal obtained from the two pMHC species is then

$$\frac{A_{n+}}{A_{n-}} = \left[\frac{C^{+}}{C^{-}}\right]^{n} \exp^{-n(k_{\mathrm{off}}^{+}-k_{\mathrm{off}}^{-})\tau} \tag{2.1}$$

where τ is the lag phase required for signalling. The equation is further simplified by assuming 1) both pMHC are present at concentrations that exceed the TCR concentration and 2) both pMHC are present at equal concentrations. The ratio of signals generated becomes

$$\frac{A_{n+}}{A_{n-}} = \left[\frac{k_{\mathrm{on}}^{+}/k_{\mathrm{off}}^{+}}{k_{\mathrm{on}}^{-}/k_{\mathrm{off}}^{-}}\right]^{n} \exp^{-n(k_{\mathrm{off}}^{+}-k_{\mathrm{off}}^{-})\tau} .$$

Note that molecular concentrations do not appear in the equation. Based on the form of this equation the authors argue that the discriminatory capacity of T cells increases more rapidly with the oligomer size n (appearing in both exponents) than the proofreading lag τ. This result is important because increasing τ will, in general, reduce the total signals obtained by the T cells (equivalent to reducing sensitivity by additional proofreading steps). By requiring both TCR in a dimer to be bound to pMHC, they propose that a further increase in specificity can be achieved.

There are several shortcomings to the model. The most critical shortcoming is the omission of kinetic processes, mainly serial binding. We expect that pMHC with large off-rates will be less likely to transduce productive signals (as predicted by the model) but in addition, these pMHC will be able to rapidly form and reform many dimers effectively increasing the probability of transducing productive signals. We suspect that including this effect will remove the n dependence in the first fraction above, reducing specificity obtained by oligomers. Additionally, it is unclear how these results will be altered if the two pMHC species are present in unequal numbers (e.g. $[P^{+}] \ll [P^{-}]$). In the future, it will be important to extend the model to the physiological scenario where the agonist pMHC is expressed at low numbers, possibly revealing the importance of agonist-endogenous pMHC heterodimers. In addition, any future model should account for the process of aggregation as a kinetic process that allows for the formation and also disassembly of TCR oligomers.

We briefly mention that models have been proposed based on the aggregation of T cell coreceptors (CD4/CD8) and TCR [49]. T cell coreceptors are able to bind pMHC directly at a site that is independent of the TCR binding site and therefore complexes composed of TCR–pMHC–coreceptor are expected to form. This coreceptor heterodimerization model posits that only once this complex forms can a productive signal be transduced. This mechanism is possible because Lck, an important kinase that can phosphorylate the TCR signalling modules, is constitutively associated with T cell coreceptors. However, many studies have shown that TCR triggering is possible in the complete absence of the T cell coreceptors, suggesting that coreceptors may enhance triggering by recruiting additional Lck to the

TCR–pMHC complex [16] or stabilizing the TCR/pMHC interaction [63]. A model incorporating the effects of agonist-endogenous pMHC heterodimers and coreceptors has also been proposed [58]. This pseudodimer model, although plausible in principle, cannot account for TCR triggering in the complete absence of coreceptors.

In summary, studies have shown that TCR aggregation, by TCR crosslinking with soluble pMHC oligomers, can activate T cells and therefore trigger TCR. However, the mechanism of TCR crosslinking at the T cell–APC interface remains elusive because pMHC do not form oligomers. Under the assumption that TCR–pMHC complexes aggregate, mathematical modelling has attempted to determine the effects of oligomerization on T cell specificity. Future experiments and modelling is required to determine the physiological mechanism of TCR aggregation and the effects that aggregation may have on pMHC detection and discrimination.

In recent years, experiments have demonstrated that TCR rapidly aggregate into sub-micron scale clusters when stimulated by pMHC on a supported planar bilayer [29, 30, 64]. These studies demonstrated the importance of TCR clusters by showing that signalling molecules localize to them. However, it is unclear if TCR cluster formation is required for TCR triggering (via aggregation) or if TCR clusters form once triggering has already taken place, possibly allowing for signal amplification. In addition to the aggregation of TCR (and signalling molecules) in clusters, studies have also revealed that certain molecules (e.g. the membrane phosphatase CD45) are excluded from clusters [30]. This raises the possibility that TCR clusters may be important for molecular segregation. The role of molecular segregation in TCR triggering and pMHC discrimination is the topic of the next section.

Segregation

Given the inability of the conformational change and aggregation mechanisms to fully account for TCR triggering, a third type of mechanism has been proposed, namely that triggering is the result of segregation of the engaged TCR/CD3 complex from inhibitory molecules [65–67]. This kinetic-segregation model of TCR triggering was inspired by two observation. Firstly, exposure of T cells to tyrosine phosphatase inhibitors stimulates a dramatic increase in tyrosine phosphorylation on the TCR/CD3 complex and results in full T cell activation. This observation indicates that there is constitutive tyrosine phosphorylation of the TCR/CD3 complex which is normally balanced by tyrosine phosphatases (reviewed in [66]). Secondly, the most abundant receptor tyrosine phosphatase, CD45, has a much larger ectodomain than the TCR and would therefore be expected to segregate from the engaged TCR at the T cell/APC contact interface [68]. As a result of this local segregation, the kinase/phosphatase balance is shifted decisively towards phosphorylation which triggers a signal cascade.

This model of TCR triggering was explored using stochastic simulations by Burroughs et al. [69]. The simulation domain was taken to be a square of area of 1 μm^2 with periodic boundary condition. In the absence of agonist pMHC,

TCR randomly diffuse on a lattice and are continually phosphorylated (kinases) and dephosphorylated (phosphatases). Since there are multiple phosphorylation sites on the TCR they use kinetic proofreading to represent each phosphorylated state of the TCR but unlike previous schemes, they account for the ability of phosphatases to reverse individual kinetic proofreading steps. In the basal state, an individual TCR cannot reach the final step in the kinetic proofreading scheme because phosphatases rapidly reverse each step. Therefore this study accounts for phosphatase activity by implementing a dual-track (reversible) kinetic proofreading mechanism.

In the presence of agonist pMHC, TCR/pMHC complexes will form which will segregate membrane proteins with large ectodomains, such as the membrane phosphatase CD45. For simplicity the model does not account for the formation of areas depleted of CD45, herein referred to as kinase rich domains (KRD), and instead models nine identical and static KRDs within the simulation domain, Fig. 2.3a. KRDs have two major effects that collectively allow for TCR triggering. First, the rate to traverse forward steps in the reversible kinetic proofreading scheme is increased while the backward rate decreases, due to the depletion and enrichment of phosphatases and kinases, respectively. Secondly, TCR that bind pMHC within KRDs are expected to become trapped for longer periods of time due to membrane deformation. This effect is modelled by assuming that the TCR/pMHC bond is spring-like, favoring an optimal intermembrane separation. Departing a close-contact zone will stress the bond and therefore effectively confines complexes to KRDs. Longer durations in KRDs will increase the probability of TCR triggering.

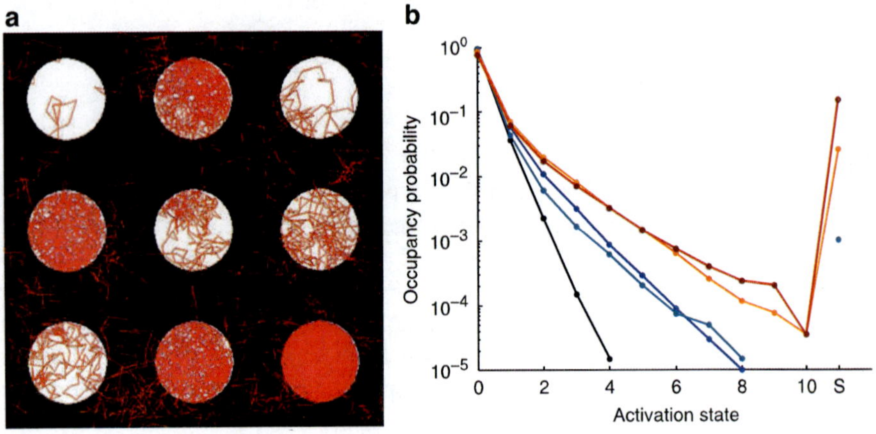

Fig. 2.3 T cell receptor triggering via kinetic segregation. (**a**) Stochastic spatial simulations of diffusing TCRs (*red traces*) are performed on a periodic boundary consisting of nine kinase rich domains (KRDs). (**b**) The probability of the TCR being in the ith proofreading step is shown without any pMHC (*black line*), low activity self pMHC with $k_{\text{off}} = 5\text{s}^{-1}$ and [pMHC]=300 μm^{-2} (*dark blue*), self pMHC with $k_{\text{off}} = 3\text{s}^{-1}$ and [pMHC] = 50 μm^{-2} (*light blue*), high density self pMHC with $k_{\text{off}} = 3\text{s}^{-1}$ and [pMHC]=300 μm^{-2} (*orange*), and high density self pMHC with agonist pMHC having $k_{\text{off}} = 0.1\text{s}^{-1}$ and [pMHC]=1 μm^{-2} (*red*). The physiologically relevant comparisons are the *light blue curve* (little productive signalling) and *red curve*, showing substantial productive signalling only when agonist pMHC are present. For further details see Burroughs et al. [69]

The critical results from the model are shown in Fig. 2.3b, where the kinetic proofreading steps are plotted against the occupancy probability. First, we see that TCR are able to reach the final activation state (S) and hence become triggered by remaining in KRDs for sufficiently long. Second, TCR triggering by self pMHC is minimal and therefore kinetic-segregation is also able to minimize noise from a high density of such pMHC. Lastly, small changes in the off-rate of agonist pMHC alters the number of triggered TCR and therefore allow for pMHC discrimination (not shown).

The model itself has several shortcomings. For simplicity, forward and reverse kinetic proofreading steps are assumed to be first order. However, these kinase/phosphatase enzymatic interactions are bimolecular two-step reactions. In the future it will be important to investigate the role of these nonlinearities in pMHC detection. The model is also sensitive to the size of KRDs, whereby KRDs with a radius of over 300 nm lead to TCR triggering from self pMHC. It will be important to determine which factors determine the size of KRDs and whether they reach this critical size on the T cell surface.

In addition to kinetic-segregation, another class of TCR triggering model based on segregation invokes lipid rafts, which are lipid microdomains thought to be enriched tyrosine kinases such as Lck and deficient in tyrosine phosphatases such as CD45. These models postulate that TCR engagement of pMHC results in association of the TCR/CD3 complex with lipid rafts, resulting in enhances phosphorylation because of the altered kinase/phosphatase balance within rafts. The main drawback of lipid raft models is that they do not provide a plausible molecular mechanism by which TCR engagement of pMHC drives association with rafts.

Concluding Remarks

Experiments have shown that T cells respond to very few and very specific pMHC presented on antigen presenting cells. In order to understand these observations various mathematical models have been formulated based on the known biophysics and biochemistry of TCR/pMHC interactions and the signalling events that are triggered within the T cell upon pMHC binding to TCR. The backbone of all mathematical models to date is the kinetic proofreading model which is able to explain pMHC discrimination based on the TCR/pMHC bond off-rate. Over the past decade, this simple model has been extended and modified to explain various experimental observations. In parallel to this research aimed at understanding antigen discrimination, molecular immunologists have been investigating the mechanism by which pMHC binding to TCR transduces a signal across the plasma membrane that initiates the very first steps intracellular signalling. This process of TCR triggering is intricately linked to antigen discrimination and we believe that additional insights can be made by future mathematical modelling that couples the two processes.

Acknowledgments DC and OD acknowledge financial support from the National Science and Engineering Research Council of Canada and the Mathematics of Information Technology and Complex Systems National Centre of Excellence. PAV is supported by the UK Medical Research Council. We also thank Salvatore Valitutti and Raibatak Das for valuable discussions.

References

1. Sykulev Y, Joo M, Vturina I, Tsomides TJ, Eisen HN (1996) Evidence that a single peptide-MHC complex on a target cell can elicit a cytolytic T cell response. Immunity 4:565–571
2. Irvine DJ, Purbhoo MA, Krogsgaard M, Davis MM (2002) Direct observation of ligand recognition by T cells. Nature 419:845–849
3. Sloan-Lancaster J, Shaw AS, Rothbard JB, Allen PM (1994) Partial T cell signaling: Altered phospho-ζ and lack of ZAP-70 recruitment in APL-induced T cell anergy. Cell 79:913–922
4. Kersh GJ, Miley MJ, Nelson CA, Grakoui A, Horvath S, Donermeyer DL, Kappler J, Allen PM, Fremont DH (2001) Structural and functional consequences of altering a peptide MHC anchor residue. J Immunol 166:3345–3354
5. Jönsson U, Fägerstam L, Ivarsson B, Johnsson B, Karlsson R, Lundh K, Löfás S, Persson B, Roos H, Rönnberg I (1991) Real-time biospecific interaction analysis using surface plasmon resonance and a sensor chip technology. BioTechniques 11:620–627
6. Malmqvist M (1993) Biospecific interaction analysis using biosensor technology. Nature 361:186–187
7. van der Merwe PA (2001) The TCR triggering puzzle. Immunity 14:665–668
8. Tolentino TP, Wu J, Zarnitsyana VI, Fang Y, Dustin ML, Zhu C (2008) Measuring diffusion and binding kinetics by contact area FRAP. Biophys J 95:920–930
9. Dushek O, Das R, Coombs D (2009) A role for rebinding in rapid and reliable T cell responses to antigen. PLoS Comput Biol 5:e1000578
10. Aleksic M, Dushek O, Zhang H, Shenderov E, Chen JL, Cerundolo V, Coombs D, van der Merwe PA (2010) Dependence of T cell antigen recognition on T cell receptor-peptide MHC confinement time. Immunity 32:1–12
11. Wofsy C, Coombs D, Goldstein B (2001) Calculations show substantial serial engagement of T cell receptors. Biophys J 80:606–612
12. Qi S, Krogsgaard M, Davis MM, Chakraborty AK (2006) Molecular flexibility can influence the stimulatory ability of receptor-ligand interactions at cell-cell junctions. Proc Natl Acad Sci USA 103:4416–4421
13. Krogsgaard M, Prado N, Adams EJ, He XL, Chow DC, Wilson DB, Garcia KC, Davis MM (2003) Evidence that structural rearrangements and/or flexibility during TCR binding can contribute to T cell activation. Mol Cells 12:1367–1378
14. van der Merwe PA, Davis SJ (2003) Molecular interactions mediating T cell antigen recognition. Annu Rev Immunol 21:659–684
15. Feinerman O, Germain RN, Altan-Bonnet G (2008) Quantitative challenges in understanding ligand discrimination by alphabeta T cells. Mol Immunol 45:619–631
16. Holler PD, Kranz DM (2003) Quantitative analysis of the contribution of TCR/pepMHC affinity and CD8 to T cell activation. Immunity 18:255–264
17. Tian S, Maile R, Collins EJ, Frelinger JA (2007) CD8+ T cell activation is governed by TCR-peptide/MHC affinity, not dissociation rate. J Immunol 179:2952–2960
18. Wooldridge L, van den Berg HA, Glick M, Gostick E, Laugel B, Hutchinson SL, Milicic A, Brenchley JM, Douek DC, Price DA, Sewell AK (2005) Interaction between the CD8 coreceptor and major histocompatibility complex class I stabilizes T cell receptor-antigen complexes at the cell surface. J Biol Chem 280:27491–27501
19. van den Berg HA, Rand DA, Burroughs NJ (2001) A reliable and safe T cell repertoire based on low-affinity T cell receptors. J Theor Biol 209:465–486

20. McKeithan K (1995) Kinetic proofreading in T-cell receptor signal transduction. Proc Natl Acad Sci USA 92:5042–5046
21. Hopfield JJ (1974) Kinetic proofreading: A new mechanism for reducing errors in biosynthetic processes requiring high specificity. Proc Natl Acad Sci USA 71:4135–4139
22. Ninio J (1975) Kinetic amplification of enzyme discrimination. Biochimie 57:587–595
23. van den Berg HA, Burroughs NJ, Rand DA (2002) Quantifying the strength of ligand antagonism in TCR triggering. Bull Math Biol 64:781–808
24. Coombs D, Kalergis AM, Nathenson SG, Wofsy C, Goldstein B (2002) Activated TCRs remain marked for internalization after dissociation from pMHC. Nat Immunol 3:926–931
25. Chan C, George AJT, Stark J (2003) T cell sensitivity and specificity - kinetic proofreading revisited. Discrete Continuous Dyn Syst - Series B 3:343–360
26. Gonzalez PA, Carreno LJ, Coombs D, Mora JE, Palmieri E, Goldstein B, Nathenson SG, Kalergis AM (2005) T cell receptor binding kinetics required for T cell activation depend on the density of cognate ligand on the antigen-presenting cell. Proc Natl Acad Sci USA 102: 4824–4829
27. Wedagedera JR, Burroughs NJ (2006) T-cell activation: A queuing theory analysis at low agonist density. Biophys J 91:1604–1618
28. George AJT, Stark J, Chan C (2005) Understanding specificity and sensitivity of T-cell recognition. Trends Immunol 26:653–659
29. Campi G, Varma R, Dustin ML (2005) Actin and agonist MHC-peptide complex-dependent T cell receptor microclusters as scaffolds for signaling. J Exp Med 202:1031–1036
30. Varma R, Campi G, Yokosuka T, Saito T, Dustin ML (2006) T cell receptor-proximal signals are sustained in peripheral microclusters and terminated in the central supramolecular activation cluster. Immunity 25:117–127
31. Valitutti S, Muller S, Cella M, Padovan E, Lanzavecchia A (1995) Serial triggering of many T-cell receptors by a few peptide-MHC complexes. Nature 375:148–151
32. Dushek O, Coombs D (2008) Analysis of serial engagement and peptide-MHC transport in T cell receptor microclusters. Biophys J 94:3447–3460
33. Torigoe C, Inman JK, Metzger H (1998) An unusual mechanism for ligand antagonism. Science 281:568–572
34. Faeder JR, Hlavacek WS, Reischl I, Blinov ML, Metzger H, Redondo A, Wofsy C, Goldstein B (2003) Investigation of early events in FceRI-mediated signaling using a detailed mathematical model. J Immunol 170:3769–3781
35. Goldstein B, Coombs D, Faeder JR, Hlavacek WS (2008) Kinetic proofreading model. Adv Exp Med Biol 640:82–94
36. Stefanová I, Hemmer B, Vergelli M, Martin R, Biddison WE, Germain RN (2003) TCR ligand discrimination is enforced by competing ERK positive and SHP-1 negative feedback pathways. Nat Immunol 4:248–254
37. Altan-Bonnet G, Germain RN (2005) Modeling T cell antigen discrimination based on feedback control of digital ERK responses. PLoS Biol 3:e356
38. Lipniacki T, Hat B, Faeder JR, Hlavacek WS (2008) Stochastic effects and bistability in T cell receptor signaling. J Theor Biol 254:110–122
39. Lauffenburger D, Linderman J (1993) Receptors: Models for binding, trafficking, and signaling. Oxford University Press, Oxford
40. Rabinowitz JD, Beeson C, Lyons DS, Davis MM, McConnell HM (1996) Kinetic discrimination in T-cell activation. Proc Natl Acad Sci USA 93:1401
41. Chan C, George AJT, Stark J (2001) Cooperative enhancement of specificity in a lattice of T cell receptors. Proc Natl Acad Sci USA 98:5758–5763
42. Chan C, Stark J, George AJT (2004) Feedback control of T-cell receptor activation. Proc R Soc B: Biol Sci 271:931–939
43. Wylie DC, Das J, Chakraborty AK (2007) Sensitivity of T cells to antigen and antagonism emerges from differential regulation of the same molecular signaling module. Proc Natl Acad Sci USA 104:5533–5538

44. Feinerman O, Veiga J, Dorfman JR, Germain RN, Altan-Bonnet G (2008) Variability and robustness in T cell activation from regulated heterogeneity in protein levels. Science 321:1081–1084
45. Casal A, Sumen C, Reddy TE, Alber MS, Lee PP (2005) Agent-based modeling of the context dependency in T cell recognition. J Theor Biol 236:376–391
46. Wylie DC, Hori Y, Dinner AR, Chakraborty AK (2006) A hybrid deterministic-stochastic algorithm for modeling cell signaling dynamics in spatially inhomogeneous environments and under the influence of external fields. J Phys Chem B 110:12749–12765
47. Dushek O (2008) Mathematical modeling in cellular immunology: T cell activation and parameter estimation. PhD thesis, University of British Columbia
48. Choudhuri K, Kearney A, Bakker TR, van der Merwe PA (2005) Immunology: How do T cells recognize antigen? Curr Opin Biol 15:R382–R385
49. Choudhuri K, van der Merwe PA (2007) Molecular mechanisms involved in T cell receptor triggering. Semin Immunol 19:255–261
50. Ma Z, Janmey PA, Finkel TH (2008) The receptor deformation model of TCR triggering. FASEB J 22:1002–1008
51. Aivazian D, Stern LJ (2000) Phosphorylation of T cell receptor zeta is regulated by a lipid dependent folding transition. Nature 7:1023–1026
52. Gil D, Schamel WWA, Montoya M, Sánchez-Madrid F, Alarcón B (2002) Recruitment of Nck by CD3ϵ reveals a ligand-induced conformational change essential for T cell receptor signaling and synapse formation. Cell 109:901–912
53. Xu C, Gagnon E, Call ME, Schnell JR, Schwieters CD, Carman CV, Chou JJ, Wucherpfennig KW (2008) Regulation of T cell receptor activation by dynamic membrane binding of the CD3epsilon cytoplasmic tyrosine-based motif. Cell 135:702–713
54. Gil D, Schrum AG, Alarcon B, Palmer E (2005) T cell receptor engagement by peptide-MHC ligands induces a conformational change in the CD3 complex of thymocytes. J Exp Med 201:517–522
55. Mingueneau M, Sansoni A, Grégoire C, Roncagalli R, Aguado E, Weiss A, Malissen M, Malissen B (2008) The proline-rich sequence of CD3epsilon controls T cell antigen receptor expression on and signaling potency in preselection $CD4^+CD8^+$ thymocytes. Nat Immunol 9:522–532
56. Heldin CH (1995) Dimerization of cell surface receptors in signal transduction. Cell 80: 213–223
57. Cochran JR, Cameron TO, Stern LJ (2000) The relationship of MHC-peptide binding and T cell activation probed using chemically defined MHC class II oligomers. Immunity 12:241–250
58. Krogsgaard M, Li QJ, Sumen C, Huppa JB, Huse M, Davis MM (2005) Agonist/endogenous peptide-MHC heterodimers drive T cell activation and sensitivity. Nature 434:238–243
59. Bachmann MF, Salzmann M, Oxenius A, Ohashi PS (1998) Formation of TCR dimers/trimers as a crucial step for T cell activation. Eur J Immunol 28:2571–2579
60. Utzny C, Coombs D, Müller S, Valitutti S (2006) Analysis of peptide/MHC-induced TCR downregulation: Deciphering the triggering kinetics. Cell Biochem Biophys 46:101–111
61. Salzmann M, Bachmann MF (1998) The role of T cell receptor dimerization for T cell antagonism and T cell specificity. Mol Immunol 35:271–277
62. Bachmann MF, Ohashi PS (1999) The role of T-cell receptor dimerization in T-cell activation. Immunol Today 20:568–576
63. Wyer J, Willcox B, Gao G, Gerth U, Davis S, Bell J, van der Merwe P, Jakobsen B (1999) T cell receptor and coreceptor CD8 alphaalpha bind peptide-MHC independently and with distinct kinetics. Immunity 10:219–225
64. Yokosuka T, Sakata-Sogawa K, Kobayashi W, Hiroshima M, Hashimoto-Tane A, Tokunaga M, Dustin ML, Saito T (2005) Newly generated T cell receptor microclusters initiate and sustain T cell activation by recruitment of Zap70 and SLP-76. Nat Immunol 6:1253–1262
65. Davis SJ, van der Merwe PA (1996) The structure and ligand interactions of CD2: Implications for T-cell function. Immunol Today 17:177–187
66. van der Merwe PA, Davis SJ, Shaw AS, Dustin ML (2000) Cytoskeletal polarization and redistribution of cell-surface molecules during T cell antigen recognition. Semin Immunol 12:5–21

67. Davis SJ, van der Merwe PA (2006) The kinetic-segregation model: TCR triggering and beyond. Nat Immunol 7:803–809
68. Springer TA (1990) Adhesion receptors of the immune system. Nature 346:425–434
69. Burroughs NJ, Lazic Z, van der Merwe PA (2006) Ligand detection and discrimination by spatial relocalization: A kinase-phosphatase segregation model of TCR activation. Biophys J 91:1619–1629

Chapter 3
Dynamic Tuning of T Cell Receptor Specificity by Co-Receptors and Costimulation

Hugo A. van den Berg and Andrew K. Sewell

Abstract Mounting evidence that each clonotype of T cell antigen receptor can productively interact with hundreds or even thousands of peptide antigens would appear to conflict, *prima facie*, with the immune system's primary task of guarding against auto-immunity and singling out harmful "non-self" epitopes against a background of "self" epitopes. This paradox dissolves somewhat once it is appreciated that, at any one time, a TCR will only have high functional sensitivity to a small subset of all its potential agonists, that is, when presented at low copy numbers, only this small subset will be able to activate the T cell. In this light, the self-nonself problem becomes a matter of keeping the TCR trained on the appropriate subset of salient epitopes. We review evidence and models that show how the co-receptors CD4 and CD8, as well as the "signal 2" costimulatory system, act to keep the TCR focussed on the appropriate agonist subset. On the theory of dynamic tuning of TCR specificity, both immune tolerance and specific reactivity against salient epitopes rely on continual regulation by other components of the immune system. We present a model of "avidity maturation" during the early phase of a T cell response.

The TCR repertoire, while vast, is thought to be too small by at least one order of magnitude to encompass a specific TCR against any pathogen-derived peptide the host may ever encounter [1, 2]. Several lines of evidence suggest that a given TCR may recognise a wide variety of peptides, a phenomenon known as *TCR degeneracy* [3, 4]. This raises the questions of how the immune system ensures that such degeneracy does not engender autoimmunity, and if any of these mechanisms may also play a role in TCR maturation. This chapter reviews models of T cell activation that propose that the degeneracy of a TCR is not merely a static property determined by the molecular identity of its CDRs, but instead a highly malleable property of the TCR/pMHC/co-receptor complex, subject to dynamic modulation that can restrict or expand the set of ligands for which the TCR has high functional

H.A. van den Berg (✉)
Warwick Systems Biology Centre, University of Warwick, Coventry CV47AL, UK
e-mail: hugo@maths.warwick.ac.uk

C. Molina-París and G. Lythe (eds.), *Mathematical Models and Immune Cell Biology*,
DOI 10.1007/978-1-4419-7725-0_3,

sensitivity, and which, moreover, can act to focus the TCR on one particular ligand, suggestive of an avidity maturation mechanism analogous to affinity maturation in humoral immunity.

A key variable in the analysis is *functional sensitivity*, a term coined by Wooldridge et al. [5]. This quantity corresponds to what has been termed *TCR avidity* by some immunologists (e.g. [6]), although use of this latter term may invite confusion, since *avidity* already has an established biochemical meaning. Quantification of TCR signalling strength has been controversial, which is perhaps not surprising since a read-out at the cellular/population level must somehow be related to biomolecular parameters. The importance of both the TCR/pMHC off-rate (equivalently, the mean TCR/pMHC interaction time) and the affinity (off-rate divided by on-rate) has been emphasised, as well as that of TCR and pMHC densities on the surfaces of, respectively, the T cell and the APC. The TCR triggering rate model of functional sensitivity due to Rand and co-workers [7–9] is perhaps the simplest mathematical model that combines all these parameters, and is briefly reviewed below.

When one considers TCR degeneracy in statistical terms, each TCR is viewed as functionally sensitive to every possible peptide ligand, and characterised by its distribution over these ligands. Almost all of the probability mass is concentrated about zero sensitivity; these *null peptides* trivially include those that cannot bind to the appropriate MHC isoform. On the order of 10^{-5} of the probability mass is concentrated over the non-null range of functional sensitivities, with $\sim 10^{-6}$–10^{-7} found at the maximum functional sensitivity end of the spectrum.

Looking at TCR degeneracy in statistical terms is necessary because a T cell can actively *alter* its functional sensitivity distribution, and hence its TCR degeneracy. That is to say that a T cell can tune itself to become more broad-spectrum promiscuous, or sensitive only to a very small set of peptide ligands.[1] Moreover, it seems possible that the T cell can select this ligand from among a set of potential strong agonists. These abilities can be understood, in mechanistic terms, on the basis of the TCR triggering rate model, as will be explained below.

Dynamic degeneracy would vastly amplify the flexibility of the T cell response and allow it to mount a specific response to virtually every antigen even with a limited repertoire. However, dynamic degeneracy might equally well allow a substantial portion of T cells to become autoreactive – even those who have been exposed to negative selection. This underlines the importance of costimulation in regulating the T cell's functional sensitivity, in both health and disease. The role of costimulation and threshold adaptation in the maintenance of the naïve T cell repertoire has been analysed elsewhere [10, 11]; we consider the role of costimulation in adjusting the functional sensitivity spectrum associated with an immune response.

[1] A residual level of promiscuity is inherent in the physico-chemical properties of the TCR/pMHC contact; one might say that a TCR tuned to minimal degeneracy is sensitive to a single *molecular fingerprint*.

The Triggering Rate Model for Functional Sensitivity

The T cell integrates, at the whole-cell level, intracellular signals emanating from individual TCR/CD3 complexes that have been triggered by a peptide/MHC molecule (the triggered TCR corresponds to a signalosome, a complex involving several kinases and adapter linkers). The philosophy of the triggering model, due to Rand and co-workers, is that any pMHC can trigger a given TCR, but will do so with widely varying likelihood. In particular, the rate at which TCRs are triggered on the surface of a given T cell is given by a simple formula:

$$\sum_j \Omega_{ij} P_{\mathrm{trig}}(i,j)$$

where the sum ranges over all pMHC species j; Ω_{ij} denotes the rate at which couplings occur between the TCR of clonotype i and a ligand of species j, whereas $P_{\mathrm{trig}}(i,j)$ denotes the probability that a given interaction between a TCR of type i and a pMHC of type j results in the TCR being triggered.

The kinetic pre-factor Ω_{ij} generally depends on several factors: the density of free (unbound) TCR molecules on the T cell surface; the density of free pMHC molecules on the APC surface; and the affinity constant of the TCR/pMHC interaction. A detailed derivation of these dependencies is given by van den Berg et al. [8]. The kinetic pre-factor depends only on the free TCR density when the latter is sufficiently low relative to the affinity constant; in this *TCR-limited triggering regime* the expression levels of ligand are immaterial. This regime may prevail during positive selection in the thymus, and is perhaps also important in CTLs. On the other hand, interactions between T cells and APCs in secondary lymphoid tissues may be expected to take place in the *MHC-limited triggering regime*, where the rate factor depends only on the pMHC levels:

$$\Omega_{ij} = \frac{Z_j}{T_{ij}} \tag{3.1}$$

where Z_j is the copy number of pMHC species j engaged by the T cell and T_{ij} denotes the mean interaction time of the TCR/pMHC interaction (that is, $1/T_{ij}$ is the off-rate). Although the affinity constant does not appear explicitly in this expression, it plays an important role in governing the change-over between TCR- and MHC-limited triggering; a theoretical prediction that was borne out by experimental observations [12]. The theory postulates that the T cell response is a functional of the triggering rate $\sum_j \Omega_{ij} P_{\mathrm{trig}}(i,j)$ (cf. [13–15]); e.g. the T cell is activated if the integrated rate $\sum_j \Omega_{ij} P_{\mathrm{trig}}(i,j)$ exceeds a certain value, called the *cellular activation threshold*, during the T cell:APC conjugation [16].

Equation (3.1) forms a link between theory and experiment. Write the TCR triggering rate due to pMHC species j as W_{ij} and assume MHC-limited triggering. Then

$$W_{ij} = Z_j \frac{P_{\mathrm{trig}}(i,j)}{T_{ij}} \,. \tag{3.2}$$

Also assume that the T cell responds when $W_{ij} > W_{\text{act}}$ where $W_{\text{act}} > 0$ denotes the cellular activation threshold. The functional sensitivity is determined experimentally by exposing T cells to APCs with various ligand levels and determining the midpoint of the dose-response curve.[2] Let Z_j^{crit} denote this midpoint level. Then the reciprocal is taken as a measure of functional sensitivity: the higher the pMHC copy number required, the lower the functional sensitivity of the TCR for that ligand. In formula:

$$\frac{1}{Z_j^{\text{crit}}} = \frac{1}{W_{\text{act}}} \cdot \frac{P_{\text{trig}}(i,j)}{T_{ij}} \tag{3.3}$$

which shows that *functional sensitivity is proportional to the MHC-specific TCR triggering rate*[3] $P_{\text{trig}}(i,j)/T_{ij}$. Moreover, (3.3) shows how functional sensitivity might be modulated: through changes of $P_{\text{trig}}(i,j)$ and T_{ij} at the receptor level and through changes of W_{act} at the cellular level. We discuss modulation of $P_{\text{trig}}(i,j)$ and T_{ij} via the co-receptor and then modulation of W_{act} via the costimulatory ligands.

Co-Receptor Tuning of Functional Sensitivity

T cells vary considerably with respect to the amount of stimulation that is required to elicit a response: naïve, nearly quiescent T cells require an interaction with a professional APC that may last for hours, whereas fully differentiated CTLs may be triggered by very few ligand molecules [17, 18]. These differences reflect differences in the expression of the TCR, various kinases, phosphorylases, including CD45 which occurs in isoforms that correlate with the cellular activation threshold of the T cell, as well as different forms of costimulation which modulate the strength of the cognate TCR/pMHC interaction. The T cell surface glycoproteins CD4 and CD8 interact with invariable regions of the MHC molecule, independently of the cognate contact [19], and hence termed *co-receptors* (see Fig. 3.1). This section focuses on the effects exerted by these co-receptors on TCR triggering.

[2] The width of the dose-response curve is attributable to natural variability in Z_j among the APCs and W_{act} among the T cells; this variation comes into play because a population-level response is measured.

[3] In fact, an additional step intervenes between experiment and theory: the immunologist does not control the presentation level Z directly, but rather the concentration at which the APC is incubated with peptide ligand. The latter may be assumed to increase with Z and will often be simply proportional.

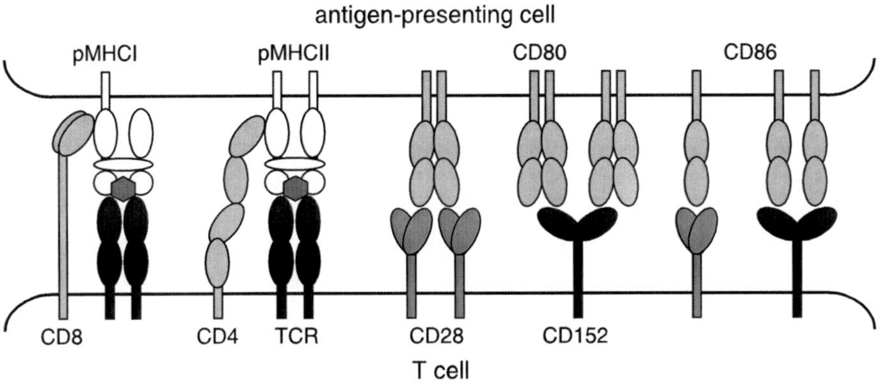

Fig. 3.1 Receptors and co-receptors at the T cell:APC interface

The TCR Triggering Probability

The probability that the TCR/CD3 complex achieves signalosome status (i.e. starts to contribute to the activation signal that is conveyed through intracellular signalling pathways) following an interaction with a pMHC molecule is given by the following formula[4]:

$$P_{\text{trig}} = \int_0^\infty P_n(t) f_T(t)\mathrm{d}t \tag{3.4}$$

where $f_T(t)$ is the probability density function of the TCR/pMHC residence time, that is, the probability that the TCR remains bound to the pMHC molecule for at least an amount of time t equals $\int_t^\infty f_T(u)\mathrm{d}u$, and $P_n(t)$ is the probability that the TCR/CD3 complex is phosphorylated at all n sites at time t. If the dissociation rate is not influenced by the age of the ternary complex, the TCR/pMHC interaction time follows the exponential distribution:

$$f_T(t) = \frac{\exp\left(-t/T_{ij}\right)}{T_{ij}} \tag{3.5}$$

where T_{ij} denotes the average docking time of a TCR/pMHC complex constituting a TCR molecule of clonotype i and a pMHC molecule of species j; the half-life of the interaction is $\log(2)T_{ij}$.

If there are a large number of individual reaction steps that all have to occur in order for the TCR to be triggered, the waiting time for completion of all steps may be expected to be well-approximated by a narrow Gaussian, in view of the Central Limit Theorem. Such steps include phosphorylations of ITAMs, binding of kinases

[4] For the sake of simplicity, the dependence on i and j will be suppressed if it is less important in the immediate context.

and linker proteins, which themselves may need to undergo phosphorylations for further steps to happen [20, 21]. The narrow Gaussian may be expected to be obtained if the steps are strictly ordered in a linear succession. For more complex transition graphs, the narrow Gaussian remains a good approximation if all steps are forward and all completion paths through the graph contain many steps; in this case the shortest path dominates. The T cell can modulate which path is shortest by diverting certain routes through the transition graph by means of ITIM phosphorylation. If backward steps, such as dephosphorylations, are allowed, the waiting time distribution widens.

In the limit for large n, the narrow Gaussian becomes degenerate. The function $P_n(t)$ is then well-approximated by the Heaviside step function:

$$P_n(t) = \begin{cases} 0 & \text{for } t \leq T_R \\ 1 & \text{for } t > T_R \end{cases} \tag{3.6}$$

where T_R denotes the receptor threshold duration: whenever the TCR remains ligated with a pMHC molecule for longer than this threshold time T_R, the TCR/CD3 complex is triggered. Equations (3.4)–(3.6) together give:

$$P_{\text{trig}} = \int_{T_R}^{\infty} \frac{\exp(-t/T_{ij})}{T_{ij}} \mathrm{d}t = \exp(-T_R/T_{ij}) . \tag{3.7}$$

Recall from (3.2) that the MHC-limited kinetic co-factor is inversely proportional to T_{ij}. The TCR triggering rate is then proportional to $\exp(-T_R/T_{ij})/T_{ij}$, which implies that the TCR triggering rate depends non-monotonically on T_{ij}, with a maximum at T_R. This is the *serial triggering* effect first proposed by Valitutti & Lanzacecchia [22] and experimentally confirmed by Kalergis et al. [23]. Clearly, the serial triggering effect will only be found under MHC-limited conditions.

The Classic Proof-Reading Model

One may question how robust approximation (3.7) is, given the simplifying assumptions. Some support is lent by the classic kinetic proof-reading model [24], which approaches the narrow Gaussian very neatly via a sequence of Gamma distributions. On this model, the TCR/CD3 complex has to undergo n ITAM phosphorylations before it acquires the ability to mediate intracellular signal transduction, leading to various cellular responses (notably gene activation in the naïve T cell) [25,26]. It is assumed that the ith phosphorylation can take place only after phosphorylations 1 through $i-1$ have taken place. Let the ITAM phosphorylation rate λ be nil if the TCR is not bound to a peptide/MHC (pMHC) complex, and take $\lambda > 0$ during TCR/pMHC ligation. The probability $P_i(t)$ that there are i phosphorylated ITAMs at time t then obeys the following differential equations during ligation:

$$\begin{cases} \frac{d}{dt} P_0 &= -\lambda P_0 \\ \frac{d}{dt} P_i &= \lambda (P_{i-1} - P_i) \quad \text{for } i = 1, \ldots, n-1 \\ \frac{d}{dt} P_n &= \lambda P_{n-1} \end{cases} \tag{3.8}$$

with initial conditions $P_0(0) = 1$, $P_i(0) = 0$ for $i > 0$. These equations are readily solved to give

$$P_i(t) = \exp(-\lambda t) \frac{(\lambda t)^i}{i!} \tag{3.9}$$

for $i = 0, \ldots, n-1$ and

$$P_n(t) = 1 - \exp(-\lambda t) \sum_{i=0}^{n-1} \frac{(\lambda t)^i}{i!} \tag{3.10}$$

for the nth phosphorylation. Substituting this result in the general expression for the triggering probability, (3.4), with an exponentially distributed dwell time, (3.5), we obtain

$$P_{\text{trig}} = \left(\frac{n}{T_R / T_{ij} + n} \right)^n \quad \text{where} \quad T_R = \frac{n}{\lambda} \tag{3.11}$$

and $\lim_{n \to \infty} P_{\text{trig}} = \exp(-T_R / T_{ij})$ in accordance with the threshold model, (3.7). The latter is already an excellent approximation to the n-step proof-reading model when n is of order 10, even if the corresponding Gamma distribution does not adequately approximate a step function for n this low.

Co-Receptor Kinetics

The co-receptors CD4 and CD8 modulate the rate of TCR triggering by pMHC engagement [27–29] and thereby the TCR's functional avidity [30]. Various distinct modulatory roles of the co-receptor, possibly acting in concert, have been proposed: (1) promoting the association of TCR and pMHCI [31]; (2) stabilising the TCR/pMHCI interaction [28, 32], thus prolonging the mean dwell time of the interaction which alters the efficacy of the pMHCI ligand [23]; and (3) enhancing the rate at which the TCR/CD3 complex attains signalling status [29, 33], by association of TCR/CD3 with protein tyrosine kinases such as $p56^{lck}$ [34] and adaptor molecules such as LAT [35] and LIME [36].

The first of these two mechanisms affects the affinity of the TCR/pMHC interaction, whereas the third affects the TCR triggering threshold. Thus, let $T^\star_{ij}$ denote the mean lifetime of the TCR/pMHC interaction when the MHC molecule is bound to the co-receptor. The stabilising effect is expressed by the inequality $T^\star_{ij} > T_{ij}$. Similarly, let $T^\star_R$ denote the TCR triggering threshold when the MHC molecule is bound to the co-receptor. Co-receptor enhancement is expressed by the inequality $T_R^\star < T_R$. There is evidence that CD8 acts through all three mechanisms [32, 37], whereas CD4 seems to act only through the third mechanism [38]. Moreover, the CD8 $\alpha\beta$ heterodimer is considerably more potent as a co-receptor than the $\alpha\alpha$ homodimer [34, 39, 40], which stresses the importance of the third function, which is strongly dependent on the presence of the CD8β chain [39, 41].

During any particular interaction between TCR/pMHC, the co-receptor may bind and unbind the MHC molecule any number of times. It will be assumed that the co-receptor-dependent effects (viz. reduced TCR/pMHC off-rate and enhanced rate of ITAM phosphorylations and so on) hold instantaneously and momentarily whenever the co-receptor is bound. Let s denote the number of transitions between the co-receptor-bound and unbound forms of the TCR/pMHC complex during the lifetime of the latter. The triggering probability, conditioned on this number of transitions, is then expressed as follows:

$$P_{\text{trig}} = \sum_{s=0}^{\infty} \mathbb{P}(\text{trig} \mid s, C^+)\mathbb{P}(s \mid C^+)\mathbb{P}(C^+) + \mathbb{P}(\text{trig} \mid s, C^-)\mathbb{P}(s \mid C^-)\mathbb{P}(C^-) \tag{3.12}$$

where $\mathbb{P}(C^+)$ and $\mathbb{P}(C^-)$ are the probabilities that the TCR/pMHC complex is bound (C^+) or unbound (C^-) to the co-receptor when the cognate contact forms. Equilibrium considerations give the following:

$$\mathbb{P}(C^+) = \frac{\gamma}{\gamma + \rho} \quad \text{and} \quad \mathbb{P}(C^-) = \frac{\rho}{\gamma + \rho} \tag{3.13}$$

where γ is the co-receptor association rate and ρ the dissociation rate. The co-receptor may engage the TCR/pMHC complex via a TCR/CD3 binding site as well as an MHC biding site, making association and dissociation two-stage processes [5, 34, 42]. Provided the kinetics of the second step are sufficiently rapid, the kinetics may be treated as first-order (see [37] for a detailed argument).

The probability of s transitions can be readily calculated by considering the geometric distribution associated with the embedded Markov chain (the "jump chain"). For s even, one finds:

$$\mathbb{P}(s \mid C^+) = \frac{1}{1 + \rho T_{ij}^\star} \left(\frac{\gamma\rho T_{ij} T_{ij}^\star}{(1 + \gamma T_{ij})(1 + \rho T_{ij}^\star)} \right)^{s/2} \tag{3.14}$$

$$\mathbb{P}(s \mid C^-) = \frac{1}{1 + \gamma T_{ij}} \left(\frac{\gamma\rho T_{ij} T_{ij}^\star}{(1 + \gamma T_{ij})(1 + \rho T_{ij}^\star)} \right)^{s/2} \tag{3.15}$$

whereas for s odd, these probabilities are:

$$\mathbb{P}(s \mid C^+) = (\gamma T_{ij})^{(s-1)/2} \left(\frac{\rho T_{ij}^\star}{(1 + \gamma T_{ij})(1 + \rho T_{ij}^\star)} \right)^{(s+1)/2} \tag{3.16}$$

$$\mathbb{P}(s \mid C^-) = \left(\rho T_{ij}^\star\right)^{(s-1)/2} \left(\frac{\gamma T_{ij}}{(1+\gamma T_{ij})(1+\rho T_{ij}^\star)}\right)^{(s+1)/2} . \tag{3.17}$$

The triggering probabilities can be computed, in general, from a Markov chain whose transition graph is obtained by doubling the original TCR/CD3 transition graph as discussed above, and inserting directed arcs linking corresponding nodes bidirectionally, expressing the processes of co-receptor association and dissociation.[5] Since the off-rate is affected by co-receptor engagement, the lifetime T of the TCR/pMHC complex is no longer an exponential variate. The mixture of two off-rates gives rise to a bi-exponential distribution:

$$\mathbb{P}(T > t) = \frac{1}{2}\left(\mathrm{e}^{-\lambda_+ t} + \mathrm{e}^{-\lambda_- t}\right) + \frac{\left((\gamma+\rho)^2 + (T_{ij}^{-1} - T_{ij}^{\star-1})(\gamma-\rho)\right)\left(\mathrm{e}^{-\lambda_+ t} - \mathrm{e}^{-\lambda_- t}\right)}{2(\gamma+\rho)\sqrt{(\gamma+\rho+T_{ij}^{-1}+T_{ij}^{\star-1})^2 - 4(\gamma/T_{ij}^\star + \rho/T_{ij} + (T_{ij}T_{ij}^\star)^{-1})}} \tag{3.18}$$

where

$$\lambda_\pm = -\frac{1}{2}(\gamma+\rho+T_{ij}^{-1}+T_{ij}^{\star-1}) \pm \frac{1}{2}\sqrt{(\gamma+\rho+T_{ij}^{-1}+T_{ij}^{\star-1})^2 - 4(\gamma/T_{ij}^\star + \rho/T_{ij} + (T_{ij}T_{ij}^\star)^{-1})} .$$

The mean lifetime is readily obtained:

$$\mathbb{E}(T) = \frac{1}{2}\left(\frac{1}{\lambda_+} + \frac{1}{\lambda_-}\right) + \frac{\left((\gamma+\rho)^2 + (T_{ij}^{-1} - T_{ij}^{\star-1})(\gamma-\rho)\right)(1/\lambda_+ - 1/\lambda_-)}{2(\gamma+\rho)\sqrt{\left(\gamma+\rho+T_{ij}^{-1}+T_{ij}^{\star-1}\right)^2 - 4\left(\gamma/T_{ij}^\star + \rho/T_{ij} + (T_{ij}T_{ij}^\star)^{-1}\right)}} . \tag{3.19}$$

[5] A simple formula is obtained using the Heaviside simplification, (3.6):

$$\mathbb{P}(\mathrm{trig} \mid s, C^\pm) = \mathbb{P}\left(\widetilde{T} + \frac{T_R}{T_R^\star}\widetilde{T}^\star > T_R \mid s, C^\pm\right)$$

where $\widetilde{T}$ is a Gamma-distributed random variable with location parameter $T_{ij}/(1+\gamma T_{ij})$ and shape parameter $\varkappa = 1 + \mathrm{ent}((s+1)/2)$ iff s is even and $C^\pm = C^-$, whereas $\varkappa = \mathrm{ent}((s+1)/2)$ otherwise (where $\mathrm{ent}(x)$ is the largest integer smaller than x), and $\widetilde{T}^\star$ is another Gamma-distributed random variable with location parameter $T_{ij}^\star/(1+\rho T_{ij}^\star)$ and shape parameter $\varkappa = 1 + \mathrm{ent}((s+1)/2)$ iff s is even and $C^\pm = C^+$, whereas $\varkappa = \mathrm{ent}((s+1)/2)$ otherwise.

Co-Receptor Kinetics in the 'Slow' and 'Fast' Limits

More insight into the above results can be gained by considering two special cases, which form the endpoints of a continuum of possibilities. The first is the case where the co-receptor kinetics is slow with respect to the TCR/pMHC kinetics, so that the MHC molecule either retains its contact with the co-receptor during the cognate interaction or remains unbound during its comparatively short docking with the TCR. Then (3.19) reduces to:

$$\mathbb{E}(T) = \frac{\rho}{\gamma + \rho} T_{ij} + \frac{\gamma}{\gamma + \rho} T^{\star}_{ij}$$

as one would expect from elementary considerations.[6] On the other hand, if the co-receptor kinetics is very rapid compared to the cognate interaction, many transitions happen during a TCR/pMHC docking, and the lifetime becomes exponential again, with an effective off-rate determined by the co-receptor binding equilibrium. Thus (3.19) reduces to

$$\mathbb{E}(T) = \left(\frac{\rho}{\gamma + \rho} \cdot \frac{1}{T_{ij}} + \frac{\gamma}{\gamma + \rho} \cdot \frac{1}{T^{\star}_{ij}} \right)^{-1}.$$

Simple expressions for the triggering probability are similarly obtained for these special cases. If the co-receptor is slow relative to the TCR, the $s = 0$ term in (3.12) dominates and one finds:

$$P_{\text{trig}} = \frac{\rho}{\gamma + \rho} \exp\left(-T_R / T_{ij}\right) + \frac{\gamma}{\gamma + \rho} \exp\left(-T^{\star}_R / T^{\star}_{ij}\right) \tag{3.20}$$

whereas for rapid co-receptor kinetics one obtains:

$$P_{\text{trig}} = \exp\left(- \frac{\rho / T_{ij} + \gamma / T^{\star}_{ij}}{\rho / T_R + \gamma / T^{\star}_R} \right).$$

The parameter γ is proportional to the surface density of co-receptor molecules; if [CD(4—8)] denotes this surface density and K_D is the dissociation constant of the co-receptor–MHC interaction, the following holds:

$$\frac{\gamma}{\gamma + \rho} = \frac{[\text{CD(4—8)}]}{[\text{CD(4—8)}] + K_D}.$$

---

[6] This presupposes a co-receptor binding equilibrium, which is realistic for CD8/TCR or CD4/TCR adducts, but may not be realistic for CD8/MHCI or CD4/MHCII couplings, whose stability may depend on the TCR/pMHC docking.

Thus, by varying the co-receptor levels, the T cell modulates interaction time and triggering probability as well as the functional sensitivity $P_{\text{trig}}/\mathbb{E}(T)$. Interestingly, the effect of increased co-receptor levels is not necessarily to increase $P_{\text{trig}}/\mathbb{E}(T)$. An increase at high co-receptor levels (i.e. [CD(4—8)] $\gg K_D$) compared to low co-receptor levels ([CD(4—8)] $\ll K_D$) is only found if the following condition is satisfied:

$$\frac{T_R^\star}{T_R} < \frac{T_{ij}^\star}{T_{ij}}\left(1 - \frac{T_{ij}}{T_R}\ln\frac{T_{ij}^\star}{T_{ij}}\right) . \tag{3.21}$$

This means that, should the co-receptor only exert the TCR/pMHC stabilisation effect (i.e. $T_R^\star = T_R$, no threshold effect), an increase of the functional sensitivity is possible only if $T_{ij} < T_R$ (ligands which satisfy this condition are *sub-optimal agonists* at [CD(4—8)] $\ll K_D$). This is illustrated in the left panel of Fig. 3.2, which shows scaled functional sensitivity $\exp[-T_R/T_{ij}](T_R/T_{ij})$ as a function of T_{ij}/T_R. The co-receptor stabilisation effect amounts to a shift to the right in this graph; it is clear that the TCR's functional sensitivity to sub-optimal agonists can be enhanced by such a shift. The opposite will be true of ligands such that $T_{ij} > T_R$ (the so-called *heteroclitic agonists*). The right panel of Fig. 3.2 shows the percentage co-receptor enhancement obtained when saturating levels of co-receptor are compared to zero levels. It is assumed that $T_{ij}^\star = 2T_{ij}$, as suggested by the data of the Sewell group [32]; the leftmost curve corresponds to the case where there is no effect on the receptor threshold, while the middle curve corresponds to $T_R^\star = T_R/2$ and the rightmost to $T_R^\star = T_R/10$. This illustrates that the receptor-level effect enlarges the range of ligands that are positively boosted by the co-receptor, and that the receptor-level boost factor $T_R^\star/T_R$ need not be much larger than the stabilisation factor $T_{ij}/T_{ij}^\star$; van den Berg et al. [37] estimate that this boost is twofold for the pMHCI/CD8 interaction.

For slow co-receptor kinetics, condition (3.21) applies at all co-receptor densities, whereas for rapid kinetics, an enhancement relative to nil co-receptor may occur even when condition (3.21) is not satisfied; in this case, the functional sensitivity $P_{\text{trig}}/\mathbb{E}[T]$ first increases with increasing co-receptor levels and then decreases at

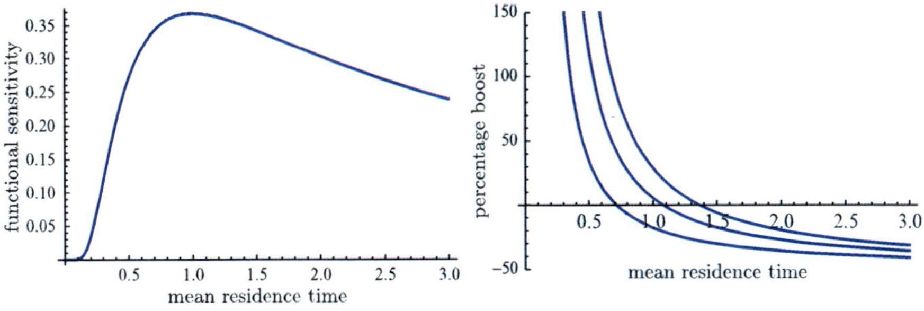

Fig. 3.2 Co-receptor-mediated enhancement of functional sensitivity

higher levels. This effect[7] is qualitatively important only for marginally sub-optimal agonists whose 'unboosted' T_{ij} is lower than, but close to T_R.

The present analysis suggests that T cells can use co-receptor expression levels not only to modulate the functional sensitivity ("avidity") to a given peptide ligand, but, moreover, that co-receptor level variation should enable *differential* regulation of TCR sensitivity. That is to say: the functional sensitivity for some ligands goes down while that for others goes up as the T cell up- or down-regulates its co-receptor. All these ligands have the potential to be good agonists, but at any one time, the T cell is tuned in on only a few, or just one, of them.

While the above theory suggests the possibility of *differential regulation of TCR avidity*, such a phenomenon remains to be experimentally observed. One would have to find a TCR X and two ligands A and B such that A is a good agonist for X at low levels of the co-receptor whereas B is a good agonist for X at high levels of the co-receptor, such that the ordering of functional sensitivities reverses as the co-receptor levels are varied. Since good agonists at nil co-receptor levels will be rare and are highly unlikely to be the index ligand through which any given clone is initially identified, it is perhaps not too surprising that the predicted effect has not yet been observed.

If it can be confirmed that differential regulation of TCR avidity occurs, there are two far-reaching consequences: (1) T cells have an – albeit limited – ability for "avidity maturation" which effectively multiplies the diversity of the repertoire (cf. [45]), in terms of enhancing its ability to match any pathogenic challenge to a highly sensitive TCR; (2) costimulation plays a pivotal role in supplying the T cell with accessory information: in the absence of disease, the T cell should be kept from tuning in (at least not too strongly) to the ligands it encounters,[8] whereas the opposite response is required when the APC carries a salient epitope. The remainder of this chapter focuses on the latter case, and describes how the unique properties of the 'signal 2' system help shape the T cell response.

Costimulatory Tuning of the Responder Spectrum

The professional APC involved in the initiation of an immune response presents the harm-associated epitope on the MHC molecule, together with an accessory costimulus ('signal 2'[9]) which it transmits via the receptors CD80 and CD86 [54]. The

---

[7] A mathematically equivalent effect is obtained by keeping co-receptor levels constant, but increasing K_D; this has been achieved by mutating the $\alpha 2$ domain of the MHCI molecule [43, 44], which interacts with the co-receptor.

[8] Since such ligands are "self" or in some cases "harmless non-self"; co-receptor expression is up- and down-regulated by, respectively, interferon-γ and interleukin-4 [46], which is thought to be directly involved in modulating the autoreactivity of the quiescent repertoire [47]. Costimulation also governs $CD25^{+}$ regulatory T cells which are important in maintaining immune tolerance to autoligands [48].

[9] Depending on the nature of this costimulus, a naïve T cell may induced to undergo cell division and differentiation, to take part in an immune response [16, 49, 50]. Alternatively, it may be

two main[10] costimulatory receptors on the T cell are CD28 and CD152 (Fig. 3.1); the former being a key activator and the latter a key attenuator for the activation of naïve T cells [49, 58]. Costimulation through CD28 and CD152 regulates the expansion of the T cell clones that take part in the immune response [59, 60].

The characteristics of the CD80/CD86–CD28/CD152 costimulatory system are discussed below. It has been observed that both CD80 and CD86 can interact with CD28 as well as CD152 [61]. One could interpret this finding as mere redundancy, or, alternatively, one could regard this as a system which exploits differences in affinities as a means to transmit information encoded in the CD80/CD86 ratio on the APC. We analyse this (relatively novel) *differential affinity* signalling paradigm.

It has been proposed that the CD80/CD86–CD28/CD152 costimulatory system modulates the spectrum of functional sensitivities for the antigenic peptide, among the responding clones [62,63]. This seems plausible, since costimulation is known to modulate the cellular activation threshold[11] [65], and (3.3) suggests a link between activation threshold and functional sensitivity; Sect. 3 analyses this idea in detail, based on the following model:

$$W_{\mathrm{act}} = m_{\mathrm{APC}} \times m_{\mathrm{T\,cell}} \times \widehat{W_{\mathrm{act}}} \tag{3.22}$$

where $\widehat{W_{\mathrm{act}}}$ expresses the baseline activation threshold (for a naïve T cell interacting with a mature professional APC) and $m_{\mathrm{APC}} \times m_{\mathrm{T\,cell}}$ is multiplier expressing modulation of the activation threshold. It is assumed here that modulation of $\widehat{W_{\mathrm{act}}}$ by the APC (m_{APC}) and by the T cell ($m_{\mathrm{T\,cell}}$) combine multiplicatively; this assumption, which forms the basis for our dynamic model, will be examined first.

Differential Affinity Signalling in the CD80/CD86–CD28/CD152 System

Engagement of CD28 potentiates T cell survival, proliferation, and activation [66]. CD28 appears to reduce the cellular activation threshold (W_{act}) in naïve T cells[12]

---

induced to enter an unresponsive anergic state [51], from which it may further differentiate into an immunoregulatory phenotype [52, 53].

[10] Another receptor important in T cell priming is CD27, which binds CD70 [55]; CD278 as well as members of the TNF receptor family are expressed in activated T cells and are important in regulating the size of the expansion [56, 57].

[11] In the absence of CD28, the activation threshold is large and can only be attained by prolonged stimulation [50] or by high presentation levels of the antigenic ligand [64].

[12] CD28-mediated costimulation regulates the number of mitotic events that responding T cells undergo [67]; acting through phosphatidylinositol 3'-kinase (PI3K) and Akt, the costimulatory receptor CD28 increases the rate of glucose uptake and glycolysis, preparing the T cell for proliferation [68]. It is unclear whether CD28 can have a direct effect on the TCR triggering rate or the rate of pMHC engagement [16,69]; CD28 could enhance TCR triggering by Itk/Emt-mediated activation of tyrosine kinase, as well as by recruitment and clustering of such kinases in membrane

while CD152 generally counteracts the effects of CD28 [58, 76]; its engagement increases the cellular activation threshold [77].[13] Up-regulation of CD152 is dependent on TCR signalling [63] as well as CD28 [85]; this constitutes a negative feedback effect which limits clonal expansion and may broaden the avidity spectrum among the responding clones by being strongest in high-avidity T cells, which receive the strongest cognate stimulation [62, 86].

The APC costimulates the T cell through two counter-receptors CD80 and CD86, each of which interact with both CD28 and CD152. CD80 binds both costimulatory receptors more strongly than CD86, and, moreover, markedly favours engagement of CD152 over CD28, whereas the more weakly binding CD86 shows much less bias [61]. While CD28 has a single binding site for CD80(86), CD152 has two identical binding sites which exhibit negative cooperativity; the site which binds first has a much higher affinity for both CD80 and CD86 than CD28's single site [61].

To represent these interactions in a simple[14] mathematical model, assume that CD28 has a single binding site for CD80 and CD86, with two distinct affinities expressed by the two-dimensional dissociation constants $K_{28/80}$ and $K_{28/86}$. For the CD152 dimer assume two identical, interdependent binding sites. The cooperativity between binding to one of the sites and the occupancy of the other is represented by the following two-dimensional dissociation constants: $K^{\circ}_{152/80}$ ($K^{\circ}_{152/86}$) for binding to CD80 (CD86) when the other site is still unoccupied, and $K^{\bullet}_{152/80}$ ($K^{\bullet}_{152/86}$) for binding to CD80 (CD86) when the other site is already occupied by either CD80 or CD86.[15] At equilibrium, the following conditions then govern the surface densities of the various species in the area of contact between the T cell and the APC:

---

microdomains [70–73]. Various T cell activation genes depend on the nuclear factor of activation in T cells (the transcription factor NFAT). The relevant active form of this intracellular messenger is unphosphorylated NFAT, localised in the nucleus. The enzyme glycogen synthase kinase-3 (GSK3) is thought to phosphorylate serine residues on NFAT, which causes it to be exported from the nucleus. It is thought that the general effect of CD28 costimulation is to depress the level of active GSK3, while CD152 costimulation may elevate GSK3 activity [62, 74, 75].

[13] Engagement of CD152, also known as CTLA-4, also restricts clonal expansion, i.e. the number of mitotic events following activation [78–81]. Furthermore, CD152 interferes with TCR signalling, possibly by a direct interaction with the TCR/CD3ζ-chain, thus negatively regulating recruitment of the TCR to kinase-rich membrane microdomains [82, 83]. CD152 can recruit PP2A-family serine/threonine phosphatases that may attenuate intracellular signalling cascades, or interfere directly with CD28 by targeting PP2A activity to CD28 [84]. While the cytoplasmic tail of CD152 contains a binding motif for Scr homology-2 domain containing tyrosine phosphatases, the involvement of such phosphatases in CD152 signalling remains unclear [77].

[14] Spatial aspects related to CD80 dimerization, as proposed by Schwartz et al. [87], are ignored.

[15] Collins et al. [61] furnish the following relations between the dissociation constants, based on 3D measurements: $K_{28/80}/K_{28/86} = 0.2$; $K^{\circ}_{152/80}/K_{28/86} = 0.0185$; $K^{\circ}_{152/86}/K_{28/86} = 0.2845$; $K^{\bullet}_{152/80}/K_{28/86} = 0.065$; $K^{\bullet}_{152/86}/K_{28/86} = 0.725$. These values indicate negative cooperativity: the dissociation constants for the second ligands are over two times higher than the corresponding first ligand value (i.e. $K^{\bullet}$ as compared to K°).

$$\left\{\begin{aligned}
[\mathrm{CD28}]_{\mathrm{free}} \times [\mathrm{CD80}]_{\mathrm{free}} &= K_{28/80} \times [\mathrm{CD28/CD80}]\\
[\mathrm{CD28}]_{\mathrm{free}} \times [\mathrm{CD86}]_{\mathrm{free}} &= K_{28/86} \times [\mathrm{CD28/CD86}]\\
[\mathrm{CD152}]_{\mathrm{free}} \times [\mathrm{CD80}]_{\mathrm{free}} &= K^{\circ}_{152/80} \times [\mathrm{CD152/CD80}]\\
[\mathrm{CD152}]_{\mathrm{free}} \times [\mathrm{CD86}]_{\mathrm{free}} &= K^{\circ}_{152/86} \times [\mathrm{CD152/CD86}]\\
[\mathrm{CD152/CD80}]_{\mathrm{free}} \times [\mathrm{CD80}]_{\mathrm{free}} &= K^{\bullet}_{152/80} \times [\mathrm{CD152/CD80/CD80}]\\
[\mathrm{CD152/CD86}]_{\mathrm{free}} \times [\mathrm{CD80}]_{\mathrm{free}} &= K^{\bullet}_{152/80} \times [\mathrm{CD152/CD86/CD80}]\\
[\mathrm{CD152/CD80}]_{\mathrm{free}} \times [\mathrm{CD86}]_{\mathrm{free}} &= K^{\bullet}_{152/86} \times [\mathrm{CD152/CD80/CD86}]\\
[\mathrm{CD152/CD86}]_{\mathrm{free}} \times [\mathrm{CD86}]_{\mathrm{free}} &= K^{\bullet}_{152/86} \times [\mathrm{CD152/CD86/CD86}]
\end{aligned}\right. \tag{3.23}$$

where square brackets indicate surface densities and 'free' means 'unoccupied'. Let A_{A} denote the total surface area of the APC, A_{T} the total surface area of the T cell, and A_{C} the surface area of the conjugate interface. Then the relevant conservation laws can be written as follows:

$$|\mathrm{CD28}| = A_{\mathrm{T}}[\mathrm{CD28}]_{\mathrm{free}} + A_{\mathrm{C}}\left([\mathrm{CD28/CD80}] + [\mathrm{CD28/CD86}]\right) \tag{3.24}$$

$$\begin{aligned}
|\mathrm{CD152}| = A_{\mathrm{T}}[\mathrm{CD152}]_{\mathrm{free}} &+ A_{\mathrm{C}}\left([\mathrm{CD152/CD80}] + [\mathrm{CD152/CD86}]\right.\\
&+ [\mathrm{CD152/CD80/CD80}] + [\mathrm{CD152/CD80/CD86}]\\
&\left.+ [\mathrm{CD152/CD86/CD80}] + [\mathrm{CD152/CD86/CD86}]\right)
\end{aligned} \tag{3.25}$$

$$\begin{aligned}
|\mathrm{CD80}| = A_{\mathrm{A}}[\mathrm{CD80}]_{\mathrm{free}} &+ A_{\mathrm{C}}\left([\mathrm{CD28/CD80}] + [\mathrm{CD152/CD80}]\right.\\
&+ [\mathrm{CD152/CD80/CD86}] + [\mathrm{CD152/CD86/CD80}]\\
&\left.+ 2\,[\mathrm{CD152/CD80/CD80}]\right)
\end{aligned} \tag{3.26}$$

$$\begin{aligned}
|\mathrm{CD86}| = A_{\mathrm{A}}[\mathrm{CD86}]_{\mathrm{free}} &+ A_{\mathrm{C}}\left([\mathrm{CD28/CD86}] + [\mathrm{CD152/CD86}]\right.\\
&+ [\mathrm{CD152/CD80/CD86}] + [\mathrm{CD152/CD86/CD80}]\\
&\left.+ 2\,[\mathrm{CD152/CD86/CD86}]\right)
\end{aligned} \tag{3.27}$$

where $|\cdot|$ denotes the total number of molecules present on the cell's surface. It is straightforward to reduce these equations to a system of only two non-linear equations, which are readily solved numerically by means of a fixed-point algorithm.[16]

The hypothesis of *differential affinity signalling* is embodied in the following three parameters:

$$\text{APC} \quad \sigma = \frac{|\mathrm{CD86}|}{|\mathrm{CD80}| + |\mathrm{CD86}|}\,; \tag{3.28}$$

$$\text{T cell} \quad \upsilon = \frac{|\mathrm{CD28}|}{|\mathrm{CD152}| + |\mathrm{CD28}|}\,; \tag{3.29}$$

$$\text{T cell:APC} \quad \gamma = \frac{(|\mathrm{CD152}| + |\mathrm{CD28}|)/A_{\mathrm{T}}}{(|\mathrm{CD80}| + |\mathrm{CD86}|)/A_{\mathrm{A}}}\,. \tag{3.30}$$

[16] The species CD152/CD86/CD80 and CD152/CD80/CD86 are physically identical, and distinguished in the calculations by the order in which they engaged their ligands; the two densities are added to give the density of the single molecular species they represent.

The parameter σ represents the balance between CD80 and CD86 on the APC. The APC can influence the nature of signal transmission between APC and T cell by adjusting σ. Similarly, υ represents the balance between CD28 and CD152, and this balance is under the control of the T cell. The third parameter γ represents the balance between CD28/152 and CD80/86; this parameter is jointly determined by the two interacting cells. The costimulatory status of the APC is expressed by the following dimensionless parameter:

$$\kappa = \frac{K_{28/86}}{(|\mathrm{CD80}| + |\mathrm{CD86}|)/A_{\mathrm{A}}} .$$

The discussion at the start of this section motivates the following assumptions: (1) both CD28 and CD152, when expressed, make a constitutive contribution to intracellular processing of the TCR-stimulus; and (2) the strength of the stimulatory effect of CD28 (and, similarly, of the inhibitory effect of CD152) depends on the relative enrichment of these receptors in the T cell:APC contact area, where the TCR signal arises. These enrichment effects can be expressed as *sequestration ratios*, defined as the contact area density of all species relative to the density prior to the cell–cell contact:

$$\varrho_{\mathrm{CD28}} \stackrel{\mathrm{def}}{=} (|\mathrm{CD28}|/A_{\mathrm{T}})^{-1} ([\mathrm{CD28}]_{\mathrm{free}} + [\mathrm{CD28/CD80}] + [\mathrm{CD28/CD86}])$$

and

$$\begin{aligned}\varrho_{\mathrm{CD152}} \stackrel{\mathrm{def}}{=} (|\mathrm{CD152}|/A_{\mathrm{T}})^{-1} & ([\mathrm{CD152}]_{\mathrm{free}} + [\mathrm{CD152/CD80}] + [\mathrm{CD152/CD86}] \\ & + [\mathrm{CD152/CD80/CD80}] + [\mathrm{CD152/CD80/CD86}] \\ & + [\mathrm{CD152/CD86/CD80}] + [\mathrm{CD152/CD86/CD86}]) .\end{aligned}$$

This sequestration effect corresponds to the increase in signalling intensity.

The effective stimulation of the T cell depends on the balance between positive stimulation (due to the combined CD28 species in the cell–cell contact area) and negative stimulation (due to the combined CD152 species in the cell–cell contact area), which is expressed by the ratio $\varrho_{\mathrm{CD28}}\upsilon/(\varrho_{\mathrm{CD152}}(1-\upsilon))$. This quantity can be identified with the multiplier in (3.22) under the following identifications:

$$m_{\mathrm{APC}} \equiv \frac{\varrho_{\mathrm{CD28}}}{\varrho_{\mathrm{CD152}}} \quad \text{and} \quad m_{\mathrm{T\,cell}} \equiv \frac{\upsilon}{(1-\upsilon)}$$

that is, the T cell sets the balance between CD28 and CD152 whereas the APC determines the sequestering ratios by varying κ and σ. Comparison of the top and bottom panels in Fig. 3.3 indicates that these two factors combine multiplicatively, as assumed by (3.22); the effect of changing $m_{\mathrm{T\,cell}}$ amounts – to a very good approximation – to a simple scaling. As the left and middle panels of Fig. 3.3 show, sequestering of both co-receptors decreases with an increasing proportion of CD86, but at different rates, which leads to either an increase or a decrease of the stimulatory balance (right panels), depending on the value of κ.

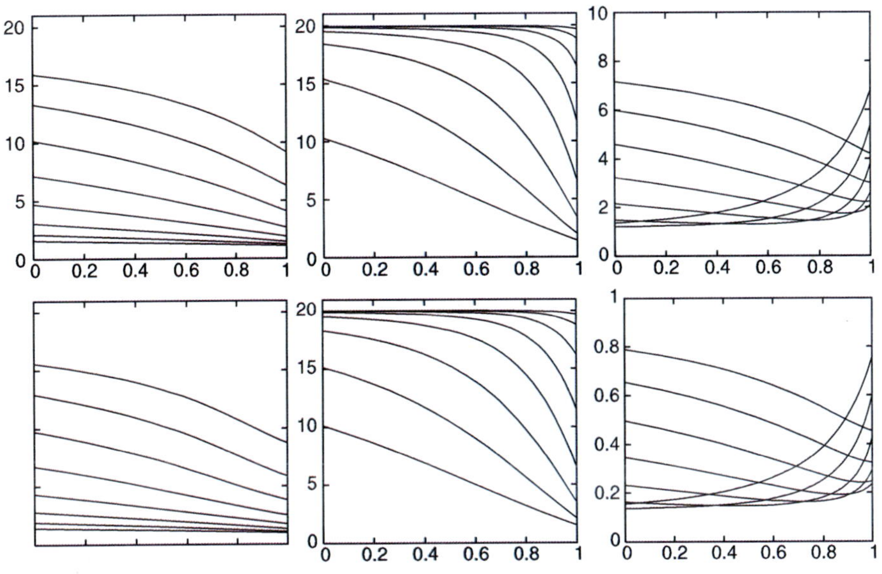

Fig. 3.3 Differential affinity signalling. *Left*: CD28 sequestering (ϱ_{CD28}); *middle*: CD152 sequestering (ϱ_{CD152}); *right*: strength of stimulation ($\varrho_{CD28}\upsilon/(\varrho_{CD152}(1-\upsilon))$) all as a function of σ. $A_C/A_T = 0.05$, $A_C/A_A = 0.005$, $\upsilon = 0.9$ (*top panels*), 0.5 (*lower panels*); $\gamma = 1$. *Curves* for (*top to bottom* in the sequestering graphs) $\kappa = \frac{1}{16}, \frac{1}{8}, \frac{1}{4}, \frac{1}{2}, 1, 2, 4, 8$ (lowest curves for lowest CD80(86) densities). The costimulatory balance (in favour of T cell activation) increases with σ when κ is large (i.e. when CD80(86) densities are low)

Evolution of the Responder Functional Sensitivity Spectrum

Clonal expansion involves antigen-driven selection of TCR clonotypes that have the highest functional sensitivity for the pathogen-derived epitopes presented on professional APCs [88]. Costimulation through CD28 and CD152 is thought to influence the shape of the TCR avidity spectrum among the clones responding to a given pMHC antigen. In some instances, the most efficacious T cell response may be composed of a few (or one) dominating clonotypes of very high functional sensitivity (a narrow spectrum response), whereas in other cases the immune response needs to maintain T cell diversity, with a broader spectrum [62]. The professional antigen presenting cells involved in priming naïve T cells may play an active role in guiding the evolution of the 'avidity spectrum' during the primary expansion, employing the differential affinity properties of the CD80/CD86–CD28/CD152 system. This section describes a model for the dynamics of the 'avidity spectrum' which elucidates the possible roles of APCs and co-stimulation.

Let w_{ij} denote the functional sensitivity of the interaction between a TCR of clonotype i and a pMHC ligand of species j:

$$w_{ij} \stackrel{\text{def}}{=} \frac{P_{\text{trig}}(i,j)}{T_{ij}}$$

(the quantities on the right are those for the prevailing levels of co-receptor, not the 'unboosted' values; a slight abuse of notation for the sake of simplicity). The aim is to describe the statistical distribution of w_{ij}-values among the TCR clones that are activated by a given pMHC ligand j. The probability that a T cell is activated during a given encounter with a professional APC is calculated as follows:

$$\mathbb{P}(Z_j w_{ij} > W_{\text{act}}) = \mathbb{P}\left(\frac{Z_j}{m_{\text{APC}}\widehat{W_{\text{act}}}} > \frac{m_{\text{T cell}}}{w_{ij}}\right) \tag{3.31}$$

(with (3.22)). For the random variable $Z_j/(m_{\text{APC}}\widehat{W_{\text{act}}})$, it is convenient and reasonable to choose a log-logistic distribution. With location parameter ϖ and scale parameter ϑ one then has:

$$\mathbb{P}\left(\frac{Z_j}{m_{\text{APC}}\widehat{W_{\text{act}}}} > \frac{m_{\text{T cell}}}{w_{ij}}\right) = \left(1 + (m_{\text{T cell}}\varpi/w_{ij})^{\vartheta}\right)^{-1}.$$

Responding T cells will experience a number of encounters with pAPCs, each of which may activate the T cell to undergo further proliferation or terminal differentiation. It is believed that a T cell responds to such an activation event by increasing its activation threshold W_{act} through up-regulation of CD152 [79]. Consider a T cell with functional sensitivity w_{ij} which has thus far experienced k activating encounters. The probability that this T cell will be activated in the next encounter is given by:

$$P_{\text{act}}(k+1; w_{ij}) = \int_1^{\infty} \mathbb{P}\left(Z_j/(m_{\text{APC}}\widehat{W_{\text{act}}}) > m/w_{ij}\right) dF_k(m; w_{ij}) \tag{3.32}$$

which is obtained by conditioning the activation probability, (3.31), on the distribution F_k, defined as follows:

$$F_k(m; w_{ij}) \stackrel{\text{def}}{=} \mathbb{P}(m_{\text{T cell}} \le m \text{ after } k \text{ encounters, with functional specificity } w_{ij}).$$

If the encounter is successful, the T cell's threshold multiplier changes as follows:

$$m_{\text{T cell}}(k+1; w_{ij}) = m_{\text{T cell}}(k; w_{ij})\, f_0\left(Z_j w_{ij}/W_{\text{act}}(k; w_{ij})\right)^{\eta}.$$

Egen et al. [62] proposed that the increment of the activation threshold depends on the strength of the TCR signal; this effect is represented in the above equation by the non-negative parameter η (for $\eta = 0$, the Egen effect is absent and the threshold increase is always by a factor $f_0 > 1$).

Equation (3.32) gives the activation probability for a T cell with a given functional sensitivity w_{ij}; the probability that any of the T cells that were activated in the k previous rounds is activated during the next encounter is found by conditioning on the distribution of functional sensitivities among T cells that have been activated k times. Denote this distribution as $G_k(w)$. Then:

$$P_{\text{act}}(k+1) = \int P_{\text{act}}(k+1; w) dG_k(w) \tag{3.33}$$

where

$$G_k(\omega) \overset{\text{def}}{=} \frac{1}{P_{\text{act}}(k)} \int_0^{\omega} P_{\text{act}}(k; w) dG_{k-1}(w) \,. \tag{3.34}$$

The initial condition for this recurrence relation is the distribution $G_0(w)$ of functional sensitivities among antigen-inexperienced T cells[17]; this distribution has been described in detail by van den Berg & Rand [89]. To complete the mathematical description, a recurrence relationship for F_k must be found. Straightforward probability considerations give the following expression:

$$\begin{aligned} F_{k+1}(m; w_{ij}) = 1 - \Bigg[&\int_1^{m/f_0} \mathbb{P}\left(\frac{Z_j}{m_{\text{APC}} \widehat{W_{\text{act}}}} > \frac{\mu}{w_{ij}} \left(\frac{\mu/f_0}{\mu} \right)^{1/\eta} \right) dF_k(\mu; w_{ij}) \\ &- \int_{m/f_0}^{\infty} \mathbb{P}\left(\frac{Z_j}{m_{\text{APC}} \widehat{W_{\text{act}}}} > \frac{\mu}{w_{ij}} \right) dF_k(\mu; w_{ij}) \Bigg] \frac{1}{P_{\text{act}}(k; w_{ij})} \end{aligned} \tag{3.35}$$

The initial condition $F_0(m; w_{ij})$ is the degenerate distribution at $m = 1$ for all values of w_{ij}. Equations (3.32), (3.33)–(3.35) constitute a system of integro-difference equations which can be solved numerically.

By the avidity principle, a T cell whose TCR has comparatively low functional sensitivity w_{ij} for the relevant ligand can still be activated if the presentation level Z_j is sufficiently high. Professional APCs (mature dendritic cells) that activate T cells from the naïve repertoire can, in principle, prevent T cells from exploiting this avidity advantage, by correlating high presentation levels with high multiplier (m_{APC}) values,[18] whereas a negative correlation would emphasise the avidity advantage. In the present model this correlation is expressed by the parameter ϑ. The variance of

[17] It is reasonable to assume that this function does not depend on the ligand at hand. This naïve distribution G_0 has the following features: most of its probability mass is concentrated near zero avidity ($w_{ij} \approx 0$ for almost all clonotypes i, as the vast majority of naïve T cells have no or very little sensitivity for any given ligand j), while a minute fraction of the probability mass is concentrated near the maximum avidity; these latter cells constitute likely responders.

[18] Such correlations require a connection between the antigen presentation pathways and co-receptor expression; such a correlation may be induced by DC maturation, which sees an increase in presentation levels as well as a shift from CD80 to CD86 expression [90, 91], which appears to depend on stimulation of CD80 [92].

the random variable $Z_j/(m_{\text{APC}}\widehat{W_{\text{act}}})$, which represents the modulated influence of the APC, is minimised (i.e. ϑ is large) when high m_{APC} correlate positively with Z_j, so that a high threshold tends to cancel out the higher number of ligand molecules presented to the T cell. By contrast, a negative correlation tends to increase this variance (i.e. to lower ϑ). The effect of this parameter is shown in Fig. 3.4. The top panel shows that large variability of $Z_j/(m_{\text{APC}}\widehat{W_{\text{act}}})$ gives rise to a broad spectrum centred on $\widehat{W_{\text{act}}}$.

During successive rounds of activation, this spectrum shifts toward the right, that is, becomes enriched with TCRs with greater functional sensitivity.[19] When the variance is low (Fig. 3.4, bottom panel) the spectrum is just the naïve spectrum,

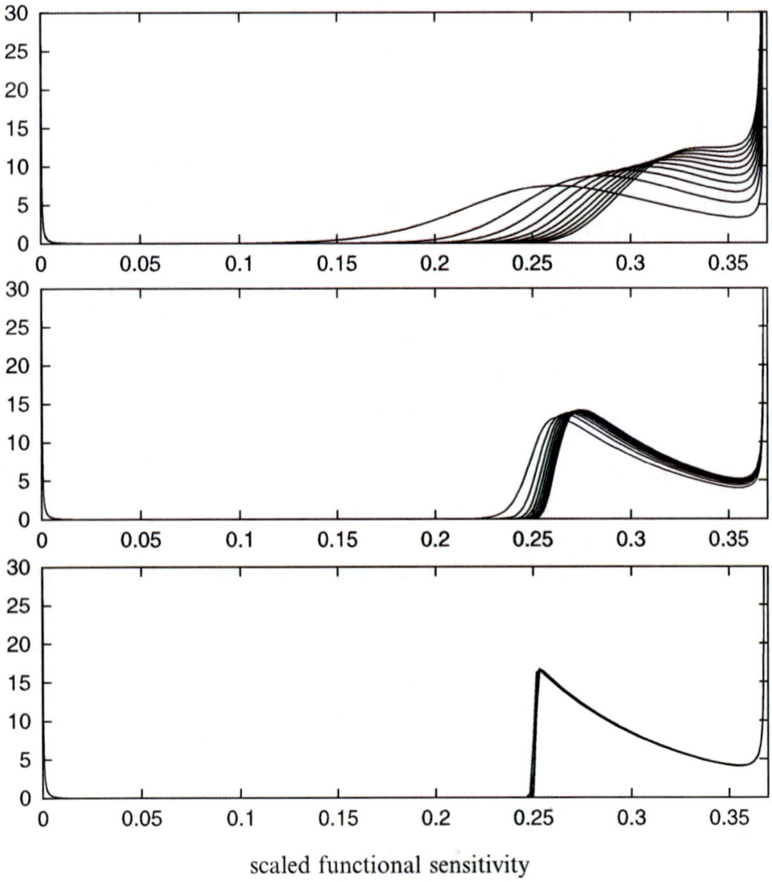

Fig. 3.4 Functional sensitivity spectra among responding clonotypes: the effect of APC variability alone. $\varpi = 0.25$; $f_0 = 1$; $\eta = 0$; $\vartheta = 10, 50, 500$ (*top to bottom*); abscissa is scaled functional sensitivity $T_R w_{ij}$; ordinate is probability density

[19] Generally, the avidity spectrum among the responders at any given moment in time is a weighted sum of the spectra for $k = 1, 2, 3, \ldots$. However, if the number of divisions between activation

left-truncated at $\widehat{W_{\rm act}}$. It is interesting that successive rounds do not lead to an enrichment of strong TCRs in this case; this is because clones just above $\widehat{W_{\rm act}}$ are likely to be re-activated in later rounds when the APC-related variance is low, whereas only those with high functional sensitivity are secure when APC-related variance is high.

Thus, selection for high functional sensitivity can be driven by APC variability alone, in the absence of threshold increases on the part of the T cell. Figure 3.5 shows the effect of changing f_0 from 1 to 1.05. As one would expect, this improves the drive to select stronger clonotypes. At low APC variability the threshold eventually becomes too stringent for any of the responding T cells to receive stimulation, whereas high APC variability ensures that a left tail of intermediate-strength TCRs is retained.

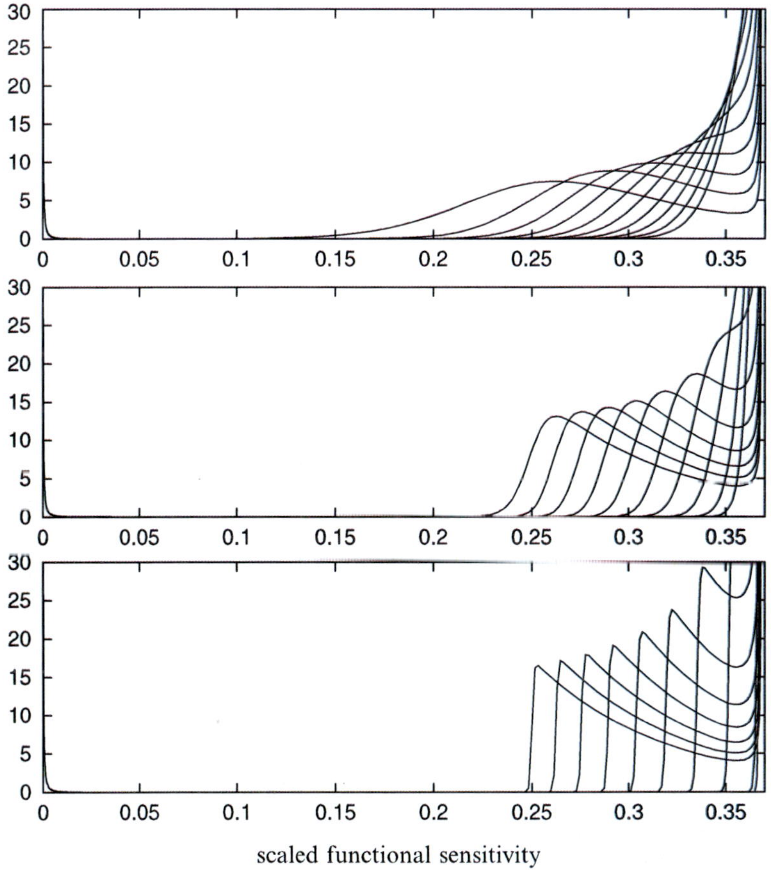

Fig. 3.5 Functional sensitivity spectra among responding clonotypes: the effect of activation-dependent threshold increments. $\varpi = 0.25$; $f_0 = 1.05$; $\eta = 0$; $\vartheta = 10, 50, 500$ (*top to bottom*); abscissa is scaled functional sensitivity $T_R w_{ij}$; ordinate is probability density

events is sufficiently large, the actual spectrum will be dominated by that corresponding to the latest activation round.

The effect of stimulation-dependent threshold increments is demonstrated in Fig. 3.6, for various positive values of η. As Egen et al. [62] predicted, the effect of 'penalising' high-strength TCRs is to make the spectrum more uniform, so that it acquires the appearance of a plateau. Moreover, the spectrum converges quickly after $\sim$10 activation rounds.

The spectrum G_k is continuous probability density function, effectively describing an ensemble of identically prepared immune systems; in any one of these systems, the spectrum will be a histogram describing a relatively small[20] number of responding clones. This number is Poisson distributed[21] with parameter

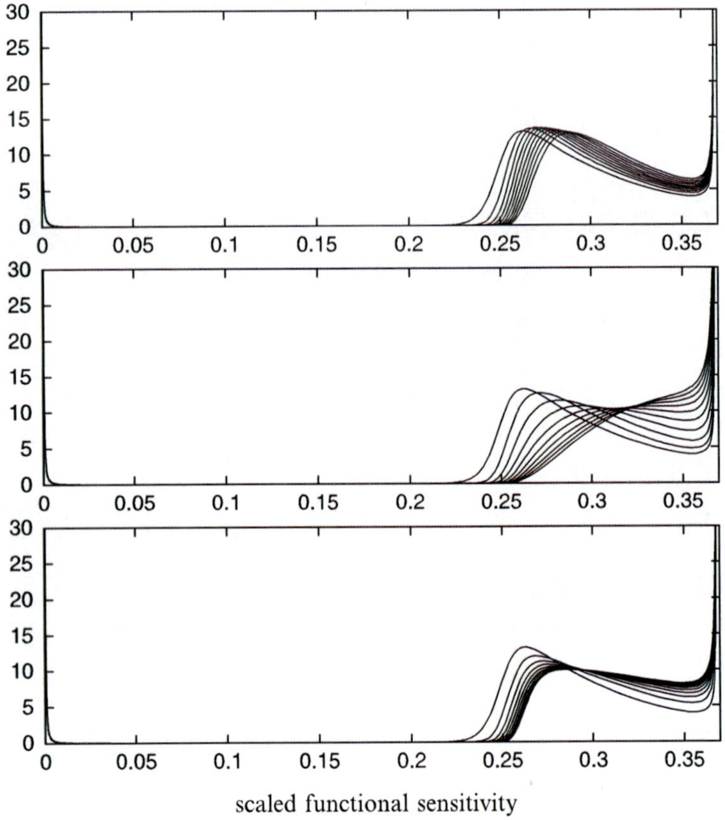

Fig. 3.6 Functional sensitivity spectra among responding clonotypes: the effect of stimulation-dependent threshold increments. $\varpi = 0.25$; $f_0 = 1$; $\eta = 0.1, 0.5, 0.9$ (*top to bottom*); $\vartheta = 50$; abscissa is scaled functional sensitivity $T_R w_{ij}$; ordinate is probability density

[20] The detectable expansion consists of on the order of 10 clones [93]; clonotypical heterogeneity typically has a "quasi-species" structure with only one or a few clones constituting the bulk [94]; the number of clones initially activated may be up to an order of magnitude larger.

[21] As may easily be shown using a technique called *compounding*.

$$\mu_k = N_C \prod_{\ell=1}^{k} P_{\text{act}}(\ell)$$

after k rounds; here, N_C denotes the number of TCR clonotypes in the naïve repertoire. The number of clones whose functional sensitivity exceeds a given value ω is similarly Poisson distributed, with parameter $\mu_k(\omega) = (1 - G_k(\omega))\mu_k$ after k rounds (clearly $\mu_k \equiv \mu_k(0)$). An open problem is to specify the parameters $\widehat{W_{\text{act}}}$, f_0, η, and ϑ so that $\mu_k(\omega)$ is maximised, given ω.

References

1. Mason D (1998) A very high level of crossreactivity is an essential feature of the T-cell receptor. Immunol Today 19:395–404
2. Rudolph MG, Wilson IA (2002) The specificity of TCR/pMHC interaction. Curr Opin Immunol 14:52–65
3. Hemmer B, Vergelli M, Pinilla C, Houghten R, Martin R (1998) Probing degeneracy in T-cell recognition using peptide combinatorial libraries. Immunol Today 19:163–168
4. Wilson DB, Wilson DH, Schroder K, Pinilla C, Blondelle S, Houghten RA, Garcia KC (2004) Specificity and degeneracy of T cells. Mol Immunol 40:1047–1055
5. Wooldridge L, Lissina A, van den Berg HA, Price DA, Sewell AK (2009) Tricks with tetramers: how to get the most from multimeric peptide-MHC. Immunology 126:147–164
6. Alexander-Miller MA, Leggatt GR, Berzofsky JA (1996) Selective expansion of high- or low avidity cytotoxic T lymphocytes and efficacy for adoptive immunotherapy. Proc Natl Acad Sci USA 93:4102–4107
7. van den Berg HA, Rand DA, Burroughs NJ (2001) A reliable and safe T cell repertoire based on low-affinity T cell receptors. J Theor Biol 209:465–486
8. van den Berg HA, Burroughs NJ, Rand DA (2002) Quantifying the strength of ligand antagonism in TCR triggering. Bull Math Biol 64:781–808
9. van den Berg HA, Rand DA (2007) Quantitative theories of T-cell responsiveness. Immunol Rev 216:81–92
10. van den Berg HA, Rand DA (2004) Dynamics of T cell activation threshold tuning. J Theor Biol 228:397–416
11. Stirk ER, Molina-París C, Berg HA van den(2008) Stochastic niche structure and diversity maintenance in the T cell repertoire. J Theor Biol 255:237–249
12. Gonzales PA, Carreño LJ, Coombs D, Mora JE, Palmeiri E, Goldstein B, Nathenson SG, Kalergis AM (2005) T cell receptor binding kinetics required for T cell activation depend on the density of the cognate ligand of the antigen-presenting cell. Proc Natl Acad Sci USA 102:4824–4829
13. Iezzi G, Karjalainen K, Lanzavecchia A (1998) The duration of antigenic stimulation determines the fate of naive and effector T cells. Immunity 8:89–95
14. Lanzavecchia A, Iezzi G, Viola A (1999) From TCR engagement to T cell activation: a kinetic view of T cell behavior. Cell 96:1–4
15. Rachmilewitz J, Lanzavecchia A (2002) A temporal and spatial summation model for T cell activation: Signal integration and antigen decoding. Trends Immunol 23:592–595
16. Bachmann MF, McKall-Faienza K, Schmits R, Bouchard D, Beach J, Speiser DE, Mak TW, Ohashi PS (1997) Distinct roles for LFA-1 and CD28 during activation of naive T cells: adhesion versus costimulation. Immunity 7:549–557
17. Lanzavecchia A, Sallusto F (2000) Dynamics of T lymphocyte responses: intermediates, effectors, and memory cells. Science 290:92–97

18. Sykulev Y, Joo M, Vturina I, Tsomides TJ, Eisen HN (1996) Evidence that a single peptide-MHC complex on a target cell can elicit a cytolytic T cell response. Immunity 4:565–571
19. Wyer JR, Willcox BE, Gao GF, Gerth UC, Davis SJ, Bell JI, Merwe PA van der, Jakobsen BK (1999) T cell receptor and co-receptor CD8$\alpha\alpha$ bind peptide-MHC independently and with distinct kinetics. Immunity 10:219–225
20. Shores EW, Tran T, Grinberg A, Sommers CL, Shen H, Love PE (1997) Role of the multiple T cell receptor (TCR)-ζ chain signaling motifs in selection of the T cell repertoire. J Exp Med 185:893–900
21. Werlen G, Palmer E (2002) The TCR signalosome: a dynamic structure with expanding complexity. Curr Opin Immunol 14:299–305
22. Valitutti S, Lanzavecchia A (1997) Serial triggering of TCRs: a basis for the sensitivity and specificity of antigen recognition. Immunol Today 18:299–304
23. Kalergis AM, Boucheron N, Doucey MA, Palmieri E, Goyarts EC, Vegh Z, Luesher IF, Nathenson SG (2001) Efficient T cell activation requires an optimal dwell-time of interaction between the TCR and the pMHC complex. Nat Immunol 2:229–234
24. McKeithan TW (1995) Kinetic proofreading in T-cell receptor signal transduction. Proc Natl Acad Sci USA 92:5042–5046
25. Itoh Y, Germain RN (1997) Single cell analysis reveals regulated hierarchical T cell antigen receptor signaling thresholds and intraclonal heterogeneity for individual cytokine responses of $CD4^+$ cells. J Exp Med 186:757–766
26. Rangel C, Angus J, Ghahramani Z, Lioumi M, Sotheran E, Gaiba A, Wild DL, Falciani F (2004) Modeling T-cell activation using gene expression profiling and state-space models. Bioinformatics 20:1361–1372
27. Janeway Jr. CA (1992) The T cell receptor as a multicomponent signalling machine: CD4/CD8 coreceptors and CD45 in T cell activation. Annu Rev Immunol 10:645–674
28. Luescher IF, Vivier E, Layer A, Mahiou J, Godeau F, Malissen B, Romero P (1995) CD8 modulation of T-cell antigen receptor-ligand interactions on living cytotoxic T lymphocytes. Nature 373:353–356
29. Purbhoo MA, Boulter JM, Price DA, Vuidepot AL, Hourigan CS, Dunbar PR, Olson K, Dawson SJ, Phillips RE, Jakobsen BK, Bell JI, Sewell AK (2001) The human CD8 coreceptor effects cytotoxic T cell activation and antigen sensitivity primarily by mediating complete phosphorylation of the T cell receptor ζ chain. J Biol Chem 276:32786–32792
30. Cawthon AG, Alexander-Miller MA (2002) Optimal colocalization of TCR and CD8 as a novel mechanism for the control of functional avidity. J Immunol 169:3492–3498
31. Pecht I, Gakamsky DM (2005) Spatial coordination of CD8 and TCR molecules controls antigen recognition by $CD8^+$ T-cells. FEBS Lett 579:3336–3341
32. Wooldridge L, van den Berg HA, Glick M, Gostick E, Brenchley JM, Douek DC, Price DA, Sewell AK (2005) Interaction between the CD8 coreceptor and MHC class I stabilizes TCR-antigen complexes at the cell surface. J Biol Chem 280:27491–27501
33. Holler PD, Kranz DM (2003) Quantitative analysis of the contribution of TCR/pepMHC affinity and CD8 to T cell activation. Immunity 18:255–264
34. Arcaro A, Grégoire C, Bakker TR, Baldi L, Jordan M, Goffin L, Boucheron N, Wurm F, Merwe PA van der, Malissen B, Luescher IF (2001) CD8β endows CD8 with efficient coreceptor function by coupling T cell receptor/CD3 to raft-associated CD8/p56^{lck} complexes. J Exp Med 194:1485–1495
35. Bosselut R, Zhang W, Ashe JM, Kopacz JL, Samuelson LE, Singer A (1999) Association of the adaptor molecule LAT with CD4 and CD8 coreceptors identifies a new coreceptor function in T cell receptor signal transduction. J Exp Med 190:1517–1525
36. Brdičková N, Brdička T, Angelisová P, Horváth O, Špiča J, Hilgert I, Pačes J, Simeoni L, Kliche S, Merten C, Schraven B, Hořejší V (2003) LIME: a new membrane raft-associated adaptor protein involved in CD4 and CD8 coreceptor signaling. J Exp Med 198:1453–1462
37. Berg HA van den, Wooldridge L, Laugel B, Sewell (2007) AK Coreceptor CD8-driven modulation of T cell antigen receptor specificity. J Theor Biol 249:395–408

38. Filipp D, Leung BL, Zhang J, Veillette A, Julius M (2004) Enrichment of lck in lipid rafts regulates colocalized fyn activation and the initiation of proximal signals through TCR$\alpha\beta$. J Immunol 172:4266–4274
39. Bosselut R, Kubo S, Guinter T, Kopacz JL, Altman JD, Feigenbaum L, Singer A (2000) Role of CD8β domains in CD8 coreceptor function: importance for MHC I binding, signaling, and positive selection of CD8$^+$ T cells in the Thymus. Immunity 12:409–418
40. Gangadharan D, Cheroute H (2004) The CD8 isoform CD8$\alpha\alpha$ is not a functional homologue of the TCR co-receptor CD8$\alpha\beta$. Curr Opin Immunol 16:264–270
41. Arcaro A, Grégoire C, Boucheron N, Stolz S, Palmer E, Malissen B, Luescher IF (2000) Essential role of CD8 palmitoylation in CD8 coreceptor function. J Immunol 165:2068–2076
42. Doucey MA, Legler D, Boucheron N, Cerottini JC, Bron C, Luescher IF (2001) CTL activation is induced by cross-linking of TCR/MHC-peptide-CD8/p56^{lck} adducts in rafts. Eur J Immunol 31:1561–1570
43. Hutchinson SL, Wooldridge L, Tafuro S, Laugel B, Glick M, Boulter JM, Jakobsen BK, Price DA, Sewell AK (2003) The CD8 T cell coreceptor exhibits disproportionate biological activity at extremely low binding affinities. J Biol Chem 278:24285–24293
44. Wooldridge L, Lissina A, Vernazza J, Gostick E, Laugel B, Hutchinson SL, Mirza F, Dunbar PR, Boulter JM, Glick M, Cerundolo V, van den Berg HA, Price DA, Sewell AK (2007) Enhanced immunogenicity of CTL antigens through mutation of the CD8 binding MHC class I invariant region. Eur J Immunol 37:1323–1333
45. Maile R, Siler CA, Kerry SE, Midkiff KE, Collins EJ, Frelinger JA (2005) Peripheral "CD8 tuning" dynamically modulates the size and responsiveness of an antigen-specific T cell pool in vivo. J Immunol 174:619–627
46. Apte SH, Baz A, Groves P, Kelso A, Kienzle N (2008) Interferon-γ and interleukin-4 reciprocally regulate CD8 expression in CD8$^+$ T cells. Proc Natl Acad Sci USA 105:17475–17480
47. Park JH, Adoro S, Lucas PJ, Sarafova SD, Alag AS, Doan LL, Erman B, Liu X, Ellmeier W, Bosselut R, Feigenbaum L, Singer A (2007) 'Coreceptor tuning': cytokine signals transcriptionally tailor CD8 coreceptor expression to the self-specificity of the TCR. Nat Immunol 8:1049–1058
48. Jordan MS, Boesteanu A, Reed AJ, Petrone AL, Holenbeck AE, Lerman MA, Naji A, Caton AJ (2001) Thymic selection of CD4$^+$CD25$^+$ regulatory T cells induced by an agonist self-peptide. Nat Immunol 2:301–306
49. Mittrücker HW, Shahinian A, Bouchard D, Kündig TM, Mak TW (1996) Induction of unresponsiveness and impaired T cell expansion by staphylococcal enterotoxin B in CD28-deficient mice. J Exp Med 183:2481–2488
50. Kündig TM, Shahinian A, Kawai K, Mittrücker HW, Sebzda E, Bachmann MF, Mak TW, Ohasi PS (1996) Duration of TCR stimulation determines costimulatory requirement of T cells. Immunity 5:41–52
51. Greenwald RJ, Boussiotis VA, Lorsbach RB, Abbas AK, Sharpe AH (2001) CTLA-4 regulates induction of anergy in vivo. Immunity 14:145–155
52. Roncarolo MG, Levings MK (2000) The role of different subsets of T regulatory cells in controlling autoimmunity. Curr Opin Immunol 12:676–683
53. Salomon B, Lenschow DJ, Rhee L, Ashourian N, Singh B, Sharpe A, Bluestone JA (2000) B7/CD28 costimulation is essential for the homeostasis of the CD4$^+$CD25$^+$ immunoregulatory T cells that control autoimmune diabetes. Immunity 12:431–440
54. Parham P (2000) The immune system. Garland Publishing, New York
55. Keller AM, Schildknecht A, Xiao Y, van den Broek M, Borst J (2008) Expression of costimulatory ligand CD70 on steady-state dendritic cells breaks CD8$^+$ T cell tolerance and permits effective immunity. Immunity 29:1–13
56. Rudd CE, Schneider H (2003) Unifying concepts in CD28, ICOS and CTLA4 co-receptor signalling. Nat Rev Immunol 3:544–556
57. Watts TH (2005) TNF/TNFR family members in costimulation of T cell responses. Annu Rev Immunol 23:23–68
58. Krummel M, Allison JP (1995) CD28 and CTLA-4 have opposing effects on the response of T cells to stimulation. J Exp Med 182:459–465

59. Salomon B, Bluestone JA (2001) Complexities of CD28/B7: CTLA-4 costimulatory pathways in autoimmunity and transplantation. Annu Rev Immunol 19:225–252
60. Davis MM, Krogsgaard M, Huppa JB, Sumen C, Purbhoo MA, Irvine DJ, Wu LC, Ehrlich L (2003) Dynamics of cell surface molecules during T cell recognition. Annu Rev Biochem 72:717–742
61. Collins AV, Brodie DW, Gilbert RJC, Iaboni A, Manso-Sancho R, Stuart DI, Merwe PA van der, Davis SJ (2002) The interaction properties of costimulatory molecules revisited. Immunity 17:201–210
62. Egen JG, Kuhns MS, Allison JP (2002) CTLA-4: new insights into its biological function and use in tumor immunotherapy. Nat Immunol 3:611–618
63. Egen JG, Allison JP (2002) Cytotoxic T lymphocyte antigen-4 accumulation in the immunological synapse is regulated by TCR signal strength. Immunity 16:23–35
64. Teh HS, Teh SJ (1997) High concentrations of antigenic ligand activate and do not tolerize naive T cells in the absence of CD28/B7 costimulation. Cell Immunol 179:74–83
65. Stipdonk MJB van, Lemmens EE, Schoenberger SP (2001) Naïve CTLs require a single brief period of antigenic stimulation for clonal expansion and differentiation. Nat Immunol 2: 423–429
66. Sansom DM (2000) CD28, CTLA-4 and their ligands: who does what and to whom? Immunology 101:169–177
67. Wells AD, Gudmundsdottir H, Turka LA (1997) Following the fate of individual T cells throughout activation and clonal expansion—signals from T cell receptor and CD28 differently regulate the induction and duration of a proliferative response. J Clin Invest 71:3173–3183
68. Frauwirth KA, Riley JL, Harris MH, Parry RV, Rathmell JC, Plas DR, Elstrom RL, June CH, Thompson CB (2002) The CD28 signalling pathway regulates glucose metabolism. Immunity 16:769–777
69. Bromley SK, Iaboni A, Davis SJ, Whitty A, Green JM, Shaw AS, Weiss A, Dustin ML (2001) The immunological synapse and CD28-CD80 interactions. Nat Immunol 2:1159–1166
70. Truitt KE, Hicks CM, Imboden JB (1994) Stimulation of CD28 triggers an association between CD28 and phosphatidylinositol 3-kinase in Jurkat T cells. J Exp Med 179:1071–1076
71. Marengére LEM, Okkenhaug K, Clavreul A, Couez D, Gibson S, Mills GB, Mak TW, Rottapel R (1997) The SH3 domain of Ikt/Emt binds to proline-rich sequences in the cytoplasmic domain of the T cell costimulatory receptor CD28. J Immunol 159:3220–3229
72. Viola A, Schroeder S, Sakakibara Y, Lanzavecchia A (1999) T lymphocyte costimulation mediated by reorganization of membrane microdomains. Science 283:680–682
73. Wülfing C, Davis MM (1998) A receptor/cytoskeletal movement triggered by costimulation during T cell activation. Science 282:2266–2269
74. Diehn M, Alizadeh AA, Rando OJ, Liu CL, Stankunas K, Botstein D, Crabtree GR, Brown PO (2002) Genomic expression programs and the integration of the CD28 costimulatory signal in T cell activation. Proc Natl Acad Sci USA 99:11796–11801
75. Riley JL, Mao M, Kobayashi S, Biery M, Burchard J, Cavet G, Gregson BG, June CH, Linsey PS (2002) Modulation of TCR-induced transcriptional profiles by ligation of CD28, ICOS, and CTLA-4 receptors. Proc Natl Acad Sci USA 99:11790–11795
76. Walunas TL, Bakker CY, Bluestone JA (1996) CTLA-4 ligation blocks CD28-dependent T cell activation. J Exp Med 183:2541–2550
77. Gajewski TF, Fallarino F, Fields PE, Rivas F, Alegre ML (2001) Absence of CTLA-4 lowers the activation threshold of primed $CD8^+$ TCR-transgenic T cells: lack of correlation with src homology domain 2-containing protein phosphatase. J Immunol 166:3900–3907
78. Brunner MC, Chambers CA, Chan FKM, Hanke J, Winoto A, Allison JP (1999) CTLA-4-mediated inhibition of early events of T cell proliferation. J Immunol 162:5813–5820
79. Doyle AM, Mullen AC, Villarino AV, Hutchins AS, High FA, Lee HW, Thompson CB, Reiner SL (2001) Induction of cytotoxic T lymphocyte antigen 4(CTLA-4) restricts clonal expansion of helper T cells. J Exp Med 194:893–902
80. Greenwald R, Oosterwegel M, Woude D van der, Kubal A, Mandelbrot DA, Boussiotis VA, Sharpe AH (2002) CTLA-4 regulates cell cycle progression during a primary immune response. Eur J Immunol 32:366–373

81. Krummel M, Allison JP (1996) CTLA-4 engagement inhibits IL-2 accumulation and cell cycle progression upon activation of resting T cells. J Exp Med 183:2533–2540
82. Chikuma S, Imboden JB, Bluestone JA (2003) Negative regulation of T cell receptor-lipid raft interaction by cytotoxic T lymphocyte-associated antigen 4. J Exp Med 197:129–135
83. Lee KM, Chuang E, Griffin M, Khattri R, Hong DK, Zhang W, Straus D, Samelson LE, Thompson CB, Bluestone JA (1998) Molecular basis of T cell inactivation by CTLA-4. Science 282:2263–2266
84. Chuang E, Fisher TS, Morgan RW, Robbins MD, Duerr JM, Vander Heiden MG, Gardner JP, Hambor JE, Neveu MJ, Thompson CB (2000) The CD28 and CTLA-4 receptors associate with the serine/threonine phosphatase PP2A. Immunity 13:313–322
85. Alegre ML, Noel PJ, Eisfelder BJ, Chuang E, Clark MR, Reiner SL, Thompson CB (1996) Regulation of surface and intracellular expression of CTLA-4 on mouse T cells. J Immunol 157:4762–4770
86. Kuhns MS, Epshteyn V, Sobel RA, Allison JP (2000) Cytotoxic T lymphocyte antigen-4 (CTLA-4) regulates the size, reactivity, and function of a primed pool of $CD4^+$ T cells. Proc Natl Acad Sci USA 97:12711–12716
87. Schwartz JCD, Zhang X, Nathenson SG, Almo SC (2002) Structural mechanisms of costimulation. Nat Immunol 3:427–434
88. Price DA, Brenchley JM, Ruff LE, Betts MR, Hill BJ, Roederer M, Koup RA, Migueles SA, Gostick E, Wooldridge L, Sewell AK, Connors M, Douek DC (2005) Avidity for antigen shapes clonal dominance in $CD8^+$ T cell populations specific for persistent DNA viruses. J Exp Med 202:1349–1361
89. Berg HA van den Rand DA (2004) Foreigness as a matter of degree: the relative immunogenicity of peptide/MHC ligands. J Theor Biol 231:535–548
90. Sansom DM, Manzotti CN, Zheng Y (2003) What's the difference between CD80 and CD86? Trends Immunol 24:313–318
91. Guermonprez P, Valladeau J, Zitvogel L, Théry C, Amigorena S (2002) Antigen presentation and T cell stimulation by dendritic cells. Annu Rev Immunol 20:621–667
92. Balkhi MY, Latchumanan VK, Singh B, Sharma P, Natarajan K (2004) Cross-regulation of CD86 by CD80 differentially regulates T helper responses from *Mycobacterium tuberculosis* secretory antigen-activated dendritic cell subsets. J Leukoc Biol 75:874–883
93. Bousso P, Levraud JP, Kourilsky P, Abastado JP (1999) The composition of a primary T cell response is largely determined by the timing of recruitment of individual T cell clones. J Exp Med 189:1591–1600
94. Sourdive DJ, Murali-Krishna K, Altman JD, Zajac AJ, Whitmire JK, Pannetier C, Kourilsky P, Evavold B, Sette A, Ahmed R (1998) Conserved T cell receptor repertoire in primary and memory CD8 T cell responses to an acute viral infection. J Exp Med 188:71–82

Chapter 4
T Cell Activation and Function: Role of Signal Strength

Asma Ahmed and Dipankar Nandi

Abstract Optimal T cell function lies at the heart of an efficient adaptive response. T cell activation is a highly regulated process and it is important to ensure that activation occurs in the proper context to prevent the development of harmful conditions such as autoimmunity and excessive inflammatory responses. One of the important factors in this process is the strength of the primary activating signal which is delivered upon ligation of the T cell receptor (TCR) with the major histocompatibility complex (MHC) encoded class I or class II molecules bearing the antigenic peptide. The strength of signal (SOS), in turn, depends on several factors: the affinity/avidity of the TCR for the MHC–peptide complex, the time of engagement, antigen concentrations, costimulatory interactions, etc. This chapter reviews the effects of SOS on thymocyte selection and education, T cell costimulation, proliferation, survival, formation of T helper T_H1 and T_H2 subsets, responses to infectious agents etc. The role of the SOS in modulating diverse T cell responses is well appreciated. However, further studies are required to understand the mechanisms by which SOS signals are relayed from the TCR to downstream effectors to modulate T cell activation and responses.

Introduction

Higher vertebrates have complex immune systems which allow them to survive in an environment teeming with infectious agents and also to combat internal abnormalities such as tumors. Immune cells, irrespective of whether they belong to the innate or adaptive arms, remain quiescent in a healthy individual and swing into action only when the body is under threat, for example during infections. Usually, the cells belonging to the innate arm, e.g. neutrophils and macrophages, are the first to come in contact with an invading microbe. These cells produce non-specific effectors,

D. Nandi (✉)
Department of Biochemistry, Indian Institute of Science, Bangalore 560012, India
e-mail: nandi@biochem.iisc.ernet.in

C. Molina-París and G. Lythe (eds.), *Mathematical Models and Immune Cell Biology*,
DOI 10.1007/978-1-4419-7725-0_4,

for example free radicals, to contain the infection. In addition, antigen processing cells (APC) process and present peptides derived from microbial proteins to initiate the T cell response. Unlike the initial non-specific innate immune response, the adaptive immune response is relatively delayed but highly antigen-specific due to the presence of clone specific B cell receptors or T cell receptors (TCR) on B cells and T cells respectively. The subsequent activation and expansion of specific T and B cell clones followed by the secretion of effector molecules, for example antibodies secreted by plasma cells, cytokines by $CD4^+$ T cells, granzymes and perforins by $CD8^+$ T cells, play important roles in protecting the host. Subsequently, immune responses are down modulated (involving death of activated cells) although some memory B cells and T cells survive. Clearly, the ability to induce a specific immune response against antigens is an important hallmark of the immune system in higher organisms.

T cells are one of the most versatile immune cells and perform several functions ranging from providing B cell help, killing infected/tumor cells and regulating immune responses. The study of T cells becomes important because any perturbation in their function leads to severe disorders. Hyperactive T cells are observed in several autoimmune diseases such as multiple sclerosis, insulin dependent diabetes mellitus etc. On the other hand, reduced T cell function makes individuals susceptible to pathogens and tumors as observed in patients with the acquired immunodeficiency syndrome. The importance of the T cell response is highlighted by the fact that immunotherapy is very often targeted towards modulating T cell function, for example suppressing the response during autoimmunity, hypersensitivity and transplantation or boosting it for successful vaccination purposes.

The adaptive immune response mediated by T cells begins in the secondary lymphoid organs with the interaction of dendritic cells (DCs) bearing surface antigen loaded MHC molecules with specific TCR bearing T cells. A productive interaction leads to clonal expansion and differentiation of naïve T cells into effector T cells. Depending on their surface homing receptors, these effectors migrate to specific tissues, under the influence of chemokine gradients, to perform their functions. There are several outcomes downstream of T cell activation and multiple factors determine the differentiation program of naïve T cells, whether they proliferate or die and, finally, whether a productive T cell response clears the infection/tumor. One of these factors is the strength of the activating signal delivered to the cell upon interaction of the TCR with the MHC–peptide complex on APCs. The strength or potency (weak or strong) of the signal is, primarily, determined by the affinity/avidity of the TCR–MHC interaction, the duration of the interaction, the dose of antigen available and presence of costimulatory signals. This chapter will give a brief introduction to T cell activation and will delve deeper into the mechanisms by which the SOS impacts cell fate decisions in the thymus and influences T cell function and death in the periphery.

The T Cell Receptor

A T cell recognizes processed antigenic peptides presented on MHC class I or class II molecules with the help of specific cell surface receptors known as the TCR. It is important to point out that the TCR is different from the B cell receptor in that it does not bind antigen alone but only does so in the context of self MHC molecules. This attribute is known as *self-MHC restriction*, which was first described by Doherty and Zinkernagel [1]. The TCR is a heterodimer of two disulphide linked chains, either $\alpha\beta$ or $\gamma\delta$. About 95% of TCRs are a heterodimer of α and β chains and a small fraction (2–5%) expresses the $\gamma\delta$ heterodimer. On account of having an immunoglobulin like domain structure, TCRs are considered members of the immunoglobulin superfamily. Each α and β chain has an amino terminal variable (V) and a constant (C) region much like an antibody. In addition, each chain has a transmembrane region of 21 or 22 amino acids and a very short 5–12 amino acid long cytoplasmic tail. In humans the α and δ gene segments are located on chromosome 14 and the β and γ segments on chromosome 7. The generation of functional TCR molecules is due to rearrangement of V and J segments of the α and γ chains and V, D and J segments of the β and δ chains. This mechanism is similar to antibody gene rearrangements and occurs in the thymus.

As TCRs contain a very short cytoplasmic tail, signal transduction after MHC–TCR binding occurs via the CD3 molecule. In fact, TCRs are associated with the CD3 complex (Fig. 4.1) consisting of five invariant chains which form two heterodimers ($\gamma\varepsilon$ and $\varepsilon\delta$) and one homodimer ($\zeta\zeta$). The cytoplasmic tails of the CD3 chains contain special immunoreceptor tyrosine-based activation motifs (ITAM) which interact with tyrosine kinases and play important roles in signal transduction. The sequence of a typical ITAM is YXXL/IX$_{6-8}$YXXL. γ, δ and ε have single ITAMs while ζ has three in a row, taking the tally up to ten ITAMs per TCR–CD3 complex.

Apart from the TCR–CD3 complex, T cells express several other important surface molecules, the most vital among them being the CD4 and CD8 co-receptors, which are also members of the immunoglobulin superfamily. T cells are, broadly, divided into two subsets: CD4$^+$ T cells recognize antigen in context of MHC class II, and CD8$^+$ in context of MHC class I (Fig. 4.2). CD4 is a monomer whereas CD8 is a heterodimer of α and β chains. These co-receptors perform two functions: first, their extracellular domains bind regions on the MHC molecule thus strengthening interaction with the TCR. Second, their intracellular domains associate with the Src family tyrosine kinase, Lck, and this association helps in signal transduction. Mice lacking CD4 generate reduced MHC class II restricted T cells, produce less Interleukin (IL)-2 upon activation and are unable to mount an efficient antibody response [2, 3].

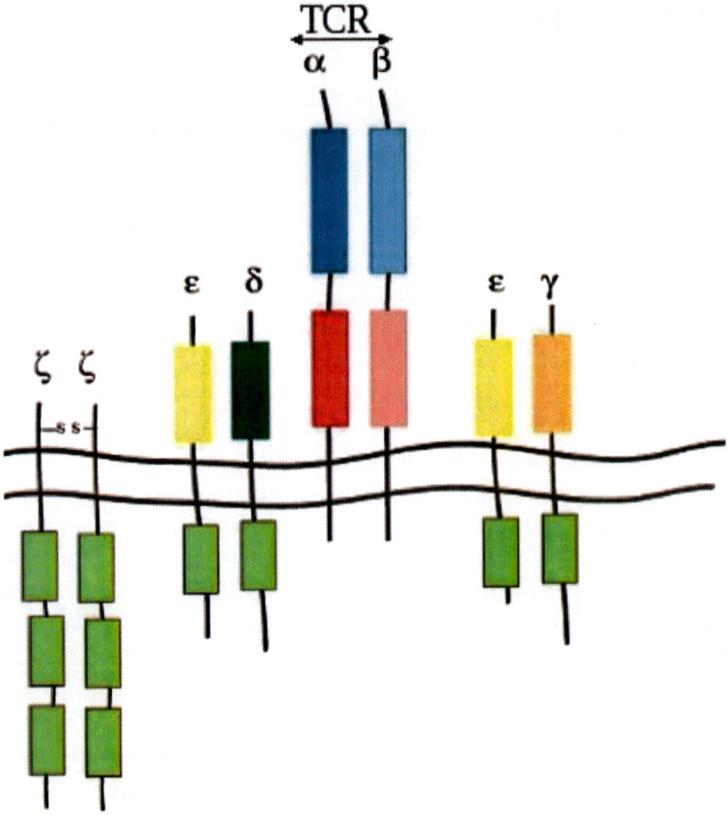

Fig. 4.1 *Structure of the T cell receptor–CD3 complex.* Most T cells express a cell surface TCR consisting of a heterodimer of α and β chains bearing variable antigen binding (*blue*) and constant (*pink*) regions. The TCR does not transduce a signal upon antigen binding and this function is performed by the associated CD3 complex comprising of ϵ, γ, δ and ζ chains. CD3 subunits have long cytoplasmic tails bearing ITAMs (*green*) which assist in signal transduction

Signalling Pathways Leading to T Cell Activation

T cell activation is initiated when a naïve circulating T cell comes in contact with an APC expressing on its surface the cognate MHC–peptide complex. The TCR binds to the MHC–peptide complex and a cascade of signalling events is triggered, the end result of which is the secretion of IL-2 and clonal expansion (Fig. 4.3). Events downstream of the binding of the TCR upon activation can be classified as early and late. Early responses include formation of the immunological synapse, a specialized structure at the contact point(s) between the T cell and APC, phosphorylation and activation of kinases and other proteins, massive remodelling of the actin cytoskeleton, changes in cytosolic Ca^{2+} concentrations and, finally, activation of transcription factors such as NF-AT and their translocation into the nucleus [4]. Subsequent events include transcription and translation of genes, such as IL-2,

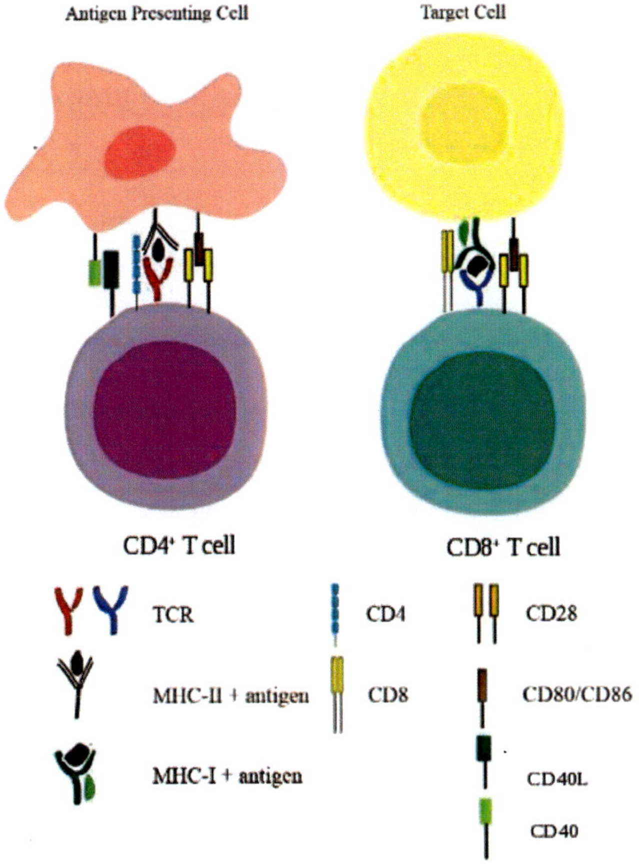

Fig. 4.2 *Optimal T cell activation requires a combination of primary and secondary signals.* CD4+ and CD8+ T cells recognize antigenic peptides in context of MHC class II and MHC class I molecules, respectively. CD4 and CD8 co-receptors stabilize the TCR–MHC peptide interaction which sends signal 1 to the T cell. Binding of CD28 on T cells with CD80/CD86 on the APCs or target cells delivers signal 2 which drives T cell cytokine secretion and clonal expansion. Also important is the interaction between CD40 on APCs and CD40L on CD4+ T cells for T dependent B cell responses and isotype switching

appearance of activation markers such as the CD25 (the high affinity IL-2 receptors), DNA replication and cell division. Further down the line are events which lead to differentiation of cells into effector and memory types and the eventual clearance of effector cells to maintain homeostasis.

The first signalling event downstream of the TCR is the phosphorylation of the ITAMs by the kinases Lck and Fyn. Multiple pathways might be operating in coupling the binding of TCR and the cognate-MHC–peptide complex to phosphorylation of ITAMs in CD3: First, the proximity of CD4 linked Lck to the TCR–CD3 complex may be responsible. Second, conformational changes in the CD3 polypeptides induced by TCR ligation may make the ITAMs more accessible to phosphorylation. Alternately, the exclusion of Csk and other phosphatases may

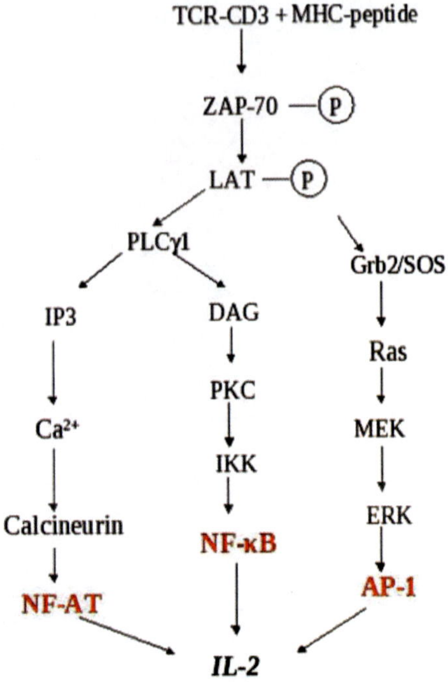

Fig. 4.3 *An overview of T cell signalling pathways leading to transcription of IL-2.* Activation via the TCR–CD3 complex stimulates multiple pathways that upregulate three major transcription factors (shown in *red*), leading to enhanced IL-2 production in T cells

push the momentum towards phosphorylation. Phosphorylated ITAMs now serve as docking sites for Src homology (SH)2 domain containing protein tyrosine kinases, e.g. Zeta associated protein 70 kDa (ZAP70), spleen tyrosine kinase (Syk), and a Tec family member, IL-2 induced tyrosine kinase (Itk) and Receptor like kinase (Rlk) [4, 5].

A dramatic increase in cytosolic Ca^{2+} due to release from ER stores is one of the early events which takes place after ligation of the TCR–CD3 complex with the cognate MHC–peptide complex. The trigger for release of Ca^{2+} from smooth endoplasmic reticulum stores is the binding of IP3 (generated by the activation of PLCγ) to its receptors (Fig. 4.4). Cytosolic Ca^{2+}amounts in a resting T cell are about 0.1 mM but increase to 0.5–2 mM as soon as the smooth endoplasmic reticulum Ca^{2+} stores are opened. However, these stores are soon exhausted and this depletion acts as a trigger for Ca^{2+} from the extracellular milieu to flow into the cell through Ca^{2+} release activated channels (CRAC). This process known as capacitative Ca^{2+} entry ensures that cytosolic Ca^{2+} levels are maintained at higher than basal amounts, e.g. ~0.2–0.5 mM, which is required for sustained T cell activation [6].

The Immunological Synapse

The immunological synapse (IS) is a highly specialized structure which is formed at the interface of a T cell and APC. Molecules present in the IS are the TCR, CD4 associated with Lck, CD28, PKCθ and LFA-1 on the T cell side and pMHC,

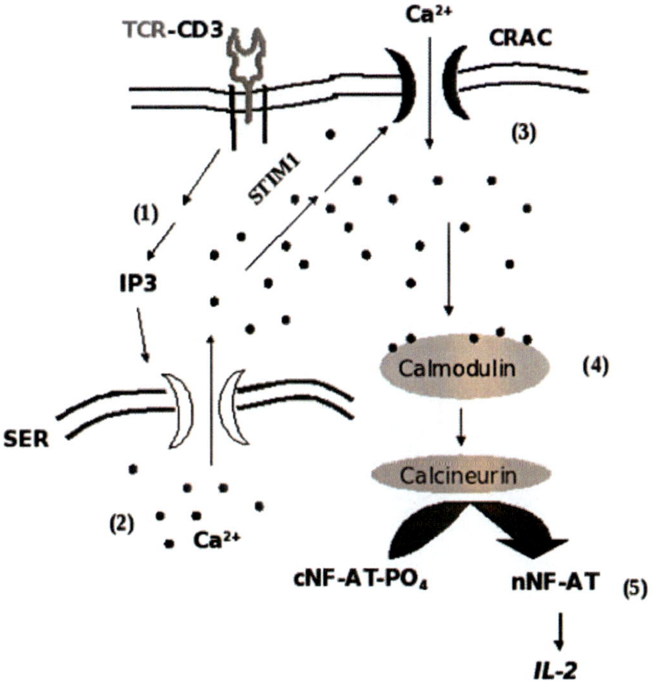

Fig. 4.4 *A simplified overview of the Ca^{2+} signalling pathway*. First, signals downstream of the TCR–CD3 complex lead to generation of InsP3. Second, the InsP3 binds to its receptors on the SER leading to release of Ca^{2+} in to the cytosol. Soon, the intracellular Ca^{2+} stores are exhausted. Third, this depletion is sensed and conveyed by stromal interaction molecule 1 (STIM1) to the plasma membrane CRAC channels, which now open and allow Ca^{2+} influx from extrcellular environment. Fourth, sufficient Ca^{2+} is now present to bind calmodulin and enable activation of phosphatase calcineurin. Finally, calcineurin dephosphorylates cytosolic NF-AT (cNF-AT), leads to its nuclear translocation and nNF-AT, along with other transcription factors, enhances transcription of *IL-2*

CD80/CD86 and ICAM-1 on the APC side. Circulating T cells which have not encountered antigen are symmetrical, round and non-polarized. Also, their adhesion molecules are in a low activity state. However, when they migrate to inflamed tissue or extravasate into secondary lymphoid tissue they become polarized, possibly under the influence of chemokine gradients. These T cells now have a leading F-actin rich lamellipodium and a myosin-II rich uropod (lagging end), migrating with a speed of 5–10 μm/min. The lamellipodium is five- to tenfold more sensitive to antigen than the uropod. Enhanced actin polymerization at the leading edge also leads to the activation of integrins. The chances of adhesion and formation of a synapse between the T cell and APC bearing the agonist MHC–peptide complex are now increased manifold. Recognition of agonist MHC–peptide complex by the TCR leads to elevation of Ca^{2+}, which serves as a stop signal for T cell migration. Formation of the IS begins with a few TCR–MHC–peptide complexes and, subsequently, these clusters come together to form a supra-molecular complex [7].

Costimulation

Optimal T cell activation, in addition to the signal generated by TCR–MHC–peptide complex ligation, requires a second costimulatory signal. This interaction involves costimulatory receptors like CD28 on the T cells and their ligands CD80/CD86 on APCs. This is known as the two signal hypothesis [8] and is required to drive T cell proliferation and effector functions. In addition, it prevents anergy, a state of unresponsiveness during a secondary encounter with antigen [9]. In vitro experiments performed to study anergy primarily characterized it as a condition of reduced proliferation and cytokine production (mainly IL-2) upon restimulation of T cells. The situation is slightly different in vivo where the phenomenon of un-responsiveness is termed as adaptive tolerance and is required to prevent autoreactivity.

CD28, a 44 kDa glycoprotein homodimer, is a member of the immunoglobulin superfamily and is constitutively expressed on T cells [10]. Structurally, it has a leader sequence, an extracellular domain comprising of three complementarity determining regions (CDRs), a transmembrane region and an intracellular domain. Its ligands are CD80 and CD86 expressed on APCs. CD86 is constitutively expressed on APCs but CD80, almost absent on resting cells, is upregulated during inflammation. The ligand binding motif of CD28 consists of MYPPY which is present in the CDR3 region and shows high conservation in CD28 from different species. CD28 is present at the immunological synapse and functions mainly as an amplifier of TCR signals [11] and enhances recruitment and redistribution of lipid rafts [12]. It boosts Ca^{2+} signalling and prolongs T cell–APC interactions [13]. The signalling molecules, Itk, PI3K and PLCγ, assemble on the CD28 intracellular tail to form a signalling unit [13]. Consequently, CD28 signalling leads to higher nuclear levels of transcription factors like NF-AT and AP-1 and enhanced transcription of *Il-2*. Also, CD28 signalling stabilizes IL-2 mRNA levels. All this leads to enhanced IL-2 levels and cell cycle progression in cells which receive both the first and the second signal [14]. CD28 prolongs T cell responses as is highlighted by the phenotype of $Cd28^{-/-}$ mice which can initiate but not sustain T cell activation [15].

CD28 also enhances T cell survival by upregulating anti-apoptotic proteins such as Bcl-2 and $BclX_L$ and inhibiting FasL expression [16]. CD28 apart from amplifying TCR signals can generate its own unique signal. After it has bound to its ligand, Tyr residues on its cytoplasmic tail get phosphorylated and lead to the recruitment of PI3K [17]. Among other molecules that get activated downstream of CD28 are Tec, Itk [18] and the guanine nucleotide exchange factor Vav-1 (involved in actin polymerization) and PLCγ1 [19]. A comparative study of expression in cells activated with TCR alone, CD28 alone or the two together showed both quantitative and qualitative differences in genes modulated by CD28 [11].

While CD28 is the prototypic positive costimulatory receptor, at the other end of the spectrum is the cytotoxic T lymphocyte associated molecule 4 (CTLA4 or CD152). Interestingly, CD28 and CTLA4 bind to the same ligands, i.e. CD80 and CD86. CTLA4, a 33–45 kDa glycoprotein, binds to CD80 and CD86 but with a manifold higher affinity than CD28. Also, CTLA4 is expressed only on activated T cells, unlike CD28. CTLA4 is a negative regulator of T cell responses as concluded from the phenotype of $Ctla4^{-/-}$ mice which die from hyperproliferation of lymphocytes by 4–6 weeks of age [20, 21]. Reducing *Ctla4* expression in mice using small interfering RNA causes onset of diabetes in mice [22]. This is important in light of the fact that several autoimmune disorders map to mutations in *CTLA4* in humans [23]. Several mechanisms have been proposed to explain the inhibition of T cell activation by CTLA4. These can be broadly classified into two categories: (1) by competing with CD28 for CD80/CD86 ligands or (2) generating its unique set of signals. CTLA4 inhibits the TCR and CD28 induced raft formation, association with the tyrosine phosphatase SHP-2 and the serine/threonie phosphatase PP2A which downregulates MAPK and ERK signalling. Increase in TGFβ secretion and increment in indoleamine dioxygenase (IDO) activity have been observed with CTLA4 ligation. More recently, CTLA4 has been shown to decrease T cell–APC dwell times by modulating intracellular Ca^{2+} levels. Overall, the net effect of CTLA4 ligation results in a decrease in IL-2 production and cell cycle progression [24–27].

Some other positive costimulatory receptors are CD40, inducible-costimulator (ICOS), OX40, 4-1BB, signalling lymphocyte activation molecule (SLAM). CD40 is expressed on B cells and other professional APCs and binds to its ligand CD40L on T cells (Fig. 4.2) [28]. Productive T dependent B cell responses and isotype switching requires CD40-CD40L interactions [29]. $Cd40l^{-/-}$ mice show reduced antigen-induced inflammatory responses, e.g. lower IgE and IgG amounts, and are unable to clear some microbial infections, e.g. *Leishmania major* [30, 31]. ICOS is a member of the B7 family and binds to its ligand ICOS-L. It is induced on activated T cells and $Icos^{-/-}$ mice have impaired germinal centre and Peyer's patch formation [32,33]. OX40 is a member of the tumor necrosis factor (TNF) family and binding to its ligand, OX40L, sustains proliferation of T cell effectors. OX40 signalling enhances IL-2 and T_H2 cytokine production. $Ox40^{-/-}$ mice have impaired effector and primary response to keyhole limpet hemocyanin [34]. Also, OX40 signalling dampens Treg function, increases symptoms of autoimmunity but lowers anti-tumor responses [35]. 4-1BB (CD137) is another tumor necrosis factor family member which binds to its ligand 4-1BBL to enhance $CD4^+$ and $CD8^+$ T cell responses and promote IFNγ production. Importantly, $4\text{-}1bb^{-/-}$ mice have defective recall $CD8^+$ responses to some viruses like influenza [36]. Binding of CD27, a member of the TNF family, to its ligand CD70 promotes proliferation and survival of activated T cells. Constitutive activation of CD27 results in immunopathology during autoimmunity and viral infections [37]. Signalling lymphocyte activation molecule (SLAM) deficiency reduces IL-4 and marginally increases IFNγ production in mice [38].

Other negative regulators of T cell activation, apart from CTLA4, are programmed death (PD)1 and B and T lymphocyte attenuator (BTLA). PD1 is a B7 family member and binds to its ligand PD-L1 and PD-L2. $Pd1^{-/-}$ mice develop lupus-like disorders [39]. BTLA is also a B7 family member and a negative regulator of B and T cell responses and $Btla^{-/-}$ mice develop exacerbated experimental autoimmune encephalomyelitis (EAE) [40]. A regulatory role has recently been assigned to the T-cell Ig domain and mucin domain (TIM) family members TIM-1, 2, 3 and 4. TIM1, 2 and 3 are expressed on T cells while TIM4 is primarily expressed on APCs [41]. TIM-3 was identified as T_H1 specific molecule responsible for regulating autoimmunity [42]. TIM-1, 3 and 4 can costimulate T cells, regulate apoptosis, tolerance and clearance of apoptotic cells [41]. Links have been found between TIM proteins and allergy, asthma and autoimmunity [43].

The proper functioning of positive and negative costimulatory molecules is crucial for an optimal T cell response and is highlighted by the phenotypes of mice deficient in any of these molecules. It is important to point out that costimulation ensures that T cell activation occurs in a proper context e.g. during an ongoing immunological response and this safeguards against development of autoimmunity.

SOS and Its Intracellular Modulators

Two important factors during T cell activation are the affinity/avidity of the TCR for the MHC–peptide complex and the time for which the two are engaged. Both these factors along with the availability of co-stimulation determine the overall SOS that is delivered to a T cell during activation. T cells express heterogeneous TCRs which may bind to the same antigen with different affinities. Therefore, T cells get activated with varying SOS, which has a profound impact during thymic education and modulate T cell responses in the periphery with respect to proliferation, cytokine secretion, survival and death. It is thought that optimal T cell activation requires sustained signalling for several hours. To achieve this, TCRs need to be continuously engaged or triggered at periodic intervals (serial triggering). The model of 'temporal summation' proposes that each short-term TCR–MHC–peptide interaction leads to accumulation of intracellular signalling intermediates until a threshold for optimal activation is reached [44].

One of the signalling intermediates robustly modulated by signal strength is intracellular Ca^{2+}. The stronger the signal, the higher is the Ca^{2+} flux [45]. Also, intracellular amounts of Ca^{2+} and ROS increase with the SOS in primary mouse $CD4^+$ T cells [46]. The TCR interacting molecule (TRIM), a transmembrane adaptor protein expressed on T cells, stabilizes TCR surface levels after engagement with MHC–peptide complexes and leads to higher intracellular Ca^{2+} amounts, leading to enhanced activation [47]. The other intracellular integrator of signal strength is ERK. There is evidence to suggest that strong signals lead to greater but transient ERK activation. On the other hand, weak signals result in lower but sustained ERK phosphorylation [48]. Proteins such as Fos which have an ERK targeting DEF motif are sensitive to transient versus sustained ERK activation and can trigger gene transcription in accordance with the SOS [49]. Disruption of *Mekk2* (an upstream MAPK kinase) in mice leads to enhanced IL-2 and IFNγ. In these mice JNK activation, but not ERK activation, is affected and it is possible that MEKK acts as a negative regulator of TCR signal strength by activating JNK [50]. Another negative regulator of TCR signal strength was identified as cyclophillin A, a peptidyl-prolyl isomerase (Ppia). Ppia interacts and inhibits Itk and $Ppia^{-/-}$ mice have elevated levels IgE, develop allergic disorders and their T cells are hypersensitive to activation [51].

Other cell surface molecules that are regulators of signal strength are CD5 and the co-receptor CD8. CD5 is a negative regulator of TCR signalling and, hence, protects against autoimmunity. T cells from $Cd5^{-/-}$ mice are hyperresponsive to anti-CD3 mediated activation [52]. CD5 expression levels are also controlled by the avidity of the TCR–MHC–peptide interaction and a strong interaction leads to higher surface amounts of CD5. Massive deletion of thymocytes in transgenic mice expressing high affinity TCR in a $Cd5^{-/-}$ background is observed due to a shift from positive to negative selection [53]. Unlike CD5, CD8 expression is inversely proportional to TCR signal strength. Experiments done to study this relationship found that, as the affinity of TCR and self MHC–peptide complexes increased, CD8 surface expression is reduced. It is possible that reducing surface CD8 amounts lowers the avidity of T cells for self ligands and prevents autoimmunity [54].

Although the molecular mechanisms and the intracellular modulators of signal strength are not completely understood, progress has been made in identifying several molecules that influence the SOS. However, the modulation of expression of these molecules under varying SOS and their interaction with other molecules affecting T cell responses need to be understood in greater detail. Nevertheless, the impact of SOS on almost every aspect of T cell life is well established and discussed below.

Role of Signal Strength in Cell Fate Decisions in the Thymus

The thymus is a bilobed structure located just above the heart comprised of supporting cells which constitute the stroma and T cells, in various stages of development, known as thymocytes. Progenitor bone marrow cells migrate to the thymus under the influence of chemotactic factors by day 11 of gestation in mice and the eighth or ninth week in humans. Once in the thymus, progenitor T cells undergo a process of maturation and education to yield mature single positive (SP) $CD4^{+}$ or $CD8^{+}$T cells that enter the peripheral circulation (Fig. 4.5). Approximately 99% of thymocytes die by apoptosis during this process and only mature T cells that recognize peptides presented in the context of either MHC class I or MHC class II are selected. The importance of the thymus is highlighted in children suffering from DiGeorge syndrome which results from deletions in chromosome 22q11.2. Similar deletions in mice chromosome have highlighted the importance of the transcription factor Tbx1 in this disorder. These children have an underdeveloped or completely absent thymus and, along with several other defects, are highly susceptible to viral, fungal and protozoan infections [55].

The development and maturation of thymocytes can be tracked by following the changes in expression of cell surface molecules. During the early stages of differentiation, thymocytes lack surface expression of CD4 and CD8 and are referred to as $CD4^{-}CD8^{-}$ or "double negative (DN)". A small fraction of $CD4^{-}CD8^{-}$

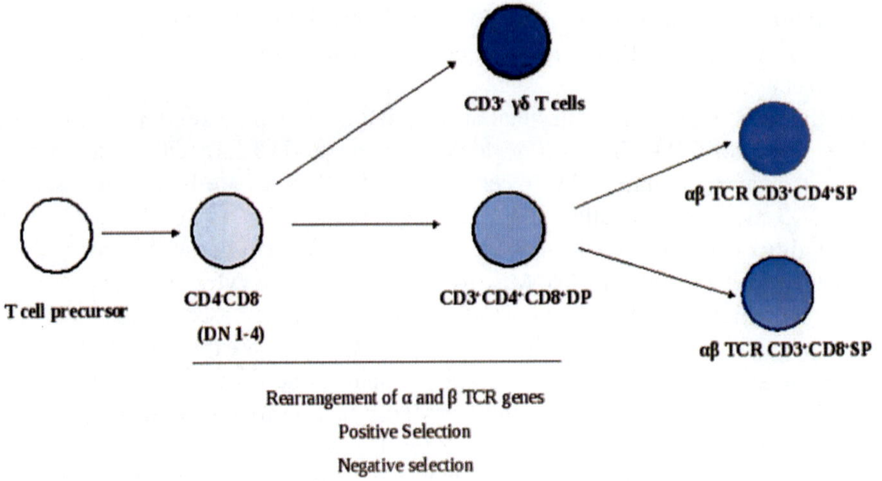

Fig. 4.5 *A simplified overview of thymic development and education.* T cell precursors that enter the thymus undergo a rigorous programme of development and education involving positive and negative selection to emerge into the periphery as $\gamma\delta$ T cells or $\alpha\beta$ TCR bearing $CD4^+$ or $CD8^+$ T cells

thymocytes make productive rearrangements of their γ and δ chain genes to express surface $\gamma\delta$ TCR together with the CD3 complex. $\gamma\delta$ T cells represent about 0.5–10% of peripheral T lymphocytes and are quite abundant in the skin, intestinal epithelium and pulmonary epithelium in mice. These cells are capable of recognizing a broad range of peptide and non-peptide antigens but not necessarily in the context of the MHC class I and class II molecules. They are believed to be members of the earliest cell-mediated immune system and may have evolved to protect the integrity of epithelial tissues [56].

A vast majority of thymocytes productively rearrange α and β TCR genes to express on their surface the $\alpha\beta$ TCR–CD3 complex. TCR gene rearrangement occurs in a manner similar to antibody gene rearrangement in B cells with the help of RAG-1 and RAG-2 and other proteins required for DNA recombination. During these stages, thymocytes also acquire surface CD4 and CD8 molecules and are referred to as $CD4^+CD8^+$ or "double positive (DP)" thymocytes. This stage is important for thymic education as only those thymocytes which recognize self MHC molecules are selected by a process known as "positive selection." Thymocytes which bind to self-MHC molecules with very high affinity are deleted by a process known as "negative selection." This process is important as it reduces the risk of autoimmunity and is well illustrated by the master transcription factor, AIRE, which is responsible for expression of several self proteins in the thymus. The deficiency of AIRE results in a multi-organ autoimmune disease known as autoimmune polyglandular syndrome, type 1 [57]. Thymocytes that fail these selection processes die by apoptosis. The MHC restricted self tolerant double-positive thymocytes lose either CD4 or CD8 co-receptor to become mature SP $CD4^+$ (T_H) or $CD8^+$ (T_C) T cells and enter the peripheral circulation [3].

Apart from positive and negative selection, two more cell fate decisions relevant in terms of the SOS in the thymus are: (1) the development of T cells expressing either $\gamma\delta$ TCR or the $\alpha\beta$ TCR and (2) differentiation of $\alpha\beta$ TCR$^+$ cells into CD4$^+$ or CD8$^+$ lineages. In both these cell fate decisions, the SOS delivered via the TCR plays crucial roles. Precursor cells in the thymus consisting of c-kit$^+$CD44$^+$CD25$^+$CD24$^+$CD4$^-$CD8$^-$can differentiate into different lineages and the DN stage is divided into 4 stages (1–4). As development progresses, thymocytes express less c-kit, CD44, CD25 and CD24, acquire Thy1, CD4, CD8 and begin to rearrange their TCR genes. Commitment to the T cell lineage takes place at the DN stage 3 once precursors acquire Thy1 and express a fully functional preTα (an invariant α chain of the TCR which is expressed early in the thymus) and rearrange their γ and δ chains. This is also the stage when the choice between the two development programs, i.e. $\alpha\beta$ and $\gamma\delta$, needs to be made. Those cells which fail to productively rearrange their TCR genes or do not express surface TCR die by apoptosis. Two models have been proposed to explain how the choice between the $\alpha\beta$ TCR versus the $\gamma\delta$ TCR is made. The first model is the stochastic/selection model according to which cell fate specification is made independent of signals induced by the TCR, i.e. fate is decided before fully functional TCRs ($\gamma\delta$ and preT$\alpha\beta$) are expressed [58]. The second is the instructive model in which fate is decided based on distinct signals generated through the preT$\alpha\beta$ and $\gamma\delta$ TCRs. One of the distinguishing factors in the generation of these two TCRs is SOS and the development of one lineage should affect the development of the other, according to this model [58–60]. The SOS model for lineage commitment was put forward by Hayes et al. [61]. Using a model where fate specification was entirely mediated by the $\gamma\delta$ TCR, this group showed that increasing signal strength through the $\gamma\delta$ TCR enhanced the development of $\gamma\delta$ T cells. Conversely, weakening of signal strength resulted in the development of T cells belonging to the $\alpha\beta$ lineage. Similar results were reported by another group [62]. One way in which signal strength through $\gamma\delta$ and preTα could be altered is through differences in their surface expression. Reduction in $\gamma\delta$ TCR surface expression skewed development towards $\alpha\beta$ T cells. Concomitantly, increasing expression of the $\gamma\delta$ TCR increased the development of $\gamma\delta$ T cells and reduced the development of $\alpha\beta$ T cells [63]. Also, reducing $\gamma\delta$ TCR signalling by reducing the number of ITAMs on the CD3ζ chain enhances $\alpha\beta$ commitment [64]. It has been shown that the magnitude of signalling downstream of the $\gamma\delta$ TCR is greater compared to that generated by the pre-TCR [65].

A biochemical basis for the lineage commitment has also been explained in terms of ERK activation. A strong signal as delivered by the $\gamma\delta$ TCR leads to sustained activation of ERK, while transient ERK activation by the weak pre-TCR signal is more conducive for proliferation and transit of cells to the DP stage [61, 62]. Sustained ERK activation leads to enhanced induction of early growth response factor (Egr), a Zn^{2+} finger transcription factor. Egr proteins are induced by both $\gamma\delta$ and pre-TCR signals; however, the level of induction is regulated by the SOS. Overexpression of Egr leads to more $\gamma\delta$ T cells and a decrease in $\alpha\beta$ T cell numbers. Downstream to Egr is Id3, a helix-loop-helix (HLH) protein. Induction of Id3 is found to be concomitant with $\gamma\delta$ T cell lineage commitment [62]. One mechanism by which Id3 nudges cells towards the $\gamma\delta$ T cell lineage is by suppressing the activity of basic helix-loop-helix E proteins which are required for VDJ recombination by the TCR β chain [60, 62].

In addition to differing SOS in determining $\alpha\beta$ 'vs' $\gamma\delta$ cell fate, a role for Notch signalling has also been proposed. According to this model, moderate pre-TCR signals synergize with Notch signals enabling generation and proliferation of $\alpha\beta$ T cells. While strong $\gamma\delta$ TCR signals are sufficient for the lineage commitment of $\gamma\delta$ T cells, weak $\gamma\delta$ TCR signals can synergize with strong Notch signals to give rise to $\alpha\beta$ cells. In other words, it appears that development of $\alpha\beta$ T cells depends more on Notch signalling compared to $\gamma\delta$ T cells [66].

After commitment to the $\alpha\beta$ lineage, DN stage 4 cells become $CD4^+CD8^+$ or DP in addition to the expression of the TCR β chain. The α TCR chain now undergoes rearrangement. In fact, DP cells continue to express RAG proteins for an extended period of time and multiple V/J recombinations occur at the same α locus so that several $\alpha\beta$ TCRs can be produced per cell. DP thymocytes in the cortex bearing the $\alpha\beta$ TCR undergo selection processes that are dependent on signals generated through the binding of TCR to the MHC–peptide complex. The affinity/avidity model for positive and negative selection is the most widely accepted. $\alpha\beta$ TCRs which have low to intermediate affinity for self MHC, undergo positive selection and differentiate into $CD4^+$ or $CD8^+$ SP cells while those TCRs which have high affinities are negatively selected and undergo clonal deletion. There are several lines of evidence that indicate that the SOS through the TCR determines the fate of DP thymocytes. Cells which are unable to engage MHC or do so with a very low affinity die by neglect, a fate which befalls most DP thymocytes. Those which engage MHC with an intermediate intensity are positively selected and those that bind with too high an intensity are either clonally deleted, become anergic or differentiate into regulatory T cells (Treg), which are peripheral immune regulators [67]. Those thymocytes that survive the rigorous selection process and are MHC restricted are present in the medulla and enter the peripheral circulation as mature T cells.

The binding affinities of TCR–MHC peptide complexes are reflected in the signal strengths generated downstream of the TCR which trigger specific programs leading to either positive or negative selection. Experiments have shown that by varying the SOS, positive and negative selection decisions can be altered. In a study to examine the role of TCR signal intensity on positive and negative selection, TCR signal potency was varied by substituting transgenic TCRs having 1, 2, 3 or no ITAMs on the CD3ζ chain. Interestingly, decreasing the signalling ability in cells with low affinity led to reduced selection. However, in cells which bound self-MHC with very high affinity and would normally have undergone clonal deletion, decreasing TCR signal potency by varying the number of ITAMs increased the chances of positive selection [68]. Low affinity ligands generate the p21 form of CD3 ζ which has only 2/3 ITAMs phosphorylated while high affinity ligands lead to phosphorylation of all 3 ζ ITAMs resulting in formation of the p23 form. The signalling molecules and adaptor proteins recruited by p21 and p23 are different and may explain the differences in the distinct signalling programs triggered during positive and negative selection [69].

The activation of ERK has been shown to integrate signal strength and cell fate decisions in the thymus. Positive selecting ligands trigger low but sustained ERK activation whereas negative selecting ligands lead to strong but transient ERK activation. Also, in foetal thymic organ cultures, negative selection stimuli can be converted to positive selection signals by use of pharmacological inhibitors of ERK. One of the reasons behind differences in ERK activation during positive and negative selection is thought to be the time for which TCRs remain on the surface. High affinity/negative selection ligands induce rapid internalization of TCR. However, during low affinity interactions TCRs stay on the surface for longer times, leading to prolonged ERK activation which is essential for survival and proliferation of positively selected DP thymocytes. Sustained ERK activation leads to phosphorylation and ubiquitin mediated proteosomal degradation of pro-apoptotic protein Bim. On the other hand strong and transient ERK activation leads to upregulation of genes such as Nur77 that lead to apoptosis [70–72]. Another line of evidence suggesting the involvement of TCR-Ras-MEK-ERK pathway is the fact that targeted deletion of RasGRP, a guanine nucleotide exchange factor for Ras, has a profound effect on positive selection by virtue of reduced ERK activation [73]. A germline knockout of ERK1 having a conditional deletion of ERK2 at the DN stage resulted in developmental arrest at the DP stage, suggesting a role for ERK in positive selection [74].

The process of commitment to either the helper ($CD4^+$) or cytotoxic ($CD8^+$) lineages is crucial and DP thymocytes downregulate either CD4 or CD8 co-receptors to become MHC class II or class I restricted. This process could be random and independent of TCR signals such that cells which down regulate the mismatched co-receptor are eliminated. However, the stochastic model could not explain why constitutive co-receptor expression could not rescue cells which had mismatched CD4 or CD8 expression. The instructive model, on the other hand, suggests that commitment to the CD4 and CD8 lineage occurs in response to specific TCR signals [75]. Lck associates with CD4 more compared to CD8 and this may lead to stronger signals being delivered by MHC class II compared to MHC class I restricted cells [76]. Strong and weak signals lead to selective downmodulation of either CD4 or CD8. Evidence suggests that weak signal stimulated cells develop into $CD8^+$ T cells whereas strong signals stimulate differentiation into $CD4^+$ T cells. Mice expressing an inactive Lck contain MHC class II restricted cells differentiating into $CD8^+$ T cells and constitutively active Lck can direct MHC-I restricted cells to differentiate into the CD4 lineage [77]. Lineage commitment can also be decided by signal duration where, longer and shorter duration signals lead to CD4 and CD8 lineage commitment respectively [78]. Another model termed the 'kinetic signalling model' proposes that the duration of TCR signalling rather than the strength plays a key role in determining the $CD4^+$ or $CD8^+$ T cell lineage. However, in this case too, the quantity of signal delivered to a cell is the deciding factor. What this model proposes is that all DP cells transiently terminate CD8 expression so that they become $CD4^+CD8^{lo}$. If this transient population of cells receives sustained TCR signalling, it develops to form $CD4^+$ cells. If the signalling is of a short duration, transcription of CD8 is reinitiated, CD4 is silenced and $CD8^+$ SP cells develop [79].

Consistent with this model, mice which expressed CD4 under the control of CD8 $E8_{III}$ enhancer, showed progressive downmodulation of CD4 after the DP stage. Positive selection of MHC-II restricted cells was normal but the differentiation of these cells was skewed towards the CD8 subtype by virtue of co-receptor downmodulation [80].

Overall, the strength or duration of signal plays crucial roles during thymic development and education: First, the conversion of DN cells to $\gamma\delta$ T cells or $\alpha\beta$ T cells; second, positive and negative selection of $\alpha\beta$ TCR^+ DP cells and, finally, the generation of mature $CD4^+$ or $CD8^+$ SP T cells.

Signal Strength Influences T_H1/T_H2 Dichotomy

The encounter of an antigenic peptide presented on MHC class II triggers a series of events leading to the differentiation of $CD4^+$ T cells into different T_H subtypes, which are characterized by distinct cytokine secretion patterns. Two primary subtypes identified are T_H1 which secrete IFNγ, IL-2, lymphotoxin β and T_H2 which secrete IL-4, IL-5, IL-6 and IL-13 [81,82]. Another recently identified effector subset is T_H17 which, primarily, produces IL-17. Their generation is regulated by IL-6 and TGFβ [83] and their development is regulated by the master transcription factors Stat3 and RORγ [84]. T_H17 cells are crucial for combating several pathogens. Mice lacking IL-17 are highly susceptible to infection with *Klebsiella* and *Candida*. More importantly, T_H17 cells have been associated with several autoimmune conditions such as multiple sclerosis and rheumatoid arthritis. Indeed IL-17 deficient mice develop less severe experimental autoimmune encephalomyelitis [85].

T_H1 responses are crucial for clearing intracellular pathogens while T_H2 responses are effective during parasite infections and antibody production. Abnormal or excessive T_H activation leads to autoimmune disorders (T_H1) or atopy (T_H2). The local cytokine milieu, co-stimulatory interactions, antigen concentration and strength of TCR signals are the factors which govern T_H differentiation. The cytokine environment existing during priming has a far reaching effect on differentiation. The presence of IL-12 skews differentiation towards T_H1 whereas IL-4 results in greater T_H2 responses [86]. The SOS, in terms of duration of engagement and antigen dose, is a key determinant in T_H fate decision [87, 88]. T cells from allergic individuals produce IL-4 only when stimulated with a low dose of allergen. When the allergen dose is increased, IL-4 production is reduced and may form the basis for allergen immunotherapy [87]. On the other hand, IL-12 coupled with a short duration TCR signal is sufficient for T_H1 polarisation; however, prolonged TCR signalling is required for T_H2 responses [88]. A weak intensity signal provided by a low affinity altered peptide ligand favors T_H2 whereas optimal concentrations of an agonist peptide which transduce a relatively strong signal favors T_H1 differentiation [89–91]. Agonist peptides trigger strong and more sustained signalling events such as CD3ζ phosphorylation, Ca^{2+} mobilization, JNK and

MAPK activation as compared to altered peptide ligands [92–95]. The magnitude of Ca^{2+} fluxes downstream of weak and strong TCR signals is an important determinant of T_H subtypes. Weak TCR stimulation induce Ca^{2+} signals that promote IL-4 production whereas strong signals induce MAPK activation that induce IFNγ [96]. Consistent with this observation, inhibition of MAPK and JNK in both human and mouse T cells promotes T_H2 and inhibits T_H1 differentiation [97]. Constitutive activation of p38 MAPK promotes T_H1 but expression of a dominant negative form of p38 MAPK inhibits T_H1 differentiation [98]. Stimulation of JNK1 deficient cells with a combination of anti-CD3 + anti-CD28 enhances T_H2 differentiation [99]. JNK2 deficiency enhances T_H2 differentiation but suppresses T_H1 primarily due to a lack of upregulation of the IL-12Rβ2 subunit [100].

The key transcription factors required for T_H1 differentiation are T-bet and Stat4 and those involved in T_H2 differentiation are Stat6, c-Maf and GATA-3. T-bet and GATA-3 are master regulators of T_H1 and T_H2 differentiation respectively [86]. Over expression of T-bet induces T_H2 cells to make IFNγ and T-bet$^{-/-}$ mice have impaired T_H1 differentiation, show increase in T_H2 cytokines and develop asthma like symptoms [101, 102]. On the other hand, deletion of GATA-3 leads to T_H2 cytokine defects [103]. Stat4 is an amplifier of IL-12 signals and can induce T_H1 differentiation even in absence of TCR signals, whereas Stat6, activated by IL-4, in turn is essential for activation of GATA-3 [82, 86]. However, the transcription factors that are primarily modulated by the SOS belong to the NF-AT class. Activation and nuclear localization of NF-AT is dependent on the Ca^{2+} signal. Stronger Ca^{2+} signals ensure longer nuclear dwell times for NF-AT. The SOS regulates amounts of NF-ATp (NF-ATc2 or NF-AT1) and NF-ATc (NF-ATc1 or NF-AT2) in the nucleus. Strong signals as opposed to weak ones lead to more NF-ATp accumulation in the nucleus, which promotes GATA-3 suppression and T_H1 differentiation. Altered peptide ligands on the other hand have low NF-ATp but sufficient NF-ATc in the nucleus to drive IL-4 transcription and T_H2 differentiation. Both IL-4 and IL-2 have been shown to be essential for T_H2 differentiation in response to low dose antigenic peptide. Cells receiving low intensity signals upregulate GATA-3 which can, in turn, enhance its own levels. GATA-3, along with IL-2 activated Stat5, drives IL-4 transcription and T_H2 differentiation [104].

Co-stimulation through CD28 has not been found to independently contribute to T_H differentiation. The addition of anti-CD28 enhances cytokine secretion but does not change the kind of cytokines secreted. However, the ability of anti-CD28 to enhance IL-4 production in cells which receive a low intensity signal results in greater differentiation of T_H2 differentiated cells [91,95]. However what does have a qualitative effect on cytokine pattern is interaction between LFA-1 and ICAM. Blockade of this interaction promotes T_H2 whereas increasing the LFA-1-ICAM interaction promotes T_H1 development due to a sustained Ca^{2+} signal [105, 106]. Overall, it appears that SOS, along with other factors, is important during T_H cell differentiation. This aspect is relevant during T_H responses as the dose and potency of the antigen clearly modulate immune responses.

Role of Signal Strength in Treg Function

Treg cells are important in the maintenance of peripheral tolerance. They have been characterized as a subset (5–10%) of $CD4^+$ T cells expressing high levels of CD25, the IL-2Rα chain, and are capable of suppressing in vivo and in vitro T cell responses [107]. The primary function of Treg cells is to regulate autoimmunity and depletion of this population in mice has been shown to precipitate several autoimmune disorders. Injection of $CD4^+CD25^+$ cells reduces symptoms of several autoimmune diseases such as colitis, gastritis, insulin dependent autoimmune diabetes and thyroiditis in mouse models [107, 108].

Treg cells do not proliferate or produce IL-2 in response to TCR mediated stimulation [109]. Most Treg cells are $CD4^+$, express the transcription factor Foxp3 and can be divided into two groups: natural Treg (nTreg) and induced Treg (iTreg). The former arise in the thymus whereas the latter are derived from peripheral $CD25^-$ SP $CD4^+$ T cells under the influence of IL-2 and TGFβ. Both express high levels of CD25, glucocorticoid-induced tumor necrosis factor inhibitor (GITR), CTLA4, CCR4, and CD62L [110]. Tregs are $CD45RB^{low}$ (mice) and $CD45RO^{low}$ (humans), a sign that they are pre-activated [110]. Apart from the site of their generation, the conditions that give rise to natural and induced Tregs are quite different. nTregs are the result of strong TCR signals delivered by self peptides in the thymus [111]. Also required for their generation are strong costimulatory signals because, in the absence of CD28, the numbers of nTreg are reduced [112–114]. On the other hand, iTregs arise in the periphery due to weak or suboptimal primary and secondary signals [115, 116]. However, absolutely essential in this case is CTLA4 signalling because mice lacking CTLA4 do not have this category of Tregs [117]. IL-2 is the cytokine that is crucial for generation and maintenance of Treg cells [118] as the expression of Foxp3 is dependent on IL-2 signalling [119]. Indeed, $Il\text{-}2^{-/-}$ mice suffer from fatal autoimmunity [120]. Treg cells are thought to mediate their suppressive effects via direct cell–cell contact. Numerous mechanisms, including secretion of suppressive cytokines IL-10 and TGFβ, repression of IL-2 transcription and induction of the tryptophan depleting enzyme indoleamine 2, 3 dioxygenase in target cells, have been proposed to explain their mode of action [121].

Tregs, when isolated from peripheral blood, are anergic. As they have immense therapeutic potential, the possibility of expanding them in vitro would be of great benefit. What has been found is that TCR and costimulatory signals that drive proliferation of naïve $CD4^+$ T cells are insufficient for propagation of Tregs bearing very high avidity TCRs for self antigens. Apart from strong TCR signals, Tregs require high intensity CD28 signals for expansion, but high amounts of IL-2 cannot substitute for CD28 signalling [122]. These results are consistent with results obtained with a superagonist anti-CD28 that selectively increased Treg numbers and their suppressive activity in rats suffering from EAE. In fact, treatment with this antibody was found to confer protection against EAE and was proposed as a therapy for treating autoimmune disorders [123].

Signal strength, apart from controlling Treg expansion, also dictates resistance or susceptibility to Treg mediated suppression. Using an in vitro model, it has been shown that human T cells became more and more resistant to Treg suppression as the SOS, in terms of anti-CD3 mode (soluble 'vs' plate-immobilised) and antigen concentration, was increased [124]. Using this information, a model was proposed according to which Tregs suppress T cells which receive low intensity signals but not those that are activated with a stronger signal. The former would be the case of autoreactive T cells binding to self-peptides and the latter of T cells binding antigenic peptides derived from pathogens. This mechanism might help Tregs to control autoimmunity and at the same time avoid suppressing pathogen specific T cell responses that are essential for the host [125]. Another mechanism by which Treg function is suppressed during infection is by IL-6 secretion by DCs in response to Toll like receptor (TLR) ligands. Depletion of $CD25^+$ cells restores effector T cell priming in *Il-6*$^{-/-}$ mice, suggesting a role for this cytokine in suppression of Treg function [126]. This aspect is important during the generation of immune responses during infections.

Role of Signal Strength in Effector and Memory Phenotype Delineation

Retaining memory long after an infection has been cleared is an important hallmark of the adaptive immune response. The frequency of a particular $CD8^+$ T cell clone in a normal healthy individual is around 1 in 100,000. Once a clone is exposed to a specific antigen, it undergoes a burst of expansion dividing several times so that its numbers increase by about 50,000-fold. After expansion these cells migrate to inflamed tissue, perform effector functions such as cytokine secretion and cytolysis. Once the infection is cleared and the antigen is no longer present most (90–95%) effectors die by apoptosis to maintain homeostasis and prevent damage to self tissue. However, some survive and develop a memory phenotype and are capable of providing protection during subsequent infections. This entire program of expansion of effector cells and commitment to the memory phenotype requires 2–3 days after first encounter with antigen. Memory T cells are much faster in responding to an antigenic challenge compared to naïve T cells. Unlike naïve T cells, a fraction of memory cells known as effector memory cells ($CD62L^{lo}CCR7^-$) resides in non-lymphoid mucosal sites so that they can encounter antigen immediately. The other subset is the central memory cells ($CD62L^{hi}CCR7^+$) that reside in secondary lymphoid tissue. The balance between effector and central memory cells is dependent on the dose of antigen and frequency of TCR stimulation [127, 128].

It is unclear whether effector and memory differentiation are coupled. Experiments where effector cells were tagged and then tracked showed that these tagged effectors appeared in the memory population [129]. However, in some cases it has been seen that conditions during priming dictate effector/memory differentiation. A combination of a primary and co-stimulatory signals in presence of cytokines such as IL-12, IL-21 and IFNα/β early on during infection promotes effector cell generation while the primary and secondary signals in absence of an inflammatory milieu, when antigen concentrations are lower, reduce the differentiation of memory cells [130, 131]. Also important during memory phenotype acquisition is help from $CD4^+$ cells. $CD8^+$ memory cells which form without $CD4^+$ help are qualitatively poor, do not respond efficiently to secondary antigenic challenge, and die by activation induced cell death (AICD). $CD4^+$ T cell interaction with DCs generates chemokines that attract $CD8^+$ cells to appropriate niches where they can receive the entire gamut of signals essential for differentiation into the memory phenotype. $CD4^+$ T cells are essential for long survival of memory cells as CD4 knock-out mice show gradual decline in memory $CD8^+$ cells [127].

Qualitatively good and functional memory cells arise only under conditions of optimal antigen dose. Chronic infection conditions, such as infection with *Mycobacterium tuberculosis* or human immunodeficiency virus (HIV), result in the persistence of antigens for a very long time. This condition leads to unresponsive effector cells that are 'exhausted' and do not give rise to functional memory cells but undergo deletion [132]. The degree of $CD8^+$ T cell exhaustion depends directly on the persisting antigen load [127, 133] and with increasing antigen dose, $CD8^+$ effector T cells undergo more rounds of cell divisions and the frequency of memory cells progressively decreases. Exhaustion results from interaction of programmed death-1 (PD-1) with its ligand PD-L1, which is highly expressed on chronically infected cells [134]. This line of therapy i.e. blockade of PD-1-PD-L1 interactions, is being pursued to boost immune responses during vaccination and immunotherapy [127, 135].

Another important factor in memory cell generation is IL-7 because, in the absence of signalling by this cytokine, memory cells do not form in mice [136]. Memory cell precursors can be identified by the presence of IL-7R [137]. What has also been found is that memory cell characteristics i.e. IL-7Rα expression, proliferation and expansion in response to IL-7 via a PI3K dependent signalling pathway are acquired as a function of increasing signal strength. However, the ability to proliferate and secrete IL-2 in response to TCR stimulation decreases when high signal strengths are used for priming. These results demonstrate that qualitatively good memory cells arise when priming is done at intermediate signal strengths [138].

Long lived $CD4^+$ memory cells are very high avidity clones suggesting that, during differentiation, cells which receive the strongest stimulus acquire the memory phenotype. The SOS during priming has a profound impact on the generation of memory $CD4^+$ T cells. This was shown using a model of adoptive transfer in which transgenic SMARTA cells specific for I-A^b-restricted GP_{61-80} epitope of lymphocytic choriomeningitis virus were injected into normal C57BL/6 mice and challenged with either the virus or *Listeria monocytogenes* expressing the viral peptide. When these mice were challenged with virus, SMARTA cells along with the endogenous viral specific $CD4^+$ T cells expanded normally, contracted and developed into long lived memory cells. However, when mice were injected with a recombinant *Listeria monocytogenes* secreting GP_{61-80}, unlike the endogenous C57BL/6 CD4 population, SMARTA cells failed to develop memory cells after the initial expansion. Also, post expansion SMARTA cells had high Bim and low Bcl-2 levels and eventually died by apoptosis. Compared to the endogenous $CD4^+$ T cell population SMARTA cells had lower affinity TCRs and the signals they received during priming were, most likely, sufficient to drive proliferation but insufficient for memory differentiation [139].

SOS Modulates the Outcome of Costimulatory Interactions

As mentioned earlier, CD28 by binding to CD80/CD86 molecules delivers the costimulatory signal essential for clonal expansion of T cells and preventing anergy. The strength of the CD28 signal can alter responses to weak and strong TCR signals. Increasing the strength of CD28 costimulation can change a weak signal and a strong signal into a strong and super strong signal respectively [140]. On the other hand, it has also been shown that absence of CD28 signals affects blastogenesis and frequency of cell division equally in T cells activated with a weak and strong signal [141].

As described earlier, several lines of evidence and the phenotype of *Ctla4*$^{-/-}$ mice suggest that CTLA4 is a negative regulator of T cell activation. On the other hand there are several reports that suggest that there could be a degree of plasticity in CTLA4 responses: First, CTLA4–CD80/CD86 interactions were found capable of costimulating T cells [142]. Second, CTLA4 could be converted from a negative to a positive costimulator by use of a recombinant single chain Fv ligand [143]. Third, CTLA4 lacking the cytoplasmic domain was found to costimulate IL-2 production in T cell hybridomas [144]. It has been found that the SOS is a key determinant of effects downstream of CTLA4. Using an in vitro system where CD4^{+} T cells were activated with varying SOS, it was found that CTLA4 function switches from positive to negative in terms of IL-2 production and cell cycle progression as the strength of the primary signal is increased [145, 146]. The downstream effectors of CTLA4 in this system of study were found to be Ca^{2+} and reactive oxygen species (ROS) whose intracellular levels were found to be directly proportional to the SOS. CTLA4, under both weak and strong activation conditions, increased intracellular Ca^{2+} and ROS levels but with distinct outcomes: T cells activated with a weak signal possess low amounts of Ca^{2+} and ROS and CTLA4–CD80/CD86 interaction enhances activation. However, cells activated with a strong signal already have substantially high levels of ROS and Ca^{2+} and their further increase by CTLA4–CD80/CD86 interactions lowers cell cycling [46].

It is possible that the plasticity displayed by CTLA4 in vitro also has physiological roles. In a pool of heterogenous T cells, CTLA4 has differential effects: T cells from multiple sclerosis patients when stimulated with myelin basic protein in the presence of CTLA4 blockade showed unequal expansion of high and low affinity clones. CTLA4 ligation enhances expansion of clones with low affinity TCRs but inhibits that of high affinity TCRs [147]. In another in vivo mouse model of EAE, immunization with a disease antagonist peptide along with CTLA4 blockade leads to lowered frequency of clones reactive to the disease agonist peptide [148]. Taking these results together, it is possible that CTLA4 inhibits expansion of high affinity

TCR clones which might dominate the T cell response. It also enhances expansion of low affinity clones which may promote the generation of a broader cross reactive immune response that may play important roles to clear rapidly mutating pathogens [149].

SOS Controls Survival and Death of T Cells in the Periphery

Circulating naïve T cells are fairly long lived owing to cytokines such as IL-7, IL-6 and IL-15. When T cells encounter antigen, they undergo a phase of rapid expansion which enables them to perform their effector functions and clear the infection. Expansion is followed by contraction where most effectors are deleted to maintain homeostasis and prevent conditions such as autoimmunity. From the expansion to the contraction state, T cells become increasingly susceptible to apoptosis. There are two apoptotic pathways operating during the contraction phase: activation induced cell death (AICD) and activated T cell autonomous death or ACAD [150]. The former relies on extrinsic death signals transmitted through surface death receptors such as Fas (CD95) and TNFR1 and occurs when pre-activated cells undergo secondary stimulation. The role of CD95 in T cell homeostasis is observed in *lpr* (mutated CD95 or Fas) and *gld* (mutated CD95L or FasL) mice which suffer from autoantibody production and excessive lymphoproliferation. On the other hand, ACAD relies on death signals arising from the mitochondria. Death in this case depends on the ratio of pro- and anti-apoptotic proteins. However, both AICD and ACAD pathways converge on caspase 3 activation and eventual cell death [150,151].

One of the factors that regulates resistance and sensitivity to death during expansion and contraction is the cytokine milieu. IL-2, which is an important cytokine during the expansion phase, is also responsible for priming cells for death by modulating expression of pro-apoptotic factors such as c-FLIP and CD95L [152]. IFNγ limits expansion of activated cells and promotes apoptosis by upregulating caspases [151, 153]. Once antigen is withdrawn, T cells become susceptible to death by neglect and their survival depends on how well they are able to respond to survival signals provided by ubiquitous cytokines IL-7 and IL-15. This ability of T cells to survive post the expansion phase is termed as 'T cell fitness' and is directly proportional to the SOS [154]. Human and mouse $CD4^+$ and $CD8^+$ cells activated with a weak activating signal expand but die by neglect as soon as the antigen is withdrawn because they are unable to respond to IL-15. On the other hand, cells which receive a relatively strong signal not only proliferate during the expansion phase but also survive in vitro and in vivo even in absence of the signal by upregulating the anti-apoptotic protein, $BclX_L$. In addition, IL-15R expression is enhanced so that the response to IL-15 survival signals continues. Together, these mechanisms ensure these cells are 'fit' [154]. However, there can be conditions where signals can be weaker or stronger than those that are used. In some cases, T cells become anergic whereas in other cases the cells die due to excessive activation [155–157].

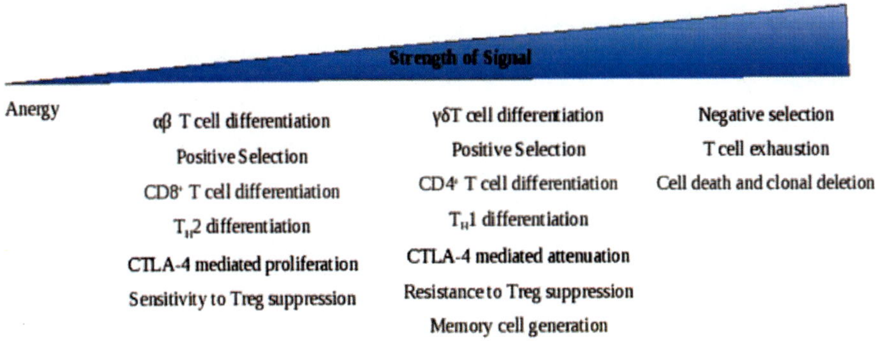

Fig. 4.6 Summary of the effects of signal strength on T cell differentiation and function

Summary

The relative SOS perceived by a T cell is translated into differential amounts of intracellular mediators, e.g. Ca^{2+}, ERK etc, which leads to distinct responses. Consequently, these changes decide the fate and nature of T cell responses in terms of differentiation into various subsets, survival and death (Fig. 4.6). During thymic differentiation, immature thymocytes which receive strong TCR signals are more likely to become $\gamma\delta$ T cells compared to $\alpha\beta$ T cells. $CD4^+CD8^+$ thymocytes that express low amounts of the $\alpha\beta$ TCR and bind to self peptide–MHC complexes with very high affinity, i.e high SOS, are negatively selected and die by apoptosis. This process reduces the risk of auto-reactivity in the periphery. Only those thymocytes that bind with the "right" affinity to self peptide–MHC complexes are positively selected. Once again, SOS plays a role and those thymocytes that receive a strong signal are likely to differentiate into $CD4^+$ T cells whereas those that receive a weaker signal differentiate into $CD8^+$ T cell. Only SP thymocytes are mature and enter the peripheral circulation.

In the periphery, a combination of TCR signal strength and local cytokine milieu determine T_H1 and T_H2 differentiation pathways. Strong signals drive T_H1 whereas weak signals result in generation of T_H2 cells. During an ongoing immune response, T cells activated with a strong signal are better equipped to survive during the contraction phase and develop into memory cells compared to cells that are primed with a comparatively weaker signal. There can be signals that are too weak or too strong to drive productive T cell responses. In case of the former, cells become unresponsive or anergic and the latter usually results in cell death. This implies that a T cell should receive the optimal amount of signal (as determined by antigen dose, receptor affinity and occupancy) for the desired response. This aspect becomes important during vaccination and autoimmune disorders. Therefore, the intensity/potency of the TCR signal is a simple and elegant mechanism to regulate several T cell processes. It is possible that modelling of T cell responses taking into account the role of SOS may be an important criterion to predict in vivo T cell responses. Further experimental verification of these modelling profiles will be important in devising better immunotherapeutic strategies to enhance T cell responses.

Acknowledgements The suggestions and comments by members of the DpN laboratory and colleagues from different institutions during our studies on T cell activation are greatly appreciated. We thank the Department of Biotechnology, Government of India and the Commission of the European Communities for financial support.

References

1. Zinkernagel R, Doherty P (1974) Immunological surveillance against altered self components by sensitised T lymphocytes in lymphocytes choriomeningitis. Nature 251:547–548
2. Goldsby RA, Kindt TJ, Osborne BA (2000) Kuby immunology. WH Freeman, New York
3. Paul WE (Ed) (2008) Fundamental immunology. Lippincott Williams and Wilkins, Philadelphia
4. Iwashima M (2003) Kinetic perspectives of T cell antigen receptor signaling. Immunol Rev 191:196–210
5. Lin J, Weiss A (2001) T cell receptor signalling. J Cell Sci 114:243
6. Lewis R (2001) Calcium signaling mechanisms in T lymphocytes. Annu Rev Immunol 19:497–521
7. Krummel M (2007) Immunological synapses: breaking up may be good to do. Cell 129: 653–655
8. Bretscher P, Cohn M (1970) A theory of self-nonself discrimination. Science 169:1042–1049
9. Harding FA, McArthur JG, Gross JA, Raulet DH, Allison JP (1992) CD28 mediated signaling co-stimulates murine T cells and prevents induction of anergy in T cell clones. Nature 356:607
10. Gross J, St John T, Allison J (1990) The murine homologue of the T lymphocyte antigen CD28. Molecular cloning and cell surface expression. J Immunol 144:3201
11. Riley J, Mao M, Kobayashi S, Biery M, Burchard J, Cavet G, Gregson B, June C, Linsley P (2002) Modulation of TCR-induced transcriptional profiles by ligation of CD28, ICOS, and CTLA-4 receptors. Proc Natl Acad Sci USA 99:11790
12. Martin M, Schneider H, Azouz A, Rudd C (2001) Cytotoxic T lymphocyte antigen 4 and CD28 modulate cell surface raft expression in their regulation of T cell function. J Exp Med 194:1675
13. Jacobelli J, Andres P, Boisvert J, Krummel M (2004) New views of the immunological synapse: variations in assembly and function. Curr Opin Immunol 16:345–352
14. Brunner-Weinzerl MC, Hoff H, Burmester GR (2004) Multiple functions for CD28 and CTLA4 during different phases of T cell responses: implications for arthritis and autoimmune disease. Arthritis Res Ther 6:45
15. Lucas P, Negishi I, Nakayama K, Fields L, Loh D (1995) Naive CD28-deficient T cells can initiate but not sustain an in vitro antigen-specific immune response. J Immunol 154:5757
16. Carreno B, Collins M (2002) The B7 family of ligands and its receptors: new pathways for costimulation and inhibition of immune responses. Annu Rev Immunol 20:29–53
17. Pagès F, Ragueneau M, Rottapel R, Truneh A, Nunes J, Imbert J, Olive D (1994) Binding of phosphatidyl-inositol-3-OH kinase to CD28 is required for T-cell signalling. Nature 369: 327–329
18. Yang W, Olive D (1999) Tec kinase is involved in transcriptional regulation of IL-2 and IL-4 in the CD28 pathway. Eur J Immunol 29:1842–1849
19. Michel F, Attal-Bonnefoy G, Mangino G, Mise-Omata S, Acuto O (2001) CD28 as a molecular amplifier extending TCR ligation and signaling capabilities. Immunity 15:935–945

20. Tivol E, Borriello F, Schweitzer A, Lynch W, Bluestone J, Sharpe A (1995) Loss of CTLA-4 leads to massive lymphoproliferation and fatal multiorgan tissue destruction, revealing a critical negative regulatory role of CTLA-4. Immunity 3:541
21. Waterhouse P, Penninger J, Timms E, Wakeham A, Shahinian A, Lee K, Thompson C, Griesser H, Mak T (1995) Lymphoproliferative disorders with early lethality in mice deficient in CTLA-4. Science 270:985
22. Chen Z, Stockton J, Mathis D, Benoist C (2006) Modeling CTLA4-linked autoimmunity with RNA interference in mice. Proc Natl Acad Sci USA 103:16400
23. Ueda H, Howson JM, Esposito L, Heward J, Snook H, Chamberlain G, Rainbow DB, Hunter KM, Smith AN, Di Genova G, Herr MH, Dahlman I, Payne F, Smyth D, Lowe C, Twells RC, Howlett S, Healy B, Nutland S, Rance HE, Everett V, Smink LJ, Lam AC, Cordell HJ, Walker NM, Bordin C, Hulme J, Motzo C, Cucca F, Hess JF, Metzker ML, Rogers J, Gregory S, Allahabadia A, Nithiyananthan R, Tuomilehto-Wolf E, Tuomilehto J, Bingley P, Gillespie KM, Undlien DE, Rnningen KS, Guja C, Ionescu-Trgoviste C, Savage DA, Maxwell AP, Carson DJ, Patterson CC, Franklyn JA, Clayton DG, Peterson LB, Wicker LS, Todd JA, Gough SC (2003) Association of the T-cell regulatory gene CTLA4 with susceptibility to autoimmune disease. Nature 423:506–511
24. Chambers C, Kuhns M, Egen J, Allison J (2001) CTLA-4-mediated inhibition in regulation of T-cell responses: mechanisms and manipulation in tumor immunotherapy. Annu Rev Immunol 19:565–594
25. Rudd C, Schneider H (2003) Unifying concepts in CD28, ICOS and CTLA4 co-receptor signalling. Nat Rev Immunol 3:544–556
26. Teft W, Kirchhof M, Madrenas J (2006) A molecular perspective of CTLA-4 function. Annu Rev Immunol 24:65–97
27. Rudd C (2008) The reverse stop-signal model for CTLA4 function. Nat Rev Immunol 8:153–160
28. Elgueta R, Benson M, de Vries V, Wasiuk A, Guo Y, Noelle R (2009) Molecular mechanism and function of CD40/CD40L engagement in the immune system. Immunol Rev 229:152–172
29. Van Kooten C, Banchereau J (1997) Functions of CD40 on B cells, dendritic cells and other cells. Curr Opin Immunol 9:330
30. Campbell K, Ovendale P, Kennedy M, Fanslow W, Reed S, Maliszewski C (1996) CD40 ligand is required for protective cell-mediated immunity to Leishmania major. Immunity 4:283–290
31. Lei X, Ohkawara Y, Stämpfli M, Mastruzzo C, Marr R, Snider D, Xing Z, Jordana M (1998) Disruption of antigen-induced inflammatory responses in CD40 ligand knockout mice. J Clin Invest 101:1342
32. Dong C, Temann U, Flavell R (2001) Cutting edge: critical role of inducible costimulator in germinal center reactions. J Immunol 166:3659
33. Iiyama R, Kanai T, Uraushihara K, Totsuka T, Nakamura T, Miyata T, Yagita H, Kushi A, Suzuki K, Tezuka K, Watanabe M (2003) The role of inducible co-stimulator (ICOS)/B7-related protein-1 (B7RP-1) interaction in the functional development of Peyer's patches. Immunol Lett 88:63–70
34. Kopf M, Ruedl C, Schmitz N, Gallimore A, Lefrang K, Ecabert B, Odermatt B, Bachmann M (1999) OX40-deficient mice are defective in Th cell proliferation but are competent in generating B cell and CTL responses after virus infection. Immunity 11:699
35. Sharpe A (2009) Mechanisms of costimulation. Immunol Rev 229:5
36. Dawicki W, Bertram E, Sharpe A, Watts T (2004) 4-1BB and OX40 act independently to facilitate robust CD8 and CD4 recall responses. J Immunol 173:5944
37. Nolte M, van Olffen R, Van Gisbergen K, van Lier R (2009) Timing and tuning of CD27-CD70 interactions: the impact of signal strength in setting the balance between adaptive responses and immunopathology. Immunol Rev 229:216–231
38. Wang N, Satoskar A, Faubion W, Howie D, Okamoto S, Feske S, Gullo C, Clarke K, Sosa M, Sharpe A, Terhorst C (2004) The cell surface receptor SLAM controls T cell and macrophage functions. J Exp Med 199:1255

39. Nishimura H, Minato N, Nakano T, Honjo T (1998) Immunological studies on PD-1 deficient mice: implication of PD-1 as a negative regulator for B cell responses. Int Immunol 10:1563
40. Watanabe N, Gavrieli M, Sedy JR, Yang J, Fallarino F, Loftin SK, Hurchla MA, Zimmerman N, Sim J, Zang X, Murphy TL, Russell JH, Allison JP, Murphy KM (2003) BTLA is a lymphocyte inhibitory receptor with similarities to CTLA-4 and PD-1. Nat Immunol 4:670–679
41. Rodriguez-Manzanet R, DeKruyff R, Kuchroo V, Umetsu D (2009) The costimulatory role of TIM molecules. Immunol Rev 229:259–270
42. Monney L, Sabatos CA, Gaglia JL, Ryu A, Waldner H, Chernova T, Manning S, Greenfield EA, Coyle AJ, Sobel RA, Freeman GJ, Kuchroo VK (2002) Th1-specific cell surface protein Tim-3 regulates macrophage activation and severity of an autoimmune disease. Nature 415:536–541
43. McIntire J, Umetsu D, DeKruyff R (2004) TIM-1, a novel allergy and asthma susceptibility gene. Springer Semin Immunopathol 25:335–348
44. Rachmilewitz J, Lanzavecchia A (2002) A temporal and spatial summation model for T-cell activation: signal integration and antigen decoding. Trends Immunol 23:592–595
45. Wulfing C, Rabinowitz J, Beeson C, Sjaastad M, McConnell H, Davis M (1997) Kinetics and extent of T cell activation as measured with the calcium signal. J Exp Med 185:1815
46. Ahmed A, Mukherjee S, Nandi D (2009) Intracellular concentrations of $Ca2^{+}$ modulate the strength of signal and alter the outcomes of cytotoxic T-lymphocyte antigen-4 (CD152)-CD80/CD86 interactions in $CD4^{+}$ T lymphocytes. Immunology 126:363
47. Kirchgessner H, Dietrich J, Scherer J, Isomaki P, Korinek V, Hilgert I, Bruyns E, Leo A, Cope A, Schraven B (2001) The transmembrane adaptor protein TRIM regulates T cell receptor (TCR) expression and TCR-mediated signaling via an association with the TCRζ chain. J Exp Med 193:1269
48. Schade A, Levine A (2004) Cutting edge: extracellular signal-regulated kinases 1/2 function as integrators of TCR signal strength. J Immunol 172:5828
49. Murphy L, Smith S, Chen R, Fingar D, Blenis J (2002) Molecular interpretation of ERK signal duration by immediate early gene products. Nat Cell Biol 4:556–564
50. Guo Z, Clydesdale G, Cheng J, Kim K, Gan L, McConkey D, Ullrich S, Zhuang Y, Su B (2002) Disruption of Mekk2 in mice reveals an unexpected role for MEKK2 in modulating T-cell receptor signal transduction. Mol Cell Biol 22:5761
51. Colgan J, Asmal M, Neagu M, Yu B, Schneidkraut J, Lee Y, Sokolskaja E, Andreotti A, Luban J (2004) Cyclophilin A regulates TCR signal strength in $CD4^{+}$ T cells via a proline-directed conformational switch in Itk. Immunity 21:189–201
52. Tarakhovsky A, Kanner S, Hombach J, Ledbetter J, Muller W, Killeen N, Rajewsky K (1995) A role for CD5 in TCR-mediated signal transduction and thymocyte selection. Science 269:535
53. Azzam H, Grinberg A, Lui K, Shen H, Shores E, Love P (1998) CD5 expression is developmentally regulated by T cell receptor (TCR) signals and TCR avidity. J Exp Med 188:2301–2311
54. Park JH, Adoro S, Lucas PJ, Sarafova SD, Alag AS, Doan LL, Erman B, Liu X, Ellmeier W, Bosselut R, Feigenbaum L, Singer A (2007) 'Coreceptor tuning': cytokine signals transcriptionally tailor CD8 coreceptor expression to the self-specificity of the TCR. Nat Immunol 8(10):1049–1059
55. Hollander G, Gill J, Zuklys S, Iwanami N, Liu C, Takahama Y (2006) Cellular and molecular events during early thymus development. Immunol Rev 209:28–46
56. Casetti R, Martino A (2008) The plasticity of $\gamma\delta$ T cells: innate immunity, antigen presentation and new immunotherapy. Cell Mol Immunol 5:161
57. Mathis D, Benoist C (2009) Aire. Annu Rev Immunol 27:287–312
58. Pennington D, Silva-Santos B, Hayday A (2005) $\gamma\delta$T cell development having the strength to get there. Curr Opin Immunol 17:108–115
59. Hayes S, Love P (2006) Strength of signal: a fundamental mechanism for cell fate specification. Immunol Rev 209:170–175

60. Lauritsen J, Haks M, Lefebvre J, Kappes D, Wiest D (2006) Recent insights into the signals that control $\alpha\beta/\gamma\delta$-lineage fate. Immunol Rev 209:176–190
61. Hayes S, Li L, Love P (2005) TCR signal strength influences $\alpha\beta/\gamma\delta$ lineage fate. Immunity 22:583–593
62. Haks M, Lefebvre J, Lauritsen J, Carleton M, Rhodes M, Miyazaki T, Kappes D, Wiest D (2005) Attenuation of $\gamma\delta$TCR signaling efficiently diverts thymocytes to the $\alpha\beta$ lineage. Immunity 22:595–606
63. Fehling H, Gilfillan S, Ceredig R (1999) $\alpha\beta/\gamma\delta$ lineage commitment in the thymus of normal and genetically manipulated mice. Adv Immunol 71:1–76
64. Hayes S, Shores E, Love P (2003) An architectural perspective on signaling by the pre-, $\alpha\beta$ and $\gamma\delta$ T cell receptors. Immunol Rev 191:28–37
65. Hayes S, Love P (2002) Distinct structure and signaling potential of the $\gamma\delta$TCR complex. Immunity 16:827–838
66. Garbe A, von Boehmer H (2007) TCR and Notch synergize in $\alpha\beta$ versus $\gamma\delta$ lineage choice. Trends Immunol 28:124–131
67. Hogquist K, Baldwin T, Jameson S (2005) Central tolerance: learning self-control in the thymus. Nat Rev Immunol 5:772–782
68. Love P, Lee J, Shores E (2000) Critical relationship between TCR signaling potential and TCR affinity during thymocyte selection. J Immunol 165:3080
69. van Oers N, Tohlen B, Malissen B, Moomaw C, Afendis S, Slaughter C (2000) The 21- and 23-kD forms of TCRζ are generated by specific ITAM phosphorylations. Nat Immunol 1:322–328
70. Werlen G, Hausmann B, Palmer E (1999) A motif in the T-cell receptor controls positive selection by modulating ERK activity. J Immunol 29:1912–1918
71. Mariathasan S, Zakarian A, Bouchard D, Michie A, Zuniga-Pflucker J, Ohashi P (2001) Duration and strength of extracellular signal-regulated kinase signals are altered during positive versus negative thymocyte selection. J Immunol 167:4966
72. McNeil L, Starr T, Hogquist K (2005) A requirement for sustained ERK signaling during thymocyte positive selection in vivo. Proc Natl Acad Sci USA 102:13574
73. Dower N, Stang S, Bottorff D, Ebinu J, Dickie P, Ostergaard H, Stone J (2000) RasGRP is essential for mouse thymocyte differentiation and TCR signaling. Nat Immunol 1:317
74. Fischer A, Katayama C, Pagès G, Pouysségur J, Hedrick S (2005) The role of ERK1 and ERK2 in multiple stages of T cell development. Immunity 23:431–443
75. He X, Kappes D (2006) CD4/CD8 lineage commitment: light at the end of the tunnel? Curr Opin Immunol 18:135–142
76. Wiest D, Yuan L, Jefferson J, Benveniste P, Tsokos M, Klausner R, Glimcher L, Samelson L, Singer A (1993) Regulation of T cell receptor expression in immature CD4$^+$ CD8$^+$ thymocytes by p56lck tyrosine kinase: basis for differential signaling by CD4 and CD8 in immature thymocytes expressing both coreceptor molecules. J Exp Med 178:1701
77. Hernández-Hoyos G, Sohn S, Rothenberg E, Alberola-Ila J (2000) Lck activity controls CD4/CD8 T cell lineage commitment. Immunity 12:313–322
78. Yasutomo K, Doyle C, Miele L, Germain R (2000) The duration of antigen receptor signalling determines CD4$^+$ versus CD8$^+$ T-cell lineage fate. Nature 404:506–510
79. Brugnera E, Bhandoola A, Cibotti R, Yu Q, Guinter T, Yamashita Y, Sharrow S, Singer A (2000) Coreceptor reversal in the thymus: signaled CD4$^+$CD8$^+$ thymocytes initially terminate CD8 transcription even when differentiating into CD8$^+$ T cells. Immunity 13:59
80. Sarafova S, Erman B, Yu Q, Van Laethem F, Guinter T, Sharrow S, Feigenbaum L, Wildt K, Ellmeier W, Singer A (2005) Modulation of coreceptor transcription during positive selection dictates lineage fate independently of TCR/coreceptor specificity. Immunity 23:75–87
81. Mosmann T, Cherwinski H, Bond M, Giedlin M, Coffman R (1986) Two types of murine helper T cell clone. I. Definition according to profiles of lymphokine activities and secreted proteins. J Immunol 136:2348
82. Mowen K, Glimcher L (2004) Signaling pathways in Th2 development. Immunol Rev 202:203–222

83. Bettelli E, Carrier Y, Gao W, Korn T, Strom T, Oukka M, Weiner H, Kuchroo V (2006) Reciprocal developmental pathways for the generation of pathogenic effector Th17 and regulatory T cells. Nature 441:235–238
84. Chen Z, O'Shea J (2008) Th17 cells: a new fate for differentiating helper T cells. Immunol Res 41:87–102
85. Bettelli E, Korn T, Kuchroo VK (2007) Th17: the third member of the effector T cell trilogy. Curr Opin Immunol 19:652–657
86. Zhu J, Paul W (2008) CD4 T cells: fates, functions, and faults. Blood 112:1557
87. Secrist H, DeKruyff R, Umetsu D (1995) Interleukin 4 production by $CD4^+$ T cells from allergic individuals is modulated by antigen concentration and antigen-presenting cell type. J Exp Med 181:1081
88. Iezzi G, Scotet E, Scheidegger D, Lanzavecchia A (1999) The interplay between the duration of TCR and cytokine signaling determines T cell polarization. Eur J Immunol 29:4092–4101
89. Constant S, Pfeiffer C, Woodard A, Pasqualini T, Bottomly K (1995) Extent of T cell receptor ligation can determine the functional differentiation of naive $CD4^+$ T cells. J Exp Med 182:1591
90. Hosken N, Shibuya K, Heath A, Murphy K, O'garra A (1995) The effect of antigen dose on $CD4^+$ T helper cell phenotype development in a T cell receptor-$\alpha\beta$-transgenic model. J Exp Med 182:1579
91. Tao X, Constant S, Jorritsma P, Bottomly K (1997) Strength of TCR signal determines the costimulatory requirements for Th1 and Th2 $CD4^+$ T cell differentiation. J Immunol 159:5956
92. Sloan-Lancaster J, Allen P (1996) Altered peptide ligand-induced partial T cell activation: molecular mechanisms and role in T cell biology. Annu Rev Immunol 14:1–27
93. Alberola-Ila J, Takaki S, Kerner J, Perlmutter R (1997) Differential signaling by lymphocyte antigen receptors. Annu Rev Immunol 15:125–154
94. Boutin Y, Leitenberg D, Tao X, Bottomly K (1997) Distinct biochemical signals characterize agonist-and altered peptide ligand-induced differentiation of naive $CD4^+$ T cells into Th1 and Th2 subsets. J Immunol 159:5802
95. Leitenberg D, Bottomly K (1999) Regulation of naïve T cell differentiation by varying the potency of tcr signal transduction. Semin Immunol 11:283–292
96. Badou A, Savignac M, Moreau M, Leclerc C, Foucras G, Cassar G, Paulet P, Lagrange D, Druet P, Guery J, Pelletier L (2001) Weak TCR stimulation induces a calcium signal that triggers IL-4 synthesis, stronger TCR stimulation induces MAP kinases that control IFN-gamma production. Eur J Immunol 31:2487–2496
97. Dumont F, Staruch M, Fischer P, DaSilva C, Camacho R (1998) Inhibition of T cell activation by pharmacologic disruption of the MEK1/ERK MAP kinase or calcineurin signaling pathways results in differential modulation of cytokine production. J Immunol 160:2579
98. Rincon M, Enslen H, Raingeaud J, Recht M, Zapton T, Su M, Penix L, Davis R, Flavell R (1998) Interferon-gamma expression by Th1 effector T cells mediated by the p38 MAP kinase signaling pathway. EMBO J 17:2817
99. Dong C, Yang D, Wysk M, Whitmarsh A, Davis R, Flavell R (1998) Defective T cell differentiation in the absence of Jnk1. Science 282:2092
100. Yang D, Conze D, Whitmarsh A, Barrett T, Davis R, Rincon M, Flavell R (1998) Differentiation of CD4 T cells to Th1 cells requires MAP kinase JNK2. Immunity 9:575–585
101. Finotto S, Neurath MF, Glickman JN, Qin S, Lehr HA, Green FH, Ackerman K, Haley K, Galle PR, Szabo SJ, Drazen JM, De Sanctis GT, Glimcher LH (2002) Development of spontaneous airway changes consistent with human asthma in mice lacking T-bet. Science 295:336
102. Szabo S, Kim S, Costa G, Zhang X, Fathman C, Glimcher L (2000) A novel transcription factor, T-bet, directs Th1 lineage commitment. Cell 100:655–669
103. Zhu J, Min B, Hu-Li J, Watson C, Grinberg A, Wang Q, Killeen N, Urban J, Guo L, Paul W (2004) Conditional deletion of Gata3 shows its essential function in Th1-Th2 responses. Nat Immunol 5:1157–1165
104. Brogdon J, Leitenberg D, Bottomly K (2002) The potency of TCR signaling differentially regulates NFATc/p activity and early IL-4 transcription in naive $CD4^+$ T cells. J Immunol 168:3825

105. Salomon B, Bluestone J (1998) Cutting edge: LFA-1 interaction with ICAM-1 and ICAM-2 regulates Th2 cytokine production. J Immunol 161:5138
106. Wülfing C, Sjaastad M, Davis M (1998) Visualizing the dynamics of T cell activation: intracellular adhesion molecule 1 migrates rapidly to the T cell/B cell interface and acts to sustain calcium levels. Proc Natl Acad Sci USA 95:6302
107. Sakaguchi S (2000) Regulatory T cells key controllers of immunologic self-tolerance. Cell 101:455–458
108. Sakaguchi S, Sakaguchi N, Asano M, Itoh M, Toda M (1995) Immunologic self-tolerance maintained by activated T cells expressing IL-2 receptor alpha-chains (CD25). Breakdown of a single mechanism of self-tolerance causes various autoimmune diseases. J Immunol 155:1151
109. Su L, Creusot R, Gallo E, Chan S, Utz P, Fathman C, Ermann J (2004) Murine CD4$^+$ CD25$^+$ regulatory T cells fail to undergo chromatin remodeling across the proximal promoter region of the IL-2 gene. J Immunol 173:4994
110. Horwitz D, Zheng S, Gray J (2008) Natural and TGF-β–induced Foxp3$^+$ CD4$^+$ CD25$^+$ regulatory T cells are not mirror images of each other. Trends Immunol 29:429–435
111. Jordan M, Boesteanu A, Reed A, Petrone A, Holenbeck A, Lerman M, Naji A, Caton A (2001) Thymic selection of cd4$^+$ cd25$^+$ regulatory T cells induced by an agonist self-peptide. Nat Immunol 2:301–306
112. Salomon B, Lenschow D, Rhee L, Ashourian N, Singh B, Sharpe A, Bluestone J (2000) B7/CD28 costimulation is essential for the homeostasis of the CD4$^+$ CD25$^+$ immunoregulatory T cells that control autoimmune diabetes. Immunity 12:431–440
113. Tai X, Cowan M, Feigenbaum L, Singer A (2005) CD28 costimulation of developing thymocytes induces Foxp3 expression and regulatory T cell differentiation independently of interleukin 2. Nat Immunol 6:152–162
114. Yu P, Haymaker CL, Divekar RD, Ellis JS, Hardaway J, Jain R, Tartar DM, Hoeman CM, Cascio JA, Ostermeier A, Zaghouani H (2008) Fetal exposure to high-avidity TCR ligand enhances expansion of peripheral T regulatory cells. J Immunol 181:73
115. Liang S, Alard P, Zhao Y, Parnell S, Clark S, Kosiewicz M (2005) Conversion of CD4$^+$ CD25-cells into CD4$^+$ CD25$^+$ regulatory T cells in vivo requires B7 costimulation, but not the thymus. J Exp Med 201:127
116. Kretschmer K, Apostolou I, Hawiger D, Khazaie K, Nussenzweig M, von Boehmer H (2005) Inducing and expanding regulatory T cell populations by foreign antigen. Nat Immunol 6:1219–1227
117. Zheng S, Wang J, Stohl W, Kim K, Gray J, Horwitz D (2006) TGF-β requires CTLA-4 early after T cell activation to induce FoxP3 and generate adaptive CD4$^+$ CD25$^+$ regulatory cells. J Immunol 176:3321
118. Malek T (2008) The biology of interleukin-2. Annu Rev Immunol 26:453–479
119. Fontenot J, Rasmussen J, Gavin M, Rudensky A (2005) A function for interleukin 2 in Foxp3-expressing regulatory T cells. Nat Immunol 6:1142–1151
120. Sadlack B, Lohler J, Schorle H, Klebb G, Haber H, Sickel E, Noelle R, Horak I (1995) Generalized autoimmune disease in interleukin-2-deficient mice is triggered by an uncontrolled activation and proliferation of CD4$^+$ T cells. Eur J Immunol 25:3053–3059
121. Beissert S, Schwarz A, Schwarz T (2006) Regulatory T cells. J Invest Dermatol 126:15–24
122. Hombach A, Kofler D, Hombach A, Rappl G, Abken H (2007) Effective proliferation of human regulatory T cells requires a strong costimulatory CD28 signal that cannot be substituted by IL-2. J Immunol 179:7924
123. Beyersdorf N, Gaupp S, Balbach K, Schmidt J, Toyka K, Lin C, Hanke T, Hunig T, Kerkau T, Gold R (2005) Selective targeting of regulatory T cells with CD28 superagonists allows effective therapy of experimental autoimmune encephalomyelitis. J Exp Med 202:445
124. Baecher-Allan C, Viglietta V, Hafler D (2002) Inhibition of human CD4$^+$ CD25$^+$ high regulatory T cell function. J Immunol 169:6210
125. Baecher-Allan C, Hafler D (2006) Human regulatory T cells and their role in autoimmune disease. Immunol Rev 212:203–216

126. Pasare C, Medzhitov R (2003) Toll pathway-dependent blockade of $CD4^+$ $CD25^+$ T cell-mediated suppression by dendritic cells. Science 299:1033
127. Kalia V, Sarkar S, Gourley T, Rouse B, Ahmed R (2006) Differentiation of memory B and T cells. Curr Opin Immunol 18:255–264
128. Williams M, Bevan M (2007) Effector and memory CTL differentiation. Annu Rev Immunol 25:171–192
129. Jacob J, Baltimore D (1999) Modelling T-cell memory by genetic marking of memory T cells in vivo. Nature 399:593–597
130. Curtsinger J, Johnson C, Mescher M (2003) CD8 T cell clonal expansion and development of effector function require prolonged exposure to antigen, costimulation, and signal 3 cytokine. J Immunol 171:5165
131. Badovinac V, Porter B, Harty J (2004) $CD8^+$ T cell contraction is controlled by early inflammation. Nat Immunol 5:809–817
132. Day C, Kaufmann D, Kiepiela P, Brown J, Moodley E, Reddy S, Mackey E, Miller J, Leslie A, DePierres C, Mncube Z, Duraiswamy J, Zhu B, Eichbaum Q, Altfel M, Wherry EJ, Coovadia HM, Goulder PJR, Klenerman P, Ahmed R, Freeman GJ, Walker BD (2006) PD-1 expression on HIV-specific T cells is associated with T-cell exhaustion and disease progression. Nature 443:350–354
133. Lim D, Höllsberg P, Hafler D (2002) Strength of prior stimuli determines the magnitude of secondary responsiveness in $CD8^+$ T cells. Cell Immunol 217:36–46
134. Wherry E, Ha S, Kaech S, Haining W, Sarkar S, Kalia V, Subramaniam S, Blattman J, Barber D, Ahmed R (2007) Molecular signature of $CD8^+$ T cell exhaustion during chronic viral infection. Immunity 27:670–684
135. Barber D, Wherry E, Masopust D, Zhu B, Allison J, Sharpe A, Freeman G, Ahmed R (2005) Restoring function in exhausted CD8 T cells during chronic viral infection. Nature 439: 682–687
136. Li J, Huston G, Swain S (2003) IL-7 promotes the transition of CD4 effectors to persistent memory cells. J Exp Med 198:807
137. Kaech S, Tan J, Wherry E, Konieczny B, Surh C, Ahmed R (2003) Selective expression of the interleukin 7 receptor identifies effector CD8 T cells that give rise to long-lived memory cells. Nat Immunol 4:1191–1198
138. Lozza L, Rivino L, Guarda G, Jarrossay D, Rinaldi A, Bertoni F, Sallusto F, Lanzavecchia A, Geginat J (2008) The strength of T cell stimulation determines IL-7 responsiveness, secondary expansion, and lineage commitment of primed human $CD4^+$ IL-7Rhi T cells. Eur J Immunol 38:30
139. Williams M, Ravkov E, Bevan M (2008) Rapid culling of the $CD4^+$ T cell repertoire in the transition from effector to memory. Immunity 28:533–545
140. Murtaza A, Kuchroo V, Freeman G (1999) Changes in the strength of co-stimulation through the B7/CD28 pathway alter functional T cell responses to altered peptide ligands. Int Immunol 11:407
141. Bonnevier J, Mueller D (2002) Cutting edge: B7/CD28 interactions regulate cell cycle progression independent of the strength of TCR signaling. J Immunol 169:6659
142. Wu Y, Guo Y, Huang A, Zheng P, Liu Y (1997) CTLA-4-B7 interaction is sufficient to costimulate T cell clonal expansion. J Exp Med 185:1327
143. Madrenas J, Chau L, Teft W, Wu P, Jussif J, Kasaian M, Carreno B, Ling V (2004) Conversion of CTLA-4 from inhibitor to activator of T cells with a bispecific tandem single-chain Fv ligand. J Immunol 172:5948
144. Hueber A, Matzkies F, Rahmeh M, Manger B, Kalden J, Nagel T (2006) CTLA-4 lacking the cytoplasmic domain costimulates IL-2 production in T-cell hybridomas. Immunol Cell Biol 84:51–58
145. Mukherjee S, Ahmed A, Malu S, Nandi D (2006) Modulation of cell cycle progression by CTLA4-CD80/CD86 interactions on $CD4^+$ T cells depends on strength of the CD3 signal: critical role for IL-2. J Leukoc Biol 80:66

146. Mukherjee S, Ahmed A, Nandi D (2005) CTLA4-CD80/CD86 interactions on primary mouse $CD4^+$ T cells integrate signal-strength information to modulate activation with Concanavalin A. J Leukoc Biol 78:144
147. Anderson D, Bieganowska K, Bar-Or A, Oliveira E, Carreno B, Collins M, Hafler D (2000) Paradoxical inhibition of T-cell function in response to CTLA-4 blockade; heterogeneity within the human T-cell population. Nat Med 6:211–214
148. Kuhns M, Epshteyn V, Sobel R, Allison J (2000) Cytotoxic T lymphocyte antigen-4 (CTLA-4) regulates the size, reactivity, and function of a primed pool of $CD4^+$ T cells. Proc Natl Acad Sci USA 97:12711
149. Egen J, Kuhns M, Allison J (2002) CTLA-4: new insights into its biological function and use in tumor immunotherapy. Nat Immunol 3:611–618
150. Hildeman D, Zhu Y, Mitchell T, Kappler J, Marrack P (2002) Molecular mechanisms of activated T cell death in vivo. Curr Opin Immunol 14:354–359
151. Krammer P, Arnold R, Lavrik I (2007) Life and death in peripheral T cells. Nat Rev Immunol 7:532–542
152. Lenardo M (1991) Interleukin-2 programs mouse $\alpha\beta$ T lymphocytes for apoptosis. Nature 353:858–861
153. Arnold R, Brenner D, Becker M, Frey C, Krammer P (2006) How T lymphocytes switch between life and death. Eur J Immunol 36:1654–1658
154. Gett A, Sallusto F, Lanzavecchia A, Geginat J (2003) T cell fitness determined by signal strength. Nat Immunol 4:355–360
155. Alexander-Miller M, Leggatt G, Sarin A, Berzofsky J (1996) Role of antigen, CD8, and cytotoxic T lymphocyte (CTL) avidity in high dose antigen induction of apoptosis of effector CTL. J Exp Med 184:485–492
156. Critchfield J, Racke M, Zuniga-Pflucker J, Cannella B, Raine C, Goverman J, Lenardo M (1994) T cell deletion in high antigen dose therapy of autoimmune encephalomyelitis. Science 263:1139
157. Lenardo M, Chan F, Hornung F, McFarland H, Siegel R, Wang J, Zheng L (1999) Mature T lymphocyte apoptosis-immune regulation in a dynamic and unpredictable antigenic environment 1. Annu Rev Immunol 17:221–253

Chapter 5
The Cyton Model for Lymphocyte Proliferation and Differentiation

Cameron Wellard, John F. Markham, Edwin D. Hawkins, and Phillip D. Hodgkin

Abstract We discuss a stochastic model for lymphocyte population dynamics based on the interaction of sub-cellular mechanisms responsible for cell death, division and differentiation. Competition between these mechanisms determines the fate of an individual, and their stochastic nature allows a range of outcomes, generating the large-scale diversity that is characteristic of an immune response.

Mathematically the function of each mechanism can be expressed as a probability density function for the time for the particular event to occur, and it is the parameters that define these distributions that are the free parameters of the model.

We formulate the problem as a branching process, and derive an expression for the probability generating function for the number of live cells of each type in all divisions, as a function of time. This allows the calculation of arbitrary moments for the live cell numbers, giving quantification not just of the mean system behaviour, but of the random fluctuations that are a consequence of the stochasticity of the underlying mechanisms.

Introduction

In this chapter we present a mathematical model for lymphocyte proliferation and differentiation based on the notion that individual cellular processes are controlled by independent molecular mechanisms. On a molecular level these mechanisms are incredibly complex, however we find that for the purposes of describing their effect on cell dynamics it is sufficient to take a statistical approach. In this way we can ignore the complex molecular pathways responsible for apoptosis and simply use an empirically derived probability density function (PDF) for the observed death

C. Wellard (✉)
The Walter and Eliza Hall Institute of Medical Research, 1G Royal Parade, Parkville, Victoria 3052, Australia
and
Department of Medical Biology, The University of Melbourne, Parkville, Victoria 3010, Australia
e-mail: wellard@wehi.edu.au

C. Molina-París and G. Lythe (eds.), *Mathematical Models and Immune Cell Biology*,
DOI 10.1007/978-1-4419-7725-0_5,

times of a particular cell type. Thus we summarise each cellular function as a PDF for the time-to-transformation, and each process is considered to act independently, competing to determine the fate of the cell.

This approach naturally encompasses the observed diversity of immune responses, with the stochastic nature of the independent mechanisms ensuring different trajectories for individual cells, generating a diverse response from the stimulation of a homogeneous population.

We begin with a discussion of the cyton model of the population dynamics of B lymphocytes. Here we use the model to derive a set of integro-differential equations for the expected number of cells in each division as a function of time. We apply the theory of branching processes to the model, and derive a recursion relation for the probability generating function (PGF) for the population size in each generation. In addition to the mean population dynamics, the PGF allows description of the fluctuations around the mean due to the stochastic nature of the cell dynamics. To illustrate this we derive a set of integral equations for both the mean, and the standard deviation, of the population size of each generation as a function of time. We introduce differentiation into the model, showing how the modules used to describe division and death can be generalised to include various differentiation processes. We then use the branching process approach to derive equations for the population dynamics of each of the subsets defined by the differentiation process. Finally, we briefly discuss some of the issues involved with direct simulation of such a system using an agent-based approach.

Proliferation

CFSE and Cell Population Dynamics

The study of cell proliferation has been facilitated by the use of the fluorescent dye Carboxyfluorescein diacetate succinimidyl ester (CFSE) [1]. CFSE is uniformly distributed throughout the cellular cytoplasm and on division is distributed approximately evenly between the two daughter cells. Thus each generation of cells has approximately half the CFSE of the previous generation, and this will appear half as bright when measured using flow-cytometry.

Stochastic Lymphocyte Division

CFSE studies have shown an element of randomness in the proliferation of lymphocytes. In a typical assay a system of lymphocytes are stained with CFSE and then stimulated, the CFSE intensity is measured at various time-points after stimulation

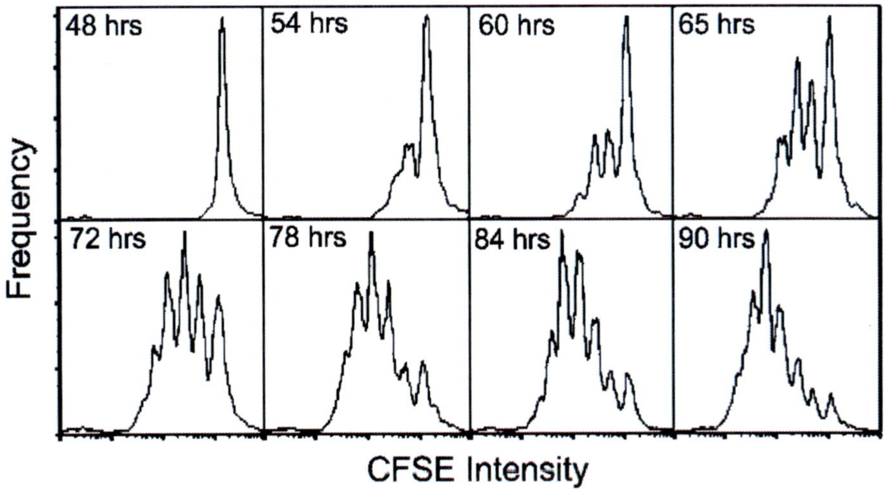

Fig. 5.1 Sample CFSE intensity profiles showing the asynchronous progression of cells through divisions

using flow-cytometry. An example profile is shown in Fig. 5.1. Immediately after stimulation the profile shows one distinct peak, indicating that none of the cells have divided. At a later time however, it is generally found that the profile shows several distinct peaks, indicating that not all of the cells in the culture have undergone the same number of divisions. This implies some element of randomness controlling the process of cell division [2, 3], with the division time distributed according to some probability density function (PDF), that may be characteristic of both the cell type, and the experimental conditions.

Death and Division as Competing Random Processes

Similar to division, cell death is a stochastic process. This has been demonstrated by measuring the survival curve of a system of cells and observing that individual cells die in a non-deterministic fashion, and that the death times of a system of cells describe some PDF. Often this PDF is, for mathematical convenience, treated as exponential however evidence suggests that a right-skewed distribution gives a better fit in many circumstances [4].

For the purposes of modelling, the processes of death and division can be treated as competing random events. On stimulation, each cell randomly draws a time-to-die and a time-to-divide from the relevant PDFs, and the earliest of these determines the fate of the cell. On division these process clocks are reset, possibly from new PDFs characteristic of the division number of the cell.

Division Destiny and Progressor Fraction

There is evidence that there exists a mechanism that intrinsically limits the maximum number of division rounds that cells will undergo in a response. Experiments performed on cells in which the apoptotic machinery has been disabled show that after a certain number of divisions, cells stop dividing and enter a state of quiescence [5]. The division destiny, the ultimate number of division reached by a cell in a response, is not uniform but is usually distributed over several generations.

Division Dependence

In general the mechanisms for death and division within a cell seem to be dependent on the number of division rounds that the cell has undergone since initial stimulation. In particular, the average times for both death and division are much longer for undivided cells, than they are for cells in subsequent generations. Further, the progressor fraction, that is the fraction of cells that are division capable, is strongly dependent on the generation number, tending to decrease as cells progress through the response.

The Cyton Model

The cyton model [4] is a stochastic model for lymphocyte proliferation based on these observations; namely that

- Death and division are stochastic processes, characterised by a PDF for the time to divide or die respectively
- These processes are independent, and compete to determine the fate of the cell
- Division events reset the mechanism responsible for these processes
- Only a fraction of cells in each division are division capable
- These mechanisms are division dependent

Let γ_i, called the progressor fraction, be the fraction of cells that, after i divisions, are capable of further division. Let $p_i^B(t)$, $p_i^D(t)$ be the PDFs for the division and death times of cells in generation i. Given that a cell arrives in generation i, the probability that it dies at a time between t and $t+dt$ after the ith division is $q_i^D(t)dt$, where $q_i^D(t)$ equals $p_i^D(t)$ multiplied by the probability that no cell division occurs before t:

$$q_i^D(t) = p_i^D(t)\left(1 - \gamma_i \int_0^t p_i^B(\tau)\mathrm{d}\tau\right); \tag{5.1}$$

Similarly, the conditional probability of cell division at a time between t and $t + dt$ after the ith division is

$$q_i^B(t) = \gamma_i p_i^B(t) \left(1 - \int_0^t p_i^D(\tau)\mathrm{d}\tau\right). \tag{5.2}$$

The cyton model can thus be expressed as a set of integro-differential equations for the expected number of cells $n_i(t)$, in division i, as a function of time:

$$\dot{n}_0(t) = -n_0(0)\left(q_0^D(t) + q_0^B(t)\right) \tag{5.3}$$

and, for $i > 1, 2, \ldots,$

$$\begin{aligned} \dot{n}_i(t) &= 2\dot{n}_{i-1}^B(t) - \dot{n}_i^B(t) - \dot{n}_i^D(t), \\ \dot{n}_i^B(t) &= n_i(0)q_i^B(t) + 2\int_0^t \dot{n}_{i-1}^B(\tau)q_i^B(t-\tau)\mathrm{d}\tau, \\ \dot{n}_i^D(t) &= n_i(0)q_i^D(t) + 2\int_0^t \dot{n}_{i-1}^B(\tau)q_i^D(t-\tau)\mathrm{d}\tau. \end{aligned} \tag{5.4}$$

In this form the model is completely general. With judicious choice of the forms of the PDFs in each division, as well as their defining parameters, and for the values of the progressor fractions, this model could describe the mean population dynamics of any cell system. Leaving all of these parameters free however, means that the model is over parameterised for most data sets. To be of any use as a predictive tool requires that we use system-specific knowledge to reduce the number of free parameters.

Right Skewed Distributions for Division and Death Times

Many models of cell division use exponential rates for cell death and division, usually for reasons of mathematical convenience rather than from any biological motivation [6]. Such a choice implies that both processes are age-independent, and are equally likely to occur in any given time-interval. Evidence suggests this is not the case. In particular cells are much less likely to either die, or to divide, immediately after a previous division event [4]. Careful examination of the data suggests that death and division times are well described by a right skewed distribution, such as lognormal or gamma distribution [7]. For example, in Fig. 5.2 we show a plot obtained from a course of tritiated thymidine pulses. The count rate is proportional to the number of cells taking up the thymidine during DNA replication in S-phase. Treatment with colcemid prevents divisions, preventing the pollution of the signal from further division. Thus the signal is proportional to the number of cells in S-phase of the first mitotic cycle after stimulation, and the distribution is well described by a lognormal PDF. Note that the proportion of cells participating in the

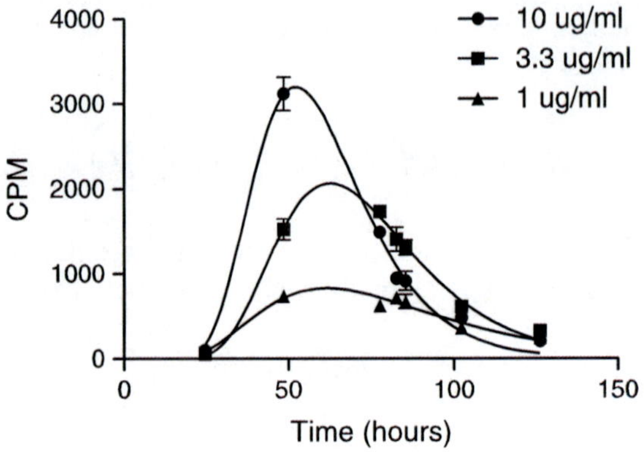

Fig. 5.2 Distribution of division times for B cells stimulated with $10\,\mu g/ml$, $3.3\,\mu g/ml$, $1.1\,\mu g/ml$ of α-CD40 in the presence of IL4. The cells were cultured in the presence of colcemid, and a time course of 1 h $\left[^{3}H\right]$thymidine pulses was conducted. The measured count rate (counts per minute) is proportional to the number of cells in S-phase. The distributions are well fit by a lognormal PDF. The figure is taken from [4]

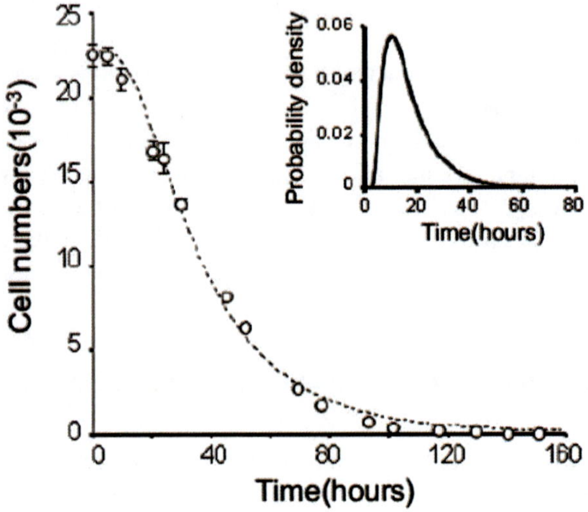

Fig. 5.3 A survival curve for B cells isolated from lymph nodes. The data have been fit to a lognormal survival *curve*, the corresponding distribution of death times is shown in the *top right*. The figure is taken from [4]

response increases with increasing concentration of the mitogen, illustrating how these PDFs can be changed by external signals. Similarly, the death times of undivided lymphocytes can also be described using a lognormal PDF. Figure 5.3 shows a survival curve for B cells isolated from a lymph node. The data have been fit to a lognormal survival curve, with the corresponding PDF shown above.

First and Subsequent Divisions and Progressor Fraction

Although the PDFs for death and division may, in principle, vary for each division, we find that for most cases it is sufficient to distinguish between undivided cells, and those in subsequent generations. Detailed analysis of a CpG-stimulated B cell approach in vitro shows that the PDFs for death and division times vary only slightly with generation after initial division [8].

We introduced the concept of the progressor fraction and discussed its division dependence. Here we apply a functional form to this division dependence, thus reducing the number of free parameters that are necessary to specify the progressor fraction in the model. The form used in [4], is based on the observation that the ultimate division number achieved by a cohort of cells is approximately normally distributed over the generations [5]. As is the case with death and division times, the undivided cells behave in a way that is different from cells in subsequent generations, and for this reason we let the progressor fraction for undivided cells remain a free parameter unconstrained by the functional form used for the subsequent generations. This implies that the fraction of cells in generation i that are division capable is given by

$$\gamma_i = \begin{cases} \gamma_0 & \text{if } i = 0, \\ \dfrac{\displaystyle\int_{i+1}^{\infty} \mathcal{N}(n, \mu_{dest}, \sigma_{dest}) dn}{\displaystyle\int_{i}^{\infty} \mathcal{N}(n, \mu_{dest}, \sigma_{dest}) dn} & \text{if } i > 0, \end{cases} \tag{5.5}$$

where $\mu_{dest}, \sigma_{dest}$ are the mean and standard deviation of the Gaussian PDF that describes division destiny:

$$\mathcal{N}(n, \mu, \sigma) = \frac{1}{\sqrt{2\pi\sigma^2}} \exp\left(-\frac{(n-\mu)^2}{2\sigma^2}\right).$$

Comparison with Experiment

This gives a model with 11 free parameters, listed in Table 5.1, which are obtained empirically by fitting to experimental data in the form of CFSE timecourses. The values of these parameters will be dependent on experimental conditions such as the strength and type of stimulus, as well as the presence of different cytokines and their concentrations. In Fig. 5.4 we show a comparison between the best-fit model prediction for the population of cells in each division compared with experimental time course. The data was obtained from CFSE analysis of a system of LPS stimulated B cells. The best-fit values for the free parameters were obtained by minimising the sum of the squared residuals between predicted and measured values.

Table 5.1 A list of free parameters used in the cyton model of lymphocyte proliferation

Parameter	Description	Parameter	Description
μ_B^0	Mean time to first division	μ_D^+	Mean time to die for cells in subsequent divisions
σ_B^0	Standard deviation of the time to first division	σ_D^+	Standard deviation of the time to die in subsequent division
μ_D^0	Mean time to die for undivided cells	γ_0	Progressor fraction for undivided cells
σ_D^0	Standard deviation of the time to die for undivided cells	μ_{dist}	Mean generation number for the PDF characterising the division destiny
μ_B^+	Mean time to subsequent division	σ_{dist}	Standard deviation of the generation number for the PDF characterising the division destiny
σ_B^+	Standard deviation of the time to subsequent division		

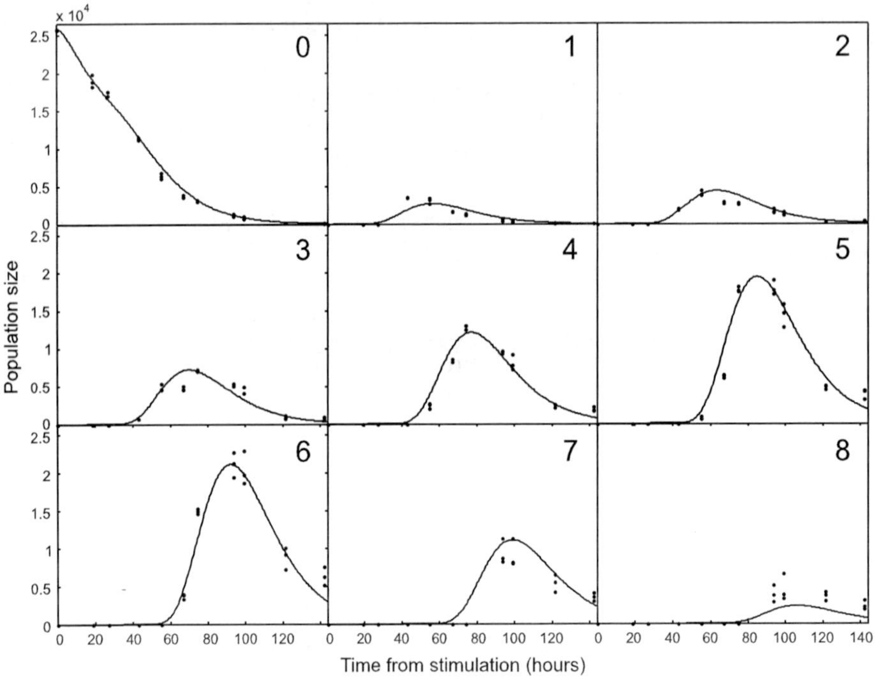

Fig. 5.4 A fit to time course data obtained from CFSE analysis of LPS stimulated B cells. Here we show the best-fit prediction for the populations of each generation as a function of time (*line*), compared to the measured values (*points*)

Stochastic Variation in Population Number

Although the cyton model is based on the idea that the fundamental cellular processes of death and division are inherently stochastic, we have only presented a solution for the average behaviour of the system. The solution of these equations gives an estimate for the number of cells that would be expected over many replicates of an identical experiment, but does not give any information about fluctuations in this result due to the stochastic nature of the underlying processes. In large systems the variation between replicates due to these fluctuations will usually be small, generally much lower than the variation due to experimental considerations, however in small systems these stochastic fluctuations can be observed. In [8], the authors present multiple time course experiments obtained from the time-lapse microscopy of CpG-stimulated B cells. In each case the clone produced from a single progenitor was followed throughout the duration of the response, and the number of cells in each generation recorded as the response progressed. In this experiment significant variation was observed in the population sizes of the different clones. While the model as formulated in (5.4) is sufficient to reproduce the average behaviour of these clones, it is not sufficient to encapsulate the stochastic variation. To do this requires a new approach.

Branching Process Models

The theory of branching process [9] has been used extensively in mathematical biology to simulate cell population dynamics, from discrete time models [10] to more complex continuous time models [11–13]. Here we use a generalisation of the Bellman–Harris model to obtain a set of recursion relations for the probability generating function (PGF) for the number of cells in each generation as a function of time. This PGF can be used to calculate the mean population dynamics, as well as the expected stochastic fluctuations around this mean.

We begin by defining the random variable $Z_i(t)$ to denote the number of live cells in generation i at time t. The PGF for the number of cells in each generation, given a single cell in generation i at time $t = 0$ is defined as

$$F_i(\mathbf{s}, t) = E\left[\prod_{i'} s_{i'}^{Z_{i'}(t)} \middle| Z_i(0) = 1, Z_j(0) = 0 \;\forall j \neq i\right], \tag{5.6}$$

where $E[x]$ denotes the expectation value of the random variable x. Following the prescription for a generalised Bellman–Harris model [12, 13] gives a recursion relation for this PGF

$$F_i(\mathbf{s}, t) = (1-Q_i^B(t))(1-Q_i^D(t))s_i + Q_i^D(t) + \int_0^t F_{i+1}(\mathbf{s}, t-\tau)^2 q_i^B(\tau)\mathrm{d}\tau. \tag{5.7}$$

Here the $q_i^B(t), q_i^D(t)$ are the effective probability distributions defined in (5.1), and $Q_i^B(t), Q_i^D(t)$ are the corresponding cumulative functions. The first term in the above equation takes into account the case in which the cell has neither died nor divided, the contribution for this case is s_i. The second term takes into account the case in which the cell has died, here the contribution is one. Finally the third term takes into account the cases in which the cell divided at a time $t - \tau$, and the contribution here is equal to the product of two PGFs for cells in the subsequent generation,born at $t - \tau$.

Mean Population Dynamics

The mean dynamics of the model can be obtained using $\mu_i^j(t) = \left.\frac{\partial F_i(s,t)}{\partial s_j}\right|_{s=0}$, where $\mu_i^j(t)$ gives the expected number of cells of generation j at time t given a single progenitor of generation i at $t = 0$. When applied to (5.7) this gives

$$\mu_i^j(t) = (1 - Q_i^B(t))(1 - Q_i^D(t))\delta_{i,j} + 2\int_0^t \mu_{i+1}^j(t-\tau)q_i^B(\tau)\mathrm{d}\tau. \tag{5.8}$$

This recursion relation can be evaluated by noting that $\mu_i^j(t) = 0, \forall j < i$, and by assuming some limit on the number of divisions that a cell has undergone. This limit may be imposed either by a progressor fraction that effectively vanishes beyond a certain generation, or by assuming a minimum division time for cells.

Stochastic Fluctuations

The advantage of the branching process approach is that, in principle, it is possible to calculate the entire probability distribution for the populations size at any point in time. In practice this is quite difficult, and usually unnecessary. A much easier prospect, and one that offers great utility, is to calculate the variance in population size. This variance will give a measure of the uncertainty in the population size resulting from the stochastic nature of the underlying processes. To calculate the variance, we begin with the second derivative of the PGF $\nu_i^{j,k}(t) = \left.\frac{\partial^2 F_i(s,t)}{\partial s_k \partial s_j}\right|_{s=0}$. When applied to (5.7) this gives

$$\nu_i^{j,k}(t) = 2\int_0^t \left(\nu_{i+1}^{j,k}(t-\tau) + \mu_{i+1}^j(t-\tau)\mu_{i+1}^k(t-\tau)\right) q_i^B(\tau)\mathrm{d}\tau. \tag{5.9}$$

From this we can use the relations

$$\sigma_i^{j,k}(t) = \begin{cases} \nu_i^{j,k}(t) - \mu_i^j(t)\mu_i^k(t) & j \neq k \\ \nu_i^{j,j}(t) + \mu_i^j(t) - \mu_i^j(t)^2 & j = k \end{cases} \tag{5.10}$$

to calculate the covariances of the population of cells in generations j, k at time t resulting from an initial cell of generation i.

Differentiation

Branching process models have been used to model differentiation in neuronal cells [11]. Here we extend our model to include possible differentiation processes for lymphocytes. The cyton model can be generalised to treat general differentiation processes in the same way as it treats death and division. To motivate this generalisation we note that various differentiation events in activated B cells, such as isotype switching and commitment to antibody-secreting cells, have been shown to be division linked [14, 15]. Similarly we assume that the death and division processes are dependent on the differentiation state, or type of the cell, and that the probability of differentiating to a particular state is dependent on the current state of the cell.

Thus we hypothesise the existence of a molecular mechanism responsible for each possible differentiation process, that acts independently of all other process, and acts in a stochastic manner. In analogy to death and division these processes can be characterised by a PDF for the differentiation time, which is dependent on the state of the cell, as well as its generation number. We define the PDF for time-to-differentiation from type j to type k for a cell in division i as $p^k_{i,j}(t)$. In analogy with progressor fraction, we define a division dependent differentiation fraction $\eta^k_{i,j}$ which gives the fraction of cells of type j that are capable of differentiation to type k in division i. In this case the effective distributions, which give the probability density at time t, are given by

$$q^B_{i,j}(t) = \gamma_{i,j} p^B_{i,j}(t)\big(1 - P^D_{i,j}(t)\big) \prod_k \big(1 - \eta^k_{i,j} P^k_{i,j}(t)\big),$$

$$q^D_{i,j}(t) = p^D_{i,j}(t)\big(1 - \gamma_{i,j} P^B_{i,j}(t)\big) \prod_k \big(1 - \eta^k_{i,j} P^k_{i,j}(t)\big),$$

$$q^k_{i,j}(t) = \eta^k_{i,j} p^k_{i,j}(t)(1 - P^D_{i,j}(t))\big(1 - \gamma_{i,j} P^B_{i,j}(t)\big) \prod_{k' \neq k, j} \big(1 - \eta^{k'}_{i,j} P^{k'}_{i,j}(t)\big).$$

Here $P(t)$ denotes the cumulative distribution of the relevant PDF.

Using the same methodology as previously we can derive a recursion relation for the PGF

$$F_{i,j}(\mathbf{s}, t) = \big(1 - Q^B_{i,j}(t)\big)\big(1 - Q^D_{i,j}(t)\big) \prod_{k \neq j} \big(1 - Q^k_{i,j}(t)\big) s_{i,j} + Q^D_{i,j}(t)$$
$$+ \int_0^t F_{i+1,j}(\mathbf{s}, t-\tau)^2 q^B_{i,j}(\tau)\mathrm{d}\tau + \sum_{k \neq j} \int_0^t F_{i,k}(\mathbf{s}, t-\tau) q^k_{i,j}(\tau)\mathrm{d}\tau. \tag{5.11}$$

As previously, we can use this relation to derive an expression for the expected population of cells of type j' in generation i' given an initial cell of type j in generation i

$$\mu_{i,j}^{i',j'}(t) = (1 - Q_{i,j}^{B}(t))(1 - Q_{i,j}^{D}(t)) \prod_{k \neq j} (1 - Q_{i,j}^{k}(t)) \delta_{i,i'} \delta_{j,j'}$$

$$+2 \int_0^t \mu_{i+1,j}^{i',j'}(t-\tau) q_{i,j}^{B}(\tau) \mathrm{d}\tau + \sum_{k \neq j} \int_0^t \mu_{i,k}^{i',j'}(,t-\tau) q_{i,j}^{k}(\tau) \mathrm{d}\tau.$$

Similarly we can use

$$\nu_{i,j}^{i',j';i'',j''}(t) = 2 \int_0^t \left(\mu_{i+1,j}^{i',j'}(t-\tau) \mu_{i+1,j}^{i'',j''}(t-\tau) + \nu_{i+1,j}^{i',j';i'',j''}(t-\tau) \right) q_{i,j}^{B}(\tau) \mathrm{d}\tau$$

$$+ \sum_{k \neq j} \int_0^t \nu_{i,k}^{i',j';i'',j''}(t-\tau) q_{i,j}^{k}(\tau) \mathrm{d}\tau,$$

to calculate values for the covariances.

Simulation

The methods mentioned so far require numerical solution of a set of deterministic integral equations for the mean number of cells or the variance. Higher order moments can also be calculated however each comes at further computational expense to the previous. An alternative approach is to simulate the cells directly. Although this approach is generally slower to calculate the mean value, there is no further cost in calculating the variance and higher moments. Further, it is more flexible and can be quickly adapted to different sorts of models. Also it can be readily extended to provide readouts of common experimentally observed quantities.

Modelling Cell Properties

Each cell being simulated carries with it a set of numbers which represent its physical state. In some cases these numbers are decided when the cell is created, in others they vary over (simulated) time. The following is a non-exhaustive list of what might be stored:

- When the cell was created. This allows age related properties of the cell to be calculated.

- A unique ID. This way relationships between cells can be kept so that correlations and clonal effects can be explored.
- Cell fate. In the case of the Cyton model, whether a cell dies or divides and when this happens can be decided at the time a cell is created.
- Cell type. Whether the cell has committed to becoming antibody secreting cell and if isotype switching has occurred.
- Stage of cell cycle and cell size. By storing where a cell is in the cell cycle the amount of DNA can be calculated. This allows comparison to be made with flow cytometry techniques that can measure DNA content such as DAPI and Hoechst stains which bind to the DNA in a stoichiometric manner. Estimating cell size allows connection to be made with forward scatter measurements from flow cytometry.
- Fluorophore content. If the stage of the cell cycle and DNA content are known then it is also possible to make connection with Bromodeoxyuridine (BrdU) experiments. BrdU is a molecule that can replace thymidine during DNA replication. In BrdU experiments, cells' exposure to BrdU is limited to certain times. Its incorporation can be measured using immunohistochemistry techniques. Normally it is used to provide confirmation that cells are dividing during a particular time (when BrdU is present) but potentially it can be used to provide more information about the parameters for a model such as Cyton [16–18].
- Affinity for antigen. This enables the models to be implemented which link a cell's ability to proliferate to its affinity, thereby addressing repertoire issues.

Implementation

In practice the simulation of cells is done using a discrete event simulator. This requires the maintenance of a list of cells and their associated states and a time-ordered list of events that can occur to them. Simulation is done asynchronously. That is, instead of stepping through time, the programme processes a sequence of queued events. In this way, time resolution can be made arbitrarily small (within machine precision) without incurring computational expense. Examples of events that need to be queued are

- Cell division and death
- Changing of some cell property such as a cell type and phase
- Making measurement corresponding to those made at experimental time points

Assuming that there are no interactions between cells then the computational cost of this method scales no worse than $O(N)$ times the cost of insertion into the event list where N is the number of cells.

References

1. Lyons A, Parish C (1994) Determination of lymphocyte division by flow cytometry. J Immunol Methods 171:131–137
2. Hasbold J, Gett A, Rush J, Deenick E, Avery D, Jun J, Hodgkin P (1999) Quantitative analysis of lymphocyte differentiation and proliferation in vitro using carboxyfluorescein diacetate succinimidyl ester. Immunol Cell Biol 77:516–522
3. Gett A, Hodgkin P (2000) A cellular calculus for signal integration by T cells. Nat Immunol 1:239–244
4. Hawkins E, Turner M, Dowling M, van Gend C, Hodgkin P (2007) A model of immune regulation as a consequence of randomized lymphocyte division and death times. Proc Natl Acad Sci USA 104:5032–5037
5. Turner M, Hawkins E, Hodgkin P (2008) Quantitative regulation of B cell division destiny by signal strength. J Immunol 181:374–382
6. Boer RD, Ganusov V, Milutinovic D, Hodgkin P, Perelson A (2006) Estimating lymphocyte division and death rates from CFSE data. Bull Math Biol 68:1011–1031
7. Leon K, Faro J, Carneiro J (2004) A general mathematical framework to model generation structure in a population of asynchronously dividing cells. J Theor Biol 229:455–476
8. Hawkins E, Markham J, McGuinness L, Hodgkin P (2009) A single-cell pedigree analysis of alternative stochastic lymphocyte fates. Proc Natl Acad Sci USA 106:13457
9. Kimmel M, Axelrod DE (2001) Branching processes in biology. Springer, New York
10. Yates A, Chan C, Strid J, Moon S, Callard R, George A, Stark J (2007) Reconstriction of cell population dynamics using CFSE. BMC Bioinformatics 8:196
11. Yakovlev A, Mayer-Proschel M, Noble M (1998) A stochastic model of brain cell differentiation in tissue culture. J Math Biol 37:49
12. Hyrien O, Mayer-Pröschel M, Noble M, Yakovlev A (2005) A stochastic model to analyze clonal data on multi-type cell populations. Biometrics 61:199–207
13. Subramanian V, Duffy K, Turner M, Hodgkin P (2008) Determining the expected variability of immune responses using the cyton model. J Math Biol 56:861–892
14. Hodgkin P, Lee J, Lyons A (1996) B cell differentiation and isotype switching is related to division cycle number. J Exp Med 184:277–281
15. Hasbold J, Corcoran L, Tarlinton D, Tangye S, Hodgkin P (2004) Evidence from the generation of immunoglobulin G-secreting cells that stochastic mechanisms regulate lymphocyte differentiation. Nat Immunol 5:55–63
16. Yanagisawa M, Dolbeare F, Todoroki T, Gray J (1985) Cell cycle analysis using numerical simulation of bivariate DNA/bromodeoxyuridine distributions. Cytometry 6:550
17. Cain S, Chau P (1997) A transition probability cell cycle model simulation of bivariate DNA/bromodeoxyuridine distributions. Cytometry 3:239
18. Hodgkin PD, Hawkins ED, Hasbold J, Gett AV, Deenick EK, Todd HF, Hommel M (2006) Monitoring T cell proliferation. In: Nagorsen D, Marincola FM (eds) Analyzing T Cell Responses: how to analyze cellular immune responses against tumor associated antigens. Springer, chapter 6, pp 123–142

Chapter 6
Modelling Intravital Two-Photon Data of Lymphocyte Migration and Interaction

Marc Thilo Figge and Michael Meyer-Hermann

Abstract Multi-photon microscopy is a powerful tool for imaging lymphocyte migration and interaction in intact biological organs. These experiments generate quantitative data on the cell motility, shape dynamics, and contact duration of cellular interactions. In this chapter we review mathematical models that have successfully contributed to the interpretation of these data with regard to lymphocyte migration and interaction. The examples involve different modelling approaches and range from T cell priming in secondary lymphoid tissue to B cell affinity maturation in germinal centres.

Introduction

Adaptive immunity implies the diligent communication of specific information that is facilitated by lymphocyte migration in lymphoid tissue and their interaction with other components of the immune system. For example, T cells scan antigen presenting cells and may become activated if the T cell receptors recognize specific foreign peptides embedded in major histocompatibility molecules. T cell priming by dendritic cells in lymph nodes occurs in subsequent phases that clearly differ in the T cell behaviour with regard to migration and interaction [1]. Differences in the migration behaviour of the distinct types of T cells in diverse lymphoid tissues remain to be explained [2].

The salient feature of B cells is that the B cell receptors undergo affinity maturation in order to optimize the immune response against pathogenic antigens. This process takes place in follicular structures of lymphoid organs, referred to as germinal centres [3, 4], and involves intercellular interactions in the selection process for high-affinity B cells. Germinal centres have a peculiar morphology consisting of

M.T. Figge (✉)
Applied Systems Biology, Friedrich Schiller University Jena, Leibniz Institute for Natural Product Research and Infection Biology – Hans Knöll Institute, Beutenbergstrasse 11a, 07745 Jena, Germany
e-mail: thilo.figge@hki-jena.de

C. Molina-París and G. Lythe (eds.), *Mathematical Models and Immune Cell Biology*,
DOI 10.1007/978-1-4419-7725-0_6,

two distinct zones, termed the dark and the light zone. The dark zone is the region of B cell proliferation, while in the light zone B cells undergo selection. How B cell migration between the zones is realized is still a matter of debate today [5–8].

For the last few years, lymphocyte migration and interaction have been extensively explored by non-invasive intravital microscopy [9]. The main achievement of this technique is that lymphocytes can be followed in real time and within their natural cellular environment. The generated data include records of cell tracks, i.e. the time-dependent position of single cells, as well as cellular contact times of interactions. The working principle of imaging by multi-photon microscopy is briefly summarized, together with an overview of dynamical cell properties that can be explored by this experimental technique. The rapidly growing body of experimental data is calling for mathematical methods that are suitable for analyzing these data in order to extract meaningful conclusions initiating new insights and experiments.

The art of mathematical modelling is to choose an adequate modelling approach. This process starts by specifying the questions to be answered and continues by identifying all variables and connections between them that are relevant for answering these questions. Each modelling approach has its assets and drawbacks. As a rule of thumb, computationally cheap methods often only give a very rough representation of reality and, therefore, they often only provide answers to questions of a very general kind. Questions with regard to specific system properties typically require detailed calculations involving sophisticated computational methods. In many cases, however, a detailed modelling approach also demands knowledge about processes and parameters that are not yet accessible experimentally. Estimating the quantitative impact of these processes ultimately leaves the questioner with the task to decide whether or not the obtained answers are reasonable. Ideally, different modelling approaches can be combined where answers to questions obtained at a certain level of resolution serve as input for modelling at a level of higher resolution.

We consider three modelling approaches of lymphocyte migration and interaction that differ in the level of resolution with respect to the functional properties of the cells. The first is a statistical modelling approach for the analysis of cell tracks as observed in two-photon experiments. Single cells are viewed as independent point particles and the only functional aspect that is taken into account is their ability to migrate. Next, we consider cellular systems in which different types of cells migrate and interact with each other. This is captured by an agent-based modelling approach, where each cell represents a discrete agent that interacts with other cells and is monitored during its whole lifetime. Finally, the agent-based modelling approach is used with a sufficiently high spatial resolution to go beyond the point particle representation of cells in order to study the dynamics of cellular shape deformations under migration and interaction.

Imaging Lymphocyte Migration and Interaction

Intravital Microscopy

Optical multi-photon microscopy is today's method of choice for in vivo imaging of single-cell dynamics and cellular interactions within intact tissue. Conventional methods employing fluorescence techniques, such as wide-field and confocal microscopy, are based on light excitations of fluorophores by single photons. A fluorophore is a functional molecule which absorbs and re-emits light at different wavelengths. The single-photon excitation typically requires photon wavelengths in the order of 400 nm corresponding to photon energies that give rise to phototoxicity and strong scattering in biological tissue.

Multi-photon microscopy is based on the virtually simultaneous light excitation by two or more photons. For example, in two-photon microscopy the fluorophore is excited by two photons, each of which contributes one half of the energy required to induce fluorescence. The required photon density at the focal spot of the microscope objective is generated by a pulsed laser, which is adjusted such that the density is sufficiently high but yet not so high as to damage the biological sample. The main advantage of this method is that using low-energetic photons with long wavelengths in the order of 800 nm strongly decreases the absorption and scattering of light in biological tissue. Moreover, since the impact of phototoxicity and photobleaching is negligible for infrared photons, imaging up to millimeter-depths into the biological tissue can be realized without inducing significant damage to the biological sample.

As with conventional microscopes, image sequences can be built up by scanning the focal spot across the sample. Since the requirement on the photon density for the induction of multi-photon fluorescence is only met within the confined volume of the focal spot, the resulting images are free of out-of-focus fluorescence and therefore have higher image contrast as compared to single-photon techniques. With the help of advanced tracking software, individual cells can be identified and followed in the three-dimensional stacks of image sequences. This makes laser scanning multi-photon microscopy the state-of-the-art technique for intravital cell imaging in biological tissue with high spatial and temporal resolution.

Exploring Dynamic Cell Properties

At each time point, multi-photon imaging yields data on the position and on the geometric shape of every cell within the three-dimensional focal spot. For an image sequence consisting of $N + 1$ measurements that are recorded with the time step Δt, the position of the ith cell, $(x_i(n), y_i(n), z_i(n))$, is determined at the time points $t_n = n\Delta t$ with $n = 0, 1, 2, \ldots, N$. Similarly, the geometric shape of the ith cell may be quantified by a shape index $s_i(n)$ characterizing the time-dependent cell geometry. The data on the cell track and cell shape can be combined into the lists

$$T_i \equiv \begin{pmatrix} x_i(0) & x_i(1) & x_i(2) & \ldots & x_i(N) \\ y_i(0) & y_i(1) & y_i(2) & \ldots & y_i(N) \\ z_i(0) & z_i(1) & z_i(2) & \ldots & z_i(N) \end{pmatrix} \text{ and } S_i \equiv \left(s_i(0)\ s_i(1)\ s_i(2) \ldots s_i(N) \right) \tag{6.1}$$

for each cell $i = 1, 2, \ldots, I$, respectively.

Various definitions of the shape index may be envisaged. For example, representing the cell geometry by an ellipsoid with equatorial radii a, b, and c that is approximated by an oblate ($a = b > c$) or a prolate ($a = b < c$) spheroid, the shape index may be defined as the ratio of the long to the short axis. This gives rise to values $s_i(n) \geq 1$, where $s_i(n) = 1$ for a spherical and $s_i(n) > 1$ for an elongated cell shape.

The cell migration is characterized by quantities that can be computed from the cell track lists T_i. These lists may be viewed as $3 \times N$ matrices, where the nth column contains the position vector of the ith cell at time t_n:

$$\boldsymbol{R}_i(n) \equiv \begin{pmatrix} x_i(n) \\ y_i(n) \\ z_i(n) \end{pmatrix}. \tag{6.2}$$

Observables of interest are often averaged over all monitored cells. This is done by summation of the corresponding quantity for each cell i and division by their total number I. In passing we note that for practical reasons the number of cells may be a function of time, $I = I(n)$, since cells may exit the focal spot at longer times and cannot be imaged anymore. The fewer remaining cells give rise to standard deviations of the averaged quantities that are increasing with the measurement time.

The time-dependent displacement vector of a cell relative to its starting point is given by

$$\boldsymbol{r}_i(n) \equiv \boldsymbol{R}_i(n) - \boldsymbol{R}_i(0) \tag{6.3}$$

and is used in the calculation of the mean cell velocity

$$\langle \boldsymbol{v}_{i,m}(n) \rangle \equiv \frac{\langle \boldsymbol{r}_i(n) \rangle - \langle \boldsymbol{r}_i(n-m) \rangle}{m \Delta t} \tag{6.4}$$

at times $t_n = n\Delta t$ with $n \geq m$. Here, the brackets $\langle \ldots \rangle$ denote the average over all monitored cells. Note that $\boldsymbol{v}_{i,m}(n)$ depends on the choice of the time interval $m\Delta t$ and the interpretation of a cell's speed, $v_{i,m}(n) = |\boldsymbol{v}_{i,m}(n)|$, may actually lead to wrong conclusions. For small time intervals $m\Delta t$ the cell speed may be overestimated due to the jitter motion of otherwise stationary cells. On the other hand, if the cells perform random walk migration, the actual cell speed may be underestimated for long time intervals.

Random walk migration refers to cells that do not move unidirectional but randomly change their migration direction. These random changes may either occur at each time step of the measurement, or only after a number of time steps have elapsed. In the latter case, a directional persistence time exists corresponding to the time interval Δt_p during which cells migrate in the same direction before randomly turning into another direction.

Unidirectional migration and random walk migration can be distinguished by analyzing the time-dependence of the mean displacement $\langle|\boldsymbol{r}_i(n)|\rangle$. For cells performing unidirectional migration, e.g. by following chemokine gradients, the mean cell velocity is a constant, such that

$$\langle|\boldsymbol{r}_i(n)|\rangle = \langle|v_{i,n}(n)|\rangle\, t_n . \tag{6.5}$$

In other words, $\langle|\boldsymbol{r}_i(n)|\rangle$ equals exactly the mean path length $\langle l_i(n)\rangle$, which is defined by:

$$\langle l_i(n)\rangle \equiv \frac{1}{n}\sum_{n'=1}^{n}\langle|\boldsymbol{r}_i(n') - \boldsymbol{r}_i(n'-1)|\rangle . \tag{6.6}$$

In contrast, for cells performing random walk migration the mean path length becomes significantly larger than the mean displacement, $\langle l_i(n)\rangle \gg \langle|\boldsymbol{r}_i(n)|\rangle$ for times $t_n \gg \Delta t$, since cells randomly change their migration direction and eventually re-approach their starting point. This is schematically shown in Fig. 6.1a. The corresponding mean displacement does not scale linearly with time, but only with the square-root of time,

$$\langle|\boldsymbol{r}_i(n)|\rangle = \sqrt{2DM}\,\sqrt{t_n}\,, \tag{6.7}$$

where the proportionality constant depends on the system's spatial dimension D and the motility coefficient M. Note that random walk migration with a directional persistence time is a combination of unidirectional migration on small time scales and random walk migration on large time scales. The time scale is set by the directional persistence time: at times $t_n < \Delta t_p$, $\langle|\boldsymbol{r}_i(n)|\rangle$ scales linear with time but assumes scaling with the square-root of time for $t_n > \Delta t_p$. The time-dependence of the mean displacement is shown in Fig. 6.1b for the three different migration behaviours.

Taken together, multi-photon imaging provides a firm data basis to characterize the migration behaviour and the shape dynamics of cells in intact biological tissue.

Statistical Cell Track Models

Statistical cell track models aim at re-constructing experimentally observed cell tracks in order to identify underlying migration mechanisms. In this modelling approach cells are treated as independently migrating point particles that are characterized by their speed and polarity. The re-construction of cell tracks is achieved by calculating the position vector $\boldsymbol{r}_i(n+1)$ of the ith cell at time $t_{n+1} = (n+1)\Delta t$ from its position $\boldsymbol{r}_i(n)$ and its velocity $v_i(n)$ at time t_n:

$$\boldsymbol{r}_i(n+1) = \boldsymbol{r}_i(n) + \Delta t\, v_i(n) . \tag{6.8}$$

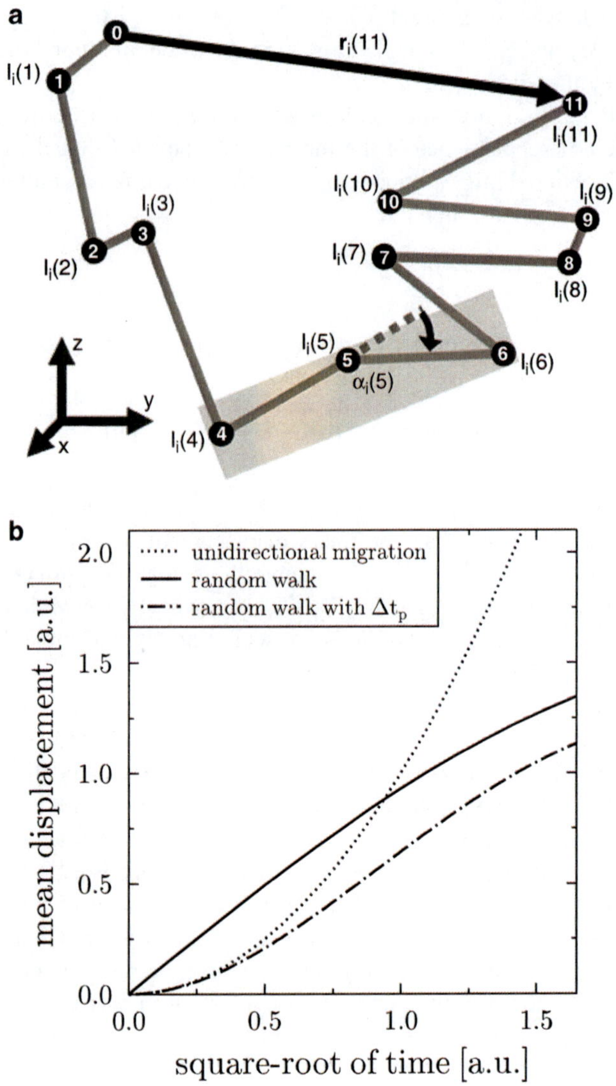

Fig. 6.1 Schematic representation of a cell track and the cellular migration behaviour in terms of the mean displacement. (**a**): The first 11 time steps of a three-dimensional cell track are shown. The cell performs random walk migration, where the length of the displacement vector $\boldsymbol{r}_i(11)$ is much smaller than the path length $l_i(11)$. Three cell positions at times t_{n-1} to t_{n+1} span a plane that defines the turning angle $\alpha_i(n)$ between the displacement vectors $\boldsymbol{r}_i(n) - \boldsymbol{r}_i(n-1)$ and $\boldsymbol{r}_i(n+1) - \boldsymbol{r}_i(n)$, as indicated for $n = 5$. (**b**): The cellular migration behaviour in terms of the mean displacement as a function of the square-root of time. Cells performing unidirectional migration scale like $\sqrt{t}^2$ (*dotted line*), whereas random walk migration scales linear with $\sqrt{t}$ (*solid line*). Random walk migration with a directional persistence time Δt_p (*dashed-dotted line*) is characterized by a scaling behaviour $\sqrt{t}^2$ at times smaller than Δt_p and $\sqrt{t}$ for larger times. If the migration is constrained to a finite volume, the mean displacement curves level off after sufficiently long times. This is depicted for the two *curves* showing random walk migration

Thus, at each time step Δt of the measurement, the new cell position is calculated by updating the velocity,

$$v_i(n) = v_i(n)\boldsymbol{e}_i(n)\,, \tag{6.9}$$

in terms of the cell speed $v_i(n)$ and the cell polarity vector $\boldsymbol{e}_i(n)$. In spherical coordinates the unit vector $\boldsymbol{e}_i(n)$ is given by

$$\boldsymbol{e}_i(n) = \begin{pmatrix} \sin(\vartheta_n)\cos(\varphi_n) \\ \sin(\vartheta_n)\sin(\varphi_n) \\ \cos(\vartheta_n) \end{pmatrix}, \tag{6.10}$$

where $\vartheta_n \in [0,\pi[$ and $\varphi_n \in [0,2\pi[$ denote the spherical angles.

So far we did not specify the procedure according to which (6.8) is updated. In fact, depending on the available data basis, different procedures may be applied in different situations. In what follows we review two procedures that have been applied to re-construct cell tracks for T cells in lymphoid tissue and for B cells in germinal centres.

T Cell Migration in Lymphoid Tissue

Imaging individual T cells within lymphoid organs has been the subject of various studies in recent years [1, 10–12]. The experimental results on T cell migration have been presented in terms of the mean displacement curve as a function of the square-root of time. In general, these curves show the characteristics of random walk migration with a directional persistence time.

Using the statistical cell track model approach, a mathematical analysis was performed with the goal to estimate values of parameters underlying T cell migration [2]. In this analysis, the three-dimensional T cell tracks were projected on a two-dimensional plane by neglecting the z-component of the polarity vector (6.10). The new polarity vector, which does not have the property of being a unit vector, reads

$$\boldsymbol{e}_i(n) = \begin{pmatrix} \sin(\vartheta_n)\cos(\varphi_n) \\ \sin(\vartheta_n)\sin(\varphi_n) \\ 0 \end{pmatrix}. \tag{6.11}$$

The re-construction of cell tracks was then performed according to (6.8) and (6.9), where the cell speed was set to a constant value,

$$v_i(n) = v\,, \tag{6.12}$$

for all cells. Furthermore, the time step Δt was decomposed into the persistence time Δt_p during which cells migrate with constant speed and the re-polarization time Δt_r during which cells do not move but are thought to reposition their lamellipods and uropod. After having paused, T cells randomly pick a new direction of migration, which was realized by choosing new angles ϑ_n and φ_n from uniform distributions.

For each parameter set $\{v, \Delta t_p, \Delta t_r\}$, a number of 10^6 individual cell tracks was generated in a randomized fashion. The parameters were explored in the range $v = [5, 50]\,\mu$m/min, $\Delta t_p = [0.5, 20]$ min, and $\Delta t_r = [0, 3.5]$ min and the optimal values were determined from a fitting procedure to the experimental data.

Combining the T cell tracks of different experimental data sets [1, 10–12] yielded the optimal values of the parameter set to be $\{v = 18.8\,\mu\text{m/min}, \Delta t_p = 2\,\text{min}, \Delta t_r = 0.5\,\text{min}\}$. Thus, T cells perform random walk migration with a persistence time of 2 min and the time for re-polarization is 30 s. Interestingly, under the constraint that the re-polarization time is neglected, $\Delta t_r = 0$ min, the optimal values changed into $\{v = 16.6\,\mu\text{m/min}, \Delta t_p = 2\,\text{min}\}$. Thus, while the T cell speed attains a somewhat smaller value, the persistence time does not change. A general conclusion of this analysis was that the re-polarization time does not play a central role in improving the agreement with the experimental data. The cell re-polarization is further discussed below and in Section "Cell Shape Dynamics of Migrating T and B Cells" within a modelling approach that resolves the shape of migrating cells.

B Cell Trans-Zone Migration in Germinal Centres

The migration of B cells between the light and dark zone in germinal centres was investigated by several experimental groups using two-photon microscopy [13–15]. All three experimental groups agree that during measurements of 1 h, 5–10% of the observed B cells will have migrated from one zone to the other. Furthermore, the three experimental groups agree on the interpretation that B cell motility follows random walk migration with a directional persistence time of about 1 min.

Based on the measured turning angle and speed distributions [13], a statistical cell track model was applied to check whether the persistent random walk hypothesis can be reconciled with the measured trans-zone migration frequency of B cells in germinal centres [7]. The persistent random walk hypothesis was incorporated in the velocity (6.9) by choosing the speed and turning angle independently. In practice, this was realized by a Monte Carlo acceptance-rejection method that generates random cell speeds $v_i(n)$ and random turning angles $\alpha_i(n)$ from the respective experimental distributions [7, 13].

It should be noted, however, that the turning angle (see Fig. 6.1a) provides information on the angle between polarity vectors $\boldsymbol{e}_i(n)$ and $\boldsymbol{e}_i(n+1)$ within the same plane:

$$\alpha_i(n) = \arccos\left(\boldsymbol{e}_i(n) \cdot \boldsymbol{e}_i(n+1)\right). \tag{6.13}$$

Thus, in three spatial dimensions, $\boldsymbol{e}_i(n+1)$ is not uniquely determined by (6.13) but may refer to any point on a cone with axis $\boldsymbol{e}_i(n)$ and radius $r_{\alpha_i(n)} = |\sin(\alpha_i(n))|$. In the spirit of the persistent random walk hypothesis, the point on this circle was chosen at random from a uniform distribution.

The turning angle and the speed are permanently changing after characteristic time steps Δt_p and Δt_v, respectively. A new speed value was randomly drawn from the corresponding distribution at every time step $\Delta t = \Delta t_v = 20\,s$, corresponding to the time interval between two consecutive speed measurements [13]. The time step Δt_p was determined such that the measured curve of the mean displacement versus the square-root of time is reproduced. It is important to note that even though the directional persistence time Δt_p is not known a priori, turning angles can nevertheless be drawn from the experimental distribution that was evaluated with a time resolution of 20 s. It has been shown that the turning angle distribution is extremely robust and represents a reasonable approximation for different values of Δt_p ranging from 20 s up to 160 s [7].

Re-constructing 10^5 B cell tracks and fitting Δt_p under the constraint that the experimentally observed mean displacement curve is recovered, yields the reasonable value $\Delta t_p = 1.24$ min for B cells in wildtype mice. Using this value, it was confirmed that the persistent random walk hypothesis can be reconciled with the trans-zone migration frequency for B cells migrating between the light and dark zone in germinal centres. This can be seen in Fig. 6.2a, where we plot the fraction of 10^5 monitored cells that migrate during a one-hour measurement across the zone boundary of 40 μm thickness within a spherical germinal centre of radius $R = 160\,\mu$m.

The position distribution $P(r, t_n)$ was calculated from the re-constructed cell tracks and represents the probability to find a cell after time t_n at a radial distance r from its initial position. The result is plotted in Fig. 6.2b at $t_n = 10$ min and $t_n = 3$ h for a spherical germinal centre of radius $R = 160\,\mu$m and cellular starting positions at the centre of the sphere. Interestingly, for B cells that perform random walk migration with a directional persistence time, it was found that within 3 h B cells will be equally distributed over the whole germinal centre. This result is in conflict with the observation of a zonal structure in germinal centres, where more B cells are found in the dark zone than in the light zone. In other words, these results suggest that additional mechanisms must be present that prohibit the quick intermixture of the germinal centre by the B cells. A natural candidate for this mechanism is chemotaxis and its implications on the germinal centre morphology will be further discussed below.

Agent-Based Models of Interacting Cell Systems

Cellular systems exhibit complex phenomena that require a modelling approach where different types of cells are represented as discrete objects with specific properties and characteristic functions. This is realized in agent-based models where each cell is an agent that can migrate and interact with other agents in space and time. It is convenient to impose a lattice of space points that mimics the spatial environment of the cells, where the distance between neighboring lattice points determines the spatial resolution. The advantage of the spatial discretization by the

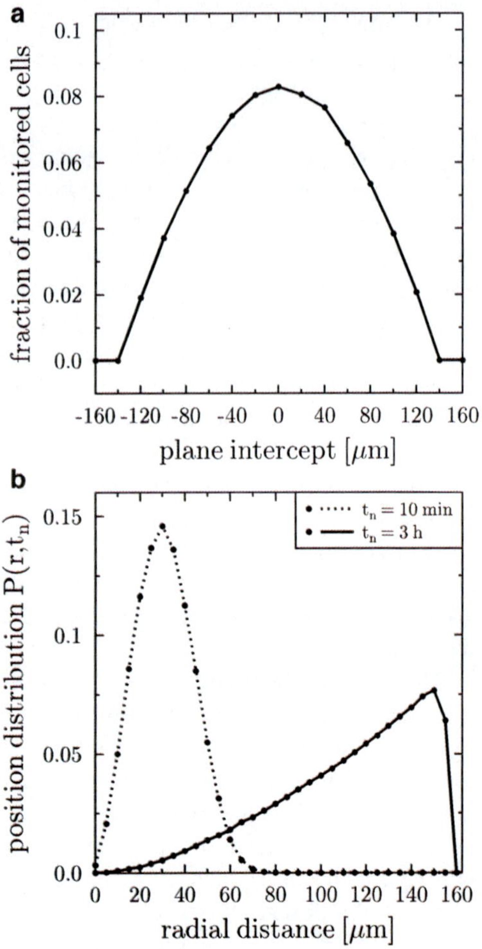

Fig. 6.2 Trans-zone migration and probability distribution of B cells in a spherical germinal centre of radius $R = 160\,\mu\text{m}$. (**a**): The fraction of 10^5 monitored cells that migrate across the zone boundary of $40\,\mu\text{m}$ thickness during a one-hour measurement. The position of the intercepting plane at the centre of the zone boundary is changed between $-R$ and $+R$ across the whole germinal centre. Over a large regime of plane intercepts the trans-zone migration frequency is found to be within the experimentally expected range of 5–10% per 1 h. (**b**): Position distribution $P(r, t_n)$ as a function of the radial distance r at times $t_n = 10$ min and $t_n = 3$ h. After 3 h, the quadratic dependence of $P(r, t_n)$ on r indicates the homogeneous distribution of B cells throughout the whole germinal centre. (This figure was modified after [7].)

lattice representation is that the neighborhood topology is kept fixed during the simulation, whereas lattice-free agent-based models with continuous space coordinates require additional measures to identify the neighbors of a cell in each time step of the simulation [16].

The procedure according to which the system evolves in time depends on the type of implemented agent-based model. In the next section, we consider a Potts-model approach, where the system dynamics is governed by minimizing a pre-defined energy functional. In contrast, we will also consider a model that is based on rules for stochastic events that determine the time-evolution on the basis of rates associated with the occurrence of these events.

T Cell Migration and Interaction with Dendritic Cells

On the basis of two-photon imaging data from lymph nodes of mice, the migration of T cells and their interaction with dendritic cells was studied using a Potts model [17, 18]. In this model, cells are represented by several connected points in a three-dimensional lattice. The cell types taken into account are T cells, dendritic cells and static fibroblastic reticular cells forming the reticular network in lymph nodes. Surface and volume energies are assigned to cells that are in contact with other cells and changes in the cellular configuration are determined by minimizing an energy functional.

The energy functional is given in terms of the Hamiltonian

$$H = \sum_{ijk} \sum_{i'j'j'} J\left[\tau(\sigma_{ijk}), \tau(\sigma_{i'j'j'})\right]\left(1 - \delta_{\sigma_{ijk},\sigma_{i'j'j'}}\right) + \sum_{\sigma}\left[\lambda_v(v_\sigma - V_\sigma)^2 + \lambda_p(p_\sigma - P_\sigma)^2\right]. \tag{6.14}$$

The first term represents the surface energy J depending on the interacting cell types $\tau(\sigma)$ with σ the identification number of the cells on the lattice. The summation runs over all lattice points $\{ijk\}$ and neighboring lattice points $\{i'j'j'\}$ with the Kronecker delta excluding self-terms of the surface energy. The volume energy is determined by λ_v and is a measure for the cell inelasticity driving the actual cell volume v_σ to a pre-defined target cell volume V_σ. For dendritic cells the large surface to volume area is imposed by the λ_p-term in the Hamiltonian with actual dendritic cell surface p_σ and target surface P_σ.

The system is advanced in time by random changes in the cellular configuration that give rise to the energy difference ΔH between the two configurations. In addition, for motile T cells the direction of migration is measured by the angle α relative to a pre-defined target direction. This gives rise to an additional contribution $\Delta H_T = -\mu \cos(\alpha)$ for T cells, such that the total energy difference is given by $\Delta H_t = \Delta H + \Delta H_T$. A Metropolis Monte Carlo algorithm is used to decide whether or not the new cellular configuration is accepted. The details of this procedure together with the chosen model parameters can be found in [17].

In the model, T cells perform random walk migration with a directional persistence time in the order of 2 min that was fitted to reproduce the experimental data [11]. This agreement was obtained without the requirement that T cells regularly pause and subsequently choose a new random direction. Instead, the model suggests that this migration behaviour is a consequence of T cells having a preferred direction of motion that is adjusted by the reticular network and dendritic cells in the nearby environment. According to the simulations, T cells migrate preferentially along the fibers of the reticular network unless they see obstacles on their migration path. It was proposed that this behaviour gives rise to small, dynamic streams of T cells through the lymph node.

The model has also been used to estimate the scanning rates of dendritic cells by T cells. The simulations revealed that T cells are able to scan 100 different dendritic cells per hour. This implies that, during negative selection in the thymic medulla, maturing T cells scan about 3.4×10^4 different dendritic cells in 14 days. This number involves many brief contacts lasting less than 1 min and the simultaneous interaction between a T cell and multiple dendritic cells. The average contact duration was between 1 and 2 min according to the simulations, with rare interactions that took up to 10 min. It should be noted, however, that the contact duration and the scanning estimate depend on the imposed densities of the different cell types.

The cellular Potts model has also been applied to investigate the three distinct phases of T cell stimulation that have been observed in two-photon experiments [18]. In the first phase, T cells rapidly migrate trough lymph nodes experiencing only brief encounters with dendritic cells. Several hours after T cells are first exposed to their cognate antigen, they enter into the second phase, which is characterized by T cells being clustered around dendritic cells for more than 30 min. It is likely that immunological synapses form during this phase that ultimately results into the stimulation of T cells. The third phase is characterized by T cells that are again rapidly migrating and proliferating in response to antigen stimulation.

Simulations suggest that, in order to enter the second phase of T cell stimulation, the assumption of adhesion between specific T cells and antigen-bearing dendritic cells alone is not sufficient. Rather, it was concluded from the simulations that stop signals have to be provided by dendritic cells that are integrated by the T cells during the first phase and enable the transition to the second phase. It was speculated that stop signals may be provided during the first and second phase of T cell priming, but are absent or ignored with the onset of the third phase, in which T cells resume rapid migration behaviour.

Transient Chemotaxis of B Cells in Germinal Centres

During the last decade, the germinal centre reaction has been simulated by various implementations of agent-based models that describe cell migration and interaction as the result of stochastic events occurring with characteristic reaction rates [8, 19–22]. Recent two-photon imaging of in vivo B cell migration in germinal centres [13–15] has re-initiated the functional analysis of the germinal centre reaction for different assumptions on the B cell migration behaviour [7].

We have discussed that the statistical cell track model, based on the hypothesis of B cells performing persistent random walk migration, reveals a conflict. On the one hand, this hypothesis is sufficient to explain the experimentally observed frequency of B cell trans-zone migration in germinal centres, on the other hand the high motility of B cells results in a quick intermixture of the germinal centre that cannot be reconciled with the observed zonal structure in germinal centres. However, performing a functional analysis within an agent-based modelling approach, this conflict can be resolved under the additional assumption that persistent random walk migration of B cells is subject to transient chemotaxis [7].

The applied agent-based model consists of the following three coupled levels:

(a) The first level contains the main lattice corresponding to the three-dimensional physical space in which cells can migrate and interact. Each lattice point can carry up to one biological cell that evolves according to reaction rates defining a probability of action or interaction with neighboring cells on the lattice. In addition, cells migrate according to a chemokine distribution that is calculated on a separate lattice (see third level below).
(b) The second level refers to the antibody shape space, which is represented by a four-dimensional lattice encoding the antibody type of each B cell. Somatic hypermutation is represented by switching the antibody type to a neighboring point in the shape space. Each point in the shape space is associated with an antibody–antigen affinity and determines the binding probability of a B cell to an antigen-presenting follicular dendritic cell.
(c) The third level deals with solving a system of reaction-diffusion equations for the chemokines CXCL12 and CXCL13. The source for CXCL12 are stromal cells at the border of the follicle in the dark zone, and follicular dendritic cells in the light zone for CXCL13.

The essential feature of this agent-based model is that it captures the whole germinal centre reaction comprising the population kinetics, cellular interactions, and affinity maturation. This level of description allows to distinguish between centroblasts, proliferating B cells that undergo somatic hypermutation, and centrocytes, nondividing B cells with activated apoptosis competing for survival signals. The included mechanisms follow, to a large extent, the classical model of the germinal centre reaction [3] and the simulations are validated by numerous experimental facts [7].

The preferred direction of B cell migration is defined by a polarity vector that is renewed after the persistence time of 1.24 min, which was obtained from the statistical cell track model that was discussed in Section "B Cell Trans-Zone Migration in Germinal Centres". The polarity vector consists of two contributions. A random contribution according to the turning angle distribution as measured in two-photon experiment [13] and a contribution that depends on the chemokine gradients.

The simulations predicted a novel migration model for B cells in germinal centres, referred to as "transient chemotaxis model" that reconciles all experimental and theoretical data [7,8]. According to this model, most of the cells in the dark and light zone actively perform persistent random walk migration. Centrocytes acquire additional sensitivity to CXCL13 after or during differentiation from centroblasts and orient toward the network of follicular dendritic cells in the light zone. In the network, the centrocytes lose CXCL13 sensitivity. This might be induced by contact with follicular dendritic cells, by overcritical CXCL13 concentrations and CXCR5 internalization, or simply by down-regulation of CXCR5 after a characteristic time. Similarly, a return to the dark zone of positively selected centrocytes is facilitated by a transient chemotactic sensitivity for CXCL12. However, random walk migration remains the dominant pathway of re-polarization since B cells undergo directed migration only temporarily. Therefore, transient chemotaxis might well be hidden in the experimental motility data that seemingly support pure persistent random walk migration.

Furthermore, the functional analysis of germinal centre simulations revealed insight about the interaction between B cells and follicular dendritic cells. Centrocytes bind antigen on follicular dendritic cells to obtain survival signals and this binding process is affinity dependent. Contacts observed in vivo have rarely been found to be longer than 5 min [14]. In the agent-based model it was assumed that, depending on the affinity of the encoded antibody, the B cells either contact follicular dendritic cells in a static condition of 30 min or in a transient manner that lasted until the B cell continued to migrate. A distribution of contact durations was obtained from the simulations and is plotted in Fig. 6.3. It was found that the majority of B cells exhibit only short contacts with follicular dendritic cells, i.e. less than 5 min (see Fig. 6.3a), and only 2–4% of the cells are in a static contact during the germinal centre reaction (see Fig. 6.3b). These results of the simulations are in agreement with the two-photon measurements [14]. Therefore, it cannot be

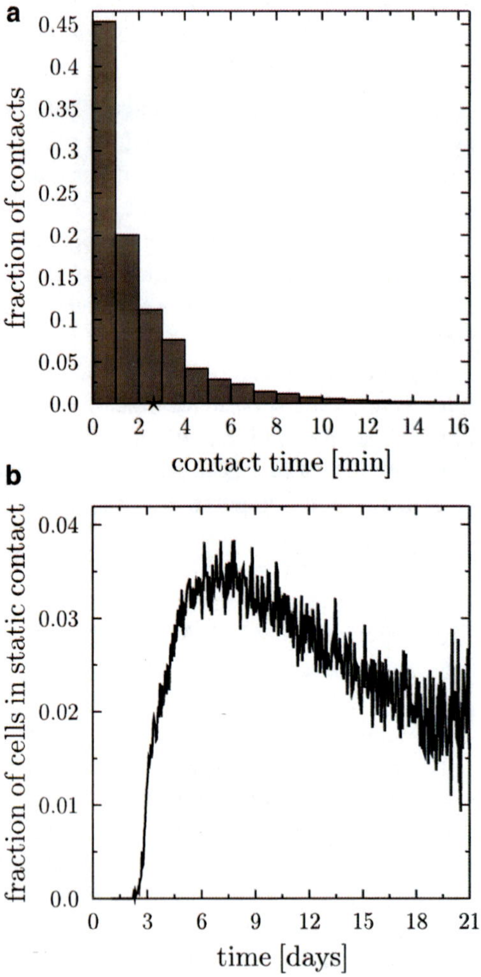

Fig. 6.3 Simulation results on the overall contact time and the number of static contacts between centrocytes and follicular dendritic cells. (**a**) The vast majority of contacts between centrocytes and follicular dendritic cells during a one-hour measurement is of dynamic nature and lasts around 3 min. (**b**) During the germinal centre reaction, the fraction of centrocytes that drive the affinity maturation and are in static contact with follicular dendritic cells for 30 min is only 2–4%. (This figure was modified after [7])

concluded from the experimental contact data that centrocytes integrate signals from short contacts with follicular dendritic cells [14]. Instead, even though signal integration cannot be ruled out, the simulations were consistent with the idea that rare static contacts are sufficient to drive affinity maturation.

Agent-Based Models of Cell Shape Dynamics

Two-photon microscopy data reveal that cell motility and cell shape are closely interlinked with each other, since the re-polarization of cells involves the repositioning of the cellular cytoskeleton. Several modelling approaches exists that capture the shape dynamics of migrating cells [17, 18, 23]. In this chapter, we concentrate on the approach that has been applied to analyze the experimentally observed shape dynamics of migrating cells in lymphoid organs [23].

Two-photon data show a stochastic variation of lymphocyte motility that is closely interlinked with the cell shape [10]. Therefore, to analyze these data, cells cannot be treated as point particles but have to be modelled in a spatially resolved fashion. This is realized by an agent-based model with a sufficiently high lattice resolution that represents each cell by a collection of subunits corresponding to connected lattice points. In addition, updating rules have to be implemented in the agent-based model that mimic realistic shape deformations of migrating cells.

Cell Shape Dynamics of Migrating T and B Cells

Following a reductionalist point of view, cells may be represented by the cell vol ume, the cell polarity and a list of connected cell subunits that constitute the cell shape on the lattice. The cell volume determines the number of cell subunits according to the chosen space resolution, while velocity states translate into probabilities of subunit movements in the direction of the cell polarity according to the chosen time resolution.

Two main contributions of cell dynamics have been identified [23]:

(1) The rearrangement of subunits with respect to a virtually shifted barycentre of the cell (active movement). A normalized polarization vector determines the direction of the active movement of a cell, which in reality is a complex function of the cytoskeleton organization as well as localized signalling pathways. This vector is considered as an approximation for the direction in which cell protrusions are developed and is assumed to change randomly with a probability that is associated with the persistence time.

(2) The rearrangement of the subunits with respect to the actual barycentre (reshaping forces). The subunit movement is realized according to heuristic rules that are interpretable as physical quantities. This ingredient of the cell motility model concerns the cell shape stability. During the procedure of active

cell movement the details of forces that reshape the cell towards a sphere, i.e. hydrostatic pressure, reduced actin filament assembly, actomyosin contraction, or membrane surface tension, are ignored. Within a phenomenological approximation, all these forces are included in a single reshaping force. This overall elastic force drives the subunits of an elongated cell back to the current barycentre and promotes a spherical shape.

The simulation of active movement was performed by the following procedure: The barycentre of the cell is virtually shifted in the direction of the polarization vector towards the membrane of the cell. Then every subunit representing a membrane point of the cell is moved in random order towards free lattice points near the virtual barycentre. Moving a border subunit which is not a direct neighbor of its target point (e.g. a subunit on the back of the cell) is considered to correspond to the cytosol shift through the whole cell. Note that in this model all subunits carried the same properties and if the movement of a membrane subunit would have caused the subunits of the cell to become disconnected this movement was suppressed. The displacement of subunits was stopped when either no membrane subunit remained to be moved or the barycentre of the cell was displaced by one lattice constant. In the new state the cell had reorganized its membrane and thus changed its shape. Thereby the total cell volume (or, the total cell area in a two-dimensional implementation) was conserved. Thus, the movement of the cell barycentre was realized by subunit rearrangements, and inherently coupled the cell movement to its deformation.

The comparison of the model results with experiment revealed that the two ingredients active movement and reshaping forces are sufficient in order to describe lymphocyte motility data as found by two-photon imaging. Here, we concentrate on the shape index that was calculated from the data obtained for T cell and B cell migration in lymph nodes of mice [10]. Projecting the cell volume on a two-dimensional plane, the area was approximated by an ellipse, which is characterized by its two axes. The shape index $s_i(n)$ of the ith cell was calculated as the ratio of the long to the short axis at each time step t_n, such that $s_i(n) > 1$ corresponds to an elongated cell shape that becomes circular for $s_i(n) = 1$.

In Fig. 6.4 the shape index distributions are plotted as obtained from the measurement of $I = 46$ T cells and B cells. The average shape index is around 2.3 for T cells and 1.6 for B cells, indicating that the cellular elongation is about a factor 1.5 larger for T cells compared to B cells. The differences in the shape index are in agreement with observations by two-photon microscopy [10]. It was speculated that this may point to cardinal differences in the cytoskeleton dynamics of migrating T and B cells [23]. Nevertheless, in the simulations both cell types dynamically round up and take a symmetric shape during the process of re-polarization. It should be noted that this result has not been enforced by assuming a pausing time in the agent-based model, as was done in the statistical cell track model combined with the assumption of a single speed state for the cells. Rather, in the agent-based model the shape index is a direct consequence of the reshaping forces that cells experience while they are continuing to migrate and the shape index is closely connected with a significant width in the underlying speed distributions [23].

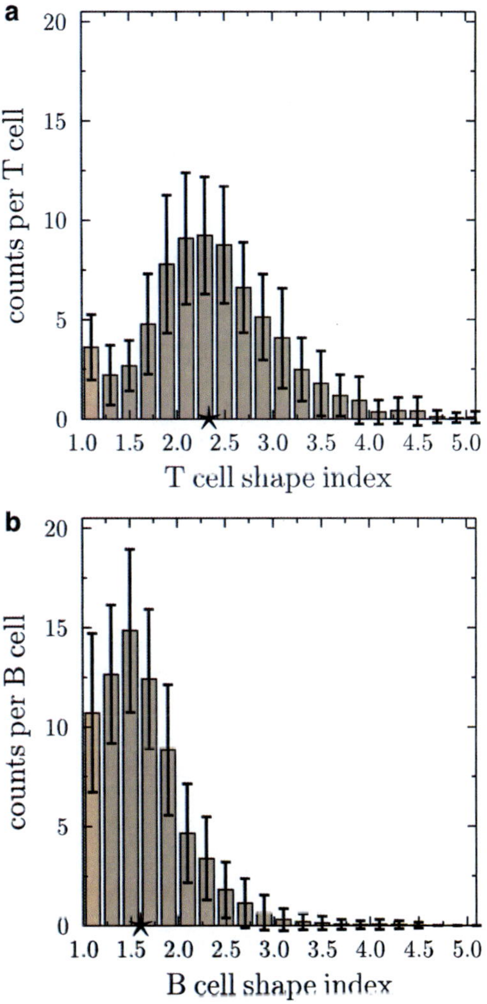

Fig. 6.4 Simulation results on the distribution of the shape index for 46 T cells and B cells. The error bars correspond to one standard deviation and stars indicate the mean values of the distributions. (**a**) The T cell shape index distribution has a mean shape index of 2.3 and shows that T cells reach peak elongations between 4 and 5. (**b**) The B cell shape index distribution has a mean shape index of 1.6, which corresponds to only 70% of the mean value for T cells. Peak elongations in B cells occur between 2.5 and 3.5

Summary

Multi-photon microscopy is a powerful tool for imaging lymphocyte migration and interaction in intact biological organs. Quantitative information on the cell motility, shape dynamics and contact durations are derived from these data by mathematical analysis that shape our interpretation of these data.

It is important to realize that mathematical analysis is not merely a descriptive approach, rather, its valuable contribution is to constitute a predictive consistency check of the working hypothesis that is underlying the interpretation of the experimental data. For example, a simple statistical cell track model was used to analyze the trans-zone migration of B cells in germinal centres. This revealed, on the one hand, that the measured trans-zone frequency is in agreement with the working hypothesis of B cells performing a persistent random walk migration in germinal centres. On the other hand, the same analysis disclosed an inconsistency with respect to the observed zonal morphology in germinal centres. This initiated further analysis using an agent-based model approach that captures the functional aspects of the germinal centre reaction. It then turned out that, hidden behind the experimental data from which the working hypothesis was derived, transient chemotaxis restores the zonal morphology in germinal centres. Since transient chemotaxis is still compatible with the B cell motility data, this implied that the initial working hypothesis of pure random walk migration with directional persistence time was actually formulated too narrow. This development of insight is a prime example for experiment and theory working hand in hand [8].

A firm experimental data basis is the mandatory prerequisite for realistic models of cellular dynamics in complex biological systems. Including more detailed aspects of cell dynamics into mathematical models reveals new system features and is becoming more feasible due to the continuously growing computer resources in terms of memory and processor speed. At the same time experimental techniques are developing fast and highly accurate data on cellular dynamics that are recorded over several hours time can be expected in the near future. The real challenge, however, is for modelers and experimentalists to keep pace in working hand in hand achieving the highest benefit from the symbiotic effects of theory and experiment.

References

1. Mempel T, Henrickson S, von Adrian U (2004) T-cell priming by dendritic cells in lymph nodes occurs in three distinct phases. Nature 427:154–159
2. Beauchemin C, Dixit NM, Perelson AS (2007) Characterizing T cell movement within lymph nodes in the absence of antigen. J Immunol 178:5505–5512
3. MacLennan ICM (1994) Germinal centers. Annu Rev Immunol 12:117–139
4. Manser T (2004) Textbook germinal centers? J Immunol 172:3369–3375
5. Hauser A, Shlomchik M, Haberman A (2007) In vivo imaging studies shed light on germinal-centre development. Nat Rev Immunol 7:499–504
6. Allen C, Okada T, Cyster J (2007) Germinal-center organization and cellular dynamics. Immunity 27:190–202
7. Figge MT, Garin A, Gunzer M, Kosco-Vilbois M, Toellner KM, Meyer-Hermann M (2008) Deriving a germinal center lymphocyte migration model from two-photon data. J Exp Med 205:3019–3029
8. Meyer-Hermann M, Figge MT, Toellner KM (2009) Germinal centres seen through the mathematical eye: B cell models on the catwalk. Trends Immunol 30:157–164
9. Cahalan M, Parker I (2008) Choreography of cell motility and interaction dynamics imaged by two-photon microscopy in lymphoid organs. Annu Rev Immunol 26:585–626
10. Miller M, Wei S, Parker I, Cahalan M (2002) Two-photon imaging of lymphocyte motility and antigen response in intact lymph node. Science 296:1869–1873
11. Miller M, Wei S, Cahalan M, Parker I (2003) Autonomous T cell trafficking examined in vivo with intravital two-photon microscopy. Proc Natl Acad Sci USA 100:2604–2609
12. Miller M, Safrina O, Parker I, Cahalan M (2004) Imaging the single cell dynamics of $CD4^+$ T cell activation by dendritic cells in lymph nodes. J Exp Med 200:847–856
13. Allen C, Okada T, Tang HL, Cyster J (2007) Imaging of germinal center selection events during affinity maturation. Science 315:528–531

14. Schwickert T, Lindquist R, Shakhar G, Livshits G, Skokos D, Kosco-Vilbois M, Dustin M, Nussenzweig M (2007) In vivo germinal center imaging reveals a dynamic open structure. Nature 446:83–87
15. Hauser A, Junt T, Mempel T, Sneddon M, Kleinstein S, Henrickson S, von Andrian U, Shlomchik M, Haberman A (2007) Definition of germinal-center B cell migration in vivo reveals predominant intrazonal circulation patterns. Immunity 26:655–667
16. Meyer-Hermann M (2008) Delaunay-object-dynamics: cell mechanics with a 3D kinetic and dynamic weighted Delaunay-triangulation. Curr Top Dev Biol 373–399
17. Beltman JB, Marée AF, Lynch JN, Miller MJ, de Boer RJ (2007) Lymph node topolgy dictates T cell migration behavior. J Exp Med 204:771–780
18. Beltman JB, Marée AF, de Boer RJ (2007) Spatial modelling of brief and long interactions between T cells and dendritic cells. Immunol Cell Biol 85:306–314
19. Meyer-Hermann M (2002) A mathematical model for the germinal center morphology and affinity maturation. J Theor Biol 216:273–300
20. Meyer-Hermann M, Maini P (2005) Back to one-way germinal centers. J Immunol 174:2489–2493
21. Figge MT (2005) Stochastic discrete event simulation of germinal center reactions. Phys Rev E 71:051907
22. Meyer-Hermann M, Maini PK, Iber D (2006) An analysis of B cell selection mechanisms in germinal centers. Math Med Biol 23:255–277
23. Meyer-Hermann M, Maini P (2005) Interpreting two-photon imaging data of lymphocyte motility. Phys Rev E 71:061912

Chapter 7
Modelling Lymphocyte Dynamics In Vivo

Becca Asquith and José A.M. Borghans

Abstract Quantification of lymphocyte dynamics contributes greatly to our understanding of many fundamental processes in immunology, including homeostasis, ageing, immunological memory and pathogenesis. Recently developed experimental techniques to label lymphocytes and mathematical models to interpret the resulting data enable us to accurately measure the rate of T and B lymphocyte proliferation and disappearance in humans in vivo. Here we describe some of the experimental and mathematical techniques involved, including a discussion of the advantages and disadvantages of the most popular methods available, as well as a practical guide to modelling labelling data, and a discussion of future challenges in this rapidly-moving area.

Why Quantify Lymphocyte Dynamics?

Lymphocytes are the key cells of our adaptive immune system and their dynamics determine our health. The correct balance between the production of lymphocytes (de novo in the thymus and by peripheral proliferation) and lymphocyte loss (by cell death or differentiation) is essential for immune function. Dysregulation of lymphocyte dynamics results in a broad spectrum of pathologies including AIDS (precipitated by CD4^{+} T lymphocyte loss), leukemia (aberrant growth of B or T lymphocytes) and autoimmune conditions such as multiple sclerosis or arthritis (inappropriate activity of self-reactive lymphocytes). Even though lymphocyte proliferation and death rates are often regarded as textbook immunology, estimates of these rates can easily vary by a 100-fold. Nevertheless, researchers are trying to understand how human diseases like HIV infection, and therapeutic interventions such

B. Asquith (✉)
Department of Immunology, Wright-Fleming Institute, Imperial College London,
Norfolk Place, London W2 1PG, UK
e-mail: b.asquith@imperial.ac.uk

C. Molina-París and G. Lythe (eds.), *Mathematical Models and Immune Cell Biology*,
DOI 10.1007/978-1-4419-7725-0_7, © Springer Science+Business Media, LLC 2011

as haematopoietic stem cell transplantation, affect lymphocyte kinetics. As long as there is controversy about lymphocyte kinetics in healthy individuals, such questions will remain hard to address.

Why Are Mathematical Models Needed?

Thanks to recent experimental advances, including the application of stable isotope labelling to label and trace cells that undergo proliferation, the time is now ripe to determine these rates of lymphocyte turnover. Proper interpretation of labelling experimental data hinges upon the use of mathematical models. Without mathematical models, labelling curves remain merely descriptive and do not yield the quantitative turnover parameters that are needed. Moreover, mathematical modelling helps to obtain insights into the complicated dynamical processes underlying the functioning of the immune system.

Mathematical models are often criticized because of the assumptions that are made. Indeed, as Lee Segel, a distinguished mathematical biologist himself, once said "mathematical biologists are people who make oversimplified models", but importantly he added "and do not even feel embarrassed". It is important to realize, that any interpretation of data incorporates assumptions. The danger with intuitive, model-free interpretations is that the assumptions are often implicit. Implicit assumptions lack transparency and researchers themselves may be unaware that they are making these assumptions and so their accuracy is never addressed. Indeed, as will be repeatedly seen in the following examples, intuitive "assumption-free" approaches that initially seem persuasive have frequently been demonstrated to give incorrect results. The fact that mathematical models make their assumptions explicit, and thereby visible, makes it easier to scrutinise them. Additionally, making simplifications and eliminating free parameters has the considerable advantage that it becomes possible to estimate key parameters such as lymphocyte proliferation and disappearance rates with limited amounts of data, which would be impossible with highly complex models (see the Section on Parameter Identifiability). Finally, simplified models can provide real insights into complex and highly dynamical systems such as the immune system.

Different Methods to Study Lymphocyte Dynamics

In recent decades, a large number of different methods have been applied to estimate the rates of production and death of different lymphocyte populations. These methods differ widely and include the interpretation of "experiments of nature", such as lymphocyte reconstitution after severe lymphopenia, as well as different markers to study lymphocyte populations that undergo proliferation and death. Some of those markers occur naturally, including the expression of the proliferation marker Ki67, or the length of a cell's telomeres. Others are administered for research purposes and include the nucleoside analogue BrdU and different stable isotopes. Here we will give an overview of these experimental methods, as well as some of the mathematical models that have been developed to interpret the data.

Experiments of Nature

In the absence of specific labelling techniques, investigators have estimated the rate of production and loss of different lymphocyte populations by following changes in the numbers of cells under specific extreme "natural" circumstances. In patients with extremely low lymphocyte numbers, e.g. because of severe regimens of chemotherapy as treatment for cancer or prior to stem cell transplantation, the reconstitution of the different lymphocyte compartments was followed over time and translated into net rates of lymphocyte production [1,2]. On the one hand, such rates may underestimate the true lymphocyte production rate, because cells may also die during the reconstitution period, a process that may go unnoticed but which will decrease the net rate of reconstitution. On the other hand, production rates in lymphopenic patients may exceed the actual rate at which cells are produced in a fully reconstituted lymphocyte compartment in steady state, because cells undergo little competition for survival factors.

Another extreme case that has been used to study the rate of lymphocyte turnover is by measuring the loss of lymphocytes with chromosomal damage in patients after treatment with radiotherapy for cancer. Assuming that no new lymphocytes with chromosomal damage are produced after stopping radiotherapy, the average loss rate of different lymphocyte populations has been estimated [3]. Again, extrapolation of these estimates to the healthy situation has to be done cautiously, because these loss rates may be influenced by the DNA damage of the cells under investigation and by the low numbers of lymphocytes in these patients.

Despite the fact that the quantitative parameters resulting from these "experiments of nature" may not translate exactly to the natural situation, these studies have had a large impact on our insights into immunological memory. Both approaches have clearly shown that the turnover rate of memory T lymphocytes exceeds that of naive T cells, and have thereby suggested that immunological memory is not maintained by a pool of long-lived memory cells, but by a highly dynamic pool of memory T cells with a rapid rate of cellular turnover.

Static Markers of Cell Proliferation and Death

Instead of measuring net production rates or net loss rates of lymphocytes under specific extreme circumstances, one can measure the natural expression of markers for cells undergoing cell proliferation or death. Ki67, for example, is a protein whose expression is restricted to proliferating cells. It is expressed during all active phases of the cell cycle, i.e. during the G_1, S, G_2 and mitotic phase, but not during the G_0 phase. Measuring the fraction of Ki67-expressing cells thus gives insights into the fraction of cells that is undergoing proliferation. Similarly, cells undergoing apoptosis can be measured by different methods. One of the plasma membrane alterations characteristic of cells in the early phase of apoptosis is the translocation of phosphatidylserines from the inner side of the membrane to the outer side. This expression can be measured by Annexin V staining, a protein that binds phosphatidylserine with high affinity. Alternatively, cells in apoptosis can be distinguished by intracellular staining for certain caspases, which are proteases involved in the cleavage of cellular proteins in cells undergoing apoptosis.

Measuring cell death and proliferation by such naturally expressed markers has the advantage that the immune system does not need to be disturbed. These measures are therefore very useful in comparisons of lymphocyte dynamics between individuals. Such static markers are, however, hard to translate into quantitative biological parameters such as the fraction of cells that proliferate or die per day. The fraction of cells expressing markers of death or proliferation reflects not only the fraction of cells undergoing death or proliferation per day, but is also influenced by the period during which these markers are expressed. The problem becomes even larger when the period of expression is not constant, e.g. because it is influenced by the immunological situation under investigation. For example, if T cells in HIV infected individuals were to die very rapidly, one might measure low fractions of Annexin V positive T cells, while in fact many cells are killed. Likewise, it has been suggested that T cells undergoing proliferation in HIV infected individuals may get stuck in the cell cycle [4], and thereby cause higher levels of Ki67 expression than one would expect from the fraction of cells that is actually producing progeny. Avoiding such problems and obtaining more quantitative biological parameters requires dynamic markers for cell proliferation and death.

Dynamic Markers of Cell Proliferation and Death

In order to study T lymphocyte production and death dynamically, a large number of markers have been used, all with their own characteristics, advantages and disadvantages. Naturally occurring dynamic markers of T cell production and death include the telomere length of a T cell population, a marker that is typically interpreted as reflecting the proliferative history of a cell population, and the expression of T cell receptor excision circles (TRECs), a marker that has been proposed for the measurement of T cell production in the thymus. Other dynamic markers are not naturally expressed and include the administration of the nucleoside analogue BrdU or stable isotopes such as deuterated water or deuterated glucose. By administration of such labels and by following their incorporation in the DNA of newly-produced cells during and after label administration, one can measure both the rate of proliferation and loss of different lymphocyte populations. Below we will provide a short summary of these different experimental techniques, as well as our favoured mathematical interpretation of the resulting data. Note that for all these techniques a number of different mathematical models have been developed. We focus on the ones that we think most clearly illustrate the impact of mathematics on our insights into lymphocyte dynamics.

T Cell Receptor Excision Circles

For decades, investigators have tried to quantify the contribution of T cells that are newly formed by the thymus to the maintenance of the peripheral T cell pool. In the late nineties a new assay was introduced to measure the number of cells that are produced by the thymus. This assay is based on the occurrence of T cell receptor excision circles (TRECs) [5,6]. TRECs are by-products of V(D)J rearrangements that occur in the thymus when the T cell receptor (TCR) is formed. During this genetic rearrangement process, parts of the TCR genome are excised and form stable DNA circles (Fig. 7.1a). These TRECs are not copied when a cell divides; they are simply passed on to one of the daughter cells (Fig. 7.1b). TRECs are thus uniquely formed in the thymus. Measurement of the average number of TRECs in a cell population by quantitative PCR was therefore proposed as a direct measure for thymus output.

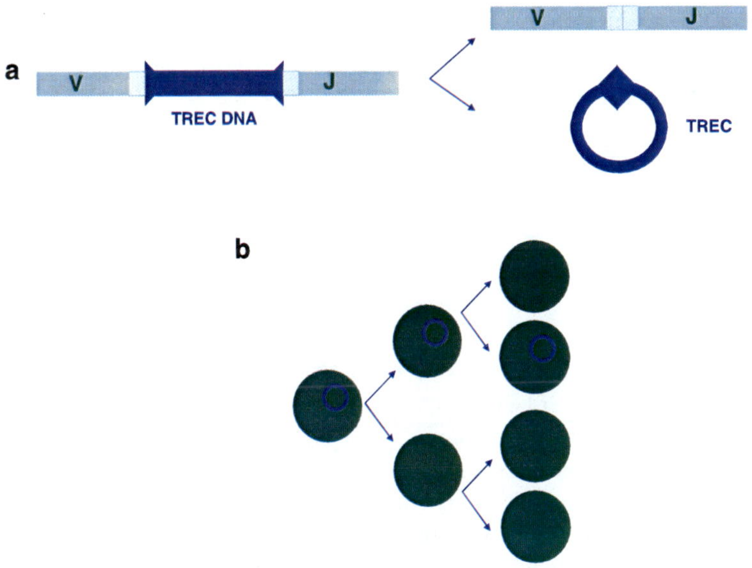

Fig. 7.1 (**a**) TRECs are excised from the DNA, during T cell receptor V(D)J gene rearrangement in the thymus. The excised pieces of DNA form stable circular DNA products, which fail to be copied during DNA duplication. (**b**) As a consequence, the average number of TRECs (in *blue*) per cell in a cell population decreases when cells proliferate

When the average number of TRECs per T cell was measured in individuals of different ages, an exponential decline was observed [6]. These findings fitted exactly to the common view that the thymus undergoes involution, and its output thereby decreases exponentially with age. Decreased TRECs per cell in HIV-infected patients were similarly interpreted as evidence for impaired thymic output [6], and increased TRECs per cell in individuals after haematopoietic stem cell transplantation as evidence for thymic rebound to compensate for the lack of T lymphocytes [7].

A simple mathematical model pointed out, however, that TRECs do not directly reflect thymic output [8–10]. The change in the average TREC content of a population of cells with age can be calculated from the change in the total number of TRECs, T, and the change in the total number of cells N with age. The total number of cells decreases by cell death at rate δ per day, and increases by T cell proliferation at rate p per day and by thymic output, $\sigma(t)$, which itself decreases exponentially with age at rate v per day:

$$\frac{\mathrm{d}N}{\mathrm{d}t} = \sigma(t) + pN - \delta N \tag{7.1}$$

where

$$\sigma(t) = \sigma(0)\mathrm{e}^{-vt}.$$

The total number of TRECs also decreases by cell death and increases by thymic output, but is not affected by T cell proliferation:

$$\frac{\mathrm{d}T}{\mathrm{d}t} = c\sigma(t) - \delta T, \tag{7.2}$$

where c is the average TREC content of a cell that leaves the thymus. The average TREC content A of a cell population can be calculated as $A = T/N$. It is easy to demonstrate that when T and N are at equilibrium, the average TREC content A is:

$$A = c(\delta - p)/\delta. \tag{7.3}$$

In other words, the average TREC content of this cell population is totally independent of the number of cells that are generated by the thymus per day. This equation demonstrates that the rate of proliferation and death of this cell population strongly influence its average TREC content. The more proliferation the cells undergo, the lower their TREC content, and the larger the death rate of the cells, the higher their TREC content. With hindsight, this can be understood intuitively, because cell proliferation increases the number of cells but not TRECs, and because cell death influences the average age of a cell population. The larger the death rate, the younger the cell population, and hence the fewer rounds of proliferation have occurred and diluted the average TREC content.

One thus has to be extremely careful when interpreting TREC data [8–10]. The fact that T cell TREC contents decline with age probably reflects a homeostatic response of the immune system to decreasing thymus output. While increased rates of T cell proliferation or decreased rates of T cell death, both possibly contributing to T cell homeostasis, strongly affect TREC contents, changes in thymic output per se will not. Likewise, low TREC contents in HIV patients are more likely to reflect increased rates of proliferation rather than decreased thymic output [8], an insight with considerable implications for the HIV field. The above equations illustrate that part of the problem can be solved by not only measuring average TREC contents of cell populations (i.e. A), but also measuring total TREC numbers per ml blood (i.e. T), a measure that is independent of the proliferation rate of the population. Nevertheless, even total TREC numbers do not directly reflect thymus output, because they are also influenced by cellular death rates. In summary, these analyses have demonstrated that a marker that is so unique for T cell generation by the thymus nevertheless also reflects all dynamical processes that T cells undergo in the periphery.

Telomere Length

Another naturally occurring dynamic marker for cell proliferation and loss is the length of a cell's telomeres. Telomeres are unique structures at the end of chromosomes, which consist of tandem DNA repeats. Because DNA polymerases fail to copy the very ends of the chromosomes during cell division, telomeres shorten with each cell division and thereby provide a record of a cell's proliferation history (Fig. 7.2) [11, 12]. Although it is very tempting to translate the decline in the average telomere length of a cell population directly into the number of cell divisions the population has undergone, describing telomere dynamics with a mathematical model has demonstrated that one has to be extremely cautious when interpreting telomere data [13]. Furthermore, the activity of the enzyme telomerase which effectively elongates telomeres can potentially destroy the relationship between telomere length and the number of divisions a cell has undergone [14, 15].

If one assumes that telomerase is not active then one can describe the shortening of telomeres in a population of cells by giving each cell an index i, depending on the number of cell divisions the cell has undergone, the so-called division index [13]. When a cell with division index i divides, it produces two daughter cells with division index $i + 1$. If the number of cells with division index i is denoted by n_i, and the total number of cells by N, the average division index μ of such a cell population is:

$$\mu = \frac{1}{N} \sum_{i=0}^{\infty} i n_i. \tag{7.4}$$

The average division index of a population increases with time. Its change over time ($\mathrm{d}\mu/\mathrm{d}t$) can be calculated by differentiating μ with respect to time, and by substituting $\mathrm{d}N/\mathrm{d}t$, the change in the total number of cells N over time, and $\mathrm{d}n_i/\mathrm{d}t$, the change in the number of cells n_i with division index i over time. If (7.4) applies

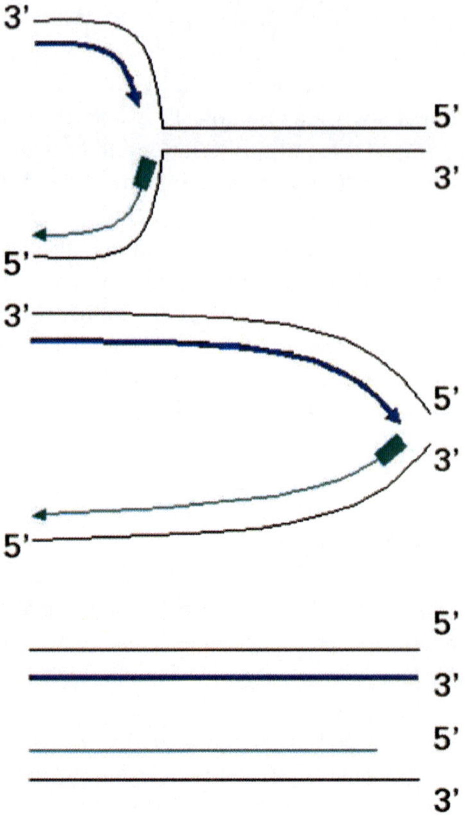

Fig. 7.2 Telomeres are repetitive DNA patterns found at the ends of the chromosomes. During DNA replication, the leading strand (*blue*) can make a full copy of the DNA, while the lagging strand (*green*) fails to copy the very end of the chromosome, leading to progressive telomere shortening

to memory T cells, the change in the total number of memory T cells ($N = M$) is described by [13]:

$$\frac{\mathrm{d}M}{\mathrm{d}t} = \sum_i \frac{\mathrm{d}m_i}{\mathrm{d}t} = M(p_M - \delta_M) + \gamma C V, \tag{7.5}$$

where p_M denotes the proliferation rate of memory T cells, δ_M denotes the rate at which memory T cells die, γ the rate at which naive T cells are primed and enter the memory compartment, C is the average number of cell divisions occurring during clonal expansion (i.e. when naive T cells are triggered to become memory T cells) and V is the number of naive T cells. See Fig. 7.3. The change in the number of memory cells m_i with division index i is described by the following differential equation [13]:

$$\frac{\mathrm{d}m_i}{\mathrm{d}t} = 2p_M m_{i-1} - (p_M + \delta_M)m_i + \gamma V n_{i-K}, \tag{7.6}$$

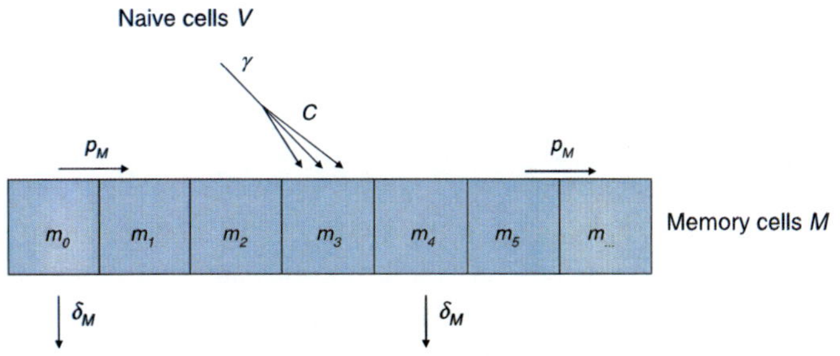

Fig. 7.3 The total memory T cell population, M, consists of sub-populations of cells m_i that have undergone i cell divisions. Memory cells in any sub-population i divide at rate p_M and produce two daughter cells with division index $i + 1$, they die at rate δ_M, and are produced when naive T cells V are activated at rate γ and clonally expand into C memory cells

where K is the number of divisions made during clonal expansion. Substituting these two equations into the derivative of (7.4), yields that the average division index μ_M of the memory T cell population changes according to:

$$\frac{d\mu_M}{dt} = 2p_M - \gamma C \frac{V}{M}(\mu_M - \mu_V - K). \tag{7.7}$$

The latter equation gives the important insight that the average telomere length of the memory T cell population does not only reflect the proliferative history of the memory T cell population, but also the transition of naive T cells into the memory compartment [13]. Telomere loss in the memory T cell pool is in part compensated by the influx of cells from the naive compartment, which – on average – have longer telomeres than memory cells. Thus, the rate of telomere loss in the memory population is to a large extent determined by the rate of telomere loss in the naive T cell pool. It is therefore perhaps not surprising that the telomere loss of naive and memory T cells was found to occur at similar rates [16], even though naive and memory T cells are thought to proliferate at different rates.

Similarly, one can argue that the rate at which naive T cells shorten their telomeres may not directly reflect their proliferation rate, but may also be influenced to a large extent by the rate of telomere loss of progenitor cells that are exported from the thymus into the periphery [13]. Since recent thymus emigrants form a permanent source of cells with relatively long telomeres, they increase the average telomere length of the naive T cell population. Ignoring this influx of cells from the thymus into the periphery may therefore lead to an underestimation of the naive T cell division rate.

In summary, although telomeric lengths may seem to be the best marker for the proliferative history of a cell population, again, just like TREC contents, they are not only influenced by T cell proliferation, but also by cellular death rates and input from the thymus. In fact, TREC contents and telomere lengths thus reflect very similar processes, even though they were originally proposed to be discriminative for thymic output and cellular proliferation, respectively.

BrdU Labelling

T cell turnover rates have extensively been studied by the use of 5-bromo-2′-deoxyuridine (BrdU), a nucleoside analogue that is incorporated into the DNA of cells that are dividing, instead of the nucleoside thymidine [17–20]. Because of its potential toxicity, BrdU has mainly been applied to rodents and non-human primates. It is typically administered via the drinking water and its incorporation in cells can be detected by flow cytometry. When cells divide in the presence of BrdU, newly-formed DNA strands build in BrdU and are subsequently randomly distributed over the two daughter cells, resulting in two BrdU-labelled cells (Fig. 7.4). When BrdU-labelled cells divide in the absence of BrdU, their label intensity decreases with every cell division. Nevertheless, BrdU labelling studies typically report the percentage of BrdU-positive cells, i.e. the percentage of cells in which the BrdU intensity exceeds a certain threshold, irrespective of their exact label intensity.

During BrdU intake, the percentage of cells that are BrdU-positive gradually increases, and after BrdU is withdrawn from the drinking water the percentage of BrdU-positive cells typically decreases. Simple, intuitive analyses interpret the rate of increase of labelled cells during the labelling period as reflecting the rate of cell proliferation and the rate of loss of labelled cells during the delabelling period as the rate of cell death. Although this may sound very persuasive, in fact it is not so straightforward. A simple mathematical model demonstrates where intuition goes wrong [21–25]. One can describe the accrual of label during BrdU administration either by the rate at which the fraction of labelled cells increases or by the rate at which the fraction of unlabelled cells decreases. Since both rates should be equal, one can simply use the easiest equation. During BrdU administration, it is easiest to describe the number of unlabelled cells, U. Unlabelled cells are lost when they

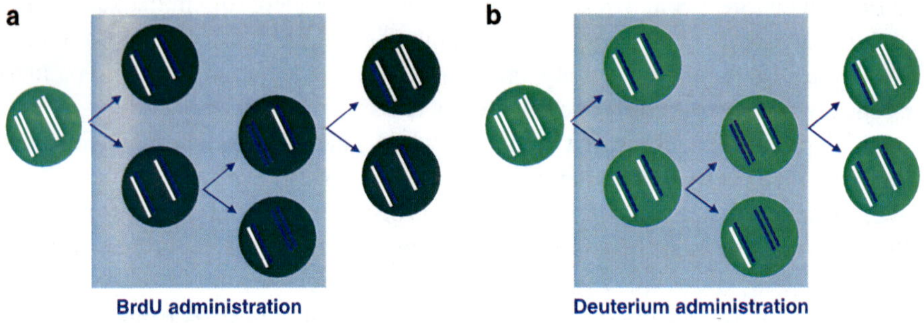

Fig. 7.4 BrdU labelling experiments follow the fraction of labelled cells, while stable isotope labelling experiments follow the fraction of labelled DNA strands. During BrdU administration (*light blue box*), dividing cells build BrdU into their newly-formed DNA strands. Labelled DNA strands (*blue*) are randomly distributed over the daughter cells. Upon BrdU-withdrawal, the label intensity of the cells decreases with every cell division, but the fraction of BrdU+ cells (*dark green*) does not decrease. During stable isotope labelling (*light blue box*), dividing cells build deuterium into their newly-formed DNA strands. Labelled DNA strands (*blue*) are randomly distributed over the daughter cells. Upon deuterium withdrawal, the fraction of labelled DNA decreases with every cell division

divide (because they become BrdU-labelled) at rate p per day, or when they die at rate δ per day:

$$\frac{dU}{dt} = -pU - \delta U. \tag{7.8}$$

This equation demonstrates that the fraction of unlabelled cells decreases, and hence the fraction of labelled cells increases, during label administration at rate $p + \delta$. Importantly, and rather counter-intuitively, label accrual during label administration thus does not only reflect proliferation but also loss of cells through death or maturation.

When BrdU is no longer administered, it is easier to follow the loss of labelled cells. Labelled cells are lost from the population by cell death at rate δ per day, but they are also gained, by cell proliferation at rate p per day, even in the absence of BrdU. Because previously labelled chromosomes are randomly distributed among daughter cells, both daughter cells will be BrdU labelled (see Fig. 7.4), albeit with lower BrdU intensity. Each BrdU labelled cell that divides thus adds another labelled cell to the population, even in the absence of BrdU:

$$\frac{dL}{dt} = pL - \delta L. \tag{7.9}$$

The latter equation demonstrates that the fraction of labelled cells decreases at rate $p - \delta$ per day when label administration is stopped [21, 24]. Importantly, the rate at which the fraction of labelled cells decreases after label cessation thus does not directly reflect the rate of cell loss. For a population at steady state, one would expect the rate of cell production p to be equal to the rate of cell loss δ, in other words $p - \delta = 0$. Consequently, one would expect the fraction of BrdU labelled cells to remain constant after label cessation. This is in contrast to what is routinely observed: that labelled cells decrease after the end of the labelling period.

It is therefore rather surprising that most BrdU studies show a significant decline in the proportion of labelled cells during de-labelling, suggesting that the average proliferation rate of lymphocytes, p, is smaller than the average loss rate of BrdU labelled cells, δ. Indeed, separate estimates of δ and p have shown up to tenfold larger rates of cell loss compared to the rate of T cell proliferation [21]. One possible explanation for this discrepancy is that dividing cells are not at equilibrium although the whole population is. Alternatively, if cell proliferation occurs in "bursts," BrdU labelling may have diluted to such an extent that cells are no longer recognized as label-positive, after extensive T cell proliferation in the absence of label. Currently ongoing research is aiming to distinguish between these options.

Stable Isotope Labelling

The most recently developed method to study lymphocyte kinetics in vivo is based on stable isotope labelling. Stable isotopes are non-radioactive variants of a chemical element, with a different mass because of the presence of extra neutrons in the nucleus. The stable isotope that is typically used in lymphocyte dynamic studies is deuterium, the heavier variant of hydrogen, which contains one proton and one neutron, and has twice the mass of hydrogen, which contains no neutrons. Deuterium has been administered in the form of deuterated glucose ($^2H_2$-glucose) [26–33] or deuterated water (^2H_2O) [34–36], and has proven to be safe and non-toxic when given in low concentrations. The deuterium atoms from these stable isotope labelled compounds are incorporated into the DNA of cells when they divide. Deuterium incorporation can be measured by a combination of gas chromatography and mass spectrometry (GC–MS) on the extracted DNA from sorted cells. In contrast to BrdU labelling studies, in which the fraction of labelled cells is measured, stable isotope labelling is measured in terms of fractions of labelled DNA fragments, not cells. The mathematics required to analyse stable isotope labelling data therefore deviate from those used in BrdU experiments [36–38].

The number of labelled DNA fragments in a population of cells increases by cell proliferation at rate p per day, independent of whether the proliferating cells had already incorporated deuterium or not. The extent to which label is incorporated during cell proliferation depends on the availability of deuterium, D. In deuterated glucose experiments, the availability is assumed to be either maximal, during label intake, or zero, after label cessation, because the turnover rate of glucose is very fast. In case of heavy water labelling, however, the supply of deuterium is more variable, and hence should be explicitly taken into account (see below). The number of labelled DNA fragments decreases by loss of labelled cells from the population at rate δ per day, through differentiation, maturation or cell death. The change in the total number of labelled DNA fragments per day can hence be described by [37]:

$$\frac{\mathrm{d}L}{\mathrm{d}t} = pcND(t) - \delta L \tag{7.10}$$

where N denotes the total number of DNA fragments in the population, and c reflects an amplification factor which needs to be introduced, because there are multiple hydrogen atoms in a single DNA fragment that can be replaced by deuterium. Typically, deuterium is given in such low concentrations that the chance to double-label a DNA fragment (and thereby miss it during mass spectrometry) is close to zero. However, the presence of multiple hydrogen atoms does increase the chance of labelling a DNA fragment at one position. Translating (7.10) into the *fraction* of labelled DNA fragments ($l = L/N$) yields:

$$\frac{\mathrm{d}l}{\mathrm{d}t} = pcD(t) - \delta l. \tag{7.11}$$

If labelling is performed with deuterated glucose, then we can assume that $D = 1$ during the labelling period and $D = 0$ during the delabelling period and (7.11) has the following solutions [37]:

$$l = \frac{pc}{\delta}(1 - \mathrm{e}^{-\delta t}), \qquad \text{during label administration } (t < \tau)$$

and

$$l = \frac{pc}{\delta}(1 - \mathrm{e}^{-\delta\tau})\mathrm{e}^{-\delta(t-\tau)} \qquad \text{after label administration } (t \geq \tau). \quad (7.12)$$

where τ represents the time point at which label administration is stopped.

If labelling is performed with deuterated water then, as mentioned above, the availability of deuterium needs to be explicitly taken into account. To this end, the heavy water enrichment in the serum of mice or the urine of humans can be measured and fitted to a simple exponential accrual and loss function:

$$D(t) = f(1 - \mathrm{e}^{-\epsilon t}) + \beta \mathrm{e}^{-\epsilon t} \qquad \text{during label administration } (t < \tau) \quad (7.13)$$

and

$$D(t) = \left(f(1 - \mathrm{e}^{-\epsilon\tau}) + \beta \mathrm{e}^{-\epsilon\tau}\right)\mathrm{e}^{-\epsilon(t-\tau)} \qquad \text{after label administration } (t \geq \tau),$$

where f represents the fraction of deuterated water in the drinking water, t denotes time in days, ϵ represents the turnover rate of body water per day, and β is the body water enrichment that is attained after a boost of label by the end of day 0. These equations can be substituted into (7.11) (and solved) to obtain a model for label enrichment in deuterated water experiments [36], which is an extension of the deuterated glucose model [37].

Equation (7.11) shows that in the absence of label, i.e. when $D = 0$, the decay of labelled DNA directly reflects the loss of labelled cells δ, and not – as in the case of BrdU – the difference between cell loss and proliferation. Typically, the loss rate of labelled cells δ appears to be several-fold higher than the estimated rate p at which cells proliferate. Although at first sight this may seem surprising for a population at steady state, the kinetics of cells that have recently divided (and hence picked up label) may be intrinsically different from those that have not [37,39,40]. It has been shown that cells that have recently divided are more likely to undergo activation-induced cell death than cells that have not. Even in the absence of such differences, the loss of label may exceed the accrual of label, because the uplabelling phase is representative of the cell population as a whole, including cells that will and cells that will not go into division during the labelling period. Loss rates, on the other hand, are based on the loss of cells that have picked up the label, and hence only involve the part of the lymphocyte pool that has recently divided. As a consequence, especially in short-term labelling experiments, rapidly turning over cells are over-represented during the downlabelling phase. Long labelling periods will give rise to lower rates of T cell loss, because the population that has picked up the label becomes more representative of the T cell population as a whole. Indeed, meta-analysis of stable isotope labelling studies with different labelling periods showed a negative correlation between the length of the labelling period and the estimated death rate [37]. Importantly, however, the average proliferation rate – which is estimated from the uplabelling phase – should in principle not be affected by the length of the labelling period.

Although stable isotope labelling has provided a large step forward in the analysis of lymphocyte dynamics, quite large discrepancies have been observed between stable isotope labelling studies from different laboratories [41]. Despite the fact that average proliferation rates should not depend on the length of the labelling period, there seems to be a tendency for deuterated glucose experiments, which typically have short labelling periods, to give rise to higher average proliferation rates than studies using deuterated water, which is typically administered for much longer periods of time. The source of this discrepancy is the subject of active research.

Mathematical and Statistical Methods of Parameter Estimation

The basis of all of the work discussed in the last Section is regression. In each case a model is formulated, usually from first principles based on an understanding of the system, and then fitted to the experimental data in order to estimate the kinetic parameters using regression. In this Section we discuss the basic techniques of model formulation and fitting for the purposes of parameter estimation focussing on the method of least squares. Our emphasis is on a practical rather than a theoretical approach. Basic techniques will be illustrated via an example taken from stable-isotope labelling studies (see the worked example later in this chapter).

Model Formulation and Selection

The fundamental requirement of a mathematical model for analysing lymphocyte kinetics is that it predicts the observed state variable(s), e.g. the fraction of labelled lymphocytes, as a function of the parameter(s) that we wish to estimate, for example, the proliferation rate. The most appropriate type of model will depend on the system being analysed: difference equations for a synchronised population, partial differential equations for a population with continuous spatial variation, multi-compartment ordinary differential equations for a population with discrete spatial variation and ordinary differential equations for a population where spatial homogeneity is assumed. It is not necessary for the model to be soluble analytically in order to fit it to experimental data. Whilst it is tempting to construct models that reflect the true complexity of the biological system, there is invariably a trade-off between the complexity of the model, in particular the number of free parameters of the model, and the ability to estimate the parameter(s) of interest. As a minimum requirement, the number of degrees of freedom D, where

$$D = \text{number of data points } N - \text{ number of free parameters } P$$

must be greater than zero and ideally should be considerably higher than zero (see the Section on Parameter Identifiability). The biological world is complex and a "true" model of an in vivo biological system would need a very large, arguably infinite, number of parameters to describe it fully. Such a model could never be used to estimate parameters from a finite data set. A model for parameter estimation should capture all phenomena likely to impact significantly on the observable being predicted but introducing further complexity is usually detrimental. When more than one model is possible, model selection should first be based on biological plausibility. If there are alternative models that are all biologically realistic then an objective selection can be made based on the goodness of fit of the models to the data. The goodness of fit of the models i.e. the discrepancy between the observed and the predicted values (often measured as the sum of squared residuals, that is the sum of the squares of the differences between the observed and predicted data) can be tested via the F test for nested models or the Akaike Information Criterion in the general case [42].

Parameter Identifiability

A parameter is identifiable if the measurements made of the state variable contain sufficient information to allow unique and accurate estimation of the parameter. Identifiability encompasses two concepts: theoretical identifiability and practical identifiability. Theoretical identifiability analyses parameter uniqueness based on the model structure and schedule of measurements, it assumes that the measurements are error free. Practical identifiability analyses parameter estimate accuracy also taking into account measurement noise. Theoretical identifiability is a necessary but not sufficient condition for practical identifiability. Theoretical identifiability (strictly speaking local theoretical identifiability) is calculated from the Jacobian or parameter sensitivity matrix. If X is the state variable observed at N timepoints $t_1, t_2, \ldots t_N$ and $\underline{\theta}$ is the vector of P parameters $\underline{\theta} = (\theta_1, \theta_2, \ldots, \theta_P)$ then the Jacobian J is the $N \times P$ matrix

$$J = \begin{bmatrix} \left.\frac{\partial X}{\partial \theta_1}\right|_{t=t_1} & \left.\frac{\partial X}{\partial \theta_2}\right|_{t=t_1} & \cdots & \left.\frac{\partial X}{\partial \theta_P}\right|_{t=t_1} \\ \left.\frac{\partial X}{\partial \theta_1}\right|_{t=t_2} & \left.\frac{\partial X}{\partial \theta_2}\right|_{t=t_2} & \cdots & \left.\frac{\partial X}{\partial \theta_P}\right|_{t=t_2} \\ \vdots & & & \\ \left.\frac{\partial X}{\partial \theta_1}\right|_{t=t_N} & \left.\frac{\partial X}{\partial \theta_2}\right|_{t=t_N} & \cdots & \left.\frac{\partial X}{\partial \theta_P}\right|_{t=t_N} \end{bmatrix}. \tag{7.14}$$

If the product of the transpose of the Jacobian with the Jacobian, $J^{\mathrm{T}}J$, is evaluated for a given choice of parameters and written in reduced row echelon form [43] then the rows that are zero except for the diagonal indicate an identifiable parameter with that row index.

Practical identifiability can be assessed via the covariance matrix C

$$C = \sigma^2 (J^{\mathrm{T}} J)^{-1}, \tag{7.15}$$

where $\sigma = (\frac{\sum_{i=1}^{N}(X_i - \hat{X}_i)^2}{N-P})^{\frac{1}{2}}$ and $\hat{X}_i$ is the predicted value of the state variable X_i at time i. The jth diagonal element of the covariance matrix, C_{jj}, approximates the variance of the estimator of the jth parameter θ_j (see the Section below on Estimating parameter errors). For the assumed noise and parameter choice this will thus provide a direct estimate of the accuracy with which a parameter can be estimated. Both theoretical and (provided a reasonable estimate of the size of the parameters and measurement error can be made) practical identifiability can be assessed prior to data collection and this should be done wherever possible to ensure that the experiment design (choice of observables and schedule of measurements) will allow identification of the parameter(s) of interest with sufficient accuracy. There are a number of factors that reduce parameter identifiability and these should be identified and minimised before conducting the experiment.

Degrees of Freedom

The variance of a parameter estimator is approximately inversely proportional to the number of degrees of freedom $N - P$. Identifiability can therefore be increased by decreasing the number of free parameters P or increasing the number of measurements N. The number of parameters can be reduced by reducing the complexity of the model or by replacing free parameters by numerical estimates where prior information is available. The number of data points can be increased by changing the experiment design to include more time points or measuring a greater number of different state variables or, if data is available for a number of individuals, using population (mixed effects) methods. The relative efficiency of the different options for increasing the effective number of degrees of freedom depends on the problem in hand. Population methods involve pooling all data from a number of individuals, which greatly increases the number of data points, and then fitting assuming that these parameters are drawn from a single distribution, so instead of needing to estimate k separate parameters for each of k individuals and for each parameter of interest (i.e. a total of kP parameters) it is only necessary to estimate one or two parameters that describe the parameter distribution, e.g. the mean and standard deviation for each parameter of interest (i.e. a total of P or $2P$ parameters). A detailed description of population methods is beyond the scope of this chapter and interested readers are referred to one of a large number of excellent textbooks including [44–46].

Sensitivity of Observations to Parameter Change

Clearly if a parameter is to be determined from the behaviour of an observable it is essential that the observable is sensitive to changes in the parameter. Sensitivity can only be optimised by a change of experiment design, e.g. using a different observable or, more realistically, using different timepoints where the observable is more sensitive to parameter changes.

Parameter Correlations

Even if an observable is sensitive to changes in a parameter it is often the case that the observable will also depend on other parameters of the model and thus there may not be a unique combination of parameters that give rise to a certain observed value. For example if

$$\text{fraction of labelled cells at time } t = (\text{proliferation rate} - \text{death rate}) \times t,$$

then a given time course of the fraction of labelled cells can be attained for a range of different combinations of proliferation and death, and the proliferation and death rates are said to be correlated. In this case it will not be possible to distinguish, based solely on the goodness of fit of the model to the data, which is the true value of proliferation and death. The correlation between the ith and jth parameters is approximated by the ith$j$th element of the correlation matrix:

$$\rho_{ij} = \frac{C_{ij}}{(C_{ii}C_{jj})^{0.5}},$$

where C is the covariance matrix defined previously. In the above example, proliferation rate and death rate would be perfectly correlated, that is the correlation would be $+1$. High (magnitude of) correlations between parameters can be reduced by simplifying the model, by changing the measurement schedule to improve parameter separability or by constructing alternative parameters that are a combination or transformation of the highly correlated parameters, e.g. in the above example neither proliferation rate nor death rate can be accurately estimated (whatever the measurement schedule in this case) but the "net growth rate"

$$\text{net growth rate} = \text{proliferation rate} - \text{death rate}$$

can be estimated.

Model Fitting

We restrict ourselves to the case where both the predictor (independent) variables and the state (dependent) variables are continuous. If a model is linear in the free parameters then analytical multiple linear regression should be used to estimate the parameters (see any introductory statistics book, e.g. [45, 47]). In general, if models cannot or should not (because of bias) be transformed to a linear form then non-linear regression should be used to the fit the model to the data. There are two basic statistical frameworks for fitting models to data: the frequentist (maximum likelihood) approach and the Bayesian approach. The well known least squares method is equivalent to a special case of the maximum likelihood approach.

Least Squares Estimate of Parameters

Consider a system with an observable X measured at time t so we have a set of N observations,

$$\{(t_1, X_1), (t_2, X_2) \cdots (t_n, X_N)\},$$

and we believe that this system is described by a model $f(t, \underline{\theta})$ where $\underline{\theta}$ is the vector of parameters of the model so that

$$X_i = f(t_i, \underline{\theta}) + \epsilon_i \qquad\qquad i = 1, \ldots N,$$

where ϵ_i is the random noise on the ith measurement. The least squares estimator of the parameters is the vector of parameters that minimises the sum of the squared differences between the observed and the predicted variables, i.e. that minimises $\sum_{i=1}^{N}(X_i f(t_i, \underline{\theta}))^2$. In general this minimum cannot be found analytically and it is necessary to use numerical searching algorithms (see the Section on Choice of software). When searching numerically for minima it is important to check that the true global minimum rather than a local minimum has been found. It is therefore advisable to start the search from a number of different initial conditions and to check that they find the same global minimum. Additionally, the convergence criteria of the search algorithms should be checked.

Maximum Likelihood Estimate of Parameters

For the system described above we wish to estimate the parameters θ given the observations, i.e. we want to find θ that is most likely to describe the system given the set of observations and the model. The likelihood of $\underline{\theta}$ given $\{t_i, X_i\}$ and f is denoted $L(\underline{\theta}|t_i, X_i, f)$ and can be calculated from:

$$L(\underline{\theta}|\{t_i, X_i\}, f) = \prod_i P(X_i|\underline{\theta}, t_i),$$

giving

$$\ln[L(\underline{\theta}|\{t_i, X_i\}, f)] = \sum_{i=1}^{N} \ln[P(X_i|\underline{\theta}, t_i)]. \tag{7.16}$$

If the errors are independent and normally distributed with mean zero and constant variance σ^2 then

$$X_i \sim \text{Normal}(f(t_i, \underline{\theta}), \sigma^2), \qquad \text{since } f(t_i, \underline{\theta}) \text{ is constant for given } i.$$

The probability density function for the Normal distribution is

$$P(X_i|\underline{\theta}, t_i) = \frac{1}{\sigma\sqrt{2\pi}} \exp\left(\frac{-(X_i - f(t_i, \underline{\theta}))^2}{2\sigma^2}\right). \tag{7.17}$$

So, substituting (7.17) into equation (7.16) yields

$$\ln[L(\underline{\theta}|\{t_i, X_i\}, f)] = N \ln \frac{1}{\sigma\sqrt{2\pi}} - \sum_i \frac{(X_i - f(t_i, \underline{\theta}))^2}{2\sigma^2}.$$

And it can be seen that the likelihood is maximised (or equivalently the $\ln[L]$ is maximised since the logarithm function is monotonic) when $\sum_i \frac{(X_i - f(t_i, \underline{\theta}))^2}{2\sigma^2}$ is minimised and the problem is reduced to the standard least squares problem. That is, when the errors are independent and normally distributed the maximum likelihood estimator of a parameter is equal to the least squares estimator.

Bayesian Parameter Inference

In the Bayesian approach parameter inference is informed not just by the data but by prior knowledge of the parameters, entered as prior distributions. The influence of prior information will depend on our confidence in the information as well as the size of the current data set. Prior data will naturally be most influential when the data set is small and the prior is strong. In the case when the prior is non-informative and the data set is infinite the Bayesian estimator is numerically equivalent to the maximum likelihood estimator. The advantages of a Bayesian approach include the ability to incorporate prior information, removal of the assumption that parameters are drawn from a normal distribution and in many cases a simpler implementation for fitting complex models with a hierarchical structure. There are numerous books that discuss Bayesian methods, those that emphasise a practical treatment include Congdon et al. [44] and Gilks et al. [46].

Choice of Software

There are numerous software packages that can be used to automate model fitting via least squares and just a few of the more popular options are listed here. As with all software there is an inverse correlation between the power and flexibility of the language and the initial investment of time to become competent in handling the software. Packages with a short learning curve include SPSS and ScoP. SPSS and SCoP are both difficult to run in batch mode so fitting a large number of datasets sequentially can involve a lot of manual data handling. Importantly, SPSS does not have a numerical differential equation solver and so cannot be used for models without an analytical solution. Probably the best balance of relatively short learning curve with software flexibility and ease of automation is offered by dedicated high level mathematical or statistical languages such as Maple (where the global optimisation package needs to be purchased as an additional add-on), Mathematica or R. For optimal flexibility and speed (balanced by a longer learning curve) the mid-level language C or C++ should be used. Software that implements a general maximum likelihood (i.e. non-normal errors) or Baysian approach are less common. Both R and C/C++ are sufficiently flexible to enable maximum likelihood or Bayesian methods, alternatively for Bayesian methods the BUGS software can be used. The software described can be downloaded from the following sites; R, C/C++ and BUGS are free.

SPSS	http://www.spss.com/
SCoP	http://www.simresinc.com/
Maple	http://www.maplesoft.com/
Mathematica	http://www.wolfram.com/
R	http://cran.r-project.org/
C / C++	http://gcc.gnu.org/
BUGS	http://www.mrc-bsu.cam.ac.uk/bugs/

Estimating Parameter Errors

There are two methods that are widely used to estimate the errors on parameter estimates: the asymptotic covariance matrix method and the bootstrap method.

Asymptotic Covariance Matrix Method

The asymptotic covariance matrix (ACM) method is based on the calculation of the covariance matrix defined in (7.15). It is only strictly true for models that are linear in the parameters of interest and when the errors on the measurements are independent and identically distributed; however if the deviation from linearity is "small" the ACM method also provides a reasonable approximation to the parameter error for nonlinear models. The covariance matrix (also known as the variance–covariance matrix or the inverse of the Fisher information matrix) is calculated from the Jacobian. The diagonal elements of the covariance matrix are the variances of the parameter estimates.

Bootstrap Method

Bootstrapping is a framework for statistical inference based on resampling and has widespread applicability beyond regression [48, 49]. In the field of regression, bootstrapping can be used to estimate the variance(s) of parameter estimates with no restrictions on the nature of the model. Here we describe case-resampling (bootstrapping the data) rather than model-based resampling (bootstrapping of the residuals).

Consider a system where there are N measurements of an observable X_i made for N predictor values t_i, that is there are N pairs (t_i, X_i). A bootstrap estimate of the variance(s) of the model parameter(s) is made by creating a bootstrap sample of size N by random sampling with replacement from the original dataset, that is from the set of N pairs (t_i, X_i). The model is fitted to the bootstrap sample and the parameters of interest estimated. This is repeated R times (typically R is of the order of 100) and the distribution of the parameter estimates constructed. The bootstrap estimate of the parameter variance is simply the variance of the bootstrapped parameters over the R runs. The number of bootstrap runs R that need to be performed can be ascertained by checking for the stabilisation of the variance with increasing R.

Choice of Method

Both the bootstrap method and the ACM method are easy to implement although the bootstrap method can be computationally intensive; which method is optimal depends on the model and data of interest. For any given model, parameter values and data schedule, the deviation from linearity can be assessed by calculating the intrinsic nonlinearity and the root mean square curvature. Details of the calculation are beyond the scope of this chapter, see [50] for an excellent description. However, whilst the curvature and intrinsic nonlinearity need to be "small" for the linear approximation of the ACM method to be applicable, how small is rather poorly defined. In practice it is often more useful to determine the optimal method of error calculation by generating "data" for known parameters using the model, adding noise by random sampling from a plausible distribution, fitting the model to the "data" and estimating the errors using both the bootstrap method and the ACM method. How often the known parameter lies within the confidence interval of the estimate of the parameter can then be assessed.

A Worked Example

As a concrete example we provide a step-by-step explanation of the fitting of the model to describe deuterated glucose labelling (see (7.11)) to experimental data taken from Macallan [29] and reproduced in Table 7.1. The data were obtained by infusing an individual with deuterated glucose for 1 day and then quantifying

Table 7.1 Deuterated glucose labelling data. The fraction of labelled DNA fragments in the $CD8^+CD45RO^+$ T cell population sorted from PBMC at successive timepoints following 1 day labelling starting at time 0. The standard deviation of three technical replicates is shown. Data is reproduced from Macallan et al. [29]

Time (days)	Fraction of labelled DNA fragments, l	Standard deviation
4	0.0141	0.0017
10	0.0093	0.0008
16	0.0096	0.0008

the fraction of labelled DNA fragments in $CD8^+CD45RO^+$ T lymphocytes from peripheral blood mononuclear cells. Measurements were made at three time points during the delabelling phase; the standard deviation of three technical replicates is provided.

Parameter Identifiability

If the measurement schedule is to be 4 days, 10 days and 16 days then, prior to data collection, we should investigate parameter identifiability to assess whether the parameters of interest can be estimated with the required accuracy.

Theoretical identifiability: Rewriting the problem in the notation of (7.14), we have $\underline{\theta} = (p, \delta)$, $t_1 = 4$, $t_2 = 10$, $t_3 = 16$ and $X = \frac{p}{\delta}(1 - e^{-\delta 1})e^{-\delta(t-1)}$. The expression for X is the solution (7.12) of the model (7.11), where the labelling period $\tau = 1$; as all the time points are taken during the delabelling phase after label administration we only use the second half of the solution. Substituting this into the expression for the Jacobian (7.14) yields

$$J = \begin{pmatrix} \left.\dfrac{(1-e^{-\delta})e^{-\delta(t-1)}}{\delta}\right|_{t=4} & \dfrac{pe^{-\delta}e^{-\delta(t-1)}}{\delta} - \dfrac{p(1-e^{-\delta})(t-1)e^{-\delta(t-1)}}{\delta} - \left.\dfrac{p(1-e^{-\delta})e^{-\delta(t-1)}}{\delta^2}\right|_{t=4} \\ \vdots & \vdots \end{pmatrix}$$

$$= \begin{pmatrix} \dfrac{(1-e^{-\delta})e^{-3\delta}}{\delta} & \dfrac{pe^{-\delta}e^{-3\delta}}{\delta} - \dfrac{3p(1-e^{-\delta})e^{-3\delta}}{\delta} - \dfrac{p(1-e^{-\delta})e^{-3\delta}}{\delta^2} \\ \vdots & \vdots \end{pmatrix}.$$

$J^{\mathrm{T}}J$ is then the 2×2 matrix with the following elements

$$J^{\mathrm{T}}J_{11} = \sum_{i=1}^{3} J_{i1}^2$$
$$= \sum_{t=4,6,10} \left(\frac{(1-\mathrm{e}^{-\delta})\mathrm{e}^{-\delta(t-1)}}{\delta} \right)^2$$
$$J^{\mathrm{T}}J_{12} = \sum_{i=1}^{3} J_{i2}J_{i1} \qquad \text{etc.}$$

Rewiting $J^{\mathrm{T}}J$ in reduced row echelon form (i.e. applying elementary matrix operations to the rows and columns of $J^{\mathrm{T}}J$ until it is transformed into a matrix with all of the nonzero rows preceding all of the zero rows, the first nonzero element in each row is 1, the first non-zero element in each row appears in the column to the right of the first nonzero element in the preceding row, and the first nonzero element in a row is the only nonzero element in its column) gives

$$\begin{pmatrix} 1 & 0 \\ 0 & 1 \end{pmatrix},$$

indicating that the parameters with index 1, i.e. p, and index 2, i.e. δ, are theoretically identifiable.

Practical identifiability: In order to assess the practical identifiability prior to data collection it is necessary to make estimates of the error σ and the parameters p and δ. If the discrepancy between the observed and predicted value of the state variable, $X_i \hat{X}_i$, is estimated to be, for example, 10% of the predicted value at each time point, then

$$\sigma = \left(\frac{1}{3-2} \sum_{i=1}^{3} (0.1\hat{X}_i)^2 \right)^{1/2}$$

and the variance of the estimates of p and δ can be calculated from the diagonal elements of the covariance matrix C (7.15).

As an example we investigate our ability to estimate p. By examining the expression for the variance of p it can be seen that the standard error (square root of the variance) is linear in p and so the coefficient of variation (standard error expressed as a fraction of p) is independent of p. The numerical value of the coefficient of variation of p is plotted as a function of δ in Fig. 7.5 (red line). Previous studies typically measure δ to be in the range 0–0.3 day^{-1} so we focus on this range. Figure 7.5 reveals that for δ in the range 0–0.3 day^{-1}, the coefficient of variation is between 20–40%. This is rather a large coefficient of variation and it is worth considering how the experimental design can be modified to reduce the error on the estimates. Of course experimental design will be constrained by logistic, financial and, in the case of human and animal experiments, ethical considerations. It is essential that these

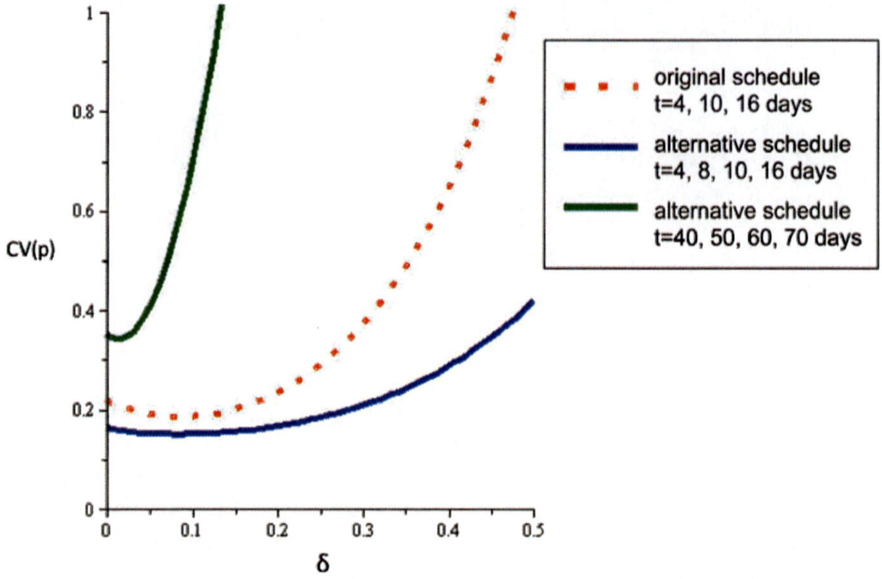

Fig. 7.5 The coefficient of variation of the estimate of p (i.e. standard error of the estimate divided by the estimate) is plotted for three different measurement schedules. *Red*: original measurement schedule t =4, 10, 16 days. *Blue*: alternative measurement schedule t =4, 8, 10, 16 days. *Green*: alternative measurement schedule t =40, 50, 60, 70 days. It can clearly be seen that the first alternative measurement schedule (t =4, 8, 10, 16 days) gives systematically lower errors in p than the original schedule for a wide range of values of δ, whilst the second alternative measurement schedule (t =40, 50 , 60, 70 days) gives systematically higher errors in p than the original schedule

points are considered when designing an experiment, because there is no point in designing the perfect experiment for parameter estimation if the experiment cannot be conducted! Conversely however, there is no point in conducting an experiment to estimate parameters if the expected error on the estimate is unacceptably high. In the example considered, the measurements are taken at days 4, 10, and 16; if the experimental design is adapted to include one extra time point at day 8, then it can be seen (Fig. 7.5, blue line) that the coefficient of variation of p falls for all values of δ considered and is between 15 and 20% for δ in the range 0–0.3 day^{-1}. This improvement in our ability to estimate p is due to the increased number of degrees of freedom as a consequence of having four data points rather than three. However, the choice of time points is also important. It is not the case that any choice of 4 time-points will give a reduced coefficient of variation of p. For example, an alternative schedule of day 40, 50, 60, and 70 increases the coefficient of variation dramatically (Fig. 7.5, green line). This is because at later time points the observed variable, i.e. the fraction of labelled DNA, is low and rather insensitive to changes in p. This preliminary analysis suggests that it may be worth considering taking an extra time point if possible, as it is likely to significantly improve the accuracy with which p can be measured. However, later time points contribute very little to the estimates and so are effectively a waste of resources. Before making any final decisions regarding experiment design, it would be prudent to consider a wider range of possible measurement schedules. Of course a comprehensive analysis is best conducted by studying the analytical expressions for the coefficient of variation.

Model Fit

We assume that in this case it was not possible to adapt the measurement schedule and the experiment proceeded with the original design yielding the data in Table 7.1. If we fit the solution (7.12) of the model (7.11) to the data by least squares regression we estimate that $p = 0.015\,\text{day}^{-1}$ and $\delta = 0.037\,\text{day}^{-1}$. So the doubling time of $CD8^+CD45RO^+$ T lymphocytes is $\ln(2)/0.015 = 46$ days and the half-life of labelled $CD8^+CD45RO^+$ T lymphocytes is $\ln(0.5)/0.037 = 18.7$ days. That is, in the absence of cell death, the $CD8^+CD45RO^+$ T-lymphocyte population would double in just over a month and (in the absence of cell proliferation) the labelled $CD8^+CD45RO^+$ T-lymphocyte population would halve in just over a fortnight. The fit of the model to the data is given in Fig. 7.6 (green line). To calculate the standard errors on the estimates via the ACM method we substitute our estimates of p and δ into the expression for the covariance matrix C (7.15). To ensure accuracy we retained a high number of decimal places on p and δ; specifically we use $p = 0.015343$ and $\delta = 0.036929$. We find that the error is $\sigma = 0.00182$ and

$$C = \begin{pmatrix} 0.00000791 & 0.0000487 \\ 0.0000487 & 0.000408 \end{pmatrix}.$$

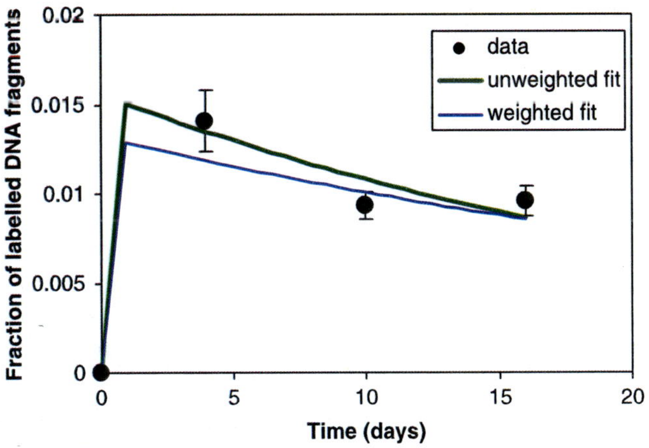

Fig. 7.6 The least squares fit of the model describing deuterated glucose labelling, (7.10), to the experimental data in Table 7.1

The standard errors on the estimates are therefore 0.003 on p and 0.020 on δ. Using the bootstrap method with 30 random samples, the errors are 0.002 and 0.016 respectively (the bootstrap method relies on random resampling, so do not be concerned if your estimates of the bootstrap errors differ slightly). Examining the covariance matrix, it can be seen that the correlation between p and δ is 0.857. The fact that the correlation is positive indicates that if p is overestimated then δ will be overestimated. The fact that the correlation is rather high indicates that it is relatively difficult to estimate p and δ uniquely as they tend to compensate for each other, i.e. a high estimate of p and a high estimate of δ will give a similar labelling pattern to a lower estimate of p and a lower estimate of δ, making it difficult to distinguish between the pairs of values. This is manifest as rather high errors on the estimates of p and δ.

Weighted Fit

In this experiment technical replicates were performed. The standard deviation of the experimental data points can be taken as a reflection of our confidence in those points. When fitting the model to the data we may wish to take into account our confidence in the points; this can be achieved by weighting the fit by an appropriate value, for instance the inverse of the variance of the replicates, i.e. instead of minimizing $\sum_{i=1}^{N}(X_i - f(t_i, \underline{\theta}))^2$, we minimize $\sum_{i=1}^{N} w_i(X_i - f(t_i, \underline{\theta}))^2$, where w_i is the inverse of the variance of the ith measurement. Performing this weighted fit to the data in Table 7.1 yields estimates of $p = 0.013\,\text{day}^{-1}$ and δ=$0.037\,\text{day}^{-1}$. The plot of this fit (Fig. 7.6, blue line) shows that, as expected, the predicted curve is weighted towards those points in which we have the most confidence, i.e. those with a lower standard deviation.

Future Challenges

Despite significant advances that have been made during the past decade in our understanding and quantification of lymphocyte dynamics, several important questions remain. For example, even though the measurement of telemore lengths and TREC contents were originally introduced to quantify cellular proliferation and thymic output, respectively, we have shown in this chapter that the kinetic behaviour of both measures is in fact determined by a combination of production of lymphocytes in the thymus and in the periphery. It is therefore very puzzling that telomeres and TREC contents of $CD4^+$ T lymphocytes are differently affected by HIV infection. While TREC contents of $CD4^+$ T lymphocytes have been shown to be reduced in HIV-infected individuals [6, 8, 51, 52], $CD4^+$ T cell telomere lengths turned out to be normal [53]. Understanding these discrepancies is the topic of ongoing research.

Also in the field of stable isotope labelling several open questions remain. Even though stable isotope labelling forms the best method that is currently available to quantify lymphocyte dynamics, several laboratories have reported quite different estimates of lymphocyte turnover. In fact, estimated proliferation and death rates based on heavy water and deuterated glucose labelling differ systematically, with higher values being obtained with deuterated glucose and with shorter labelling times [41]. The cause of these discrepancies between studies is currently being investigated.

Acknowledgements Our thinking on lymphocyte dynamics has been influenced by a large number of people. We would particularly like to acknowledge the significant impact of Charles Bangham, Rob de Boer, Vitaly Ganusov, Derek Macallan, Frank Miedema, Kiki Tesselaar, Rodolphe Thiébaut and Luc Willems. We are grateful to the Wellcome Trust, the Medical Research Council, Research Councils UK, the Engineering and Physical Sciences Research Council, the Leverhulme Trust, and the Netherlands Organisation for Scientific Research for financial support.

References

1. Mackall C, Fleisher T, Brown M, Andrich M, Chen C, Feuerstein I, Horowitz M, Magrath I, Shad A, Steinberg S, Wexler L, Gress R (1995) Age, thymopoiesis, and CD4$^+$ T-lymphocyte regeneration after intensive chemotherapy. N Engl J Med 332:143
2. Weinberg K, Annett G, Kashyap A, Lenarsky C, Forman S, Parkman R (1995) The effect of thymic function on immunocompetence following bone marrow transplantation. Biol Blood Marrow Transplant 1:18
3. Michie C, McLean A, Alcock C, Beverley P (1992) Lifespan of human lymphocyte subsets defined by CD45 isoforms. Nature 360:264–265
4. Combadière B, Blanc C, Li T, Carcelain G, Delaugerre C, Calvez V, Tubiana R, Debre P, Katlama C, Autran B (2000) CD4 Ki67 lymphocytes in HIV-infected patients are effector T cells accumulated in the G_1 phase of the cell cycle. Eur J Immunol 30:3598–3603
5. Kong F, Chen C, Cooper M (1998) Thymic function can be accurately monitored by the level of recent T cell emigrants in the circulation. Immunity 8:97–104
6. Douek D, McFarland R, Keiser P, Gage E, Massey J, Haynes B, Polis M, Haase A, Feinberg M, Sullivan J, Jamieson B, Zack J, Picker L, Koup R (1998) Changes in thymic function with age and during the treatment of HIV infection. Nature 396:690–695
7. Douek D, Vescio R, Betts M, Brenchley J, Hill B, Zhang L, Berenson J, Collins R, Koup R (2000) Assessment of thymic output in adults after haematopoietic stem cell transplantation and prediction of T-cell reconstitution. Lancet 355:1875–1881
8. Hazenberg M, Otto S, Stuart J, Verschuren M, Borleffs J, Boucher C, Coutinho R, Lange J, de Wit T, Tsegaye A, van Dongen J, Hamann D, de Boer R, Miedema F (2000) Increased cell division but not thymic dysfunction rapidly affects the T-cell receptor excision circle content of the naive T cell population in HIV-1 infection. Nat Med 6:1036–1042
9. Dutilh B, de Boer R (2003) Decline in excision circles requires homeostatic renewal or homeostatic death of naive T cells. J Theor Biol 224:351–358
10. Hazenberg M, Borghans J, de Boer R, Miedema F (2003) Thymic output: a bad TREC record. Nat Immunol 4:97–99
11. Harley C, Futcher A, Greider C (1990) Telomeres shorten during ageing of human fibroblasts. Nature 345:458–460
12. Blackburn E (1991) Structure and function of telomeres. Nature 350:569–573
13. De Boer R, Noest A (1998) T cell renewal rates, telomerase, and telomere length shortening. J Immunol 160:5832
14. Weng N, Levine B, June C, Hodes R (1995) Human naive and memory T lymphocytes differ in telomeric length and replicative potential. Proc Natl Acad Sci USA 92:11091

15. Counter C, Gupta J, Harley C, Leber B, Bacchetti S (1995) Telomerase activity in normal leukocytes and in hematologic malignancies. Blood 85:2315
16. Weng N, Hathcock K, Hodes R, Wolthers K, Miedema F (1998) Regulation of telomere length and telomerase in T and B cells. Immunity 9:151–157
17. Lempicki R, Kovacs J, Baseler M, Adelsberger J, Dewar R, Natarajan V, Bosche M, Metcalf J, Stevens R, Lambert L, Alvord W, Polis M, Davey R, Dimitrov D, Lane H (2000) Impact of HIV-1 infection and highly active antiretroviral therapy on the kinetics of $CD4^+$ and $CD8^+$ T cell turnover in HIV-infected patients. Proc Natl Acad Sci USA 97:13778
18. Kovacs J, Lempicki R, Sidorov I, Adelsberger J, Herpin B, Metcalf J, Sereti I, Polis M, Davey R, Tavel J, Falloon J, Stevens R, Lambert L, Dewar R, Schwartzentruber D, Anver M, Baseler M, Masur H, Dimitrov D, Lane H (2001) Identification of dynamically distinct subpopulations of T lymphocytes that are differentially affected by HIV. J Exp Med 194:1731
19. Kovacs J, Lempicki R, Sidorov I, Adelsberger J, Sereti I, Sachau W, Kelly G, Metcalf J, Davey Jr R, Falloon J, Polis M, Tavel J, Stevens R, Lambert L, Hosack D, Bosche M, Issaq H, Fox S, Leitman S, Baseler M, Masur H, Di Mascio M, Dimitrov D, Lane H (2005) Induction of prolonged survival of CD4 T lymphocytes by intermittent IL-2 therapy in HIV-infected patients. J Clin Invest 115:2139–2148
20. Di Mascio M, Sereti I, Matthews L, Natarajan V, Adelsberger J, Lempicki R, Yoder C, Jones E, Chow C, Metcalf J, Sidorov I, Dimitrov D, Polis M, Kovacs J (2006) Naive T-cell dynamics in human immunodeficiency virus type 1 infection: effects of highly active antiretroviral therapy provide insights into the mechanisms of naive T-cell depletion. J Virol 80:2665
21. Mohri H, Bonhoeffer S, Monard S, Perelson A, Ho D (1998) Rapid turnover of T lymphocytes in SIV-infected rhesus macaques. Science 279:1223
22. De Boer R, Mohri H, Ho D, Perelson A (2003) Estimating average cellular turnover from 5-bromo-2'-deoxyuridine (BrdU) measurements. Proc R Soc B Biol Sci 270:849
23. De Boer R, Mohri H, Ho D, Perelson A (2003) Turnover rates of B cells, T cells, and NK cells in simian immunodeficiency virus-infected and uninfected rhesus macaques. J Immunol 170:2479
24. Bonhoeffer S, Mohri H, Ho D, Perelson A (2000) Quantification of cell turnover kinetics using 5-bromo-2'-deoxyuridine. J Immunol 164:5049
25. Debacq C, Asquith B, Kerkhofs P, Portetelle D, Burny A, Kettmann R, Willems L (2002) Increased cell proliferation, but not reduced cell death, induces lymphocytosis in bovine leukemia virus-infected sheep. Proc Natl Acad Sci USA 99:10048
26. Hellerstein M, Neese R (1992) Mass isotopomer distribution analysis: a technique for measuring biosynthesis and turnover of polymers. Am J Physiol Endocrinol Metab 263:E988
27. Macallan D, Fullerton C, Neese R, Haddock K, Park S, Hellerstein M (1998) Measurement of cell proliferation by labeling of DNA with stable isotope-labeled glucose: studies in vitro, in animals, and in humans. Proc Natl Acad Sci USA 95:708
28. Hellerstein M, Hanley M, Cesar D, Siler S, Papageorgopoulos C, Wieder E, Schmidt D, Hoh R, Neese R, Macallan D, Deeks S, McCune J (1999) Directly measured kinetics of circulating T lymphocytes in normal and HIV-1-infected humans. Nat Med 5:83–89
29. Macallan D, Asquith B, Irvine A, Wallace D, Worth A, Ghattas H, Zhang Y, Griffin G, Tough D, Beverley P (2003) Measurement and modeling of human T cell kinetics. Eur J Immunol 33:2316–2326
30. Macallan D, Wallace D, Zhang Y, de Lara C, Worth A, Ghattas H, Griffin G, Beverley P, Tough D (2004) Rapid turnover of effector-memory $CD4^+$ T cells in healthy humans. J Exp Med 200:255
31. Wallace D, Zhang Y, Ghattas H, Worth A, Irvine A, Bennett A, Griffin G, Beverley P, Tough D, Macallan D (2004) Direct measurement of T cell subset kinetics in vivo in elderly men and women. J Immunol 173:1787
32. McCune J, Hanley M, Cesar D, Halvorsen R, Hoh R, Schmidt D, Wieder E, Deeks S, Siler S, Neese R, Hellerstein M (2000) Factors influencing T-cell turnover in HIV-1-seropositive patients J Clin Invest 105(5):R1–R8

33. Mohri H, Perelson A, Tung K, Ribeiro R, Ramratnam B, Markowitz M, Kost R, Hurley A, Weinberger L, Cesar D, Hellerstein M, Ho D (2001) Increased turnover of T lymphocytes in HIV-1 infection and its reduction by antiretroviral therapy. J Exp Med 194:1277
34. Neese R, Misell L, Turner S, Chu A, Kim J, Cesar D, Hoh R, Antelo F, Strawford A, McCune J, Christiansen M, Hellerstein M (2002) Measurement in vivo of proliferation rates of slow turnover cells by 2H2O labeling of the deoxyribose moiety of DNA. Proc Natl Acad Sci USA 99:15345
35. Hellerstein M, Hoh R, Hanley M, Cesar D, Lee D, Neese R, McCune J (2003) Subpopulations of long-lived and short-lived T cells in advanced HIV-1 infection. J Clin Invest 112:956
36. Vrisekoop N, den Braber I, de Boer A, Ruiter A, Ackermans M, van der Crabben S, Schrijver E, Spierenburg G, Sauerwein H, Hazenberg M, de Boer R, Miedema F, Borghans J, Tesselaar K (2008) Sparse production but preferential incorporation of recently produced naive T cells in the human peripheral pool. Proc Natl Acad Sci USA 105:6115
37. Asquith B, Debacq C, Macallan D, Willems L, Bangham C (2002) Lymphocyte kinetics: the interpretation of labelling data. Trends Immunol 23:596–601
38. Ribeiro R, Mohri H, Ho D, Perelson A (2002) In vivo dynamics of T cell activation, proliferation, and death in HIV-1 infection: why are $CD4^+$ but not $CD8^+$ T cells depleted? Proc Natl Acad Sci USA 99:15572
39. Grossman Z, Herberman R, Dimitrov D (1999) T cell turnover in SIV infection. Science 284:555
40. Asquith B, Bangham C (2003) An introduction to lymphocyte and viral dynamics: the power and limitations of mathematical analysis. Proc R Soc B Biol Sci 270:1651
41. Borghans J, de Boer R (2007) Quantification of T-cell dynamics: from telomeres to DNA labeling. Immunol Rev 216:35–47
42. Burnham K, Anderson D (2002) Model selection and multimodel inference: a practical information-theoretic approach. Springer, New York
43. Boas M (1983) Mathematical methods in physical sciences. John Wiley & Sons, New York
44. Congdon P (2003) Applied bayesian modelling. John Wiley and Sons Inc, Chichester, UK
45. Dobson A (2001) An introduction to generalized linear models. CRC, Boca Raton, FL
46. Gilks W, Spiegelhalter D (1996) Markov chain monte carlo in practice. Chapman and Hall/CRC, London
47. Armitage P, Berry G, Matthews J (2002) Statistical methods in medical research. Blackwell, Oxford, UK
48. Chernick M (1999) Bootstrap methods: a practitioner's guide. Wiley, New York
49. Efron B, Tibshirani R (1997) An introduction to the bootstrap. Chapman and Hall, New York
50. Bates D, Watts D (1988) Nonlinear regression analysis. J R Stat Soc B 22:41–88
51. Zhang L, Lewin S, Markowitz M, Lin H, Skulsky E, Karanicolas R, He Y, Jin X, Tuttleton S, Vesanen M, Spiegel H, Kost R, van Lunzen J, Stellbrink H, Wolinsky S, Borkowsky W, Palumbo P, Kostrikis L, Ho D (1999) Measuring recent thymic emigrants in blood of normal and HIV-1-infected individuals before and after effective therapy. J Exp Med 190:725
52. Harris J, Hazenberg M, Poulin J, Higuera-Alhino D, Schmidt D, Gotway M, McCune J (2005) Multiparameter evaluation of human thymic function: interpretations and caveats. Clin Immunol 115:138–146
53. Wolthers KC, Bea G, Wisman A, Otto SA, de Roda Husman AM, Schaft N, de Wolf F, Goudsmit J, Coutinho RA, van der Zee AG, Meyaard L, Miedema F (1996) T cell telomere length in HIV-1 infection: no evidence for increased $CD4^+$ T cell turnover. Science 274:1543

Chapter 8
Continuous-Time Birth and Death Processes: Diversity Maintenance of Naïve T Cells in the Periphery

Carmen Molina-París, Emily Stirk, Katie Quinn, and Grant Lythe

Abstract We construct a birth and death process for the number of T cells belonging to one clonotype. Cells are released from the thymus into the peripheral lymphoid organs. We assume that after this time, no more T cells of this clonotype are exported by the thymus, so that further T cells of this clonotype can only be generated by homeostatic proliferation, when a T cell receives a survival signal and undergoes a single round of cell division. We show that eventual extinction is guaranteed. The late-time behaviour of the process before extinction takes place is described by the limiting conditional probability distribution (LCD), which we prove exists. We show how approximations are related to the LCD of the original process and use them to study the LCD in two special cases.

T cells mature in the thymus, a small organ close to the heart, where they undergo two tests called positive and negative selection [1]. Cells that survive then migrate to the peripheral lymphoid organs such as the spleen and the lymph nodes where they become part of the naïve T cell repertoire, so called because the T cells have not yet encountered their specific antigen. T cells have molecules on their surface called T cell receptors (TCRs) which are responsible for recognising antigen. Each T cell expresses many identical copies of the TCR and T cells with differing TCR structures are said to belong to different clonotypes, i.e., each clonotype has one specificity of receptor. The number of T cells in a healthy adult human is approximately 10^{11} [2] and this total includes around 10^7–10^8 different clonotypes [3]. This diverse repertoire is produced by random rearrangements of the TCR genes during the T cell maturation process in the thymus [4].

Peptides are presented to T cells by specialised cells called antigen presenting cells (APCs). These peptides are displayed on the major histocompatibility complex (MHC) molecules that are found on the surface of the APC. The peptide–MHC (pMHC) complex interacts with the T cell via the TCR (see Fig. 8.1) and this encounter occurs in the lymph nodes when the T cell and the APC come into contact.

G. Lythe (✉)
Department of Applied Mathematics, University of Leeds, Leeds LS2 9JT, UK
e-mail: grant@maths.leeds.ac.uk

C. Molina-París and G. Lythe (eds.), *Mathematical Models and Immune Cell Biology*, DOI 10.1007/978-1-4419-7725-0_8,

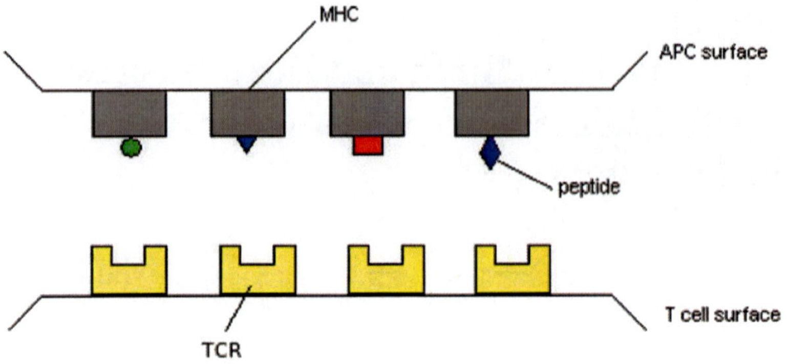

Fig. 8.1 APC and T cell surfaces. These cells interact in the lymph nodes

An APC can present around 10^3 different peptides at any given instant [5], each occurring in differing numbers, and the peptides displayed change over time. The array of peptides presented at a single point in time is referred to as an antigen presentation profile (APP) [6]. The peptides presented are mostly self-peptides, i.e., peptides derived from the body's own proteins. However, in the case of infection, foreign peptides will also be displayed. Each peptide may be capable of interacting with several T cell clonotypes and, in turn, each TCR can interact with many different peptides [7]. (If there were a one-to-one correspondence between peptides and TCRs then a very large number of T cells would be required, far greater than can be accommodated in the body.)

A diverse and balanced repertoire of T cells is essential for a functional adaptive immune response [8]. This is because the immune system is unable to predict which pathogens the organism will be exposed to during its lifetime. How is this large diversity of naïve T cells maintained over the lifetime of the host?

In the periphery, the number of naïve T cells is subject to homeostatic control [10]. This means that the number of T cells remains approximately constant and returns to the steady state following a perturbation, e.g., following infection. However, homeostatic mechanisms appear to break down in old age and the diversity of the naïve T cell repertoire declines as T cell clonotypes become extinct [2]. This may leave gaps in the repertoire and result in increased susceptibility to infection.

Peripheral naïve T cells can be produced in two ways: (1) output of new cells from the thymus and (2) homeostatic proliferation of existing cells in the periphery. Experimental evidence suggests that this proliferation occurs after the T cell receives a survival signal from an APP presenting self pMHC complexes [11]. In this chapter we use a continuous-time Markov chain to model the number of T cells belonging to a particular clonotype, which includes T cell division after receiving a survival signal from an APP and cell death. A stochastic model is more appropriate than a deterministic model because it enables us to study the probability of clonotype extinction which, as described above, is important biologically. Also, the number of T cells belonging to a given clonotype may be small, in which case stochastic fluctuations are important.

Continuous-Time Markov Chains

A stochastic process $\{\mathbf{X}_t : t \in \mathcal{T}\}$ is a collection of random variables indexed by a set $\mathcal{T}$, the elements of which usually correspond to time values. The state-space $\mathbb{S}$ of the process is the range of all possible values that the random variables $\mathbf{X}_t$ can take. In this chapter, the values of the random variables will represent the number of T cells belonging to a particular clonotype and so the state-space will be discrete. Production of a new cell or death of an existing cell may occur at any time and so the set $\mathcal{T}$ will be continuous.

A stochastic process $\{\mathbf{X}_t : t \geq t_0\}$ on the state-space $\mathbb{S} = \{0, 1, \ldots\}$ is called a continuous-time Markov chain if it satisfies the following condition. For any $0 \leq t_0 < t_1 < \ldots < t_j < t_{j+1}$,

$$\mathbb{P}(\mathbf{X}_{t_{j+1}} = n_{j+1} | \mathbf{X}_{t_0} = n_0, \ldots, \mathbf{X}_{t_j} = n_j) = \mathbb{P}(\mathbf{X}_{t_{j+1}} = n_{j+1} | \mathbf{X}_{t_j} = n_j).$$

This is the Markov property. It says that, given the current state of the process, the probability of future behaviour is not influenced by any additional knowledge of the past history of the process.

The transition probabilities are denoted by

$$p_{nm}(t_1, t_2) = \mathbb{P}(\mathbf{X}_{t_2} = m | \mathbf{X}_{t_1} = n)$$

for $t_1 < t_2$ and $n, m \in \mathbb{S}$. That is, $p_{nm}(t_1, t_2)$ is the probability that the process is in state m at time t_2 given that it was in state n at time t_1. We will say that the transition probabilities are stationary (homogeneous) if the transition probabilities depend, not on both t_1 and t_2, but only on $t_2 - t_1$. Then the notation becomes

$$p_{nm}(t_2 - t_1) = \mathbb{P}(\mathbf{X}_{t_2} = m | \mathbf{X}_{t_1} = n).$$

The transition probabilities have the property

$$\sum_{m=0}^{+\infty} p_{nm}(t) = 1 \quad \text{for } t \geq t_0, \quad n \in \mathbb{S},$$

and satisfy the equations

$$p_{nm}(t + s) = \sum_{k=0}^{+\infty} p_{nk}(t) p_{km}(s), \tag{8.1}$$

for all $s, t \in [t_0, +\infty)$ and all $n, m \in \mathbb{S}$, which are known as the Chapman–Kolmogorov equations.

Birth and Death Processes

A birth and death process is a continuous-time Markov chain where transitions are only allowed to adjacent states. The state-space of the process may be finite i.e., $\mathbb{S} = \{0, 1, \ldots, N\}$ or infinite i.e., $\mathbb{S} = \{0, 1, \ldots\}$. A birth and death process has the following transition probabilities as $\Delta t \to 0^+$:

$$\begin{aligned} p_{nm}(\Delta t) &= \mathbb{P}(\mathbf{X}_{t+\Delta t} = m | \mathbf{X}_t = n) \\ &= \begin{cases} \lambda_n \Delta t + o(\Delta t) & m = n+1 \\ \mu_n \Delta t + o(\Delta t) & m = n-1 \\ 1 - (\lambda_n + \mu_n)\Delta t + o(\Delta t) & m = n \\ o(\Delta t) & \text{otherwise,} \end{cases} \end{aligned} \tag{8.2}$$

where $f(\Delta t) = o(\Delta t)$ as $\Delta t \to 0^+$ if $\lim_{\Delta t \to 0^+} \frac{f(\Delta t)}{\Delta t} = 0$. The birth rate, λ_n, is the rate of transition from state n to $n+1$ while the death rate, μ_n, is the rate of transition from state n to $n-1$. The birth and death rates satisfy $\lambda_n \geq 0$, $\mu_n \geq 0$ for $n = 0, 1, 2, \ldots$, but $\mu_0 = 0$ so that transitions outside of the state-space cannot occur. The process can be represented as

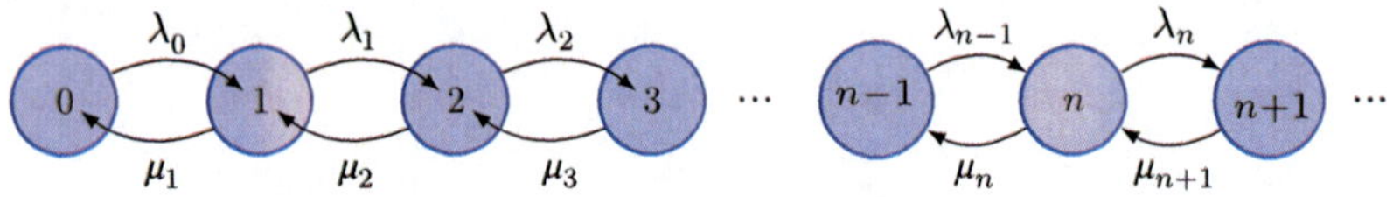

Let

$$p_n(t) = \mathbb{P}(\mathbf{X}_t = n | \mathbf{X}_{t_0} = n_0), \quad n \in \mathbb{S}, \tag{8.3}$$

which is the probability that the birth and death process is in state n at time t, given that the initial state of the process is n_0. If the state-space of the process is finite and $\lambda_N = 0$ so that transitions outside of the state-space cannot occur, the equation for the endpoint, $n = N$, is

$$p_N(t + \Delta t) = \lambda_{N-1} \Delta t p_{N-1}(t) + (1 - \mu_N \Delta t) p_N(t) + o(\Delta t). \tag{8.4}$$

Taking the limit $\Delta t \to 0^+$, results in

$$\frac{\mathrm{d}p_n(t)}{\mathrm{d}t} = \lambda_{n-1} p_{n-1}(t) + \mu_{n+1} p_{n+1}(t) - (\lambda_n + \mu_n) p_n(t), \quad 1 \leq n \leq N-1. \tag{8.5}$$

Similarly,

$$\frac{\mathrm{d}p_0(t)}{\mathrm{d}t} = \mu_1 p_1(t) - \lambda_0 p_0(t), \tag{8.6}$$

$$\frac{\mathrm{d}p_N(t)}{\mathrm{d}t} = \lambda_{N-1} p_{N-1}(t) - \mu_N p_N(t). \tag{8.7}$$

The differential equations (8.5)–(8.7) are known as the forward Kolmogorov equations.

Stationary Probability Distribution

A stationary probability distribution $\pi = (\pi_0, \pi_1, \ldots)$, where $\pi_n \geq 0$ for $n \in \mathbb{S}$, $\pi_n = 0$ for $n \notin \mathbb{S}$ and $\sum_{n=0}^{\infty} \pi_n = 1$, satisfies the equations

$$\begin{aligned} 0 &= \mu_1 \pi_1 - \lambda_0 \pi_0, \\ 0 &= \lambda_{n-1}\pi_{n-1} + \mu_{n+1}\pi_{n+1} - (\lambda_n + \mu_n)\pi_n, \qquad 1 \leq n \leq N-1, \\ 0 &= \lambda_{N-1}\pi_{N-1} - \mu_N \pi_N, \end{aligned}$$

which are the forward Kolmogorov equations with the time derivatives set to zero. If the state-space of the birth and death process $\{\mathbf{X}_t : t \geq t_0\}$ is infinite i.e., $\mathbb{S} = \{0, 1, \ldots\}$, a unique positive stationary probability distribution, π, exists if and only if [12]

$$\lambda_{n-1} > 0 \quad \text{and } \mu_n > 0 \quad \text{for } n = 1, 2, \ldots \tag{8.8}$$

and

$$\sum_{n=1}^{+\infty} \frac{\lambda_0 \lambda_1 \ldots \lambda_{n-1}}{\mu_1 \mu_2 \ldots \mu_n} < +\infty. \tag{8.9}$$

If these conditions are satisfied, the stationary probability distribution is given by

$$\pi_0 = \frac{1}{1 + \sum_{n=1}^{+\infty} \frac{\lambda_0 \lambda_1 \ldots \lambda_{n-1}}{\mu_1 \mu_2 \ldots \mu_n}}, \tag{8.10}$$

$$\pi_n = \frac{\lambda_0 \lambda_1 \ldots \lambda_{n-1}}{\mu_1 \mu_2 \ldots \mu_n} \pi_0 \quad \text{for } n = 1, 2, \ldots. \tag{8.11}$$

If the state space of the birth and death process is finite i.e., $\mathbb{S} = \{0, 1, 2, \ldots, N\}$, then a unique positive stationary probability distribution π exists if and only if

$$\lambda_{n-1} > 0 \quad \text{and } \mu_n > 0 \quad \text{for } n = 1, 2, \ldots, N.$$

Then the stationary probability distribution is given by (8.10)–(8.11), where the index n and the summation on n extend from 1 to N.

Continuous-Time Birth and Death Process with Absorbing States

If $\lambda_0 = \mu_0 = 0$ then $n = 0$ is an absorbing state, meaning that once the process reaches this state, it remains there forever. The process can be represented as

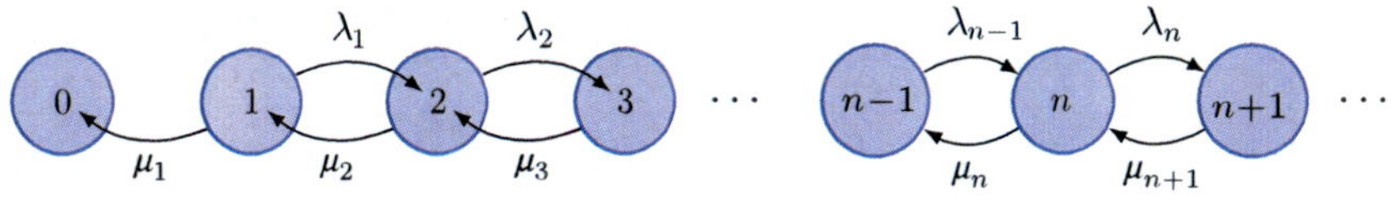

The stationary probability distribution of the process has all its mass at the absorbing state, i.e., $\pi = (1, 0, 0, \ldots)^T$. The probability of the process reaching the absorbing state is given by $\lim_{t \to +\infty} p_0(t)$. If the state-space of the process is infinite i.e., $\mathbb{S} = \{0, 1, \ldots\}$ and

$$\sum_{n=1}^{+\infty} \frac{\mu_1 \mu_2 \ldots \mu_n}{\lambda_1 \lambda_2 \ldots \lambda_n} = +\infty, \tag{8.12}$$

then $\lim_{t \to +\infty} p_0(t) = 1$, which means that eventual absorption at $n = 0$ is guaranteed [13]. On the other hand, if (8.12) does not hold, then

$$\lim_{t \to +\infty} p_0(t) = \frac{\sum_{n=m}^{+\infty} \frac{\mu_1 \mu_2 \ldots \mu_n}{\lambda_1 \lambda_2 \ldots \lambda_n}}{1 + \sum_{n=1}^{+\infty} \frac{\mu_1 \mu_2 \ldots \mu_n}{\lambda_1 \lambda_2 \ldots \lambda_n}}, \tag{8.13}$$

where m is the initial state of the process.

The Limiting Conditional Probability Distribution

Prior to extinction at $n = 0$ occurring, the probability distribution of a birth and death process may adopt a stationary shape for a long period of time, especially if the expected time until absorption is relatively large. In order to study the behaviour of the process before extinction occurs, we define the following conditional probabilities:

$$q_n(t) = \mathbb{P}(\mathbf{X}_t = n \mid \mathbf{X}_t \neq 0) = \frac{p_n(t)}{1 - p_0(t)} \quad \text{for} \quad n = 1, 2, \ldots, \tag{8.14}$$

which is the probability that the process is in state n at time t, given that absorption has not yet occurred. These conditional probabilities satisfy:

$$\begin{aligned} \frac{\mathrm{d}q_n(t)}{\mathrm{d}t} &= \frac{1}{1 - p_0(t)} \frac{\mathrm{d}p_n(t)}{\mathrm{d}t} + \frac{p_n(t)}{1 - p_0(t)} \frac{1}{1 - p_0(t)} \frac{\mathrm{d}p_0(t)}{\mathrm{d}t} \\ &= \lambda_{n-1} q_{n-1}(t) - (\lambda_n + \mu_n) q_n(t) \\ &\quad + \mu_{n+1} q_{n+1}(t) + \mu_1 q_1(t) q_n(t), \end{aligned} \tag{8.15}$$

for $n \geq 2$ and

$$\begin{aligned} \frac{\mathrm{d}q_1(t)}{\mathrm{d}t} &= \frac{1}{1 - p_0(t)} \frac{\mathrm{d}p_1(t)}{\mathrm{d}t} + \frac{p_1(t)}{1 - p_0(t)} \frac{1}{1 - p_0(t)} \frac{\mathrm{d}p_0(t)}{\mathrm{d}t} \\ &= \mu_2 q_2(t) - (\lambda_1 + \mu_1) q_1(t) + \mu_1 q_1(t) q_1(t). \end{aligned} \tag{8.16}$$

If the state space is finite, we have

$$\begin{aligned}\frac{\mathrm{d}q_N(t)}{\mathrm{d}t} &= \frac{1}{1-p_0(t)}\frac{\mathrm{d}p_N(t)}{\mathrm{d}t} + \frac{p_N(t)}{1-p_0(t)}\frac{1}{1-p_0(t)}\frac{\mathrm{d}p_0(t)}{\mathrm{d}t} \\ &= \lambda_{N-1}q_{N-1}(t) - \mu_N q_N(t) + \mu_1 q_1(t) q_N(t). \end{aligned} \tag{8.17}$$

A distribution $\bar{q}$ is called a quasi-stationary probability distribution (QSD) if it is a solution of (8.15)–(8.17) where the time derivatives are set to zero. The limiting conditional probability distribution (LCD) of the process is defined as

$$\lim_{t\to+\infty} q_n(t), \quad n \geq 1. \tag{8.18}$$

Since the LCD is independent of time, it is also a QSD. If the state-space of the process is finite, there is a unique QSD, which is also the LCD of the process. However, if the process has an infinite state-space, there may be no QSD and, if a QSD does exist, it is not necessarily unique [14]. In most cases, it is not possible to find an explicit solution for the QSD or the LCD, but numerical methods can be used (see [15] for an iterative procedure).

The LCD can be approximated analytically by setting $\mu_1 = 0$ [15]. This results in a new birth and death process which has no absorbing state and can be represented as

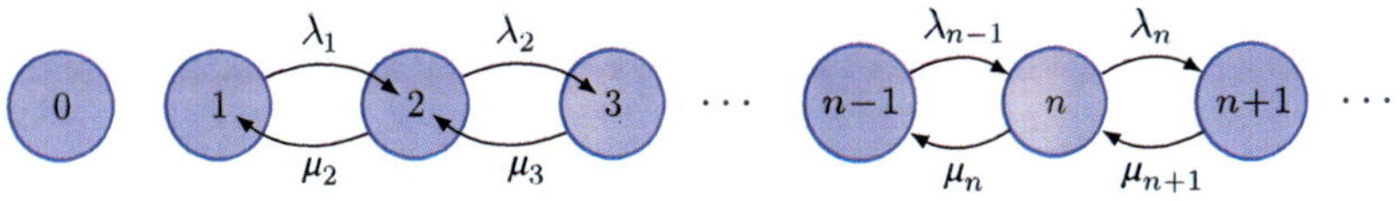

If $\lambda_{n-1} > 0$ and $\mu_n > 0$ for $n = 2, 3, \ldots,$ and

$$\sum_{n=2}^{+\infty} \frac{\lambda_1\lambda_2\ldots\lambda_{n-1}}{\mu_2\mu_3\ldots\mu_n} < +\infty \tag{8.19}$$

the new process has a unique positive stationary probability distribution which is given by

$$\begin{aligned}\pi_1^{(1)} &= \frac{1}{1+\sum_{n=2}^{+\infty}\frac{\lambda_1\lambda_2\ldots\lambda_{n-1}}{\mu_2\mu_3\ldots\mu_n}}, \\ \pi_n^{(1)} &= \frac{\lambda_1\lambda_2\ldots\lambda_{n-1}}{\mu_2\mu_3\ldots\mu_n}\pi_1^{(1)} \quad \text{for } n \geq 2. \end{aligned} \tag{8.20}$$

This distribution is an approximation to the true LCD of the process.

The LCD may also be approximated by the stationary distribution of a birth and death process which has the same birth rates as the original process, but where the death rates μ_n are replaced by μ_{n-1} to allow for one immortal individual [15]. This can be represented as

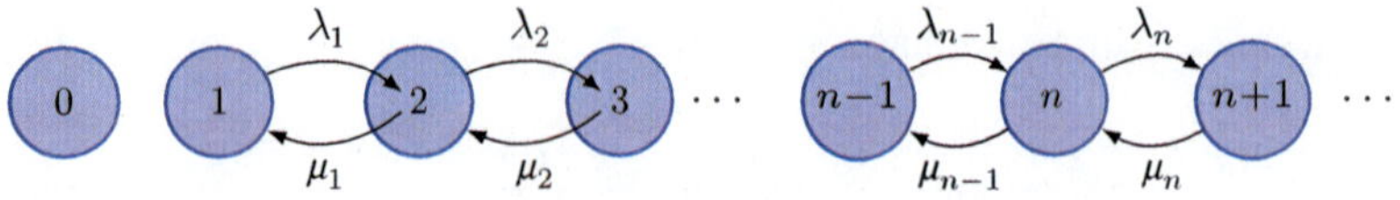

If $\lambda_n > 0$ and $\mu_n > 0$ for $n = 1, 2, \dots,$ and

$$\sum_{n=2}^{+\infty} \frac{\lambda_1 \lambda_2 \dots \lambda_{n-1}}{\mu_1 \mu_2 \dots \mu_{n-1}} < +\infty \tag{8.21}$$

this process has a unique positive stationary probability distribution which is given by

$$\pi_1^{(2)} = \frac{1}{1 + \sum_{n=2}^{+\infty} \frac{\lambda_1 \lambda_2 \dots \lambda_{n-1}}{\mu_1 \mu_2 \dots \mu_{n-1}}},$$

$$\pi_n^{(2)} = \frac{\lambda_1 \lambda_2 \dots \lambda_{n-1}}{\mu_1 \mu_2 \dots \mu_{n-1}} \pi_1^{(2)} \quad \text{for } n \geq 2. \tag{8.22}$$

In general, $\pi^{(1)}$ is a better approximation to the LCD when the mean time until extinction is long, whereas $\pi^{(2)}$ provides a better approximation when the mean extinction time is short [15].

Mathematical Model of Peripheral Maintenance

We will use the index i to label T cell clonotypes and the index q to label APPs. Let $\mathbb{C}$ be the set of all T cells in the naïve repertoire. Whether or not a T cell of clonotype i can receive a survival signal from a given APP q depends on the particular TCR it expresses on its surface. The subset $\mathbb{C}_q$ is defined to be the set of all T cells that are capable of receiving a survival signal from APP q. Also, $\mathbb{Q}$ is the set of all APPs which may occur in the periphery and $\mathbb{Q}_i$ is the subset of APPs from which T cells of clonotype i can receive a survival signal. We define $n_q = |\mathbb{C}_q|$, which is the total number of T cells that can receive a survival signal from APP q. These sets are illustrated in Fig. 8.2.

Let γ_q be the rate of survival signals from all the APCs presenting APP q, which we assume remains constant in time. We will also assume that the survival signals from any APP q are shared equally among all the T cells capable of receiving them. Now define $\lambda^{(i)}$ to be the per cell birth rate for T cells of clonotype i. Then

$$\lambda^{(i)} = \sum_{q \in \mathbb{Q}_i} \frac{\gamma_q}{|\mathbb{C}_q|}. \tag{8.23}$$

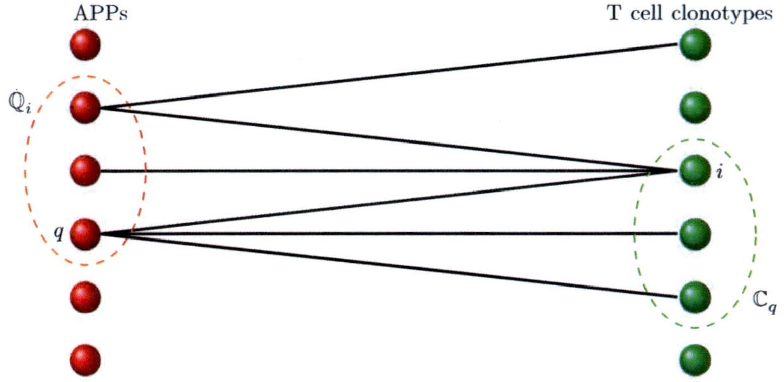

Fig. 8.2 The sets of APPs and T cell clonotypes. Each *circle* on the *left* represents an APP, while each *circle* on the *right* represents a T cell clonotype (containing all the T cells with an identical TCR). A *line* between the two *circles* indicates that T cells of this clonotype receive a survival signal from the APP

Let n_i be the number of T cells belonging to clonotype i and n_{iq} be the number of T cells not of clonotype i that receive a survival signal from an APP $q \in \mathbb{Q}_i$. Then $n_q = n_i + n_{iq}$ so that

$$\lambda^{(i)} = \sum_{q \in \mathbb{Q}_i} \frac{\gamma}{n_i + n_{iq}}, \tag{8.24}$$

where we have assumed that $\gamma_q = \gamma$ for simplicity. Next, we partition the set $\mathbb{Q}_i$ into disjoint subsets as follows:

$$\mathbb{Q}_i = \bigcup_{r=0}^{+\infty} \mathbb{Q}_{ir}, \tag{8.25}$$

where $\mathbb{Q}_{ir}$ is the set of APPs which provide survival signals to T cells of clonotype i and to r other distinct clonotypes in the repertoire. Then $\mathbb{Q}_{ir} \cap \mathbb{Q}_{ir'} = \emptyset$ for $r \neq r'$, and hence,

$$\lambda^{(i)} = \gamma \sum_{r=0}^{+\infty} \sum_{q \in \mathbb{Q}_{ir}} \frac{1}{n_i + n_{iq}}. \tag{8.26}$$

The previous equation implies that the birth rate per T cell of clonotype i depends not only on the number of T cells of clonotype i, but also on the number of T cells of any other clonotype that receives a survival signal from an APP $q \in \mathbb{Q}_i$ through the term n_{iq}.

We proceed to develop a mean field approximation to decouple the birth rates so that the expression for $\lambda^{(i)}$ depends only on the number of T cells of clonotype i.

Treating the mean number of T cells per clonotype as a parameter, $\langle n \rangle$, the approximation is [9]

$$\lambda^{(i)} = \gamma \sum_{r=0}^{+\infty} \frac{|\mathbb{Q}_{ir}|}{r\langle n \rangle + n_i}. \tag{8.27}$$

Let ν_i be the number of clonotypes that compete with T cells of clonotype i for survival signals from an APP, with the average taken over all the APPs belonging to the set $\mathcal{Q}_i$. Assuming that

$$|\mathbb{Q}_{ir}| = |\mathbb{Q}_i| \frac{\nu_i^r e^{-\nu_i}}{r!} \tag{8.28}$$

completes the model.

Summary of the Model

Using the mean field approximation, we have modelled the number of T cells belonging to a given clonotype, i, as a continuous-time birth and death process $\{\mathbf{X}_t : t \geq t_0\}$ on the state-space $\mathbb{S} = \{0, 1, 2, \ldots\}$ with the birth and death rates

$$\lambda_0 = 0, \tag{8.29}$$

$$\lambda_n = \varphi n e^{-\nu} \sum_{r=0}^{+\infty} \frac{\nu^r}{r!} \frac{1}{r\langle n \rangle + n}, \qquad n \geq 1, \tag{8.30}$$

$$\mu_n = \mu n\,, \qquad n \geq 0, \tag{8.31}$$

where the clonotype label i has been dropped for notational convenience. The model has four parameters:

(1) φ is a parameter proportional to the number of APPs which can provide survival stimuli to T cells of the fixed clonotype i. Then φ^{-1} is proportional to the mean time until a T cell of this clonotype receives a survival signal from an APP in the absence of competition with T cells of other clonotypes.
(2) ν is the "mean niche overlap" and encodes competition for survival stimuli between T cells of the fixed clonotype, i, and T cells of other clonotypes.
(3) $\langle n \rangle$ is the average clonotype size over the naïve T cell repertoire.
(4) μ is the death rate per T cell of clonotype i.

Special Cases

We introduce two special cases of the model which are defined by the value of the mean niche overlap parameter ν, which encodes competition. If $\nu \ll 1$ we say that the clone occupies a "hard niche", while if $\nu \gg 1$ we say that it occupies a "soft niche". Biologically, a clonotype with $\nu \ll 1$ possesses a TCR that is very different from the other clonotypes in the repertoire, while a clonotype with $\nu \gg 1$ has a TCR that is similar to the TCRs of many other clonotypes in terms of the APPs from which it is able to receive survival signals. We now derive expressions for the forms of the birth rates in these special cases.

For $\nu \ll 1$, the first term in the sum in (8.30) dominates so that

$$\lambda_n \approx \varphi n \mathrm{e}^{-\nu} \frac{\nu^0}{0!} \frac{1}{n} \approx \varphi, \quad \text{for } n \geq 1,$$

which means that the birth rate is approximately constant in this case. On the other hand, for $\nu \gg 1$ we have

$$\lambda_n = \varphi n \sum_{r=0}^{+\infty} \frac{\mathrm{e}^{-\nu} \nu^r}{r!} \frac{1}{r\langle n \rangle + n} = \varphi n \mathbb{E}\left[\frac{1}{\mathbb{Y}\langle n \rangle + n} \right], \tag{8.32}$$

where $\mathbb{Y}$ is Poisson distributed with mean ν. Carrying out a Taylor expansion about the mean gives

$$\lambda_n = \varphi n \left(\frac{1}{\nu \langle n \rangle + n} + \frac{\langle n \rangle^2 \nu}{(\nu \langle n \rangle + n)^3} + \ldots \right). \tag{8.33}$$

Since $\nu \gg 1$, the second and subsequent terms in the above expansion may be neglected, which results in

$$\lambda_n = \frac{\varphi n}{\nu \langle n \rangle + n}. \tag{8.34}$$

Clonal Extinction and Mean Extinction Times

Since $\mu_0 = \lambda_0 = 0$, the birth and death process has an absorbing state at $n = 0$. Clonal extinction occurs if the process reaches this state. We now calculate the probability of extinction using condition (8.12).

Firstly, note that the birth rate is bounded from above as follows:

$$\begin{aligned} \lambda_n &= \varphi n \mathrm{e}^{-\nu} \sum_{r=0}^{+\infty} \frac{\nu^r}{r!} \frac{1}{r\langle n \rangle + n} \\ &\leq \varphi n \mathrm{e}^{-\nu} \sum_{r=0}^{+\infty} \frac{\nu^r}{r!} \frac{1}{n} \\ &= \varphi \mathrm{e}^{-\nu} \mathrm{e}^{\nu} \\ &= \varphi. \end{aligned} \tag{8.35}$$

Therefore

$$\sum_{n=1}^{+\infty} \frac{\mu_1\mu_2\ldots\mu_n}{\lambda_1\lambda_2\ldots\lambda_n} \geq \sum_{n=1}^{+\infty} \frac{\mu^n n!}{\varphi^n}, \tag{8.36}$$

using the upper bound (8.35). Let $a_n = \frac{\mu^n n!}{\varphi^n}$. Then

$$\frac{a_{n+1}}{a_n} = \frac{\mu(n+1)}{\varphi} \to +\infty \text{ as } n \to +\infty,$$

and so $\sum_{n=1}^{+\infty} a_n$ diverges by the ratio test. Hence, condition (8.12) holds by the comparison test and we conclude that absorption at $n = 0$ is certain for all values of the parameters. This means that the eventual fate of any clone is extinction and disappearance from the repertoire.

We now compute the mean times to extinction. The mean time to extinction from initial state m is given by [12]

$$\tau_m = \begin{cases} \frac{1}{\mu_1} + \sum_{n=2}^{+\infty} \frac{\lambda_1\lambda_2\ldots\lambda_{n-1}}{\mu_1\mu_2\ldots\mu_n}, & m = 1, \\ \tau_1 + \sum_{k=1}^{m-1} \left[\frac{\mu_1\ldots\mu_k}{\lambda_1\ldots\lambda_k} \sum_{n=k+1}^{+\infty} \frac{\lambda_1\ldots\lambda_{n-1}}{\mu_1\ldots\mu_n} \right], & m = 2, 3, \ldots. \end{cases} \tag{8.37}$$

We first calculate the mean time to extinction for the hard niche case ($\nu \ll 1$). Substituting the birth and death rates $\lambda_n = \varphi$ and $\mu_n = \mu n$ into the above expressions results in

$$\tau_1 = \frac{1}{\varphi}\left(e^{\frac{\varphi}{\mu}} - 1\right), \tag{8.38}$$

$$\tau_m = \tau_1 + \sum_{k=1}^{m-1} \frac{k!}{\mu}\left(\frac{\mu}{\varphi}\right)^{k+1} \left[e^{\frac{\varphi}{\mu}} - \sum_{n=0}^{k} \left(\frac{\varphi}{\mu}\right)^n \frac{1}{n!} \right] \text{ for } m \geq 2. \tag{8.39}$$

We now consider the most general form of the birth and death rates given by (8.30)–(8.31). Then

$$\begin{aligned} \tau_1 &= \sum_{n=1}^{+\infty} \frac{\lambda_1\lambda_2\ldots\lambda_{n-1}}{\mu_1\mu_2\ldots\mu_n} \\ &\leq \sum_{n=1}^{+\infty} \frac{\varphi^{n-1}}{\mu^n n!} \quad \text{using the bound } \lambda_n \leq \varphi \\ &= \frac{1}{\varphi}\left(e^{\frac{\varphi}{\mu}} - 1\right), \end{aligned} \tag{8.40}$$

and

$$\begin{aligned}\tau_m &= \tau_1 + \sum_{k=1}^{m-1} \frac{\mu_1\mu_2\ldots\mu_k}{\lambda_1\lambda_2\ldots\lambda_k} \sum_{n=k+1}^{+\infty} \frac{\lambda_1\lambda_2\ldots\lambda_{n-1}}{\mu_1\mu_2\ldots\mu_n} \\ &\leq \tau_1 + \sum_{k=1}^{m-1} \frac{\mu_1\mu_2\ldots\mu_k}{\lambda_1\lambda_2\ldots\lambda_k} \sum_{n=k+1}^{+\infty} \frac{1}{\varphi n!}\left(\frac{\varphi}{\mu}\right)^n \\ &\leq \tau_1 + \sum_{k=1}^{m-1} \sum_{n=k+1}^{+\infty} \frac{k!}{\varphi n!}\left(\frac{\varphi}{\mu}\right)^{n-k} \\ &= \tau_1 + \sum_{k=1}^{m-1} \frac{k!}{\mu}\left(\frac{\mu}{\varphi}\right)^{k+1}\left[e^{\frac{\varphi}{\mu}} - \sum_{n=0}^{k}\left(\frac{\varphi}{\mu}\right)^n \frac{1}{n!}\right] \quad \text{for } m \geq 2, \end{aligned} \tag{8.41}$$

so the mean extinction times in the most general case are bounded by (8.38)–(8.39). Therefore, the hard niche case provides an upper bound on the mean extinction times. This is shown in Fig. 8.3 where τ_1 (calculated numerically) is plotted as a function of ν for fixed φ, μ and $\langle n \rangle$. For $\nu \ll 1$, τ_1 is independent of ν, in agreement with (8.38). Clones with $\nu \ll 1$ have the longest lifetimes in the repertoire. These are the clonotypes that are most different from other T cells in the repertoire in terms of the APPs they are able to receive survival signals from and so their long residence times in the repertoire ensure that diversity is maintained.

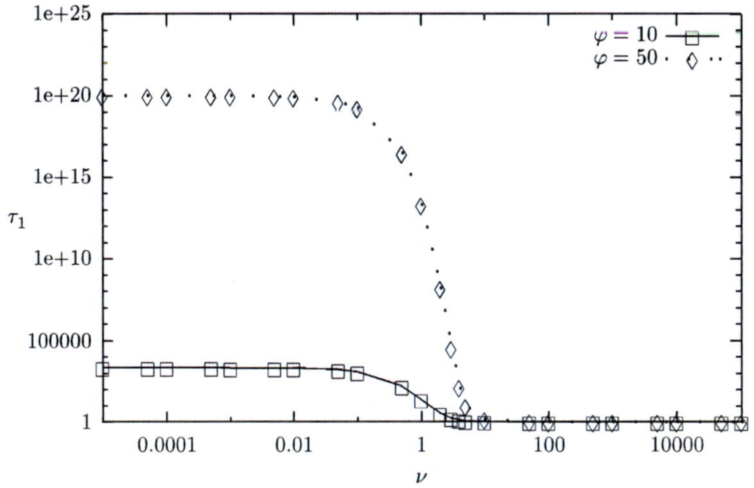

Fig. 8.3 τ_1 as a function of ν for $\varphi = 10, 50$, $\langle n \rangle = 10$ and $\mu = 1$

Homeostasis and the Limiting Conditional Distribution

Since extinction is guaranteed with probability one, the limiting distribution of the process is given by $(1, 0, 0, \ldots)^T$. To investigate the behaviour of the system before extinction takes place, we consider the limiting conditional probability distribution (LCD) of the process, which represents the homeostatic distribution of T cells.

Case $\nu \ll 1$

In the hard niche case, the birth and death rates are given by $\lambda_0 = 0$, $\lambda_n = \varphi$, for $n \geq 1$ and $\mu_n = \mu n$, for $n \geq 0$. Then

$$\sum_{n=2}^{+\infty} \frac{\lambda_1 \lambda_2 \ldots \lambda_{n-1}}{\mu_2 \mu_3 \ldots \mu_n} = \sum_{n=2}^{+\infty} \left(\frac{\varphi}{\mu}\right)^{n-1} \frac{1}{n!}$$

$$= \frac{\mu}{\varphi} \sum_{n=2}^{+\infty} \left(\frac{\varphi}{\mu}\right)^{n} \frac{1}{n!}$$

$$= \frac{\mu}{\varphi} \left(\mathrm{e}^{\frac{\varphi}{\mu}} - 1 - \frac{\varphi}{\mu}\right).$$

The first approximation (8.20) to the LCD is given by

$$\pi_1^{(1)} = \frac{1}{1 + \sum_{n=2}^{+\infty} \frac{\lambda_1 \lambda_2 \ldots \lambda_{n-1}}{\mu_2 \mu_3 \ldots \mu_n}} = \frac{\varphi}{\mu(\mathrm{e}^{\frac{\varphi}{\mu}} - 1)},$$

$$\pi_n^{(1)} = \frac{\lambda_1 \lambda_2 \ldots \lambda_{n-1}}{\mu_2 \mu_3 \ldots \mu_n} \pi_1^{(1)} = \left(\frac{\varphi}{\mu}\right)^n \frac{1}{n!(\mathrm{e}^{\frac{\varphi}{\mu}} - 1)} \quad \text{for } n \geq 2.$$

This distribution has mean

$$\sum_{n=1}^{+\infty} n \pi_n^{(1)} = \frac{\varphi}{\mu\left(1 - \mathrm{e}^{-\frac{\varphi}{\mu}}\right)},$$

and variance

$$\sum_{n=1}^{+\infty} n^2 \pi_n^{(1)} - \left(\sum_{n=1}^{+\infty} n \pi_n^{(1)}\right)^2 = \frac{1}{\left(1 - \mathrm{e}^{-\frac{\varphi}{\mu}}\right)} \left(\frac{\varphi}{\mu}\right)^2 \left(1 + \frac{\mu}{\varphi} - \frac{1}{1 - \mathrm{e}^{-\frac{\varphi}{\mu}}}\right).$$

The second approximating distribution (8.22) to the LCD is:

$$\sum_{n=2}^{+\infty} \frac{\lambda_1 \lambda_2 \ldots \lambda_{n-1}}{\mu_1 \mu_2 \ldots \mu_{n-1}} = \sum_{n=2}^{+\infty} \left(\frac{\varphi}{\mu}\right)^{n-1} \frac{1}{(n-1)!} = \mathrm{e}^{\frac{\varphi}{\mu}} - 1.$$

This means that

$$\pi_1^{(2)} = \frac{1}{1 + \sum_{n=2}^{+\infty} \frac{\lambda_1 \lambda_2 \ldots \lambda_{n-1}}{\mu_1 \mu_2 \ldots \mu_{n-1}}} = \mathrm{e}^{-\frac{\varphi}{\mu}},$$

$$\pi_n^{(2)} = \frac{\lambda_1 \lambda_2 \ldots \lambda_{n-1}}{\mu_1 \mu_2 \ldots \mu_{n-1}} \pi_1^{(2)} = \left(\frac{\varphi}{\mu}\right)^{n-1} \frac{1}{(n-1)!} \mathrm{e}^{-\frac{\varphi}{\mu}}, \quad \text{for } n \geq 2.$$

This distribution has mean

$$\sum_{n=1}^{+\infty} n \pi_n^{(2)} = \frac{\varphi}{\mu} + 1,$$

and variance

$$\sum_{n=1}^{+\infty} n^2 \pi_n^{(2)} - \left(\sum_{n=1}^{+\infty} n \pi_n^{(2)}\right)^2 = \frac{\varphi}{\mu}.$$

The mean time to extinction increases as φ increases. Hence, $\pi^{(2)}$ is the best approximation when the mean time to extinction is short, while the accuracy of approximation $\pi^{(1)}$ improves with increasing mean time to extinction.

Case $\nu \gg 1$

In the soft niche case, the birth rates are given by (8.34) and the first approximating distribution to the LCD is given by

$$\pi_1^{(1)} = \frac{1}{1 + \sum_{n=2}^{+\infty} \frac{1}{n} \left(\frac{\varphi}{\mu}\right)^{n-1} \prod_{k=1}^{n-1} \frac{1}{k + \nu\langle n \rangle}},$$

$$\pi_n^{(1)} = \frac{1}{n} \left(\frac{\varphi}{\mu}\right)^{n-1} \pi_1^{(1)} \prod_{k=1}^{n-1} \frac{1}{k + \nu\langle n \rangle} \quad \text{for } n \geq 2,$$

while the second approximating distribution is given by

$$\pi_1^{(2)} = \frac{1}{1 + \sum_{n=2}^{+\infty} \left(\frac{\varphi}{\mu}\right)^{n-1} \prod_{k=1}^{n-1} \frac{1}{k + \nu\langle n \rangle}},$$

$$\pi_n^{(2)} = \left(\frac{\varphi}{\mu}\right)^{n-1} \pi_1^{(2)} \prod_{k=1}^{n-1} \frac{1}{k + \nu\langle n \rangle} \quad \text{for } n \geq 2.$$

If φ is large enough, the clonotype behaves as a hard niche clone even though $\nu \gg 1$, and $\pi^{(1)}$ is a good approximation to the LCD for long mean extinction times, while $\pi^{(2)}$ is a better approximation than $\pi^{(1)}$ when the mean time to extinction is short, as for the case $\nu \ll 1$.

References

1. Starr TK, Jameson SC, Hogquist KA (2003) Positive and negative selection of T cells. Annu Rev Immunol 21:139–176
2. Goronzy JJ, Lee WW, Weyand CM (2007) Aging and T-cell diversity. Exp Gerontol 42: 400–406
3. Arstila TP, Casrouge A, Baron V, Even J, Kanellopoulos J, Kourilsky P (1999) A direct estimate of the human $\alpha\beta$ T cell receptor diversity. Science 286:958–961
4. Davis MM, Bjorkman PJ (1988) T cell antigen receptor genes and T cell recognition. Nature 334:395–402
5. Mahajan VS, Leskov IB, Chen J (2005) Homeostasis of T cell diversity. Cell Mol Immunol 2:1–10
6. van den Berg HA, Burroughs NJ, Rand DA (2002) Quantifying the strength of ligand antagonism in TCR triggering. Bull Math Biol 64:781–808
7. Mason D (1998) A very high level of crossreactivity is an essential feature of the T cell receptor. Immunol Today 19:395–404
8. Davenport MP, Price DA, McMichael AJ (2007) The T cell repertoire in infection and vaccination: implications for control of persistent viruses. Curr Opin Immunol 19:294–300
9. Stirk E, Molina-París C, van den Berg H (2008) Stochastic niche structure and diversity maintenance in the T cell repertoire. J Theor Biol 255:237–249
10. Tanchot C, Rosado MM, Agenes F, Freitas AA, Rocha B (1997) Lymphocyte homeostasis. Semin Immunol 9:331–337
11. Goldrath AW, Bevan MJ (1999) Selecting and maintaining a diverse T cell repertoire. Nature 402:255–262
12. Allen LJS (2003) An introduction to stochastic processes with applications to biology. Pearson Education, New Jersey, USA
13. Taylor HM, Karlin S (1998) An introduction to stochastic modeling, 3rd edn. Academic, San Diego
14. Van Doorn EA (1991) Quasi-stationary distributions and convergence to quasi-stationarity of birth-death processes. Adv Appl Probab 23:683–700
15. Nasell I (2001) Extinction and quasi-stationarity in the Verhulst logistic model. J Theor Biol 211:11–27

Chapter 9
Multivariate Competition Processes: A Model for Two Competing T Cell Clonotypes

Carmen Molina-París, Grant Lythe, and Emily Stirk

Abstract Diversity in the naïve T cell repertoire is maintained throughout the majority of an individual's lifetime. The homeostatic mechanisms involved include competition for survival stimuli furnished by antigen-presenting cells, dependent on weak recognition of arrays of self-peptides by the T cell antigen receptor. We study the dynamics of this process from the point of view of stochastic competitive exclusion between a pair of T cell clonotypes which are similar in terms of the specific survival stimuli which they are able to receive. This dynamics is formulated as a bivariate continuous-time Markov process for the number of T cells belonging to the pair of clonotypes. We prove that the ultimate fate of both clonotypes is extinction and provide a bound on mean extinction times. We concentrate mainly on the case where the two clonotypes exhibit little competition with other T cell clonotypes in the repertoire, since this case provides an upper bound on the mean extinction times. As the two clonotypes become more similar in terms of the proportion of their resources which are shared between them, one clonotype quickly becomes extinct in a process resembling the ecological principle of classical competitive exclusion. We consider the limiting probability distribution for the bivariate process, conditioned on non-extinction of both clonotypes.

Introduction: Bivariate Competition Processes

A healthy adult human possesses approximately 10^{11} naïve T cells [1], which belong to around 10^7–10^8 different clonotypes (all T cells with identical T cell antigen receptor) [2]. The number of naïve T cells is controlled homeostatically, meaning that it remains approximately constant throughout much of an individual's lifetime [3]. Experimental evidence suggests that T cell homeostasis is driven by interactions with self-antigens (antigens derived from the body's own proteins)

G. Lythe (✉)
Department of Applied Mathematics, University of Leeds, Leeds LS2 9JT, UK
e-mail: grant@maths.leeds.ac.uk

C. Molina-París and G. Lythe (eds.), *Mathematical Models and Immune Cell Biology*,
DOI 10.1007/978-1-4419-7725-0_9,

displayed on the surface of antigen-presenting cells [4–6]. Naïve T cells undergo infrequent cell divisions after receiving a survival signal from an antigen-presenting cell (APC). Whether or not a particular T cell can receive a survival signal from a given APC depends on the TCR it expresses and the array of self-antigens presented on the APC surface [7]. Competition amongst T cells for these interactions regulates the diversity of the T cell repertoire [8].

In Chap. 9, we constructed a univariate birth and death process to model the number of T cells belonging to a particular clonotype. The model relied upon a mean field approximation of the competition between different clonotypes. It was assumed that, even though a given T cell may compete with T cells of many other clonotypes for access to survival signals from an antigen-presentation profile (APP), individual competitive interactions between pairs of clonotypes are weak and do not have a significant impact on the fate of either clonotype. In terms of the sets defined in the previous chapter, this means that $|\mathbb{Q}_i \cap \mathbb{Q}_j| \ll |\mathbb{Q}_i|$ for $i \neq j$ (see Fig. 9.1). In this chapter, we introduce a model for the case when this condition does not hold. This means that the two clonotypes i and j overlap significantly in terms of the APPs from which they can receive survival signals and so the number of T cells of one clonotype affects the level of survival signals available to T cells of the other clonotype. We will assume that $|\mathbb{Q}_i \cap \mathbb{Q}_k| \ll |\mathbb{Q}_i|$, $|\mathbb{Q}_j \cap \mathbb{Q}_k| \ll |\mathbb{Q}_j|$ for $k \neq i, j$ and so competition between pairs of clonotypes (other than the pair i and j) is small. Therefore, the birth rates for the two competing clonotypes are coupled and we require a bivariate continuous-time Markov process to model the number of T cells belonging to both clonotypes [9, 10].

In the next section, a type of bivariate continuous-time Markov chain called a competition process is defined. This type of process is then used to model the number of T cells belonging to two closely competing clonotypes. It is then shown

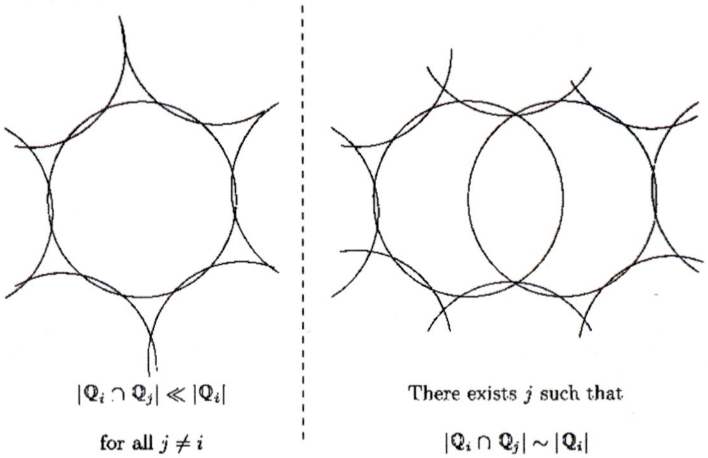

Fig. 9.1 The diagram on the *left* represents pairs of clonotypes for which the mean field approximation, described in the previous chapter, is reasonable. In this chapter, we consider a pair of clonotypes such as that represented on the *right*

that both clonotypes eventually become extinct with probability one. The model is extended to include any number of clonotypes for which the mean field approximation introduced in the previous chapter does not hold. Finally, we conclude the chapter with a numerical simulation of the full model using the Gillespie algorithm, where it is not necessary to make any mean field assumptions regarding the competition between different clonotypes.

We will construct a mathematical model for the number of T cells belonging to two different clonotypes, which we refer to as clonotypes 1 and 2, for which $|\mathbb{Q}_1 \cap \mathbb{Q}_2| \sim |\mathbb{Q}_1|$. This means that the two clonotypes overlap significantly in terms of the APPs from which they are able to receive survival signals. The random variable $\mathbb{X}(t)$ describes the number of T cells of clonotype 1 at time t, while the random variable $\mathbb{Y}(t)$ describes the number of T cells of clonotype 2 at time t. At time $t = \tilde{t}_1$, $\tilde{n}_1$ cells of clonotype 1 are released from the thymus and at time $t = \tilde{t}_2$, $\tilde{n}_2$ cells of clonotype 2 are released from the thymus. It is assumed that after $\tilde{t}_1$ no T cells of clonotype 1 are produced from the thymus and similarly, after $\tilde{t}_2$ no T cells of clonotype 2 are produced from the thymus. Without loss of generality we may assume that $\tilde{t}_1 \leq \tilde{t}_2 \stackrel{\text{def}}{=} t_0$. For $t < t_0$, only one of the clonotypes is present in the T cell repertoire and so the univariate model introduced in the previous chapter may be used. After this time we model the number of T cells belonging to clonotype 1 and 2 as a bivariate competition process.

Competition Processes

Let $\{(\mathbb{X}(t), \mathbb{Y}(t)) : t \geq t_0\}$ be a continuous-time bivariate Markov process on the state-space $\mathbb{S} = \{(n_1, n_2) : n_1, n_2 = 0, 1, \ldots\}$. The transition probabilities are defined by

$$p_{\mathbf{nm}}(\Delta t) = \mathbb{P}\{\mathbb{X}(t + \Delta t) = m_1, \mathbb{Y}(t + \Delta t) = m_2 | \mathbb{X}(t) = n_1, \mathbb{Y}(t) = n_2\} \quad (9.1)$$

for $\mathbf{n} = (n_1, n_2) \in \mathbb{S}$ and $\mathbf{m} = (m_1, m_2) \in \mathbb{S}$, where it is assumed that the transition probabilities are stationary (they only depend on the time interval between transitions and not the time at which transitions occur), as in the previous chapter.

A bivariate competition process is a continuous-time Markov process where transitions are only allowed to adjacent states. Then the transition probabilities satisfy $p_{\mathbf{nn}}(\Delta t) = 1 - (\lambda^{(1)}_{n_1,n_2} + \lambda^{(2)}_{n_1,n_2} + \mu^{(1)}_{n_1,n_2} + \mu^{(2)}_{n_1,n_2})\Delta t + o(\Delta t)$ and

$$p_{\mathbf{nm}}(\Delta t) = \begin{cases} \lambda^{(1)}_{n_1,n_2}\Delta t + o(\Delta t) & \mathbf{m} = (n_1 + 1, n_2) \\ \lambda^{(2)}_{n_1,n_2}\Delta t + o(\Delta t) & \mathbf{m} = (n_1, n_2 + 1) \\ \mu^{(1)}_{n_1,n_2}\Delta t + o(\Delta t) & \mathbf{m} = (n_1 - 1, n_2) \\ \mu^{(2)}_{n_1,n_2}\Delta t + o(\Delta t) & \mathbf{m} = (n_1, n_2 - 1) \\ o(\Delta t) & \text{otherwise,} \end{cases} \quad (9.2)$$

where $f(\Delta t) = o(\Delta t)$ if $\lim_{\Delta t \to 0+} \frac{f(\Delta t)}{\Delta t} = 0$.

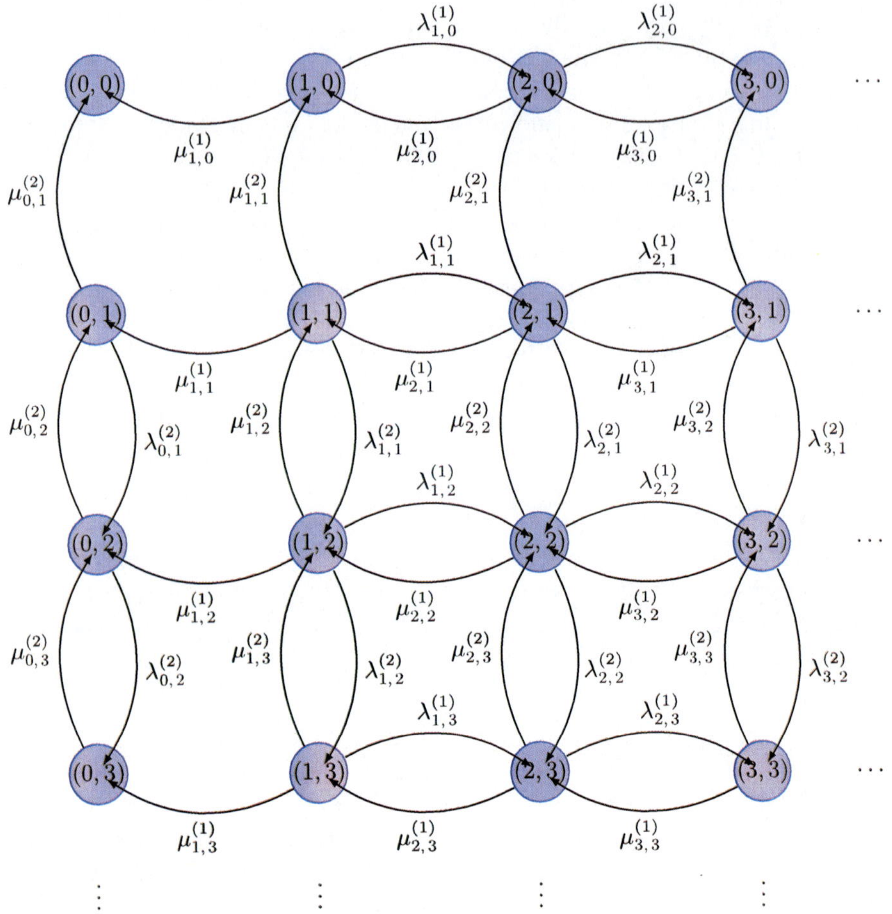

Fig. 9.2 A schematic representation of the bivariate competition process and the transitions between different states

A schematic representation of the process is given in Fig. 9.2. The quantity $\lambda^{(1)}_{n_1,n_2}$ is the birth rate of T cells of clonotype 1 and is the rate of transition from state (n_1, n_2) to $(n_1 + 1, n_2)$. Similarly, the birth rate of T cells of clonotype 2, denoted $\lambda^{(2)}_{n_1,n_2}$, is the rate of transition from state (n_1, n_2) to $(n_1, n_2 + 1)$. The death rate for T cells of clonotype 1 is given by $\mu^{(1)}_{n_1,n_2}$ and this is the rate of transition from state (n_1, n_2) to $(n_1 - 1, n_2)$. The death rate of T cells of clonotype 2, $\mu^{(2)}_{n_1,n_2}$, is the rate of transition from state (n_1, n_2) to $(n_1, n_2 - 1)$. We set $\mu^{(1)}_{0,j} = \mu^{(2)}_{j,0} = 0$ for all $j \geq 0$ so that transitions outside of the state space $\mathbb{S}$ cannot occur.

Let p_{n_1,n_2} be the probability that the process is in state (n_1, n_2) at time t. These probabilities satisfy the forward Kolmogorov equations, which are given by

$$\frac{dp_{n_1,n_2}(t)}{dt} = \lambda^{(1)}_{n_1-1,n_2} p_{n_1-1,n_2}(t) + \lambda^{(2)}_{n_1,n_2-1} p_{n_1,n_2-1}(t) + \mu^{(1)}_{n_1+1,n_2} p_{n_1+1,n_2}(t)$$
$$+ \mu^{(2)}_{n_1,n_2+1} p_{n_1,n_2+1}(t) - \left(\lambda^{(1)}_{n_1,n_2} + \lambda^{(2)}_{n_1,n_2} + \mu^{(1)}_{n_1,n_2}\right.$$
$$\left. + \mu^{(2)}_{n_1,n_2}\right) p_{n_1,n_2}(t), \tag{9.3}$$

for $n_1 \geq 1,\ n_2 \geq 1$, plus

$$\frac{dp_{0,0}(t)}{dt} = \mu^{(1)}_{1,0} p_{1,0}(t) + \mu^{(2)}_{0,1} p_{0,1}(t) - \left(\lambda^{(1)}_{0,0} + \lambda^{(2)}_{0,0}\right) p_{0,0}(t),$$
$$\frac{dp_{0,n_2}(t)}{dt} = \lambda^{(2)}_{0,n_2-1} p_{0,n_2-1}(t) + \mu^{(1)}_{1,n_2} p_{1,n_2}(t) + \mu^{(2)}_{0,n_2+1} p_{0,n_2+1}(t)$$
$$- \left(\lambda^{(1)}_{0,n_2} + \lambda^{(2)}_{0,n_2} + \mu^{(2)}_{0,n_2}\right) p_{0,n_2}(t), \quad n_2 \geq 1,$$
$$\frac{dp_{n_1,0}(t)}{dt} = \lambda^{(1)}_{n_1-1,0} p_{n_1-1,0}(t) + \mu^{(1)}_{n_1+1,0} p_{n_1+1,0}(t) + \mu^{(2)}_{n_1,1} p_{n_1,1}(t)$$
$$- \left(\lambda^{(1)}_{n_1,0} + \lambda^{(2)}_{n_1,0} + \mu^{(1)}_{n_1,0}\right) p_{n_1,0}(t), \quad n_1 \geq 1. \tag{9.4}$$

If the state-space of the process is finite, i.e., $\mathbb{S} = \{(n_1, n_2) : n_1, n_2 = 0, 1, \ldots, N\}$ and $\lambda^{(1)}_{N,n_2} = \lambda^{(2)}_{n_1,N} = 0$ for $n_1, n_2 = 0, 1, \ldots, N$ so that transitions outside of the state-space cannot occur, there are additional equations for the states on the boundary which are given by

$$\frac{dp_{N,0}(t)}{dt} = \lambda^{(1)}_{N-1,0} p_{N-1,0}(t) + \mu^{(2)}_{N,1} p_{N,1}(t) - \left(\lambda^{(2)}_{N,0} + \mu^{(1)}_{N,0}\right) p_{N,0}(t),$$
$$\frac{dp_{0,N}(t)}{dt} = \lambda^{(2)}_{0,N-1} p_{0,N-1}(t) + \mu^{(1)}_{1,N} p_{1,N}(t) - \left(\lambda^{(1)}_{0,N} + \mu^{(2)}_{0,N}\right) p_{0,N}(t),$$
$$\frac{dp_{N,n_2}(t)}{dt} = \lambda^{(1)}_{N-1,n_2} p_{N-1,n_2}(t) + \lambda^{(2)}_{N,n_2-1} p_{N,n_2-1}(t) + \mu^{(2)}_{N,n_2+1} p_{N,n_2+1}(t)$$
$$- \left(\lambda^{(2)}_{N,n_2} + \mu^{(1)}_{N,n_2} + \mu^{(2)}_{N,n_2}\right) p_{N,n_2}(t), \quad 1 \leq n_2 \leq N-1,$$
$$\frac{dp_{n_1,N}(t)}{dt} = \lambda^{(1)}_{n_1-1,N} p_{n_1-1,N}(t) + \lambda^{(2)}_{n_1,N-1} p_{n_1,N-1}(t) + \mu^{(1)}_{n_1+1,N} p_{n_1+1,N}$$
$$- \left(\lambda^{(1)}_{n_1,N} + \mu^{(1)}_{n_1,N} + \mu^{(2)}_{n_1,N}\right) p_{n_1,N}(t), \quad 1 \leq n_1 \leq N-1,$$
$$\frac{dp_{N,N}(t)}{dt} = \lambda^{(1)}_{N-1,N} p_{N-1,N}(t) + \lambda^{(2)}_{N,N-1} p_{N,N-1}(t)$$
$$- \left(\mu^{(1)}_{N,N} + \mu^{(2)}_{N,N}\right) p_{N,N}(t). \tag{9.5}$$

Mathematical Model for Two Competing Clonotypes

The transition probabilities defining the competition process take the form of those in (9.2) and are now derived. Recall that $\mathbb{C}$ is the set of all T cells in the naïve repertoire and $\mathbb{C}_q$ is the set of all T cells that are capable of receiving a survival signal from APP q. Also, $\mathbb{Q}$ is the set of all APPs which may occur in the periphery and $\mathbb{Q}_i$ is the subset of APPs from which T cells of clonotype i can receive a survival signal. We define $n_q = |\mathbb{C}_q|$, which is the total number of T cells that can receive a survival signal from APP q. Let γ_q be the rate of survival signals from all the APCs presenting APP q, and $\lambda^{(1)}$ be the per-cell birth rate of T cells of clonotype 1. Then, assuming that $\gamma_q = \gamma$ and that the survival signals from APP q are shared equally among all the T cells capable of receiving them, we have that

$$\lambda^{(1)} = \sum_{q \in \mathbb{Q}_1} \frac{\gamma}{|\mathbb{C}_q|} = \sum_{q \in \mathbb{Q}_1} \frac{\gamma}{n_q}, \tag{9.6}$$

as for the univariate case. Let n_1 be the number of T cells belonging to clonotype 1, and n_{1q} be the number of T cells not belonging to clonotype 1 that receive a survival signal from an APP $q \in \mathbb{Q}_1$. Hence, $n_q = n_1 + n_{1q}$.

We next divide the set $\mathbb{Q}_1$ into two disjoint subsets by defining

$$\mathbb{Q}_{12} = \mathbb{Q}_1 \cap \mathbb{Q}_2, \tag{9.7}$$

which is the set of APPs from which T cells of both clonotype 1 and clonotype 2 can receive a survival signal, and

$$\mathbb{Q}_{1/2} = \mathbb{Q}_1 \cap \bar{\mathbb{Q}}_2, \tag{9.8}$$

which is the set of APPs from which T cells of clonotype 1 receive a survival signal, but T cells of clonotype 2 do not. Here, $\bar{\mathbb{Q}}_2$ denotes the complement of the set $\mathbb{Q}_2$. Then, $\mathbb{Q}_1 = \mathbb{Q}_{12} \cup \mathbb{Q}_{1/2}$ and $\mathbb{Q}_{12} \cap \mathbb{Q}_{1/2} = \emptyset$. Therefore,

$$\lambda^{(1)} = \sum_{q \in \mathbb{Q}_{12}} \frac{\gamma}{n_1 + n_2 + n_{12q}} + \sum_{q \in \mathbb{Q}_{1/2}} \frac{\gamma}{n_1 + n_{1q}},$$

where n_2 denotes the number of T cells belonging to clonotype 2 and $n_{12q} = n_q - n_1 - n_2$.

Now, we partition the sets $\mathbb{Q}_{12}$ and $\mathbb{Q}_{1/2}$ into disjoint subsets as follows:

$$\mathbb{Q}_{12} = \bigcup_{r=0}^{+\infty} \mathbb{Q}_{12r} \quad \text{and} \quad \mathbb{Q}_{1/2} = \bigcup_{r=0}^{+\infty} \mathbb{Q}_{1r/2},$$

where $\mathbb{Q}_{12r}$ is the set of APPs which provide survival signals to T cells of clonotype 1 and clonotype 2 and to r other distinct clonotypes in the repertoire, and $\mathbb{Q}_{1r/2}$

is the set of APPs which provide survival signals to T cells of clonotype 1 and to r other distinct clonotypes, none of which is clonotype 2. Hence,

$$\lambda^{(1)} = \gamma \sum_{r=0}^{+\infty} \left(\sum_{q \in \mathbb{Q}_{12r}} \frac{1}{n_1 + n_2 + n_{12q}} + \sum_{q \in \mathbb{Q}_{1r/2}} \frac{1}{n_1 + n_{1q}} \right). \tag{9.9}$$

Note that the birth rate of T cells of clonotype 1 depends on the number of T cells of any other clonotype that competes with clonotype 1 for access to survival signals from an APP, through the terms n_{1q} and n_{12q}.

We now introduce a mean field approximation so that this birth rate depends only on the number of T cells belonging to clonotypes 1 and 2. We denote by

- p_1 the probability that a randomly chosen APP provides a survival stimulus to T cells of clonotype 2, given that it provides a survival stimulus to T cells of clonotype 1
- p_2 the probability that a randomly chosen APP provides a survival stimulus to T cells of clonotype 1, given that it provides a survival stimulus to T cells of clonotype 2
- $p_{\cdot|i/j}$ the probability that an APP chosen at random from $\mathbb{Q}_{i/j}$ will belong to the set $\mathbb{Q}_k$ of a different clonotype k selected at random

With the mean field approximation we write

$$\lambda^{(1)} = \gamma \sum_{r=0}^{+\infty} \left(\frac{|\mathbb{Q}_{12r}|}{n_1 + n_2 + r\langle n \rangle} + \frac{|\mathbb{Q}_{1r/2}|}{n_1 + r\langle n \rangle} \right), \tag{9.10}$$

where

$$|\mathbb{Q}_{12r}| = p_1 |\mathbb{Q}_1| \frac{\nu_{12}^r e^{-\nu_{12}}}{r!} \quad \text{and} \quad |\mathbb{Q}_{1r/2}| = (1 - p_1)|\mathbb{Q}_1| \frac{\nu_1^r e^{-\nu_1}}{r!}.$$

The parameter ν_{12} is the mean niche overlap for APPs in the set $\mathbb{Q}_{12}$, which is the average number of clonotypes that are competing with clonotype 1 and clonotype 2 for an APP in the set $\mathbb{Q}_{12}$, whereas ν_1 is the mean niche overlap for APPs in the set $\mathbb{Q}_{1/2}$, which is the average number of clonotypes that are competing with clonotype 1 for an APP in the set $\mathbb{Q}_{1/2}$. Hence, these parameters represent the strength of competition between T cells of clonotype 1 and T cells of other clonotypes in the repertoire (other than clonotype 2) for APPs in these sets, $\mathbb{Q}_{12}$ and $\mathbb{Q}_{1/2}$, respectively. These parameters are not the same as the mean niche overlap parameter for all APPs in the set $\mathbb{Q}_1$, ν, which is a parameter of the corresponding univariate model described in the previous chapter. However, these parameters are related by

$$\nu = \nu_{12} p_1 + \nu_1 (1 - p_1). \tag{9.11}$$

Hence, the per-cell birth rate of T cells of clonotype 1 is given by

$$\lambda^{(1)} = \varphi_1 \left(p_1 e^{-\nu_{12}} \sum_{r=0}^{+\infty} \frac{\nu_{12}^r}{r!} \frac{1}{n_1 + n_2 + r\langle n\rangle} + (1-p_1) e^{-\nu_1} \sum_{r=0}^{+\infty} \frac{\nu_1^r}{r!} \frac{1}{n_1 + r\langle n\rangle} \right), \tag{9.12}$$

where $\varphi_1 = \gamma|\mathbb{Q}_1|$ is a parameter proportional to the number of APPs from which T cells of clonotype 1 receive a survival signal. By a similar derivation, the per-cell birth rate of T cells of clonotype 2 is given by

$$\lambda^{(2)} = \varphi_2 \left(p_2 e^{-\nu_{12}} \sum_{r=0}^{+\infty} \frac{\nu_{12}^r}{r!} \frac{1}{n_1 + n_2 + r\langle n\rangle} + (1-p_2) e^{-\nu_2} \sum_{r=0}^{+\infty} \frac{\nu_2^r}{r!} \frac{1}{n_2 + r\langle n\rangle} \right), \tag{9.13}$$

where $\varphi_2 = \gamma|\mathbb{Q}_2|$ and $p_2 = \mathbb{P}(q \in \mathbb{Q}_1 | q \in \mathbb{Q}_2)$. The per-cell death rate of a given clonotype is assumed to be constant and is given by μ_1 for T cells of clonotype 1 and μ_2 for T cells of clonotype 2. Therefore, the birth and death rates of the bivariate competition process are given by

$$\lambda^{(1)}_{n_1,n_2} = \lambda^{(1)} n_1, \quad \lambda^{(2)}_{n_1,n_2} = \lambda^{(2)} n_2, \quad \mu^{(1)} = \mu_1 n_1 \quad \text{and} \quad \mu^{(2)} = \mu_2 n_2.$$

Summary of the Model

We have constructed a bivariate competition process $\{\mathbb{X}(t), \mathbb{Y}(t) : t \geq t_0\}$ on the state-space $\{(n_1, n_2) : n_1, n_2 = 0, 1, \ldots\}$ which models the number of T cells belonging to clonotypes 1 and 2 at time t. The birth rates are given by (9.12), (9.13) plus $\lambda^{(1)}_{0,n_2} = 0$ for $n_2 \geq 0$ and $\lambda^{(2)}_{n_1,0} = 0$ for $n_1 \geq 0$. The death rates are given by $\mu^{(1)}_{n_1,n_2} = \mu_1 n_1$ and $\mu^{(2)}_{n_1,n_2} = \mu_2 n_2$ for $n_1, n_2 \geq 0$. By definition, $\mathbb{Q}_1 \cap \mathbb{Q}_2 = \mathbb{Q}_{12} = \mathbb{Q}_{21} = \mathbb{Q}_2 \cap \mathbb{Q}_1$. We also have that $\gamma|\mathbb{Q}_{12}| = \gamma p_1 |\mathbb{Q}_1| = \varphi_1 p_1$ and, similarly, $\gamma|\mathbb{Q}_{21}| = \gamma p_2 |\mathbb{Q}_2| = \varphi_2 p_2$. This leads to the constraint on the parameters

$$\varphi_1 p_1 = \varphi_2 p_2, \tag{9.14}$$

where $0 \leq p_1 \leq 1$ and $0 \leq p_2 \leq 1$, as the parameters p_1 and p_2 represent probabilities. Hence, the model has nine independent parameters,

$$\varphi_1, \varphi_2, p_1, \nu_{12}, \nu_1, \nu_2, \mu_1, \mu_2, \langle n\rangle,$$

compared to the four parameters of the univariate model described in the previous chapter.

Clonal Extinction and Mean Extinction Times

Since we have that $\mu^{(1)}_{0,j} = \mu^{(2)}_{j,0} = \lambda^{(1)}_{0,j} = \lambda^{(2)}_{j,0} = 0$ for all $j \geq 0$, the set of states $\mathbb{A} = \{(n_1, n_2) : n_1 = 0 \text{ or } n_2 = 0\}$ forms an absorbing set. (Once the process enters this set, it will never leave.) This corresponds to extinction of one of the clonotypes from the T cell repertoire. The state $(n_1, n_2) = (0, 0)$ is an absorbing state corresponding to extinction of both clonotypes. In this section, we prove that the probability of the process eventually reaching this absorbing state is one and show that the mean time until the absorbing state is reached is finite. In order to do this, we require the following upper bound on the birth rates:

$$\begin{aligned}
\lambda^{(1)}_{n_1,n_2} &\leq \varphi_1 n_1 \left(p_1 e^{-\nu_{12}} \sum_{r=0}^{+\infty} \frac{\nu_{12}^r}{r!} \frac{1}{n_1} + (1 - p_1) e^{-\nu_1} \sum_{r=0}^{+\infty} \frac{\nu_1^r}{r!} \frac{1}{n_1} \right) \\
&= \varphi_1 n_1 \left(p_1 e^{-\nu_{12}} \frac{e^{\nu_{12}}}{n_1} + (1 - p_1) e^{-\nu_1} \frac{e^{\nu_1}}{n_1} \right) \\
&= \varphi_1 \, . \qquad (9.15)
\end{aligned}$$

Similarly, we can show that $\lambda^{(2)}_{n_1,n_2} \leq \varphi_2$.

Both Clonotypes Become Extinct with Probability One

In order to prove that the absorbing state at $(n_1, n_2) = (0, 0)$ is reached with probability one, we use the method of Iglehart [11] and bound the bivariate competition process with a univariate birth and death process which moves towards the absorbing state at the origin at a slower rate than the bivariate process. We then show that the univariate process reaches the absorbing state with certainty, and hence we may conclude that the bivariate competition process also reaches state $(0, 0)$ with probability one. We first divide the state-space $\mathbb{S}$ of the process into the following disjoint subsets:

$$\mathbb{S}_k = \{(n_1, n_2) : n_1 + n_2 = k\} \text{ for } k \geq 0.$$

A schematic diagram of these sets is given in Fig. 9.3. The birth and death rates for the univariate birth and death process are now defined in such a way that this new process moves towards the origin at a slower rate than the bivariate process. Let

$$\lambda'_k = \max_{(n_1,n_2) \in \mathbb{S}_k} \left\{ \lambda^{(1)}_{n_1,n_2} + \lambda^{(2)}_{n_1,n_2} \right\} \quad \text{and} \quad \mu'_k = \min_{(n_1,n_2) \in \mathbb{S}_k} \left\{ \mu^{(1)}_{n_1,n_2} + \mu^{(2)}_{n_1,n_2} \right\},$$

with $\lambda'_k = \mu'_k = 0$ when $k = 0$. If the process is in the set $\mathbb{S}_k$, at the next transition it moves to either $\mathbb{S}_{k-1}$ or $\mathbb{S}_{k+1}$. The rate λ'_k is the maximum rate at which the bivariate process moves upwards from the set $\mathbb{S}_k$ to $\mathbb{S}_{k+1}$ and the rate μ'_k is the

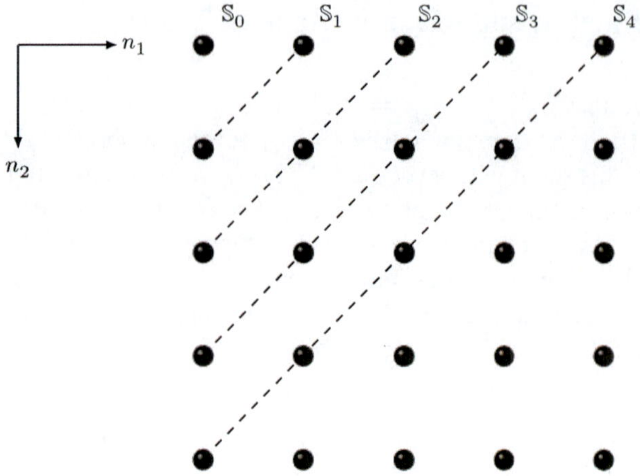

Fig. 9.3 Diagram of the sets $\mathbb{S}_k$ for $k \geq 0$

minimum rate at which the bivariate process moves downwards from the set $\mathbb{S}_k$ to $\mathbb{S}_{k-1}$. Hence, these rates define a univariate birth and death process on the state space $\{\mathbb{S}_0, \mathbb{S}_1, \mathbb{S}_2, \ldots\}$ where $\mathbb{S}_0$ is an absorbing state and $\mathbb{S}_k$ is now treated as an element rather than a set. This process moves towards the absorbing state at a slower rate than the bivariate process and can be represented by the following diagram:

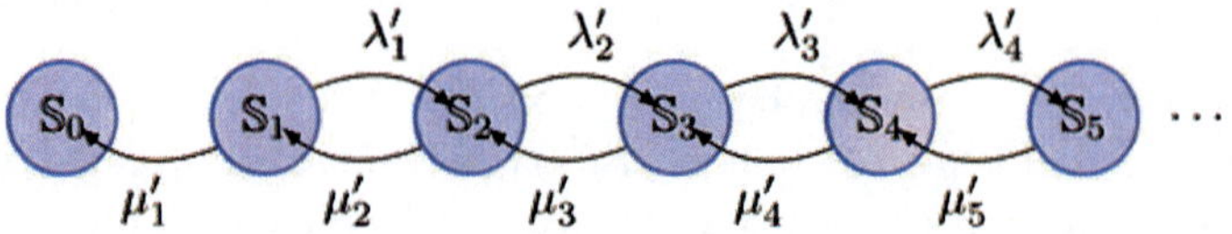

A sufficient condition for this process to reach the absorbing state with probability one is that the series

$$\sum_{k=1}^{+\infty} \frac{\mu'_1 \mu'_2 \ldots \mu'_k}{\lambda'_1 \lambda'_2 \ldots \lambda'_k} \tag{9.16}$$

diverges (see previous chapter). In order to prove this, we first observe that

$$\lambda'_k = \max_{(n_1,n_2)\in\mathbb{S}_k} \left\{ \lambda^{(1)}_{n_1,n_2} + \lambda^{(2)}_{n_1,n_2} \right\} \leq \varphi_1 + \varphi_2, \tag{9.17}$$

from (9.15) and

$$\mu'_k = \min_{(n_1,n_2)\in\mathbb{S}_k} \{\mu_1 n_1 + \mu_2 n_2\} = k \min(\mu_1, \mu_2). \tag{9.18}$$

Therefore,

$$\sum_{k=1}^{+\infty} \frac{\mu'_1\mu'_2\dots\mu'_k}{\lambda'_1\lambda'_2\dots\lambda'_k} \geq \sum_{k=1}^{+\infty} \frac{k![\min(\mu_1,\mu_2)]^k}{(\varphi_1+\varphi_2)^k}. \tag{9.19}$$

Let

$$a_k = \frac{k![\min(\mu_1,\mu_2)]^k}{(\varphi_1+\varphi_2)^k}, \tag{9.20}$$

so that

$$\frac{a_{k+1}}{a_k} = \frac{(k+1)\min(\mu_1,\mu_2)}{(\varphi_1+\varphi_2)} \to +\infty \text{ as } k \to +\infty. \tag{9.21}$$

Hence, the series $\sum_{k=1}^{+\infty} a_k$ diverges by the ratio test and so the series (9.16) diverges by comparison. Therefore, the univariate process reaches the origin with certainty and so we conclude that the bivariate competition process ultimately reaches the absorbing state $(n_1, n_2) = (0, 0)$ with probability one for all values of the parameters, corresponding to extinction of both clonotypes from the repertoire.

Finite Mean Extinction Times

The mean time until both clonotypes become extinct, when the initial state of the process is (n_1, n_2), is denoted by τ_{n_1,n_2}. This quantity is finite for all $(n_1, n_2) \in \mathbb{S} \setminus \{(0,0)\}$ if the series

$$\sum_{k=1}^{+\infty} \frac{\lambda'_1\lambda'_2\dots\lambda'_{k-1}}{\mu'_1\mu'_2\dots\mu'_k} \tag{9.22}$$

converges [11]. For the rates (9.17)–(9.18),

$$\sum_{k=1}^{+\infty} \frac{\lambda'_1\lambda'_2\dots\lambda'_{k-1}}{\mu'_1\mu'_2\dots\mu'_k} \leq \sum_{k=1}^{+\infty} \frac{(\varphi_1+\varphi_2)^{k-1}}{k![\min(\mu_1,\mu_2)]^k}. \tag{9.23}$$

Let

$$b_k = \frac{(\varphi_1+\varphi_2)^{k-1}}{k![\min(\mu_1,\mu_2)]^k}. \tag{9.24}$$

Then,

$$\frac{b_{k+1}}{b_k} = \frac{(\varphi_1+\varphi_2)}{(k+1)\min(\mu_1,\mu_2)} \to 0 \text{ as } k \to +\infty, \tag{9.25}$$

which means that the series $\sum_{k=1}^{+\infty} b_k$ converges by the ratio test. Hence, the series (9.22) converges by comparison and therefore the mean time to reach the absorbing state from all initial states $(n_1, n_2) \in \mathbb{S} \setminus \{(0,0)\}$ is finite for all values of the parameters. This is because the convergence of the series (9.22) is sufficient to prove that the univariate process with the birth and death rates λ'_k and μ'_k reaches the absorbing state in finite time from all initial states, and this process moves towards the absorbing state at a slower rate than the bivariate competition process.

The Limiting Conditional Probability Distribution

The limiting conditional probability distribution of the process represents the late-time behaviour of the system before extinction occurs. Since extinction of both clonotypes occurs with certainty in finite time, the limiting probability distribution of the process, $\lim_{t\to+\infty} p_{n_1,n_2}(t)$, has all its mass at the state $(n_1, n_2) = (0, 0)$. In order to study the behaviour of the process before extinction takes place, we condition on the event that the process has not yet reached the absorbing set $\mathbb{A}$. Let $p_{\mathbb{A}}(t)$ denote the probability that the process is not in set $\mathbb{A}$ at time t. We define

$$q_{n_1,n_2}(t) = \frac{p_{n_1,n_2}(t)}{p_{\mathbb{A}}(t)}, \tag{9.26}$$

which is the probability that the process is in state (n_1, n_2) at time t, given that extinction of either clonotype has not yet occurred. Then

$$\frac{\mathrm{d}q_{n_1,n_2}(t)}{\mathrm{d}t} = \frac{1}{p_{\mathbb{A}}(t)}\frac{\mathrm{d}p_{n_1,n_2}(t)}{\mathrm{d}t} - \frac{p_{n_1,n_2}(t)}{(p_{\mathbb{A}}(t))^2}\frac{\mathrm{d}p_{\mathbb{A}}(t)}{\mathrm{d}t}. \tag{9.27}$$

Using the fact that $\sum_{n_1=0}^{+\infty}\sum_{n_2=0}^{+\infty} p_{n_1,n_2} = 1$, we have

$$p_{\mathbb{A}}(t) = 1 - \sum_{n_2=0}^{+\infty} p_{0,n_2}(t) - \sum_{n_1=0}^{+\infty} p_{n_1,0}(t) + p_{0,0}(t), \tag{9.28}$$

and hence

$$\frac{\mathrm{d}p_{\mathbb{A}}(t)}{\mathrm{d}t} = -\frac{\mathrm{d}}{\mathrm{d}t}\sum_{n_2=0}^{+\infty} p_{0,n_2}(t) - \frac{\mathrm{d}}{\mathrm{d}t}\sum_{n_1=0}^{+\infty} p_{n_1,0}(t) + \frac{\mathrm{d}p_{0,0}(t)}{\mathrm{d}t}. \tag{9.29}$$

We now derive expressions for the terms on the right hand side of this equation.

$$\frac{\mathrm{d}}{\mathrm{d}t}\sum_{n_2=0}^{+\infty} p_{0,n_2}(t) = \mu_1 \sum_{n_2=0}^{+\infty} p_{1,n_2}(t) \quad \text{and} \quad \frac{\mathrm{d}}{\mathrm{d}t}\sum_{n_1=0}^{+\infty} p_{n_1,0}(t) = \mu_2 \sum_{n_1=0}^{+\infty} p_{n_1,1}(t).$$

Therefore,

$$\frac{\mathrm{d}p_{\mathbb{A}}(t)}{\mathrm{d}t} = -\mu_1 \sum_{n_2=1}^{+\infty} p_{1,n_2}(t) - \mu_2 \sum_{n_1=1}^{+\infty} p_{n_1,1}(t),$$

and

$$\begin{aligned}\frac{\mathrm{d}q_{n_1,n_2}(t)}{\mathrm{d}t} &= \lambda^{(1)}_{n_1-1,n_2} q_{n_1-1,n_2}(t) + \lambda^{(2)}_{n_1,n_2-1} q_{n_1,n_2-1}(t) + \mu^{(1)}_{n_1+1,n_2} q_{n_1+1,n_2}(t) \\ &\quad + \mu^{(2)}_{n_1,n_2+1} q_{n_1,n_2+1}(t) - \Big(\lambda^{(1)}_{n_1,n_2} + \lambda^{(2)}_{n_1,n_2} + \mu^{(1)}_{n_1,n_2} \\ &\qquad + \mu^{(2)}_{n_1,n_2}\Big) q_{n_1,n_2}(t) \\ &\quad + \mu_1 q_{n_1,n_2}(t) \sum_{n_2=1}^{+\infty} q_{1,n_2}(t) + \mu_2 q_{n_1,n_2}(t) \sum_{n_1=1}^{+\infty} q_{n_1,1}(t), \qquad (9.30)\end{aligned}$$

for $n_1 \geq 1, n_2 \geq 1$. A distribution, $\bar{q}$, is called a quasi-stationary probability distribution (QSD) if it is a solution of the equations obtained from (9.30) by setting the time derivatives to zero. The limiting conditional probability distribution (LCD) is defined by

$$\lim_{t \to +\infty} q_{n_1,n_2}(t), \qquad (9.31)$$

where $n_1 \geq 1, n_2 \geq 1$. This distribution is independent of time and so is a QSD. For a process where the state-space is finite, the QSD of the process is unique and is equal to the LCD [12]. This will always be the case when the LCD of the process is calculated numerically.

A Process with *m* Competing Clonotypes

In previous sections we have studied a process modelling the number of T cells belonging to two competing clonotypes. In this section we extend the model to include m competing clonotypes for which the mean field approximation does not hold, and prove that all clonotypes eventually become extinct with probability one and that the mean time until extinction is finite. In this case, we do not make use of a mean field approximation.

The number of T cell of clonotypes $1, 2, \ldots, m$ at time t is modelled as a continuous-time multivariate Markov process $\{\mathbb{X}_1(t), \mathbb{X}_2(t), \ldots, \mathbb{X}_m(t) : t \geq t_0\}$ on the state-space $\mathbb{S} = \{(n_1, n_2, \ldots, n_m) : n_1, n_2, \ldots, n_m = 0, 1, 2, \ldots\}$. As before, transitions are only allowed to adjacent states and so we have an m-dimensional generalisation of the birth and death process, which we call a multivariate competition process [13].

For $\mathbf{n} = (n_1, n_2, \ldots, n_m)$, $\mathbf{n}' = (n'_1, n'_2, \ldots, n'_m)$ we introduce the transition probabilities

$$p_{\mathbf{nn}'}(\Delta t) = \mathbb{P}\{\mathbb{X}_1(t+\Delta t) = n'_1, \mathbb{X}_2(t+\Delta t) = n'_2, \ldots,$$
$$\mathbb{X}_m(t+\Delta t) = n'_m | \mathbb{X}_1(t) = n_1, \mathbb{X}_2(t) = n_2, \ldots, \mathbb{X}_m(t) = n_m\}$$

for $\mathbf{n}$ and $\mathbf{n}' \in \mathbb{S}$. As $\Delta t \to 0^+$ these probabilities satisfy

$$p_{\mathbf{nn}}(\Delta t) = 1 - \left(\lambda_{\mathbf{n}}^{(1)} + \ldots + \lambda_{\mathbf{n}}^{(m)} + \mu_{\mathbf{n}}^{(1)} + \ldots + \mu_{\mathbf{n}}^{(m)}\right) \Delta t + o(\Delta t)$$

and

$$p_{\mathbf{nn}'}(\Delta t) = \begin{cases} \lambda_{\mathbf{n}}^{(1)} \Delta t + o(\Delta t) & \mathbf{n}' = (n_1+1, n_2, \ldots, n_m) \\ \lambda_{\mathbf{n}}^{(2)} \Delta t + o(\Delta t) & \mathbf{n}' = (n_1, n_2+1, \ldots, n_m) \\ \vdots & \vdots \\ \lambda_{\mathbf{n}}^{(m)} \Delta t + o(\Delta t) & \mathbf{n}' = (n_1, n_2, \ldots, n_m+1) \\ \mu_{\mathbf{n}}^{(1)} \Delta t + o(\Delta t) & \mathbf{n}' = (n_1-1, n_2, \ldots, n_m) \\ \mu_{\mathbf{n}}^{(2)} \Delta t + o(\Delta t) & \mathbf{n}' = (n_1, n_2-1, \ldots, n_m) \\ \vdots & \vdots \\ \mu_{\mathbf{n}}^{(m)} \Delta t + o(\Delta t) & \mathbf{n}' = (n_1, n_2, \ldots, n_m-1) \\ o(\Delta t) & \text{otherwise} \end{cases}$$

and $p_{\mathbf{nn}'}(\Delta t) = 0$ for $\mathbf{n} \notin \mathbb{S}$ or $\mathbf{n}' \notin \mathbb{S}$. Here, $\lambda_{\mathbf{n}}^{(s)}$ is the birth rate of T cells of clonotype s and is the rate of transition from state $(n_1, n_2, \ldots, n_s, \ldots, n_m)$ to state $(n_1, n_2, \ldots, n_s + 1, \ldots, n_m)$ for $1 \leq s \leq m$. Similarly, $\mu_{\mathbf{n}}^{(s)}$ is the death rate of T cells of clonotype s and is the rate of transition from state $(n_1, n_2, \ldots, n_s, \ldots, n_m)$ to state $(n_1, n_2, \ldots, n_s - 1, \ldots, n_m)$ for $1 \leq s \leq m$. Setting $\mu_{\mathbf{n}}^{(s)} = 0$ for $n_s = 0$ and $1 \leq s \leq m$ ensures that transitions outside of the state-space cannot occur. We will assume that, after the initial time, the thymus does not produce any more T cells of clonotypes $1, 2, \ldots, m$. Hence, $\lambda_{\mathbf{n}}^{(s)} = 0$ for $n_s = 0$ where $1 \leq s \leq m$. This means that the set of states $\mathbb{A} = \{(n_1, n_2, \ldots, n_m) : n_s = 0 \text{ for } 1 \leq s \leq m\}$ is an absorbing set and the state $(n_1, n_2, \ldots, n_m) = (0, 0, \ldots, 0)$ is an absorbing state.

Rather than explicitly deriving an expression for the birth rate of each clonotype in the multivariate competition process, we instead formulate a bound on each of these rates. Let $\mathbb{Q}_s$ be the set of all APPs from which T cells of clonotype s can receive a survival signal. We define $\lambda^{(s)}$ to be the per-cell birth rate for T cells of clonotype s, where $1 \leq s \leq m$. Then

$$\lambda^{(s)} = \sum_{q \in \mathbb{Q}_s} \frac{\gamma}{|\mathbb{C}_q|} = \sum_{q \in \mathbb{Q}_s} \frac{\gamma}{n_s + n_{sq}}, \tag{9.32}$$

where n_s is the number of T cells belonging to clonotype s and n_{sq} is the number of T cells in $\mathbb{C}_q$ that are not of clonotype s. Then

$$\lambda^{(s)} \leq \gamma \sum_{q \in \mathbb{Q}_s} \frac{1}{n_s} = \frac{\gamma|\mathbb{Q}_s|}{n_s} = \frac{\varphi_s}{n_s},$$

where $\varphi_s = \gamma|\mathbb{Q}_s|$ is a parameter representing the strength of the stimulation. Therefore,

$$\lambda_{\mathbf{n}}^{(s)} = \lambda^{(s)} n_s \leq \varphi^{(s)}, \tag{9.33}$$

for $1 \leq s \leq m$. To obtain the death rates, we assume there is a constant per-cell death rate, which for clonotype s is denoted by μ_s. Hence,

$$\mu_{\mathbf{n}}^{(s)} = \mu_s n_s, \tag{9.34}$$

for $1 \leq s \leq m$.

Extinction Occurs with Probability One

Using the bound on the birth rates derived in the previous section, we now show that the probability of eventual absorption at state $(n_1, n_2, \ldots, n_m) = (0, 0, \ldots, 0)$ is 1 for all values of the parameters. We first introduce the following disjoint subsets of $\mathbb{S}$:

$$\mathbb{S}_k = \{(n_1, n_2, \ldots, n_m) : n_1 + \ldots + n_m = k\} \text{ for } k \geq 0.$$

We now define the birth rates for a univariate birth and death process on the state-space $\{\mathbb{S}_0, \mathbb{S}_1, \ldots\}$ in such a way that it moves towards the absorbing state at the origin at a slower rate than the multivariate process. Let

$$\lambda'_k = \max_{(n_1,\ldots,n_m)\in\mathbb{S}_k} \left\{\lambda_{\mathbf{n}}^{(1)} + \lambda_{\mathbf{n}}^{(2)} + \ldots + \lambda_{\mathbf{n}}^{(m)}\right\}, \tag{9.35}$$

$$\mu'_k = \min_{(n_1,\ldots,n_m)\in\mathbb{S}_k} \left\{\mu_{\mathbf{n}}^{(1)} + \mu_{\mathbf{n}}^{(2)} + \ldots + \mu_{\mathbf{n}}^{(m)}\right\}. \tag{9.36}$$

From (9.33) we have that

$$\lambda'_k = \max_{(n_1,\ldots,n_m)\in\mathbb{S}_k} \left\{\lambda_{\mathbf{n}}^{(1)} + \lambda_{\mathbf{n}}^{(2)} + \ldots + \lambda_{\mathbf{n}}^{(m)}\right\} \leq \varphi_1 + \varphi_2 + \ldots + \varphi_m,$$

$$\mu'_k = \min_{(n_1,\ldots,n_m)\in\mathbb{S}_k} \{\mu_1 n_1 + \mu_2 n_2 + \ldots + \mu_m n_m\} = k \min(\mu_1, \mu_2, \ldots, \mu_m).$$

As in the bivariate case, λ'_k is the maximum rate at which the multivariate process moves from the set $\mathbb{S}_k$ to $\mathbb{S}_{k+1}$ and μ'_k is the minimum rate at which the multivariate process moves from the set $\mathbb{S}_k$ to $\mathbb{S}_{k-1}$. Hence, the univariate birth and death process with rates λ'_k and μ'_k on the state-space $\{\mathbb{S}_0, \mathbb{S}_1, \ldots\}$ moves towards the absorbing state at $\mathbb{S}_0$ at a slower rate than the multivariate one. The univariate process reaches the absorbing state with certainty if

$$\sum_{k=1}^{+\infty} \frac{\mu'_1 \mu'_2 \ldots \mu'_k}{\lambda'_1 \lambda'_2 \ldots \lambda'_k} \tag{9.37}$$

diverges. We have

$$\sum_{k=1}^{+\infty} \frac{\mu'_1 \mu'_2 \ldots \mu'_k}{\lambda'_1 \lambda'_2 \ldots \lambda'_k} \geq \sum_{k=1}^{+\infty} \frac{k![\min(\mu_1, \mu_2, \ldots, \mu_m)]^k}{(\varphi_1 + \varphi_2 + \ldots + \varphi_m)^k}.$$

Let

$$c_k = \frac{k![\min(\mu_1, \mu_2, \ldots, \mu_m)]^k}{(\varphi_1 + \varphi_2 + \ldots + \varphi_m)^k},$$

so that

$$\frac{c_{k+1}}{c_k} = \frac{(k+1)\min(\mu_1, \mu_2, \ldots, \mu_m)}{\varphi_1 + \varphi_2 + \ldots + \varphi_m} \to +\infty \text{ as } k \to +\infty,$$

which means that $\sum_{k=1}^{+\infty} c_k$ diverges by the ratio test. Hence, the series (9.37) diverges by comparison and so the univariate process reaches the absorbing state with probability one. Since this process moves towards the absorbing state at a slower rate than the multivariate competition process, we conclude that absorption at the origin is also certain for this process for all values of the parameters.

Let $\tau_{\mathbf{n}}$ be the mean extinction time if the initial state of the process is $\mathbf{n} = (n_1, n_2, \ldots, n_m)$. This time is finite for all $\mathbf{n} \in \mathbb{S} \setminus \{(0, 0, \ldots, 0)\}$ if the series

$$\sum_{k=1}^{+\infty} \frac{\lambda'_1 \lambda'_2 \ldots \lambda'_{k-1}}{\mu'_1 \mu'_2 \ldots \mu'_k} \tag{9.38}$$

converges [11]. We have

$$\sum_{k=1}^{+\infty} \frac{\lambda'_1 \lambda'_2 \ldots \lambda'_{k-1}}{\mu'_1 \mu'_2 \ldots \mu'_k} \leq \frac{(\varphi_1 + \varphi_2 + \ldots + \varphi_m)^{k-1}}{k![\min(\mu_1, \mu_2, \ldots, \mu_m)]^k}.$$

Let

$$d_k = \frac{(\varphi_1 + \varphi_2 + \ldots + \varphi_m)^{k-1}}{k![\min(\mu_1, \mu_2, \ldots, \mu_m)]^k}.$$

Then

$$\frac{d_{k+1}}{d_k} = \frac{\varphi_1 + \varphi_2 + \ldots + \varphi_m}{(k+1)\min(\mu_1, \mu_2, \ldots, \mu_m)} \to 0 \text{ as } k \to +\infty,$$

and so $\sum_{k=1}^{+\infty} d_k$ converges by the ratio test. Hence, the series (9.38) converges by comparison and so $\tau_{\mathbf{n}}$ is finite for all $\mathbf{n} \in \mathbb{S} \setminus \{(0, 0, \ldots, 0)\}$ for all values of the parameters.

Numerical Results

We have carried out numerical solutions of the competition model, without any type of mean field approximation, using the Gillespie algorithm [14]. A number of T cell clonotypes, N_c, a number of APPs, $|\mathbb{Q}|$, and a set of connections between them, are input at the start of a numerical run. In the run used to produce the data in Figs. 9.4 and 9.5, the initial number of T cell clonotypes is 1,000 and there are 100 cells of each clonotype at $t = 0$. Connections between T cell clonotypes and APPs, meaning that the clonotype receives a signal from the APP, are assigned independently with probability p at the beginning of the run and not changed thereafter. The death rate μ is constant; the birth rate of each clonotype is calculated using (9.6) at each time during the numerical run. Each clonotype's value of ν is a function of time that is the average, over the APPs from which it receives a stimulus, of the number of other surviving clonotypes also stimulated by an APP. A clonotype that only receives signals from APPs that do not signal to any other clonotype will have $\nu = 0$; a clonotype that receives signals from two APPs, one of which signals to one other clonotype and one of which signals to two other clonotypes, will have $\nu = 1.5$.

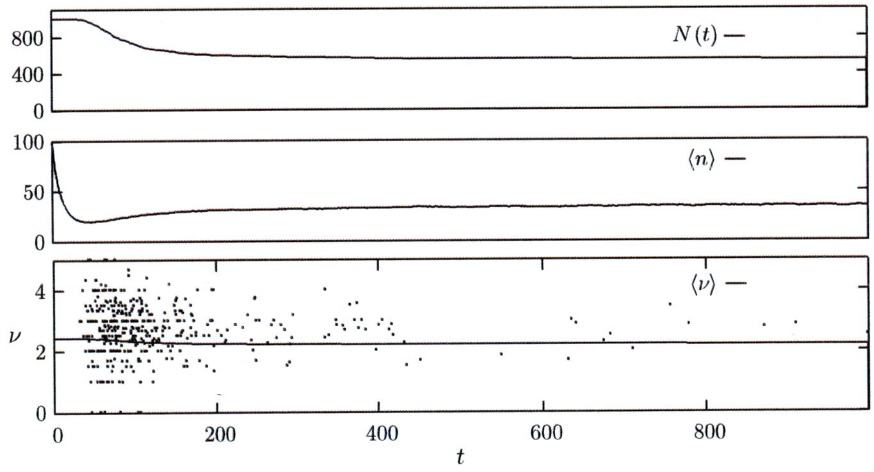

Fig. 9.4 Numerical results, showing the time evolution of the number of clonotypes with at least one living cell, the mean number of cells per clonotype, and the mean value of ν. The T cell clonotype–APP connections are randomly assigned at the beginning of the numerical run. The parameters for the run are $N_c = 1{,}000$, $|\mathbb{Q}| = 2{,}000$, $\mu = 0.1$, $\gamma = 1$ and $p = 0.0025$

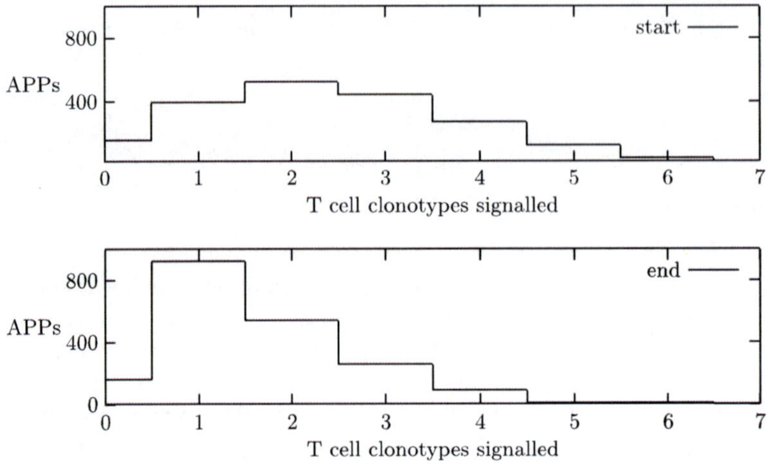

Fig. 9.5 Competition, and extinction of some clonotypes, leads to more uniform coverage of the set of APPs by the repertoire of T cell clonotypes. The *upper panel* shows a histogram of the number of clonotypes that recognise an APP at the beginning of the numerical run. The *lower panel* shows a histogram of the number of clonotypes that recognise an APP at $t = 1{,}000$

Figure 9.4 shows several properties of the repertoire of T cells as a function of time. The upper panel shows the number of surviving clonotypes (those with at least one living cell) as a function of time. The central panel shows the mean number of cells per surviving clonotype and the lower panel the mean value of ν in the set of surviving clonotypes. Each dot in the lower panel indicates the time of extinction of a clonotype (horizontal coordinate) and its ν value. Note that the majority of the clonotypes that die out have above-average values of ν. The hypothesis of the existence of a well-defined mean number of cells per clonotype, necessary for the approximate mean field model, is supported by numerical results on this "exact" model. The conjecture that competition is unfavourable to clonotypes with large values of ν is also supported.

It is illuminating to examine the dynamics from the point of view of the set of APPs. This set does not change with time, but the number of T cell clonotypes in the repertoire that recognise any one APP (the number of clonotypes in the set $\mathbb{C}_q$) does change over time because some clonotypes die out. In Fig. 9.5 we illustrate the effect of competition in producing an increasingly uniform coverage of the set, which may be envisaged as a space of epitopes. At the start of the numerical run, the distribution of the number of clonotypes that recognise an APP is wide and the most common value is 2; at the end of the run, it is much narrower, and the most common value is 1. Competition for survival signals forces T cell clonotypes and their APPs to be increasingly monogamous.

References

1. Goronzy JJ, Weyand CM (2001) Thymic function and peripheral T cell homeostasis in rheumatoid arthritis. Trends Immunol 22:251–255
2. Arstila TP, Casrouge A, Baron V, Even J, Kanellopoulos J, Kourilsky P (1999) A direct estimate of the human $\alpha\beta$ T cell receptor diversity. Science 286:958–961
3. Freitas AA, Rocha B (2000) Population biology of lymphocytes: the flight for survival. Annu Rev Immunol 18:83–111
4. Ernst B, Lee DS, Chang JM, Sprent J, Surh CD (1999) The peptide ligands mediating positive selection in the thymus control T cell survival and homeostatic proliferation in the periphery. Immunity 11:173–181
5. Freitas AA, Rocha B (1999) Peripheral T cell survival. Curr Opin Immunol 11:152–156
6. Goldrath AW, Bevan MJ (1999) Selecting and maintaining a diverse T cell repertoire. Nature 402:255–262
7. Ferreira C, Barthlott T, Garcia S, Zamoyska R, Stockinger B (2000) Differential survival of naïve CD4 and CD8 T cells. J Immunol 165:3689–3694
8. Troy AE, Shen H (2003) Cutting edge: homeostatic proliferation of peripheral T lymphocytes is regulated by clonal competition. J Immunol 170:672–676
9. Stirk E, Molina-París C, van den Berg H (2008) Stochastic niche structure and diversity maintenance in the T cell repertoire. J Theor Biol 255:237–249
10. Stirk E, Lythe G, van den Berg H, Hurst G, Molina-París C (2010) The limiting conditional distribution in a stochastic model of T cell repertoire maintenance. Math Biosci 224:74–86
11. Iglehart DL (1964) Multivariate competition processes. Ann Math Stat 35:350–361
12. Darroch JN, Seneta E (1965) On quasi-stationary distributions in absorbing discrete-time finite Markov chains. J Appl Probab 2:88–100
13. Reuter GEH (1961) Competition processes. In: Proc. Fourth Berkeley Symp. Math. Statist. Prob. II. Univ. of California Press, pp 421–430
14. Gillespie DT (1977) Exact stochastic simulation of coupled chemical reactions. J Phys Chem 81:2340–2361

Chapter 10
Stochastic Modelling of T Cell Homeostasis for Two Competing Clonotypes Via the Master Equation

Shev MacNamara and Kevin Burrage

Abstract Stochastic models for competing clonotypes of T cells by multivariate, continuous-time, discrete state, Markov processes have been proposed in the literature by Stirk, Molina-París and van den Berg (2008). A stochastic modelling framework is important because of rare events associated with small populations of some critical cell types. Usually, computational methods for these problems employ a trajectory-based approach, based on Monte Carlo simulation. This is partly because the complementary, probability density function (PDF) approaches can be expensive but here we describe some efficient PDF approaches by directly solving the governing equations, known as the Master Equation. These computations are made very efficient through an approximation of the state space by the Finite State Projection and through the use of Krylov subspace methods when evolving the matrix exponential. These computational methods allow us to explore the evolution of the PDFs associated with these stochastic models, and bimodal distributions arise in some parameter regimes. Time-dependent propensities naturally arise in immunological processes due to, for example, age-dependent effects. Incorporating time-dependent propensities into the framework of the Master Equation significantly complicates the corresponding computational methods but here we describe an efficient approach via Magnus formulas. Although this contribution focuses on the example of competing clonotypes, the general principles are relevant to multivariate Markov processes and provide fundamental techniques for computational immunology.

K. Burrage (✉)
The Oxford Computing Laboratory and Oxford Centre for Integrative Systems Biology, University of Oxford, Oxford OX1 2JD, UK
and
The Institute for Molecular Biosciences, The University of Queensland, Queensland, Brisbane, QLD 4072, Australia
e-mail: kevin.burrage@comlab.ox.ac.uk

C. Molina-París and G. Lythe (eds.), *Mathematical Models and Immune Cell Biology*, DOI 10.1007/978-1-4419-7725-0_10,

Introduction

In order to survive, an organism must constantly monitor itself for invading entities or cells. This monitoring role is fulfilled by the immune system, which responds to infections from a variety of pathogens, such as viruses, bacteria, protozoa, and fungi. Immunology is a broad field but this contribution focuses on one important part of the immune system, namely T cells. These cells choreograph what is known as the adaptive immune response [1,2].

Briefly, T cells are released from the thymus after undergoing a series of positive and negative selection processes. They recognize foreign epitopes present in an organism by scanning the organism's own cells – which display signature peptide fragments of what is inside on special membrane proteins, together known as Major Histocompatibility Complex (MHC) – with T cell receptors (TCR) on the T cell membrane. The T cell population can be partitioned into clonotypes with the same TCR. Different T cell clonotypes have different receptors that are capable of recognizing peptide fragments. The T cell population is the subject of homeostatic regulation and a particular clonotype population may rise or fall depending on the stimulus that it receives. The unique peptide fragments recognized by a particular TCR are known as the epitopes for that clonotype. There is also some overlap, in the sense that different clonotypes are cross-reactive with other epitopes to varying extents. Clonotype populations receive a stimulus to increase their population from epitopes that they recognize, so this overlap leads to competition amongst clonotypes for survival stimulus. When a pathogen is detected through one of these recognition events, the T cell involved becomes activated and, through a complex series of events, can trigger an immune response [2,3].

The space of all possible epitopes that an organism may potentially be challenged with is enormous so an organism needs to maintain diversity amongst its T cell population so that it is capable of recognizing as many different types of pathogens as possible. Also, while scanning itself, the larger the clonotype population, the sooner the invading pathogen is found. Thus, on the one hand, the more different clonotypes, and the larger the populations, the better; on the other hand, there are limits to the size of the population of T cells that an organism can maintain. In order to cover as much as possible of this epitope space with only a limited population of T cells an organism must minimize the overlap of different clonotypes. As an indication of the magnitude of the numbers involved it has been estimated that the human repertoire has up to $\approx 3 \times 10^7$ distinct TCR clonotypes but that the complexity of the space of all possible peptide-MHC 11mers is $\approx 6 \times 10^{12}$ [4]. This large difference in numbers shows that the correspondence cannot be one-to-one and that there is a need for great diversity and for some cross-reactivity in the T cell repertoire. This motivates the competitive models of T cell clonotype homeostasis considered here but see Stirk, Molina-París and van den Berg for more background to this problem [5].

Mathematical Background

Mathematical models in immunology often consist of ordinary differential equations (ODEs), based on rates of production and decay of pertinent species. However, when some species are present in small numbers, such as T cell clonotypes, a discrete and stochastic framework is more appropriate. In particular, Stirk et al. argue that the appearance and disappearance of clonotypes in the peripheral pool of naïve T cells is an inherently stochastic phenomena and that maintenance functional diversity depends critically on these chance events [5, Sect. 2]. Markov processes provide such a stochastic framework [6] and Stirk et al. employ a continuous-time, discrete state, multivariate Markov process to model T cell populations [5]. Allen provides a nice survey of the different modelling approaches [7].

In our framework, a Markov process consists of N species and the state of the system, $\boldsymbol{x} = [x_1, \ldots, x_N]$, records the integer population of each. Each state may transition to M other states, $\boldsymbol{x} + \boldsymbol{\nu}_j$, for $j = 1, \ldots, M$. Here the $\boldsymbol{\nu}_j$ are a set of M vectors that define the geometry of the Markov process. Associated with each state is a set of M *propensities*, $\alpha_j(\boldsymbol{x}) \geq 0$, that determine the relative chance of each transition occurring. The propensities are defined by the requirement that, given $\boldsymbol{x}(t) = \boldsymbol{x}$, $\alpha_j(\boldsymbol{x})dt$ is the probability of transition j, in the next infinitesimal time interval $[t, t + dt)$. This contribution focuses on numerical methods for studying such immunological models and we now describe two complementary approaches.

Trajectory Approaches

The Stochastic Simulation Algorithm (SSA) is a statistically exact algorithm for simulating strong trajectories of discrete stochastic Markov processes. Algorithm 1 summarizes the SSA. The inputs to the algorithm are: t_f, the amount of time that the simulation should run for; $\boldsymbol{x}_0$, the initial state of the process; $\boldsymbol{\nu}$, N and M as defined above. The output of the algorithm is the state of the system at time t_f. The algorithm needs a way to compute the values of the propensities of each of the M possible transitions. In Algorithm 1 this is done by making a function call. Often these propensities are functions of the current state, parameterized by some constants determined from the application being modelled. We denote the vector of these parameters by $\boldsymbol{c}$, which must also be given as input to the Algorithm. At each step, the SSA samples the waiting time until the next change occurs from an exponential distribution, and samples from a uniform distribution to determine which of the M possible changes occurs, based on the relative sizes of the propensity functions [8, 9]. Note that in Algorithm 1, r_1, r_2, denote random numbers from the uniform distribution on $[0, 1]$, and $\log(\frac{1}{r_1})$ arises because we employ the inverse-transform method for sampling from the exponential distribution. If an absorbing state is reached then $\boldsymbol{\alpha}$ is zero and the algorithm may terminate and simply return the absorbing state. On average, the time step, denoted by τ, is of the order of the reciprocal of the sum of the propensity functions, which may be very small if either

```
ALGORITHM 1: SSA (t_f, x_0, c, v, N, M)
x ← x_0
t ← 0
while t < t_f do
    α ← propensities(x, c)
    if (α == 0)
        t ← t_f
        break
    end if
    α_0 ← Σ_{j=1}^{M} α_j
    r_1, r_2 ~ U[0, 1]
    τ ← (1/α_0) log(1/r_1)
    if (t + τ > t_f)
        t ← t_f
        break
    end if
    t ← t + τ
    choose j such that Σ_{k=1}^{j-1} α_k < α_0 r_2 ≤ Σ_{k=j}^{M} α_k
    x ← x + v_j
end while
return x
```

some of the rate constants are large or some of the species occur in large numbers. Since many thousands, even hundreds of thousands, of simulations may be necessary to compute statistics about the dynamics it may be computationally cheaper to directly compute the probability density function (PDF).

PDF Approaches

Associated with the Markov process is a PDF that evolves according to the forward Kolmogorov equations, which, in this setting, are known as the Master Equations (MEs) [10]. Given an initial condition $\boldsymbol{x}(t_0) = \boldsymbol{x}_0$, the probability of being in state $\boldsymbol{x}$ at time t, $P(\boldsymbol{x}; t)$, satisfies the following discrete partial differential equation (PDE),

$$\frac{\partial P(\boldsymbol{x};t)}{\partial t} = \sum_{j=1}^{M} \alpha_j(\boldsymbol{x} - \boldsymbol{v}_j) P(\boldsymbol{x} - \boldsymbol{v}_j; t) - P(\boldsymbol{x};t) \sum_{j=1}^{M} \alpha_j(\boldsymbol{x}).$$

This ME may be written in an equivalent matrix-vector form so that the evolution of the probability density $\boldsymbol{p}(t)$ (which is a vector of probabilities $P(\boldsymbol{x};t)$, indexed by the states $\boldsymbol{x}$) is described by a system of linear, constant coefficient, ordinary differential equations,

$$\dot{\boldsymbol{p}}(t) = \boldsymbol{A}\boldsymbol{p}(t), \tag{10.1}$$

where the matrix $\boldsymbol{A} = [a_{ij}]$ is populated by the propensities and represents the *infinitesimal generator* of the Markov process, with $a_{jj} = -\sum_{i \neq j} a_{ij}$ [11].

Remark. It is common in the literature to work with the $\boldsymbol{Q}$-matrix: $\boldsymbol{Q} = \boldsymbol{A}^T$.

Given an initial distribution $\boldsymbol{p}(0)$, the solution at time t is

$$\boldsymbol{p}(t) = \exp(t\boldsymbol{A})\boldsymbol{p}(0), \tag{10.2}$$

where the exponential of a bounded operator is usually defined via a Taylor series: $\exp(t\boldsymbol{A}) = \boldsymbol{I} + \sum_{n=1}^{\infty} \frac{(t\boldsymbol{A})^n}{n!}$. The numerical solution of (10.2), for the special class of matrices arising in immunology applications, is the focus of this work. In this context, the matrices often represent birth-death processes and are large and sparse. The matrix exponential is well studied [12, 13] and numerical methods for linear ODEs [14] are closely related. There are some technical considerations when the system is infinite, as noted in [15–17]. Recently, the Finite State Projection (FSP) algorithm was suggested as a way of handling the large matrices that arise in MEs associated with chemical kinetic processes [18].

The FSP algorithm In the FSP algorithm the matrix in (10.2) is replaced by $\boldsymbol{A}_k$ where

$$\boldsymbol{A} = \left(\begin{array}{c|c} \boldsymbol{A}_k & * \\ \hline * & * \end{array}\right) \tag{10.3}$$

i.e. $\boldsymbol{A}_k$ is a $k \times k$ submatrix of the true operator $\boldsymbol{A}$. The states indexed by $\{1, \ldots, k\}$ then form the *finite state projection*. The FSP algorithm replaces (10.2) with the approximation

$$\boldsymbol{p}(t_f) \approx \begin{pmatrix} \exp(t_f \boldsymbol{A}_k)\boldsymbol{p}_k(0) \\ 0 \end{pmatrix},$$

which, by [18, Theorem 2.1], is nonnegative. The subscript k denotes the truncation just described and we note that a similar truncation is applied to the initial distribution. Consider the column sum $\Gamma_k = \mathbb{1}^T \exp(t_f \boldsymbol{A}_k)\boldsymbol{p}_k(0)$, where $\mathbb{1} = (1, ..., 1)^T$ with appropriate length. Normally the exact solution (10.2) is a probability vector with unit column sum, however due to the truncation, the sum Γ_k may be less than one, because in the approximate system, probability is no longer conserved. However, as k increases, Γ_k increases too, so that the approximation is gradually improved [18]. Additionally it is shown in [18, Theorem 2.2] that if $\Gamma_k \geq 1 - \epsilon$ for some pre-specified tolerance ϵ, then

$$\begin{pmatrix} \exp(t_f \boldsymbol{A}_k)\boldsymbol{p}_k(0) \\ \boldsymbol{0} \end{pmatrix} \leq \boldsymbol{p}(t_f) \leq \begin{pmatrix} \exp(t_f \boldsymbol{A}_k)\boldsymbol{p}_k(0) \\ \boldsymbol{0} \end{pmatrix} + \epsilon\mathbb{1}.$$

For simplicity we described the algorithm as if it merely increases k but it can be generalized so that the projection is expanded around the initial state in a way that respects the *reachability* [18] of the Markov model.

The Krylov FSP Algorithm The FSP method was recently improved to a Krylov-based approach [15, 19, 20] by adapting Sidje's Expokit codes [21, 22]. The Krylov FSP converts the problem of exponentiating a large sparse matrix to that of exponentiating a small, dense matrix in the Krylov subspace. Given an initial vector and matrix the Krylov subspace of dimension m is

$$\mathcal{K}_m = \mathcal{K}_m(\boldsymbol{A}, \boldsymbol{v}) = span\left\{\boldsymbol{v}, \boldsymbol{A}\boldsymbol{v}, \boldsymbol{A}^2\boldsymbol{v}, \ldots, \boldsymbol{A}^{m-1}\boldsymbol{v}\right\}.$$

The dimension m of the Krylov subspace is typically small and $m = 30$ was used in this implementation. The Krylov approximation to $\exp(\tau \boldsymbol{A})\boldsymbol{v}$ is

$$\beta \boldsymbol{V}_{m+1} \exp\left(\tau \overline{\boldsymbol{H}}_{m+1}\right) \boldsymbol{e}_1,$$

where $\beta \equiv \|\boldsymbol{v}\|_2$, $\boldsymbol{e}_1$ is the first unit basis vector, and $\boldsymbol{V}_{m+1}$ and $\overline{\boldsymbol{H}}_{m+1}$ are the orthonormal basis and upper Hessenberg matrix, respectively, resulting from the well-known Arnoldi process. The exponential in the smaller subspace is computed via the diagonal Padé approximation with degree $p = 6$, together with scaling and squaring.

A description of the Arnoldi process may be found in the classic text of Golub and Van Loan [23] and is often employed as a numerical scheme for eigenvalue problems, for example. In our context, we employ the Arnoldi process to build a basis for the Krylov subspace $\mathcal{K}_m$. This is similar to the usual mathematical Gram–Schmidt process for building a basis but involves a small change which makes the Arnoldi process preferable numerically. It results in a matrix $\boldsymbol{V}_m$ whose columns form an orthonormal basis of $\mathcal{K}_m$.

The Krylov FSP is a general purpose ME-solver. In this work we describe how to apply it to immunological models. Also, we will demonstrate how to generalize the algorithm to the case of time-dependent or age-dependent propensities.

Numerical experiments All numerical experiments were performed in `MATLAB` on a 2 GHz processor running the `Windows XP` operating system. For the numerical purposes here it is enough to report the values of the parameters used but for other applications appropriate scalings and units of measurement would be required.

T Cell Homeostasis for Two Competing Clonotypes

We now review the Stirk et al. model and adopt their notation [5]. The model consists of N T cell clonotypes and the state of the system records the nonnegative population of each. The model is a multivariate birth-death process [6]: the population of a particular clonotype may change by an increase or decrease of precisely one cell at a time. Thus associated with each state is a set of $M = 2N$ propensities and corresponding vectors. The vectors, previously denoted $\boldsymbol{v}_j$, are all of the form $[0, \ldots, \pm 1, 0, \ldots, 0]$, where the ± 1 occurs in the ith component to denote an increase or decrease of the ith species. Here we consider a two-dimensional version so $N = 2$, $M = 4$ and $\boldsymbol{x} = [n, n']$. With n T cells of clonotype i and n' T cells of clonotype i', the birth and death rate of cells of clonotype n are denoted $\lambda_{nn'}$ and $\mu_{nn'}$, respectively. These propensities were previously denoted $\alpha_j([n, n'])$. A similar notion denotes the corresponding rates for cells of clonotype i'. Naturally, when there are zero cells of a clonotype the birth and death rates for that clonotype are both zero, i.e., $\lambda_{0n'} = \mu_{0n'} = \lambda_{n0} = \mu_{n0} = 0$. Otherwise, for $n, n' = 1, 2 \ldots$, we define the birth and death rates to have the following functional form:

$$
\begin{aligned}
\lambda_{nn'} &= \phi n f(n, n', \nu_{ii'}, \nu_{i/i'}, p) \\
\lambda'_{nn'} &= \phi' n' f(n', n, \nu_{i'i}, \nu_{i'/i}, p') \\
\mu_{nn'} &= \mu n \\
\mu'_{nn'} &= \mu' n',
\end{aligned}
$$

where

$$
f(n, n', x, y, p) = p\mathrm{e}^{-x} \sum_{r=0}^{\infty} \frac{x^r}{r!} \frac{1}{r\,\langle n \rangle + n + n'} + (1-p)\mathrm{e}^{-y} \sum_{r=0}^{\infty} \frac{y^r}{r!} \frac{1}{r\,\langle n \rangle + n}. \tag{10.4}
$$

The form of the birth rates reflects the competition between clonotypes for survival stimulus, so that clonotypes that overlap more with other clonotypes in terms of the set of APPs that they depend on, receive less stimulus overall than clonotypes that overlap less with other clonotypes. The summation over r arises because of the way that the sets of APPs are partitioned. In general the rth term corresponds to the stimulus that the clonotype receives from those APPs that it shares with r other clonotypes. For example, the $r = 0$ term corresponds to the stimulus that the clonotype receives from APPs that it shares with no other clonotypes, the $r = 1$ term corresponds to the stimulus that the clonotype receives from one other clonotype, etc. In any particular organism the number of distinct clonotypes will be finite, albeit possibly very large, so after finitely many terms in the summation, the rest of the terms are zero. This function may be approximated, for example, by truncating each of the two summations, or by employing a Poisson approximation [5]. Notice that the model has an absorbing state at [0, 0]. Here $\langle n \rangle$ is the average clone size over the naïve repertoire, $\nu_{ii'}$ is the mean niche overlap for antigen presentation profiles (APPs) that can provide survival signals to T cells of clonotype i and clonotype i', and $\nu_{i/i'}$ is the mean niche overlap for APPs that provide survival signals to T cells of clonotype i but not T cells of clonotype i'. See Stirk et al. for more details.

This describes the model in its full generality but we consider the special case

$$
\nu_{i/i'} \gg 1, \qquad \nu_{i'/i} \gg 1.
$$

In this case the authors [5] observe that a good approximation to (10.4) is

$$
f(n, n', x, y, p) = \frac{p}{n + n'} + (1-p)\frac{1}{n + \langle n \rangle\, y}. \tag{10.5}
$$

Parameters We fix $\langle n \rangle = 10$. We begin with 10 cells of each clonotype and consider the solution at $t_f = 30$. We choose:

$$\begin{aligned} \phi = \phi' &= 60 \\ \mu = \mu' &= 1 \\ \nu_{i/i'} = \nu_{i'/i} &= 100 \\ p = p' &= 0.5 \,. \end{aligned} \tag{10.6}$$

Thus the model is symmetric.

Simulations

Figure 10.1 shows an SSA simulation of the model with these parameters. One clonotype reaches extinction before $t = 30$, while the other clonotype remains and fluctuates about an average of approximately 32 cells for a very long time. At first we may think that the simulation has settled down to its stationary distribution but this is not the case because extinction of both species is guaranteed in this model. Thus we instead associate the long period with $n > 0$ shown in Fig. 10.1 with a quasi-stationary distribution (QSD).

Here we give a simple argument to show that the simulation in Fig. 10.1 is indeed eventually absorbed. Note that Stirk et al. give a more sophisticated analysis and establish that absorption is guaranteed in general. After extinction of one of the clonotypes, we are left with a one-dimensional, infinite state, birth-death process for the remaining clonotype. There is an absorbing state at 0, and an irreducible class $\{1, 2, \ldots\}$, from which the absorbing state is accessible. Following standard notation and standard theory for birth-death processes [6], absorption with probability one is equivalent to divergence of the series

$$\sum_{i=1}^{\infty} \frac{\mu_1 \mu_2 \ldots \mu_i}{\lambda_1 \lambda_2 \ldots \lambda_i}.$$

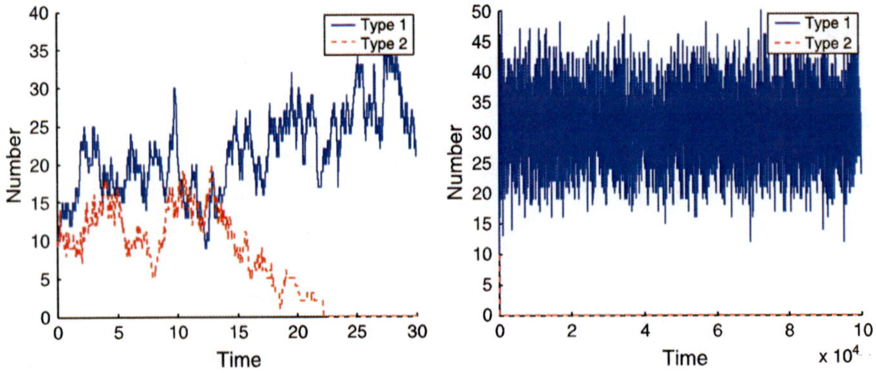

Fig. 10.1 A simulation of the T cell clonotype populations

In the simulation in Fig. 10.1, after extinction of clonotype n', the death rate is $\mu_n = \mu n$ and the function in (10.5) is

$$\frac{p}{n} + \frac{1-p}{n + \langle n \rangle \nu_{i/i'}} \approx \frac{1}{n},$$

for large n, so the birth rate $\lambda_n \approx \phi n(1/n) = \phi$ for large n. With these observations it is straight-forward to show that the series diverges, for example by the *nth* term test.

Although absorption is guaranteed, we can see from Fig. 10.1 that the time to extinction for both species is very large (at least $t = 10^5$ in this case). The length of time that a clonotype spends in a QSD before being absorbed is identified as an important issue and discussed in depth by Stirk et al. [5]. Bounds for the mean time until absorption are given [5, Sect. 3]. Also, the work of Nåsell [24] is applied to this process to find the form of the QSD [5, Sect. 3.2.2]. Significantly, it is shown that clones with a smaller niche overlap last longer, because they face less competition for survival stimuli. This mechanism drives the T cell repertoire towards greater diversity and thus endows the immune system with a greater functional capacity for recognizing a larger variety of foreign pathogens.

Figure 10.2 shows the evolution of the PDF associated with the model, computed via a Krylov FSP algorithm for the solution of the Master Equation [20]. The PDF starts as a delta distribution on the initial state, as seen in the figure labelled $t = 0$. However, after a very short time, $t = 1$, it spreads out to resemble a Gaussian distribution. By $t = 5$, it is trimodal. The middle peak then gradually disappears so that by $t = 30$, only two of the original peaks remain. The remaining peaks correspond to the QSD. Each peak corresponds to moderate levels of one T cell clonotype and extinction of the other. The distribution is symmetric about the line $n = n'$, reflecting the symmetry of this model.

Time-Dependent Propensity Functions

Thus far ϕ has been a constant but Stirk et al. identify this as an issue for future research and discuss the desire to consider it as a time-dependent parameter, because ϕ, which represents the strength of the stimulation that a T cell receives from the relevant APPs, depends on the T cell repertoire, which changes with time as some clones go extinct and new clones are released from the thymus [5]. It is well-recognized that older individuals often have a weaker immune system, as evidenced by the greater susceptibility of the aged to infection [2]. Very complicated processes in immunobiology underly this aging effect but it may be due in part to a decline in the strength of the birth rates. For example, there are limits to how many times that a T cell may replicate, perhaps partly due to the shortening of the telomeres with each replication [5, 25, 26], and the level of stimulation felt by a T cell clonotype may decline with age [27]. This motivates us to consider a variation of the original

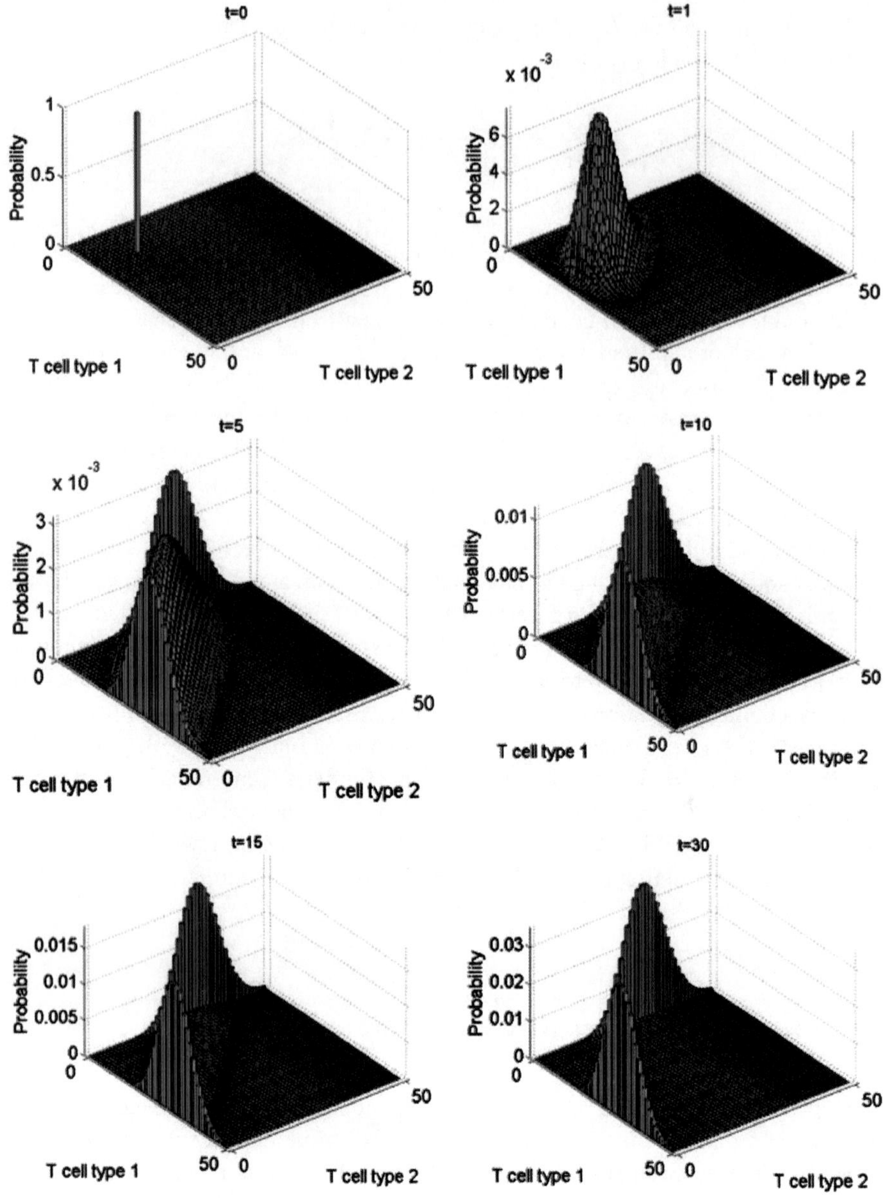

Fig. 10.2 The initial delta distribution, at $t = 0$, on the initial state [10, 10], and the solution to the Master Equation at times $t = 1, 5, 10, 15, 30$

model in which the birth rates decline with time. As Stirk et al. and others point out, similar numbers of T cells are found in older and younger individuals [28]. Thus we must bear in mind with what follows that we consider a reduction in birth rates for some clonotypes but not all. Here we consider just two clonotypes, which is only a small fraction of the total number of clones in the full T cell repertoire.

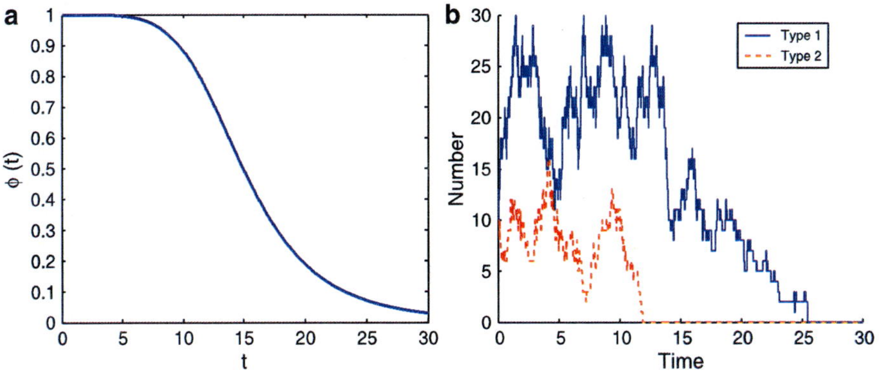

Fig. 10.3 (**a**) The time dependence of $\phi(t)$ in (10.7). (**b**) A simulation of the T cell clonotype populations, with the time-dependent propensities scaled as described near (10.7)

We represent time dependence by the hill function

$$\phi(t) = \frac{1}{1 + \left(\frac{t}{K}\right)^m}, \tag{10.7}$$

where $K = 15$ and $m = 5$. The function declines with time, with a steep sigmoidal shape, as shown in Fig. 10.3a. We scale the propensities for the birth rates by $\phi(t)$. For this example, we replace ϕ and ϕ' in (10.6) with $60\phi(t)$. Thus at $t = 0$ the new birth rates are the same as the old ones and remain similar for small values of t, but for large values of t the new birth rates are much less than the old birth rates.

Figure 10.3b shows a simulation of the time-dependent process. This simulation is obtained by a slight variation of the algorithm for the time-independent case [29]. The main subtlety here is in choosing the time step based on the exponential of the integral of the propensity functions. We see the populations decline sharply at $t \approx 15$, corresponding to the decline in $\phi(t)$. In contrast to the time-independent case, both populations are extinct by $t = 30$.

Numerical Solution of the Time-Dependent Master Equation

Having considered a trajectory approach we now discuss complementary PDF approaches to the time-dependent case. Consider the system

$$p'(t) = \boldsymbol{A}(t)p(t),$$

which is the same as in (10.1) except that here the matrix $\boldsymbol{A}(t)$ is time-dependent. In analogy with the constant-coefficient case, we would like to find a solution of the form

$$p(t) = \exp\Big(\sigma(t)\Big)p(0).$$

If, $\forall t, t'$, $\boldsymbol{A}(t)$ commutes with $\boldsymbol{A}(t')$, we have

$$\sigma(t) = \int_0^t \boldsymbol{A}(s)\mathrm{d}s.$$

In the constant-coefficient case, the integral is $\sigma(t) = t\boldsymbol{A}$, thus recovering (10.2) as an important special case. Magnus [30] derived a formula for the general case:

$$\begin{aligned}\sigma(t) = & \int_0^t \boldsymbol{A}(s_1)ds_1 - \frac{1}{2}\int_0^t\int_0^{s_1}[\boldsymbol{A}(s_2), \boldsymbol{A}(s_1)]\mathrm{d}s_2\mathrm{d}s_1 \\ & +\frac{1}{4}\int_0^t\int_0^{s_1}\int_0^{s_2}\Big[[\boldsymbol{A}(s_3), \boldsymbol{A}(s_2)], \boldsymbol{A}(s_1)\Big]\mathrm{d}s_3\mathrm{d}s_2\mathrm{d}s_1 \\ & +\frac{1}{12}\int_0^t\int_0^{s_1}\int_0^{s_1}\Big[\boldsymbol{A}(s_3), [\boldsymbol{A}(s_2), \boldsymbol{A}(s_1)]\Big]\mathrm{d}s_3\mathrm{d}s_2\mathrm{d}s_1 + \ldots \quad (10.8)\end{aligned}$$

Here the commutator $[\boldsymbol{A}, \boldsymbol{B}]$ is defined as $\boldsymbol{AB} - \boldsymbol{BA}$. This infinite series continues with higher order terms containing progressively higher order commutators and higher order multiple integrals. Burrage has discussed the Magnus formula in the context of stochastic differential equations [31]. It has been widely used for theoretical purposes but for numerical purposes, it is awkward to evaluate in this form. For example, one may consider the approximation obtained by truncating the series after suitably many terms. However, it is expensive to evaluate the many multiple integrals, and higher commutators, and it is not trivial to decide where to truncate the series, or which terms should be selected to obtain the highest order approximation for the least computational effort. Fortunately, in a series of papers on numerical methods based on the Magnus formula, Iserles et al. [32] showed that the scheme

$$\boldsymbol{A}_1 = \boldsymbol{A}\left(t_N + \left(\frac{1}{2} - \frac{\sqrt{3}}{6}\right)h\right), \qquad \boldsymbol{A}_2 = \boldsymbol{A}\left(t_N + \left(\frac{1}{2} + \frac{\sqrt{3}}{6}\right)h\right),$$

$$\sigma_{N+1} = \frac{1}{2}h(\boldsymbol{A}_1 + \boldsymbol{A}_2) + \frac{\sqrt{3}}{12}h^2[\boldsymbol{A}_2, \boldsymbol{A}_1], \quad (10.9)$$

$$\boldsymbol{p}(t_{N+1}) = \mathrm{e}^{\sigma_{N+1}}\boldsymbol{p}(t_N), \quad (10.10)$$

where $t_{N+1} = t_N + h$, is an order four approximation. Notice $\boldsymbol{A}_1$ and $\boldsymbol{A}_2$ are the evaluation of the matrix at Gaussian quadrature points.

For this work, we implement this scheme in combination with a Krylov FSP approach. The algorithm repeats two steps until the desired time point is reached: first, with a fixed step size of $h = 0.5$, we form $\boldsymbol{A}_1$, $\boldsymbol{A}_2$ and σ_{N+1} in (10.9); then pass σ_{N+1} to the Krylov FSP solver and apply the update rule (10.10). We refer to this procedure as the *Magnus Krylov FSP algorithm*. It is summarized in Algorithm 2.

The algorithm requires the following inputs: the initial distribution, $\boldsymbol{p}(0)$; the time t_f, at which the solution is required; and the matrix $\boldsymbol{A}$ representing the Markov

ALGORITHM 2: Magnus Krylov FSP($\boldsymbol{A}$, $\boldsymbol{p}(0)$, t_f, h)
$t \leftarrow 0$;
$\boldsymbol{p}(t) \leftarrow \boldsymbol{p}(0)$;
while $t < t_f$ **do**
    Compute $\boldsymbol{A}_1$, $\boldsymbol{A}_2$ and $\boldsymbol{\sigma}_{N+1}$ in (10.9) with $t_N = t$;
    Compute $\mathrm{e}^{\boldsymbol{\sigma}_{N+1}} \boldsymbol{p}(t)$ by *Krylov FSP*
    $\boldsymbol{p}(t) \leftarrow \mathrm{e}^{\boldsymbol{\sigma}_{N+1}} \boldsymbol{p}(t)$ as in (10.10);
    $t \leftarrow t + h$;
end while
return $\boldsymbol{p}(t)$;

chain. The initial distribution will often be a delta distribution on the initial state, as in the examples in this paper. The algorithm also requires a way to evaluate the matrix $\boldsymbol{A}(t)$ at various time points. Although Algorithm 2 is described as if it requires the full matrix $\boldsymbol{A}$, in fact it does not need this. A function representing the action of the matrix on a vector is sufficient for a Krylov method. At any instant the algorithm only requires a finite principal submatrix of the full matrix $\boldsymbol{A}$. This finite state projection can be dynamic so that it tracks most of the support of the distribution [18, 20]. The algorithm also requires a step size h and for this implementation a fixed step size of $h = 0.5$ was used. However this is a simplification and in future work adaptive step size strategies will be investigated.

Remarks. – In general Algorithm 2 expands the projection at each step as necessary. However for the particular application in this paper we implemented a simplified version of the algorithm. First, we truncate the state space to a size of approximately 10,000. We use this same state space to form $\hat{\boldsymbol{A}}(t)$ $(\forall t)$. Note that $\hat{\boldsymbol{A}}(t)$ is only an approximation to $\boldsymbol{A}(t)$ but we will abuse notation and denote both by $\boldsymbol{A}(t)$.
- Due to the expansion around the initial state in whole numbers of steps of reachability [20], the precise size of the state space employed by the algorithm is 10,076. Repeating the experiments with a smaller truncation size of 5,000 gives results that are visually indistinguishable from the larger system.
- The sum of the conserved probability at each step is monitored and remains $> 1 - 10^{-5}$. In the constant coefficient case this would guarantee accuracy, with the computed solution being a lower bound on the true solution, within $\epsilon = 10^{-5}$ in a component-wise sense. This suggests that the solution in our time-dependent case is also very accurate. However, this is only a heuristic algorithm and a formal analysis of the error behaviour is still to be considered.
- In the current implementation, at each step, $\boldsymbol{A}_1$, $\boldsymbol{A}_2$, $[\boldsymbol{A}_1, \boldsymbol{A}_2]$ and finally $\boldsymbol{\sigma}$ are formed as sparse matrices and then $\boldsymbol{\sigma}$ is passed to the Krylov solver, which employs an Arnoldi process together with an adaptive time step integration scheme. The matrices $\boldsymbol{A}_1$, $\boldsymbol{A}_2$ share the same sparsity pattern and are very sparse: the density of non zeros is $\approx 5 \times 10^{-4}$. The commutator is about twice as dense, at $\approx 10^{-3}$, but this is still very sparse. So the problem is well-suited to a Krylov approach.

- The solutions are computed within a few minutes. Matrices such as $\boldsymbol{A}_1$ can be formed in less than one second.
- In many applications $\boldsymbol{A}(t) \to \boldsymbol{A}_\infty$ as $t \to \infty$ so that for sufficiently large t the problem reduces to (10.1) with $\boldsymbol{A}$ replaced by the constant equilibrium matrix $\boldsymbol{A}_\infty$. In these situations it is desirable to combine the Magnus Krylov FSP with the standard Krylov FSP described in the previous section: initially, we employ the Magnus method but after sufficiently large t is reached, we switch to the cheaper, standard Krylov FSP. The successful application of this combination requires a way to recognize when sufficiently large t has been reached. For the application at hand, $\phi(t) \to 0$ for large t so we could employ this strategy for, say $t > 100$, for example.

Figure 10.4 shows the evolution of the PDF in the time-dependent case. Initially, the distribution is similar to the time-independent case: compare, for example, Fig. 10.4 at $t = 5$ with Fig. 10.2. However, at $t \approx 8$ the evolution starts to diverge and by $t = 13$ the PDF is noticeably different. In particular we see the distribution gradually moving towards the origin at the snapshots in $t = 13, 17, 23, 25$, and by $t = 30$ about 80% of the distribution is in the absorbing state. This is in contrast to the time-independent case for which the QSD lasts for a very long time. This attenuation of the time spent in the QSD is significant in understanding the immunological implications of the age-dependent case. As noted in above, diversity in the T cell repertoire is maintained because clonotypes with a smaller niche overlap tend to last longer because they face less competition for survival stimuli. However for clonotype populations that show an age-dependent decline in ϕ as in (10.7), the QSD may not last very long even if the clonotype has a small niche overlap. This results in an abnormal loss of diversity from the T cell repertoire. corresponding decrease in the functional capacity of the immune system to recognize foreign epitopes.

The Mean Time to Extinction of Both Clonotypes

In this section we quantify how the survival time of a clonotype depends on the time-dependence of the birth-rates. Assuming that the process begins in the quasi-stationary distribution, the mean time until absorption is given by $1/\lambda_C$, where λ_C is the decay parameter associated with the Markov chain. For many applications, after a brief initial transient the process is approximately in the QSD so this is a reasonable assumption. The theory of the decay parameter and QSDs is presented in [33]. Thus we use $1/\lambda_C$ as an indication of the survival time of a clonotype. For the bivariate clonotype model, we compute the eigenvalue, λ, with smallest magnitude, of the matrix $\boldsymbol{A}_C$. Here $\boldsymbol{A}_C$ is same as the FSP (10.3), except that the absorbing state has also been removed. One way to compute λ is to use the inverse power method, or to call the MATLAB sparse eigenvalue routine, `eigs`$(\boldsymbol{A}_C, 1, 0)$. We approximate the decay parameter by $\lambda_C \approx -\lambda$. Note that there are some technical issues with this approximation for infinite models [33–36]. Although the clonotype models are

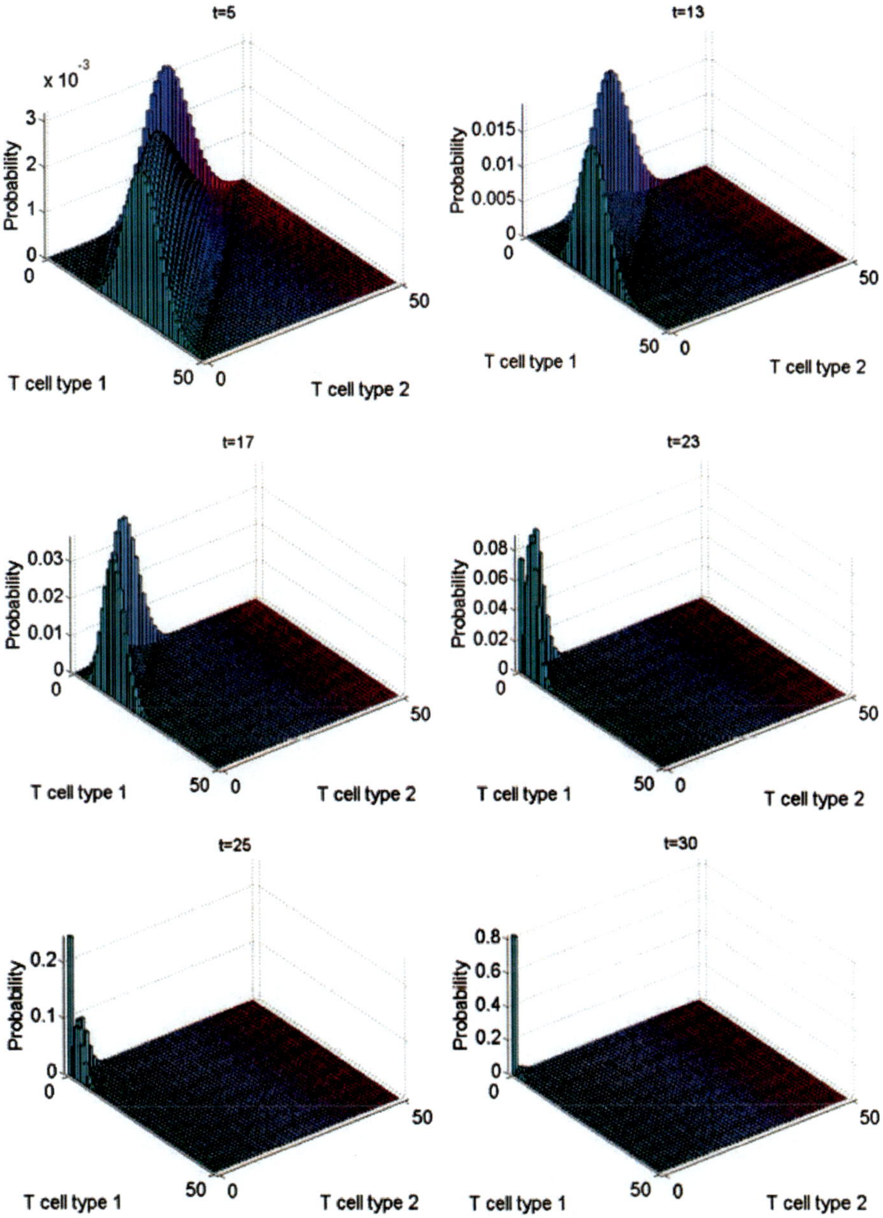

Fig. 10.4 The solution to the ME at $t = 5, 13, 17, 23, 25, 30$ in the time-dependent case

infinite this is because bounds are not known and may be large. Real applications will have finite numbers of clonotypes so it is reasonable to focus on the finite case.

Figure 10.5 shows the change in the mean time to absorption, from the QSD, as a function of time. It can be seen that the time-dependence resembles that of the time-dependent birth rates in Fig. 10.3a, although on a log scale. At $t = 0$, the average

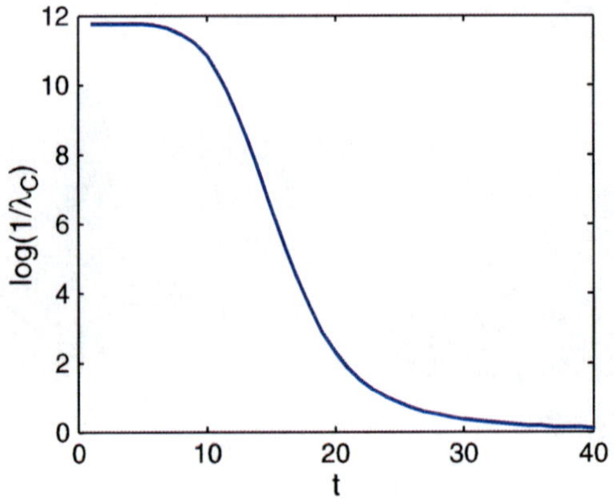

Fig. 10.5 The mean time until extinction of both clonotypes in the two-dimensional model

lifetime of a clonotype is $\approx 10^{11}$ but at $t = 20$ the lifetime is only $\approx 10^2$. This shows that the simulation in Fig. 10.1 would probably not go extinct until $\approx 10^{11}$. Also, this shows that the reduction in the birth rates over time, by about an order of magnitude, leads to a much larger effect on the life expectancy of a clonotype, which decreases by several orders of magnitude.

Discussion

This work has focused on the implications of a particular functional form of the time-dependence for diversity maintenance but future work will consider the significance of different functional forms for $\phi(t)$. Furthermore it would be interesting to explore parameter space and models that have more clonotypes. For example this work focused on the case that $\nu \gg 1$ but it would be interesting to relax this assumption and consider the model with appropriate approximations to the birth rates. We identify a number of important areas for refining the numerical methods described here:

- Generalizing the error analysis of the Krylov FSP approximation to:
  - A time-dependent matrix
  - A combination of the Krylov FSP with the order 4 approximation
- Employing adaptive time-steps in the Magnus Krylov FSP integrator.
- Identifying a more sophisticated embedding of the Magnus formula in the Krylov methods, for improved computational efficiency. For example, as a Krylov method, only a function that returns the action of σ on a given vector, $\mathbf{v}$, is required, so we do not need to form the matrix, or the commutator in (10.9). Instead, one possibility is to employ the following scheme. First form the vectors,

$$\mathbf{w}_1 = \mathbf{A}_1\mathbf{v}, \quad \mathbf{w}_2 = \mathbf{A}_2\mathbf{v}, \quad \mathbf{w}_3 = \mathbf{A}_2\mathbf{w}_1, \quad \mathbf{w}_4 = \mathbf{A}_1\mathbf{w}_2,$$

then form $\boldsymbol{\sigma v}$ as

$$\frac{h}{2}\left(w_1 + w_2 + \frac{\sqrt{3}h}{6}(w_3 - w_4)\right).$$

- However a more advanced embedding is desired. series (10.8) by a vector. For example, perhaps the Magnus formula can be more deeply incorporated into the fabric of the numerical time stepping procedure; the Arnoldi process; or the approximation of the exponential of the small, dense, matrix $\boldsymbol{H}_m$. This will be important for larger scale problems arising in models of more than two clonotypes.
- Finally, in the time-dependent case, the PDF approach can be extremely competitive with the trajectory approach [29]. For example, we could run only a few 100 simulations in the same time that it takes for the Magnus Krylov FSP to compute the PDF. However no attempt was made to optimize the trajectory algorithm, which is significantly slower than its constant-coefficient counterpart, so a thorough investigation of the relative efficiency of the approaches is deferred to future work.

This chapter is based on a technical computing article in SIAM Multiscale Modelling and Simulation by the same authors that would normally only be read by the computational mathematics community but which is not accessible to the immunology community. More technical details can be found there, where some of the issues are addressed. It is the authors' hope that by appearing in the present context, this chapter may raise awareness amongst the immunology community of the potential of computational and interdisciplinary techniques to contribute to immunology.

Conclusions

We have demonstrated a number of novel numerical strategies to be relevant to computational immunology. First, it has been demonstrated that the Krylov FSP can be applied to models in immunology, so that PDF approaches to the governing master equations are now feasible. Second, Magnus formulas allow the Krylov FSP to be applied to systems with time or age-dependent rates. Previously, Iserles, Nørsett and Rasmussen demonstrated a clever order four approximation based on the Magnus expansion. We have combined the approximation with a Krylov method, allowing us to handle matrices of larger size. For example, we have dealt with 10,000 × 10,000 matrices. This has allowed us to handle interesting applications in computational immunology with age-dependent birth rates. Furthermore, it will be straight-forward to apply the numerical methods developed here to other applications in computational immunology and also to the chemical master equation for biochemical kinetics with time-dependent propensities.

We have applied the novel computational methods, to investigate one model of aging effects on the immune system. In some cases, quasi-stationary distributions for the populations of a particular clonotype – that would otherwise be sustained for extended periods – may have their life spans significantly attenuated by a decline in the birth rates due to aging. This would seem to be a very important factor in understanding the effects of aging on health. Finally, this work concentrated on a special case of the Stirk, Molina-París and van den Berg model [5] of T cell homeostasis but in future work it is planned to investigate the model more generally.

References

1. Lodish H, Berk A, Zipursky L, Matsudaira P, Baltimore D, Darnell J (2008) Molecular cell biology. W.H. Freeman, New York
2. Janeway C, Travers P, Walport M, Shlomchik M (2005) Immunobiology: the immune system in health and disease. Garland Science/Churchill Livingstone, New York
3. Davis SJ, van der Merwe PA (2006) The kinetic-segregation model: TCR triggering and beyond. Nat Immunol 7:803–809
4. Nikolich-Zugich J, Slifka M, Messaoudi L (2004) The many important facets of T-cell repertoire diversity. Nature 4:123–132
5. Stirk ER, Molina-París C, van den Berg HA (2008) Stochastic niche structure and diversity maintenance in the T cell repertoire. J Theor Biol 255:237–249
6. Bremaud P (1999) Markov chains: Gibbs fields, Monte Carlo simulation and queues. Springer, New York
7. Allen LJS (2003) An introduction to stochastic processes with applications to biology. Prentice Hall, Upper Saddle River, NJ
8. Gillespie D (1992) Markov processes: an introduction for physical scientists. Academic, San Diego
9. Gillespie DT (1977) Exact stochastic simulation of coupled chemical reactions. J Phys Chem 81:2340–2361
10. van Kampen NG (2001) Stochastic processes in physics and chemistry. Elsevier Science, Amsterdam
11. Stewart WJ (1994) Introduction to the numerical solution of Markov chains. Princeton University Press, Princeton, NJ
12. Moler C, Van Loan C (1978) Nineteen dubious ways to compute the exponential of a matrix. SIAM Rev 20:801–836
13. Moler C, Van Loan C (2003) Nineteen dubious ways to compute the exponential of a matrix, 25 years later. SIAM Rev 45:3–49
14. Burrage K (1995) Parallel and sequential methods for ordinary differential equations. Oxford University Press, Oxford
15. MacNamara S, Burrage K, Sidje RB (2007) An improved finite state projection algorithm for the numerical solution of the chemical master equation with applications. In: Read W, Roberts AJ (eds) Proceedings of the 13th biennial computational techniques and applications conference, CTAC-2006. ANZIAM J 48:C413–C435
16. Kato T (1976) Perturbation theory for linear operators. Springer, New York
17. Norris JR (1997) Markov chains. Cambridge University Press, Cambridge
18. Munsky B, Khammash M (2006) The finite state projection algorithm for the solution of the chemical master equation. J Chem Phys 124:044104
19. Burrage K, Hegland M, MacNamara S, Sidje R (2006) A Krylov-based finite state projection algorithm for solving the chemical master equation arising in the discrete modelling of biological systems. In: Langville A, Stewart W (eds) 150th Markov Anniversary Meeting, Charleston, SC, USA. Boson Books, pp 21–38
20. MacNamara S, Burrage K, Sidje R (2008) Multiscale modeling of chemical kinetics via the master equation. Multiscale Model Simul 6:1146–1168

21. Sidje RB (1998) Expokit: a software package for computing matrix exponentials. ACM Trans Math Softw 24:130–156
22. Sidje R, Stewart W (1999) A numerical study of large sparse matrix exponentials arising in Markov chains. Comput Stat Data Anal 29:345–368
23. Golub GH, Van Loan CF (1996) Matrix computations. Johns Hopkins, Baltimore
24. Nåsell I (2001) Extinction and quasi-stationarity in the verhulst logistic model. J Theor Biol 211:11–27
25. Kilpatrick R, Rickabaugh T, Hultin L, Hultin P, Hausner M, Detels R, Phair J, Jamieson B (2008) Homeostasis of the naive $CD4^{+}$ T cell compartment during aging. J Immunol 180:1499–1507
26. Goronzy J, Lee WW, Weyand C (2007) Aging and T-cell diversity. Exp Gerontol 42:400–406
27. Sansoni P, Vescovini R, Fagnoni F, Biasini C, Zanni F, Zanlari L, Telera A, Lucchini G, Passeri G, Monti D, Franceschi C, Passeri M (2008) The immune system in extreme longevity. Exp Gerontol 43:61–65
28. Aspinall R, Andrew D (2000) Thymic involution in aging. J Clin Immunol 20:250–256
29. Anderson DF (2007) A modified next reaction method for simulating chemical systems with time dependent propensities and delays. J Chem Phys 127:214107
30. Magnus W (1954) On the exponential solution of differential equations for a linear operator. Commun Pure Appl Math 7:649–673
31. Burrage PM (1999) Runge-Kutta methods for stochastic differential equations. PhD thesis, The University of Queensland
32. Iserles A, Nørsett SP, Rasmussen AF (1998) Time-symmetry and high-order Magnus methods. Technical report, University of Cambridge
33. Anderson WJ (1991) Continuous-time Markov chains: an applications orientated approach. Springer, New York
34. Pollett P (1995) The determination of quasistationary distributions directly from the transition rates of an absorbing markov chain. Math Comput Model 22:279–287
35. Sirl D, Zhang H, Pollett P (2007) Computable bounds for the decay parameter of a birth-death process. J Appl Probab 44:476–491
36. Breyer LA, Hart AG (2000) Approximations of quasi-stationary distributions for markov chains. Math Comput Model 31:69–79

Chapter 11
Dendritic Cell Migration in the Intestinal Tract

Rowann Bowcutt and Sheena Cruickshank

Abstract T cells are critical cells for the development of adaptive immune responses and protective immunity. However, before T cells can act, they must be switched on which is done by professional antigen presenting cells such as dendritic cells (DCs). DCs are found in all parts of the body and their main role is to detect pathogens or injury, take up antigens, process them and present them to T cells. Thus, DCs represent the beginning of a well orchestrated immune response. In order for effective immune function, DCs need to both get to the sites of injury or pathogen invasion as well as the sites of T cell activation. DC movement or migration is a complex multi-step process which also involves phenotypic and functional changes to the DC itself to facilitate movement. In response to an injury or infection, DCs are recruited both locally and from the blood to the site of injury. As DCs exit the tissue they also must be replenished. Dysregulation of the DC migratory response can result in chronic inflammation and the development of inappropriate immune responses. DC migratory behaviour varies depending on the anatomical location. One of the most significant areas of antigen uptake in the body is the intestinal tract. Here we address the process of DC migration within the large and small intestine.

Dendritic cells (DCs) act as a bridge between the innate and the adaptive immune response. DCs are professional antigen presenting cells that are unique in their ability to educate or prime naive T cells in order to generate effector T cells and establish adaptive immunity. DCs arise from progenitor cells in the bone marrow and migrate via the blood to tissues in the body. DCs are found in almost all tissues, for example the intestinal tract, skin and lungs. Such tissue-resident DCs are termed immature DCs. Immature DCs are able to recognise and ingest pathogens and are efficient at taking up antigens via several mechanisms including phagocytosis. Once the immature DC has sensed and taken up an antigen, the DC goes through a well-characterised process of maturation. During DC maturation,

S. Cruickshank (✉)
Faculty of Life Sciences, AV Hill Building, University of Manchester, Oxford Road, Manchester M13 9PL, UK
e-mail: sheena.cruickshank@manchester.ac.uk

C. Molina-París and G. Lythe (eds.), *Mathematical Models and Immune Cell Biology*, DOI 10.1007/978-1-4419-7725-0_11,

the DC is re-programmed and becomes less efficient at taking up antigen and better at presenting antigen to T cells. Following maturation, the DC then migrates via the lymphatics to secondary lymphoid structures such as the lymph nodes where it can interact with T cells and prime them for effector T cell responses. Thus, DCs are highly motile cells which enable them to acquire the antigens needed to prime T cells, interact with other innate cells such as natural killer cells and migrate to secondary lymphoid organs where T cell priming occurs. DC migration is of special significance in the gastrointestinal tract (GIT) in particular the small and large intestine. The GIT has an enormous antigenic burden, the biggest in the body, consisting of dietary food antigens and bacterial antigens which derive from the large resident, commensal bacteria population. The GIT also represents a major site of pathogen attack with gastrointestinal infections amongst the most prevalent in the world. Thus, the immune system of the GIT must be adapted to ignore beneficial antigens from food and friendly bacteria whilst recognising and eradicating pathogens. Furthermore, there is significant diversity of DC function within the GIT which may be related to differences in antigenic burden. For example some DCs in the small intestine are actively involved with the maintenance of oral tolerance and may be exposed to a wide array of dietary antigens. In contrast, DCs appear to be more sequestered away in the large intestine. This difference may be due to the large population of resident commensal bacteria present in the large intestine which is largely absent from the small intestine.

Factors Mediating DC Migration

The GIT is essentially a long tube approximately 9 m in length, the inner lining of which is covered by a continuous layer of epithelial cells (Fig. 11.1). The small intestine extends from the stomach to the ileocecal junction. In humans it is about 720 cm in length and is divided into three parts: the duodenum which is fixed to the abdominal wall and only comprises about 20 cm of the small intestine, the jejunum (approximately the next two fifths) and the ileum (the remaining three fifths). The large intestine of the human is much shorter and is approximately 180 cm in length and consists of the caecum (which is continuous with the ileum), the appendix and the colon (which is divided into the ascending, transverse and descending colon), the rectum and anal canal and terminates as the anus. The epithelium acts as a barrier separating the antigenic contents of the intestinal lumen from the immune cells found underneath the epithelium within the sub-mucosa. Some DCs and immune cells can also be found within the intestinal epithelial layer [1,2] and they also help maintain the barrier integrity of the intestine. The epithelium by virtue of its position is therefore the most common initial site of pathogen invasion. DC migration is triggered by a number of events such as infection or damage to the epithelium. Infection and damage result in the release of mobilisation signals that can be detected by receptors on DCs. These signals include pathogen derived factors, soluble proteins called chemokines and even some anti-bacterial peptides [3,4] (Fig. 11.2).

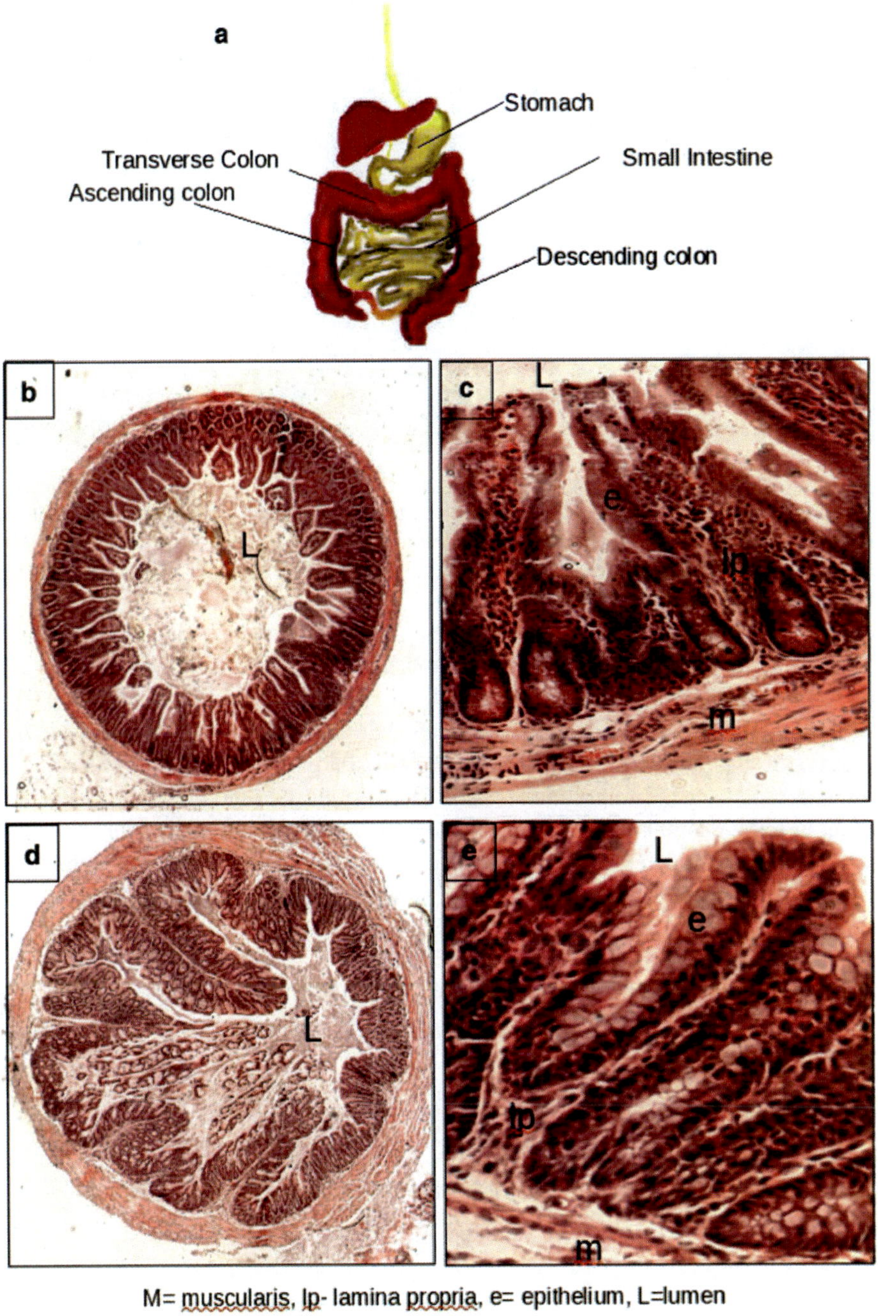

Fig. 11.1 Structure of the gut. (**a**) Cartoon showing the different regions of the gut. The images in (**b–e**) are histology micrographs showing the small intestine (**b** and **c**) and colon (**d** and **e**) at low magnification (**b**, **d**) and higher magnification (**c**, **e**). The position of the gut lumen is indicated with L. Note the presence of faecal debris within the lumen. The epithelial layer is indicated with e, lamina propria with lp and the muscularis with m in the higher magnification images

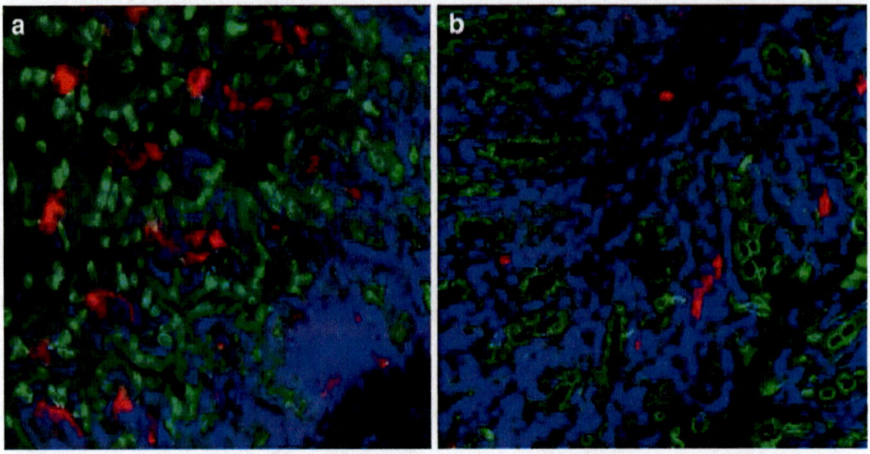

Fig. 11.2 Distribution of Dendritic Cells in the Gut. Distribution of DCs in the intestine. Immunohistochemistry images demonstrating the relative abundance of DCs in the different regions of the gut. The images are taken from normal small intestine and show the villi (**a**) and the colon (**b**) and show the crypts. The sections have been stained so that the epithelium is *green* , the dendritic cells are *red* and the cell nuclei are *blue*

Highly conserved structures on bacteria and pathogens are recognised by families of pattern recognition receptors (PRR) that are expressed by DCs as well as epithelial cells. Once the PRR has been triggered, cell signalling pathways are triggered that favour the production of chemokines and cytokines [5–7]. Although PRR ligation does trigger DC migration [8, 9], the most important mobilisation factors are thought to be chemokines [10–14]. The expression profile of chemokine receptors on DCs changes upon DC maturation; immature DCs typically express high levels of CCR2, CCR5 and CCR6 [4]. Upon DC maturation these chemokine receptors are down-regulated and the DC expresses the chemokine receptor CCR7. Even though CCR7 has been identified as being crucial in DC migration to the MLN other mediators such as cysteinyl leukotrienes and prostaglandin E2 are needed to make CCR7 responsive to its ligands [15, 16]. However, CCR7 is also expressed on so- called semi-mature DCs that for example have phagocytosed apoptotic cells [12]. Lymphatic endothelial cells express high levels of the CCR7 ligands CCL19 and CCL21, and CCR7 is important in migration of mature and possibly semi-mature DCs to lymph nodes for T cell priming [15, 16]. This means that the type of signal recognised by mature and immature DCs is potentially different. Overall, the mobilisation signals will help:

1. Maintain immune homeostasis by preventing inappropriate immune responses to beneficial antigens
2. Alert DCs to the site of damage/infection
3. Recruit DC progenitors and DCs in order to maintain the resident population or respond to damage or infection [16]

In response to the mobilisation signals cell signalling cascades are triggered in the DC resulting in multiple internal and external phenotypic and functional changes

[8,16]. For example, the interaction with pathogen- derived antigens not only causes the DCs to mature but also increases their motility [8, 17]. In order for the DC to physically move, the DC interacts with the gut substrate, blood vessels or lymphatics. These interactions are facilitated by adhesion molecules on both the DCs and the substrate it is moving through. Examples include adhesion molecules such as ICAM-1, JAM-1 and integrins which recognise substrate components such as collagen. The adhesion molecules act to enable the DC to either anchor to or detach from the cell substrate it is in contact with [16]. Changes in the expression of adhesion molecules can affect how DCs move including the speed and direction of migration [16].

Another factor that may influence the speed of DC migration to lymph nodes is the rate of the lymph flow within the lymphatic vessels. Histamine and cysteinyl leukotrienes promote DC migration and can also increase lymph flow rate. Furthermore factors that inhibit DC migration have also been shown to decrease lymph flow rate [12, 18].

DC Localisation Within the GIT

The GIT has a large population of DCs which are located in the small intestine, large intestine and the MLNs (Fig. 11.3). The majority of DCs are found in the small intestine. DCs in the small intestine are found distributed throughout the lamina propria and in defined lymphoid structures called Peyers patches. Notably, a proportion of small intestinal DCs expressing CX3CR1 are found at the epithelial layer and extend dendrites between epithelial cells and into the gut lumen. These dendrite structures are called transepithelial dendrites and may represent primary routes of antigen uptake for DCs [19]. DC with transepithelial dendrites are rare or absent in the ileum and terminal ileum of the small intestine.

Generally, DCs in the large intestine are much rarer than the small intestine. Here, the majority of resident DCs are found in small isolated lymphoid follicle structures (ILF) and colonic patches (CP) [20] with occasional DCs within the lamina propria [4,21,22]. In health there are few if any DCs in the colonic epithelial layer [21]. The formation of transepithelial dendrites in the large intestine has not been observed in healthy mice and is rare even in infection [4, 23].

DC Antigen Sampling and Migration in the Normal Non-Inflamed Intestine

Two of the main reasons DCs need to migrate are to acquire antigens or prime T cells. In the small intestine, there are three major routes by which DCs can acquire antigens: via specialised epithelial cells called microfold or M cells which are mostly found in the Peyers patches, via the intestinal epithelium within the lamina

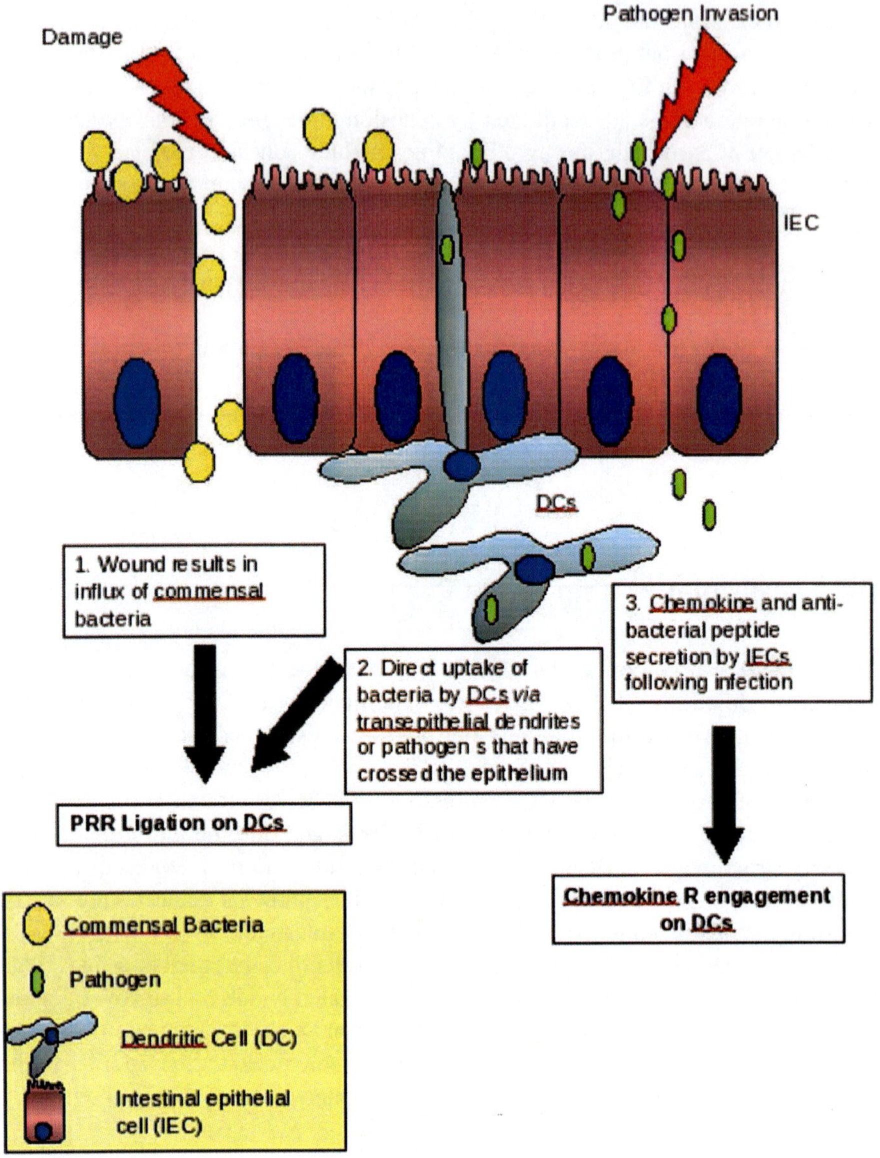

Fig. 11.3 Epithelial barrier function and Trigger for DC recruitment. Cartoon showing the main pathways that can trigger DC mobilisation in the intestine

propria or via transepithelial dendrites [2, 19, 24]. In the large intestine, the pathways of DC antigen uptake are not well-defined as there are no Peyers patches, few DCs in the epithelium and no obvious population of DCs with transepithelial dendrites [3, 23]. However, the ILFs and CPs may have an analogous role to the Peyers patches thus representing a primary site of antigen uptake for DCs in the normal un-inflamed large intestine [25, 26].

As stated previously the intestine has a considerable antigenic burden consisting of dietary antigens and bacterial antigens from the commensal bacteria. The lack of a pro-inflammatory immune response to such antigens is called immunological tolerance. It is thought that in the normal gut, DCs contribute to immunological tolerance. As well as local mechanisms of tolerance within the small intestine [27–32], it has been shown that there is a constant low level of DC traffic from the small intestinal lamina propria via the lymphatics to the MLNs [33]. Such trafficking is referred to as steady state migration. In addition it is thought that DCs within the Peyers patches are recruited in steady state migration. The chemokines CCL9 and CCL20 and the mucosal homing integrin MAdCAM-1, which are recognised by receptors on immature DCs, are thought to be critical for steady state sampling [34–36].

That intestinal DCs are involved in sampling luminal antigens is supported by the observation that non-invasive and non-pathogenic Escherichia coli can be cultured from the MLN of mice that have intestinal DCs with transepithelial dendrites [19]. The chemokine receptor CCR7 on DCs is thought to be important in this steady-state DC trafficking. In mice lacking CCR7, there is impaired trafficking of DCs and defective induction of tolerance [37]. The majority of DCs in the normal intestine are immature and do not express CCR7 whereas DCs in the MLN have been shown to be more mature [38, 39]. As the immune response primed by these steady state DCs is a tolerogenic one, it is not clear whether this is because the DCs respond differently to beneficial antigens, do not fully mature e.g. are semi-mature or only a specific subset of DCs respond. Recently, it has been shown that a subset of DCs expressing CD103 represent the motile DCs that migrate to the secondary lymphoid structures and initiate immune responses [40].

There is less data on the function of DCs in the normal non-inflamed colon. Although ILFs and CPs may represent a site of luminal sampling, it is worth noting that their formation is unaffected in the absence of the commensal bacteria [25]. DCs have not been observed with transepithelial dendrites in the normal large intestine therefore the transepithelial dendrite route is highly unlikely to be involved in steady state sampling. Furthermore, there are very few DCs at the epithelial layer; on average between zero and 1 DCs are observed per crypt adjacent to the epithelium [7]. However, a small subset of DCs express CCR7 (<5%) in the normal large intestine [7] so it is possible that there is some trafficking of DCs to the MLNs in the steady state. This small subset of CCR7$^+$ DCs may be semi-mature [7]. It does appear that in contrast to the small intestine DCs are sequestered away from potential contact with luminal antigens. This may be due to the fact that the large intestine has a significantly higher antigenic burden with the bulk of the commensal flora residing there, therefore it is beneficial to keep the most significant antigen presenting cell (APC) away from this major source of antigen. Similarly, in the lower small intestine which has the most bacteria, DCs with transepithelial dendrites are less frequently observed in normal healthy mice [2]. Certainly, inappropriate immune responses to the commensal bacteria have been shown to lead to the development of inflammatory bowel disease (reviewed in [41]). Furthermore, DCs have been strongly implicated in this dysregulated immune response in the large intestine [21, 42].

DC Migration in Response to Infection or Injury

DC migration within the gut in response to infection or damage is critical for resistance to infection. This migration may represent the movement of resident DCs to the site of injury of infection (usually at or near the epithelial layer) as well as the recruitment of DCs from the periphery and bone marrow that will move through the tissue to the site of damage. The latter cells are often termed inflammatory DCs and express CD103 [40].

In the small intestine, as we have discussed, many of the DCs are situated where they are readily able to access antigens such as those with transepithelial dendrites. In the terminal ileum there are much fewer DCs with transepithelial dendrites constitutively. However, the number of DCs with transepithelial dendrites increases dramatically upon infection in the small intestine and in particular the ileum [2]. This is thought to occur as a consequence of recruitment of inflammatory DCs [2]. The large intestine does not have normally have DCs in the epithelial layer. However, it is known that DCs are critical for immunity to a strain of colitogenic *Salmonella typhimurium*. Experiments in which DCs were depleted revealed that DCs were needed for invasion of *S. typhimurium* across the epithelium. The mechanisms by which the bacteria targeted the DCs was not clear as DCs were not observed with transepithelial dendrites in the large intestine [23]. Currently it is unclear whether DCs ever extend transepithelial dendrites in the large intestine in response to *S. typhimurium*. However, as DCs are normally found distant from the large intestinal epithelium the immunity to *S.typhimurium* must occur due to migration of DCs to the epithelium. The mouse model of the human whipworm infection *Trichuris trichiura* (*Trichuris muris*) was used to investigate DC interactions with the epithelium in the large intestine. *T. muris* resides only in the large intestine. The studies revealed that, 24 h after infection, in hosts that would be resistant to infection, there were occasional DCs ($< 5\%$) with transepithelial dendrites in the colon [7]. This analysis was performed using confocal and electron microscopy which effectively analyses only a snapshot of the immune response so is somewhat limited. With the advent of better techniques to visualise immune responses in vivo such as multiphoton microscopy, it may be possible to resolve the issue of transepithelial dendrites in the large intestine. For example, we will be able to determine the frequency and duration with which large intestinal DCs can form transepithelial dendrites in response to infection and whether there are ever DCs with transepithelial dendrites in the steady state.

Although, there are DCs present at the epithelial layer in the small intestine constitutively, a number of infections and pathogen derived antigens have been shown to promote further DC migration to the Peyers patches [43] and lamina propria epithelium of the small intestine [33, 44–47]. One chemokine receptor has been specifically linked with homing to the small intestine and this is CCL25 which is recognised by the receptor CCR9 [47]. CCL25 is expressed by small intestinal epithelial cells. CCR9 has been shown to recruit a subset of DCs called plasmacytoid DCs to the small intestine in response to PRR triggers [48]. Interestingly, this subset of DCs appeared to be necessary for the recruitment of other DC types to the small intestine epithelium but only in response to a select set of PRRs [48]. A number of other chemokines are upregulated by epithelial cells in response to infection including CCL20, CCL5 and CCL2 [4, 45, 49, 50]. In particular CCR6-dependent recruitment of DCs has been shown to facilitate better immunity to *S. typhimurium*. The mechanism for improved immunity was due to increased trafficking of antigen carried by the phagocytic DCs to the lymph nodes which then increased the efficiency of antigen presentation to T cells [42]. The migratory DCs were not necessarily involved directly in antigen presentation. The migratory DCs appeared to be blood-derived rather than tissue derived and were therefore of the inflammatory DC subset.

There is relatively little literature on DC trafficking within the large intestine but DC mobilisation and trafficking to the epithelium has been shown to be vital for effective immunity to infection [4]. In mice that are susceptible to the mouse parasitic worm infection *T. muris* it has been shown that delayed DC mobilisation in the large intestine is strongly associated with susceptibility to infection whereas resistant hosts have a rapid DC migratory response. This delayed response in susceptible hosts was linked to reduced secretion of chemokines by the intestinal epithelium within the first twenty four hours of infection [7]. The most important chemokines were CCL5 and CCL20 and blocking of these chemokines prevented DC mobilisation in response to infection [7]. Similarly, in mice with a defect in the secretion of epithelial chemokines, there is reduced DC migratory response which is associated with increased susceptibility to *S. typhimurium* infection (Cruickshank, unpublished observations). Unlike, the small intestine however, where there is a rapid trafficking of DCs to the MLNs following infection, few DCs were found in the MLNs post-infection within the first 7 days. However, the size and frequency of the DC-rich lymphoid aggregates (the isolated lymphoid follicles and colonic patches) increased suggesting that DCs can be retained in the colon following antigen stimulation [4, 51]. Similarly, the colonic patches and isolated lymphoid follicles of the colon have been confirmed as important sites for the generation of protective mucosal IgA induction following intrarectal administration of antigen [26]. In this study, CCR7 was involved in homing to another lymph node site, the iliac lymph nodes which are found below the mesenteric lymph nodes closer to the rectum (the terminal part of the large intestine).

Rapid DC mobilisation in the intestine is not always associated with a positive outcome however. Some pathogens utilise DCs to disseminate around the body such as *Toxoplasma gondii* which can directly exploit the way in which DCs move. *T.gondii* secretes chemokine mimics that are strongly chemotactic to intestinal DCs [52] as well as directly targeting the mechanisms of DC migration in order to promote faster cell migration [53]. The accumulation of DCs in the large intestine, particularly mature DCs, has also been associated with the development of chronic inflammation seen in inflammatory bowel disease [21, 42, 54, 55]. All the factors underlying the accumulation of mature DCs in IBD are not elucidated currently but clearly this demonstrates an alteration in migratory behaviour.

Recruitment of DC Progenitors

DC precursors such as monocytes are rapidly recruited from the vasculature to the site of injury or infection (reviewed in [56]). Monocytes have been shown to give rise to inflammatory DCs and actually are derived from a common precursor to conventional DCs. Within the first few hours and days of infection it has been shown that there is a massive wave of CCR2 dependent recruitment of $Gr1^+$ monocytes from the bone marrow to the bloodstream and thence to the tissues [57–59]. Under the influence of pathogen derived factors and cytokines, these cells are able to differentiate into inflammatory DCs [59, 60]. As well as acting as DC progenitors, these monocyte cells may have anti-microbial functions (reviewed in [56]).

In addition to the inflammatory subset of DCs, it is likely that monocytes may be involved in steady state replenishment of local resident DCs. There is some debate about whether the $Gr1^+$ monocytes that act as inflammatory DCs progenitors are also progenitor cells for conventional tissue resident DCs. It is generally believed that is not the case however. Gr1lo monocytes may act as progenitors for resident conventional DCs although the chemokines involved in their recruitment are not fully elucidated and there is still relatively little information on this subset (reviewed in [61]).

It is apparent that the migration of DCs both to the sites of injury or infection as well as sites of T cell priming represents critical stages in the development of effective immunity. Not only is DC migration essential for effector immunity, it also appears to be vital to maintain normal homeostasis in the gut by preventing inappropriate immune responses to normal antigens. Dysregulation of the DC migratory response can also result in significant problems leading to chronic inflammation and the development of inappropriate immune responses. Another important consideration is that the immune responses within the small intestine and large intestine are not the same as is commonly assumed. DC migration is a complex process that involves multiple factors all of which can have profound effects on the outcome of the immune response. This area of research is still in its infancy and represents an area of rapid development in the immunological field.

Acknowledgements We would like to acknowledge L. Logunova for the images in Fig. 11.2 and Dr J. Pennock for image (a) in Fig. 11.1 and helpful discussions and critical feedback.

References

1. Dalton JE, Cruickshank SM, Egan CE, Mears R, Newton DJ, Andrew EM, Lawrence B, Howell G, Else KJ, Gubbels MJ, Striepen B, Smith JE, White SJ, Carding SR (2006) Intraepithelial $\gamma\delta^+$ lymphocytes maintain the integrity of intestinal epithelial tight junctions in response to infection. Gastroenterology 131:818–829
2. Chieppa M, Rescigno M, Huang AY, Germain RN (2006) Dynamic imaging of dendritic cell extension into the small bowel lumen in response to epithelial cell TLR engagement. J Exp Med 203:2841–2852

3. Yang D, Biragyn A, Kwak LW, Oppenheim JJ (2002) Mammalian defensins in immunity: more than just microbicidal. Trends Immunol 23:291–296
4. Cruickshank SM, DeSchoolmeester ML, Svensson M, Howell G, Bazakou A, Logunova L, Little MC, English N, Grencis RK, Else KJ, Carding SR (2009) Rapid dendritic cell mobilisation to the large intestinal epithelium is associated with resistance to trichuris muris infection. J Immunol 182:3055–3062
5. Lan JG, Cruickshank SM, Singh JC, Farrar M, Lodge JP, Felsburg PJ, Carding SR (2005) Different cytokine response of primary colonic epithelial cells to commensal bacteria. World J Gastroenterol 11:3375–3384
6. Singh JC, Cruickshank SM, Newton DJ, Wakenshaw L, Graham A, Lan J, Lodge JP, Felsburg PJ, Carding SR (2005) Toll-like receptor-mediated responses of primary intestinal epithelial cells during the development of colitis. Am J Physiol – Gastrointest Liver Physiol 288:G514–G524
7. Cruickshank SM, Wakenshaw L, Cardone J, Howdle PD, Murray PJ, Carding SR (2008) Evidence for the involvement of nod2 in regulating colonic epithelial cell growth and survival. World J Gastroenterol 14:5834–5841
8. Watts C, Zaru R, Prescott AR, Wallin RP, West MA (2007) Proximal effects of toll-like receptor activation in dendritic cells. Curr Opin Immunol 19:73–78
9. West MA, Prescott AR, Chan KM, Zhou Z, Rose-John S, Scheller J, Watts C (2008) TLR ligand-induced podosome disassembly in dendritic cells is adam17 dependent. J Cell Biol 182:993–1005
10. Kupiec-Weglinski JW, Austyn JM, Morris PJ (1988) Migration patterns of dendritic cells in the mouse. traffic from the blood, and T cell-dependent and -independent entry to lymphoid tissues. J Exp Med 167:632–645
11. Cyster JG (1999) Chemokines and the homing of dendritic cells to the T cell areas of lymphoid organs. J Exp Med 189:447–450
12. Randolph GJ, Angeli V, Swartz MA (2005) Dendritic-cell trafficking to lymph nodes through lymphatic vessels. Nature 5:617–628
13. Sánchez-Sánchez N, Riol-Blanco L, Rodríguez-Fernández JL (2006) The multiple personalities of the chemokine receptor CCR7 in dendritic cells. J Immunol 176:5153–5159
14. Randolph GJ, Ochando J, Partida-Sánchez S (2008) Migration of dendritic cell subsets and their precursors. Annu Rev Immunol 26:293–316
15. MartIn-Fontecha A, Sebastiani S, Hopken UE, Uguccioni M, Lipp M, Lanzavecchia A, Sallusto F (2003) Regulation of dendritic cell migration to the draining lymph node: impact on T lymphocyte traffic and priming. J Exp Med 198:615–621
16. Alvarez D, Vollmann EH, von Andrian UH (2008) Mechanisms and consequences of dendritic cell migration. Immunity 29:325–342
17. Svensson HG, West MA, Mollahan P, Prescott AR, Zaru R, Watts C (2008) A role for ARF6 in dendritic cell podosome formation and migration. Eur J Immunol 38:818–828
18. von der Weid PY (2001) Review article: lymphatic vessel pumping and inflammation–the role of spontaneous constrictions and underlying electrical pacemaker potentials. Aliment Pharmacol Ther 15:1115–1129
19. Niess JH, Brand S, Gu X, Landsman L, Jung S, McCormick BA, Vyas JM, Boes M, Ploegh HL, Fox JG, Littman DR, Reinecker HC (2005) CX3CR1-mediated dendritic cell access to the intestinal lumen and bacterial clearance. Science 307:254–258
20. Chang SY, Cha HR, Uematsu S, Akira S, Igarashi O, Kiyono H, Kweon MN (2008) Colonic patches direct the cross-talk between systemic compartments and large intestine independently of innate immunity. J Immunol 180:1609–1618
21. Cruickshank SM, English NR, Felsburg PJ, Carding SR (2005) Characterization of colonic dendritic cells in normal and colitic mice. World J Gastroenterol 11:6338–6347
22. Takenaka S, Safroneeva E, Xing Z, Gauldie J (2007) Dendritic cells derived from murine colonic mucosa have unique functional and phenotypic characteristics. J Immunol 178: 7984–7993

23. Hapfelmeier S, Muller AJ, Stecher B, Kaiser P, Barthel M, Endt K, Eberhard M, Robbiani R, Jacobi CA, Heikenwalder M, Kirschning C, Jung S, Stallmach T, Kremer M, Hardt WD (2008) Microbe sampling by mucosal dendritic cells is a discrete, myd88-independent step in deltainvg S. typhimurium colitis. J Exp Med 205:437–450
24. Rescigno M, Urbano M, Valzasina B, Francolini M, Rotta G, Bonasio R, Granucci F, Kraehenbuhl JP, Ricciardi-Castagnoli P (2001) Dendritic cells express tight junction proteins and penetrate gut epithelial monolayers to sample bacteria. Nat Immunol 2:361–367
25. Kweon MN, Yamamoto M, Rennert PD, Park EJ, Lee AY, Chang SY, Hiroi T, Nanno M, Kiyono H (2005) Prenatal blockage of lymphotoxin beta receptor and tnf receptor p55 signalling cascade resulted in the acceleration of tissue genesis for isolated lymphoid follicles in the large intestine. J Immunol 174:4365–4372
26. Lee AY, Chang SY, Kim JI, Cha HR, Jang MH, Yamamoto M, Kweon MN (2008) Dendritic cells in colonic patches and iliac lymph nodes are essential in mucosal iga induction following intrarectal administration via CCR7 interaction. Eur J Immunol 38:1127–1137
27. Mowat AM, Donachie AM, Parker LA, Robson NC, Beacock-Sharp H, McIntyre LJ, Millington O, Chirdo F (2003) The role of dendritic cells in regulating mucosal immunity and tolerance. Novartis Found Symp 252:291–302; discussion 302–305
28. Mowat AM, Parker LA, Beacock-Sharp H, Millington OR, Chirdo F (2004) Oral tolerance: overview and historical perspectives. Ann N Y Acad Sci 1029:1–8
29. Coombes JL, Siddiqui KR, Arancibia-Carcamo CV, Hall J, Sun CM, Belkaid Y, Powrie F (2007) A functionally specialized population of mucosal $CD103^+$ DCs induces $Foxp3^+$ regulatory T cells via a TGF-β and retinoic acid- dependent mechanism. J Exp Med 204:1757–1764
30. Annacker O, Coombes JL, Malmstrom V, Uhlig HH, Bourne T, Johansson-Lindbom B, Agace WW, Parker CM, Powrie F (2005) Essential role for CD103 in the T cell-mediated regulation of experimental colitis. J Exp Med 202:1051–1061
31. Coombes JL, Powrie F (2008) Dendritic cells in intestinal immune regulation. Nat Rev Immunol 8:435–446
32. Monteleone I, Platt AM, Jaensson E, Agace WW, Mowat AM (2008) Il-10-dependent partial refractoriness to toll-like receptor stimulation modulates gut mucosal dendritic cell function. Eur J Immunol 38:1533–1547
33. Turnbull EL, Yrlid U, Jenkins CD, Macpherson GG (2005) Intestinal dendritic cell subsets: differential effects of systemic TLR4 stimulation on migratory fate and activation in vivo. J Immunol 174:1374–1384
34. Cook DN, Prosser DM, Forster R, Zhang J, Kuklin NA, Abbondanzo SJ, Niu XD, Chen SC, Manfra DJ, Wiekowski MT, Sullivan LM, Smith SR, Greenberg HB, Narula SK, Lipp M, Lira SA (2000) CCR6 mediates dendritic cell localization, lymphocyte homeostasis, and immune responses in mucosal tissue. Immunity 12:495–503
35. Iwasaki A, Kelsall BL (2000) Localization of distinct Peyer's patch dendritic cell subsets and their recruitment by chemokines macrophage inflammatory protein MIP-3α, MIP-3β, and secondary lymphoid organ chemokine. J Exp Med 191:1381–1394
36. Zhao X, Sato A, Dela Cruz CS, Linehan M, Luegering A, Kucharzik T, Shirakawa AK, Marquez G, Farber JM, Williams I, Iwasaki A (2003) CCL9 is secreted by the follicle-associated epithelium and recruits dome region Peyer's patch $CD11b^+$ dendritic cells. J Immunol 171:2797–2803
37. Worbs T, Bode U, Yan S, Hoffmann MW, Hintzen G, Bernhardt G, Forster R, Pabst O (2006) Oral tolerance originates in the intestinal immune system and relies on antigen carriage by dendritic cells. J Exp Med 203:519–527
38. Liu LM, MacPherson GG (1995) Antigen processing by rat lymph-borne dendritic cells. Adv Exp Med Biol 378:215–217
39. Liu L, MacPherson GG (1995) Dendritic cells "in vivo": their role in the initiation of intestinal immune responses. Adv Exp Med Biol 371A:271–274
40. Schulz O, Jaensson E, Persson EK, Xiasun L, Worbs T, Agace WW, Pabst O (2009) Intestinal $CD103^+$, but not $CX3CR1^+$, antigen sampling cells migrate in lymph and serve classical dendritic cells functions. J Exp Med 206:3101–3114

41. Strober W, Fuss IJ, Blumberg RS (2002) The immunology of mucosal models of inflammation. Annu Rev Immunol 20:495–549
42. Ashcroft AJ, Cruickshank SM, Croucher PI, Perry MJ, Rollinson S, Lippitt JM, Child JA, Dunstan C, Felsburg PJ, Morgan GJ, Carding SR (2003) Colonic dendritic cells, intestinal inflammation, and T cell- mediated bone destruction are modulated by recombinant osteoprotegerin. Immunity 19:849–861
43. Fleeton MN, Contractor N, Leon F, Wetzel JD, Dermody TS, Kelsall BL (2004) Peyer's patch dendritic cells process viral antigen from apoptotic epithelial cells in the intestine of reovirus-infected mice. J Exp Med 200:235–245
44. Yrlid U, Cerovic V, Milling S, Jenkins CD, Klavinskis LS, MacPherson GG (2006) A distinct subset of intestinal dendritic cells responds selectively to oral TLR7/8 stimulation. Eur J Immunol 36:2639–2648
45. Auray G, Lacroix-Lamande S, Mancassola R, Dimier-Poisson I, Laurent F (2007) Involvement of intestinal epithelial cells in dendritic cell recruitment during c. parvum infection. Microbes Infect 9:574–582
46. Wick MJ (2007) Monocyte and dendritic cell recruitment and activation during oral salmonella infection. Immunol Lett 112:68–74
47. Kunkel EJ, Campbell JJ, Haraldsen G, Pan J, Boisvert J, Roberts AI, Ebert EC, Vierra MA, Goodman SB, Genovese MC, Wardlaw AJ, Greenberg HB, Parker CM, Butcher EC, Andrew DP, Agace WW (2000) Lymphocyte cc chemokine receptor 9 and epithelial thymus-expressed chemokine (TECK) expression distinguish the small intestinal immune compartment: epithelial expression of tissue-specific chemokines as an organizing principle in regional immunity. J Exp Med 192:761–768
48. Wendland M, Czeloth N, Mach N, Malissen B, Kremmer E, Pabst O, Forster R (2007) CCR9 is a homing receptor for plasmacytoid dendritic cells to the small intestine. Proc Natl Acad Sci USA 104:6347–6352
49. Sierro F, Dubois B, Coste A, Kaiserlian D, Kraehenbuhl JP, Sirard JC (2001) Flagellin stimulation of intestinal epithelial cells triggers ccl20-mediated migration of dendritic cells. Proc Natl Acad Sci USA 98:13722–13727
50. Ravindran R, Rusch L, Itano A, Jenkins MK, McSorley SJ (2007) Ccr6-dependent recruitment of blood phagocytes is necessary for rapid cd4 t cell responses to local bacterial infection. Proc Natl Acad Sci USA 104:12075–12080
51. Little MC, Bell LV, Cliffe LJ, Else KJ (2005) The characterization of intraepithelial lymphocytes, lamina propria leukocytes, and isolated lymphoid follicles in the large intestine of mice infected with the intestinal nematode parasite trichuris muris. J Immunol 175:6713–6722
52. Aliberti J, Valenzuela JG, Carruthers VB, Hieny S, Andersen J, Charest H, Reis e Sousa C, Fairlamb A, Ribeiro JM, Sher A (2003) Molecular mimicry of a CCR5 binding-domain in the microbial activation of dendritic cells. Nat Immunol 4:485–490
53. Lambert H, Hitziger N, Dellacasa I, Svensson M, Barragan A (2006) Induction of dendritic cell migration upon toxoplasma gondii infection potentiates parasite dissemination. Cell Microbiol 8:1611–1623
54. Bell SJ, Rigby R, English N, Mann SD, Knight SC, Kamm MA, Stagg AJ (2001) Migration and maturation of human colonic dendritic cells. J Immunol 166:4958–4967
55. Hart AL, Al-Hassi HO, Rigby RJ, Bell SJ, Emmanuel AV, Knight SC, Kamm MA, Stagg AJ (2005) Characteristics of intestinal dendritic cells in inflammatory bowel diseases. Gastroenterology 129:50–65
56. Geissmann F, Auffray C, Palframan R, Wirrig C, Ciocca A, Campisi L, Narni-Mancinelli E, Lauvau G (2008) Blood monocytes: distinct subsets, how they relate to dendritic cells, and their possible roles in the regulation of T-cell responses. Immunol Cell Biol 86:398–408
57. Drevets DA, Dillon MJ, Schawang JS, Van Rooijen N, Ehrchen J, Sunderkotter C, Leenen PJ (2004) The Ly-6C^{high} monocyte subpopulation transports listeria monocytogenes into the brain during systemic infection of mice. J Immunol 172:4418–4424
58. Sunderkotter C, Nikolic T, Dillon MJ, Van Rooijen N, Stehling M, Drevets DA, Leenen PJ (2004) Subpopulations of mouse blood monocytes differ in maturation stage and inflammatory response. J Immunol 172:4410–4417

59. Serbina NV, Pamer EG (2006) Monocyte emigration from bone marrow during bacterial infection requires signals mediated by chemokine receptor ccr2. Nat Immunol 7:311–317
60. Varol C, Landsman L, Fogg DK, Greenshtein L, Gildor B, Margalit R, Kalchenko V, Geissmann F, Jung S (2007) Monocytes give rise to mucosal, but not splenic, conventional dendritic cells. J Exp Med 204:171–180
61. Tacke F, Randolph GJ (2006) Migratory fate and differentiation of blood monocyte subsets. Immunobiology 211:609–618

Chapter 12
Reassessing Germinal Centre Reaction Concepts

Jose Faro and Michal Or-Guil

Abstract To determine the number of B cells seeding germinal centres, different authors have used immunohistology of germinal centre sections in conjunction with assuming a binomial distribution of the fractions of two phenotypically distinct B cell populations participating in a given immune response. This approach further assumed that germinal centres are closed to continuous B cell entry. Using such a model, it has been concluded that germinal centres contain two to eight clones, a figure that is usually taken and cited as being essentially correct. The present re-evaluation of those and related experiments lead to an extended mathematical model. This model includes an estimation of errors created by data sampling, two new parameters that take into account possible mistakes in classification of single population GC sections, and the likely variability in the number of seeding B cells. Fitting this new model to experimental data resulted in an estimated mean number of $\langle n \rangle = 23$–37 seeder B cells.

Introduction

Vertebrates have evolved a complex immune system (IS) that efficiently contributes to protecting them from many infectious and toxic agents. To cope with such a huge variety of agents the IS generates a large diversity of lymphocyte receptors. This occurs throughout life by various mechanisms that, in higher vertebrates, are activated in two waves. The first one takes place during lymphocyte development, is antigen (Ag) independent, and comprises the random recombination of relatively few gene segments into a fully variable (V) region exon of immunoglobulin (Ig)

J. Faro (✉)
Edificio de Ciencias Experimentáis, Universidade de Vigo, Campus As Lagoas-Marcosende, 36310 Vigo, Spain
and
Estudos Avançados de Oeiras, Instituto Gulbenkian de Ciência, Apartado 14, 2781-901 Oeiras, Portugal
e-mail: jfaro@uvigo.es; jfaro@igc.gulbenkian.pt

C. Molina-París and G. Lythe (eds.), *Mathematical Models and Immune Cell Biology*, DOI 10.1007/978-1-4419-7725-0_12,

heavy or light chains and T cell receptor chains [1]. In birds and mammals the relevance of this mechanism for generating primary B-cell repertoire diversity varies with different species. Thus other mechanisms such as gene conversion and somatic hypermutation (SHM) which act on rearranged V-region exons may contribute to most of the B-cell repertoire diversity [1, 2]. The second wave of repertoire diversification in B cells is Ag dependent and triggered during immune responses (IRs) to protein-containing Ags, the major mechanism responsible being SHM. This process takes place, in higher vertebrates, in germinal centres (GCs) of secondary lymphoid organs [2, 3].

Germinal centres are dynamic, short-lived anatomical structures generated within primary follicles during humoral IRs. Since their discovery in 1884 as sites of intense mitosis [4, 5] they have been increasingly implicated in IRs [6–9]. In the 1960s studies were initiated into their cellular and anatomical dynamics during IRs [10–13] leading to current terminology. It was then also shown that they are special areas where protein Ag accumulates by an active transport mechanism [14–19]. Within the next decade the so-called follicular dendritic cell (FDC) was confirmed as a key GC cell component, essential for the follicular Ag-trapping process. It was also shown that Ag remained in GCs in a native form, in large complexes with antibodies (Ab) and complement factors, and associated with the membrane of the long dendritic arms of FDCs [16, 17, 20, 21]. At the beginning of the 1980s the role of GCs in memory B-cell generation was well established [22], but a few years later it appeared that GCs might be more profoundly involved in IRs than previously thought. It was proposed [5, 23, 24] that they could be the functional environment for two processes, well-known at that time, namely somatic hypermutation of Abs [25] and affinity maturation [26]. Their involvement in SHM was confirmed in 1991 [27, 28]. Since then, accumulated data clearly indicate that a selection process between Ag-specific B cells takes place in GCs, accounting at least partially for affinity maturation [29–31].

Current View of GCs

A consensus picture of evolving GCs in primary IRs emerged from the mid 1980s to the 1990s. This picture, still dominant in textbooks [3] and most reviews, can be described in a roughly temporal fashion as follows:

1. Foci of Ag-reactive B cells are formed
2. Seeding and initiation of GCs by B cells: this seeding is oligoclonal
3. Seeding of GCs by T cells
4. Founding centroblasts start proliferating with division times of 6–12 h
5. SHM starts
6. Some centroblasts stop dividing and become centrocytes. These cells accumulate in the apical zone, creating the light zone, while centroblasts remain mostly in the basal, dark zone. SHM takes place only in centroblasts. Ag-driven selection affects mostly centrocytes

7. Recycling: the original idea, proposed by MacLennan et al. in 1991 [32], suggested that centroblasts migrate from the dark to the light zone, become centrocytes and as such stop SHM. They are then submitted to selection; surviving cells migrate from the light to the dark zone where they again become centroblasts activating the SHM mechanism. This cycle is supposed to be repeated several times. A later theoretical analysis of the affinity maturation process, based on an optimization criterion [33], was taken as support for this hypothetical view (see discussion). The original recycling idea was later modified so that no physical migration between zones was required
8. High(er) affinity centrocytes are induced to differentiate into either memory B cells or long-lived plasma cells, exiting the GC
9. Gradual decrease of the dark zone, until a single, smaller zone remains where centrocytes and centroblasts are mixed
10. Gradual decrease by death/migration of centrocytes and centroblasts, so that the secondary follicle gradually recovers the phenotype of a primary follicle
11. The whole process takes about 3 weeks

The population dynamics and transient nature of GCs as well as the mutation-selection processes taking place there, raise very interesting problems, amenable to analysis by mathematical and computational tools. Such analyses can focus on different points of view, for instance, on developmental (tissue organization and remodelling), ecological, microevolutionary or health aspects.

However, empirical descriptions of the kinetics applying to different aspects of GC dynamics, such as initiation, maturation and decline, compartmentalization or B-cell selection, are based on experimental procedures that are technically limited with respect to the kinetics of individual GCs. This means many relevant processes can either not be observed directly or not in the relevant time scales. Thus, most if not all of the dynamic data relates to the global GC reaction and not to the average individual GC [34], leaving essential questions open [3].

In the ideal case, questions are answered directly from experimental data ("Do GCs contain T cells? Yes, immunohistology shows that they do"). If direct experimental observation is not possible, then assumptions have to be made in order to draw inferences from indirect experimental data. Often, mathematical models are used to obtain indirect estimates from these assumptions. For instance, assuming that B cells do not enter the GC after its formation, and proliferate at a constant rate, with certain assumptions on the selection process, one can set up a mathematical model where the results are compared to V region Ig sequences from GC B cells in order to estimate the rate of SHM [35].

However, some problems can arise in interpreting data. Firstly, measurement errors can be large enough to give misleading results. Secondly, if data only comprises a small sample, effects of sampling have to be considered. Thirdly, if the conclusion stems from indirect experimental observation, then the validity of the assumptions has to be assessed. Finally, averages over many events can be erroneously taken to reflect the behaviour of individual events [34].

An additional problem is that concepts resulting from data observation are usually presented without clearly distinguishing between truly established facts and those that are hypotheses, observational interpretations, assumptions or even "reasonable" beliefs. As a consequence, scientists can be easily mislead, particularly newcomers to the field. This is highly relevant because such potential interpretations or assumptions may impinge on models of GC dynamics or affinity maturation. Presenting those assumptions as facts prevents their own analysis and constrains future analysis of the models.

As a way to clarify this conundrum, we wanted to critically reassess key papers that contributed to the dominant picture of GC dynamics. We sought to disclose potential interpretations, hypotheses or assumptions that may have been accepted later as facts.

In this paper, we focus on a particular issue, namely, the proposal that GCs are founded oligoclonally. This is to exemplify how apparently good and simple quantitative experimental designs can provide results that are in fact ambiguous with a large margin of error, thus leaving the original question unresolved.

An oligoclonal GC is one that was founded and then populated by only a few clones. A clone comprises cells that originate from the same precursor cell. Hence, a B cell clone will consist of B cells with the same rearranged Ig heavy (*Igh*) and light (*Igl*) V-region sequences, although these sequences might have acquired mutations due to SHM.

Clonal diversity and clonal size distribution of GCs are a consequence of the interplay of migration, proliferation, selection, and differentiation processes, and hence provide information that can help unveil underlying dynamic processes. Therefore, investigators initially focused on determining the clonal composition of GCs. Both direct and indirect methods were used to assess this question. The first experimental approach to the question of clonal diversity that comes to mind is directly measuring *Igh* and/or *Igl* V-region sequence diversity in individual GCs. However, probably due to technical limitations at the time, this approach was not the first to be developed. A second, clever approach, and historically the first to be followed, is to estimate the clonal diversity indirectly by measuring only a qualitative GC trait. The conclusion drawn from these analyses is that GCs are founded oligoclonally by around two to eight precursor cells. Reading the original papers, however, one realizes that the data included potentially important sources of error, thus casting reasonable doubts on the reliability of the authors' interpretations. We will present the data in question and discuss their interpretation in the following sections.

Analysis of the Experimental Basis for the Concept of Oligoclonal Seeding of GCs

Indirect Experimental Evidence

The concept of oligoclonal GCs originated from evaluating indirect evidence based on the assumption that all GCs are founded by n B cells, and become virtually closed structures for Ag-specific B cells arriving later. Within this closed structure, the founder B cell populations expand at the same rate. Under these assumptions, the theoretical approach behind the indirect method is the following. Assume there are two B cell populations named A and B with, in principle, the same diversity, but with two different phenotypic traits – for instance, a plasma membrane allelic marker without functional differences between alleles. Let us call q and p $(=1-q)$ the fractions corresponding to populations A and B, respectively. If GCs are seeded by a similar number n of cells then the expected fraction of GCs seeded by k cells of type A and $n-k$ cells of type B is:

$$\binom{n}{k} p^{n-k} q^{k}. \tag{12.1}$$

In particular, the expected fractions of GCs seeded by only A cells (f_A), only B cells (f_B) or a mixture of A and B cells (f_{Mix}) are, respectively,

$$f_A = q^n \tag{12.2}$$
$$f_B = p^n \tag{12.3}$$
$$f_{Mix} = 1 - q^n - p^n. \tag{12.4}$$

The number n of founder B cells can then be calculated using (12.2), (12.3) or (12.4) alone from knowing the fraction p and the observed value for f_A, f_B or f_{Mix} (denoted here as f_A^{ob}, f_B^{ob} and f_{Mix}^{ob}). Two experimental system variants have been used that allow identification of two B cell populations and hence measurement of p, f_A^{ob}, f_B^{ob} and f_{Mix}^{ob}:

(1) *Immunization with two Ags.* This strategy involves analyzing the GC reactions triggered by two different Ags injected simultaneously [36, 37]. In the paper by Liu et al. [36] rats primed with spider crab hemocyanin (MSH) were boosted 4 weeks later with a mixture of dinitrophenyl (DNP) and 2-phenyloxazolone (Ox) coupled independently to MSH. The fractions of splenic GCs monospecific for either Ox or DNP were determined by immunohistology 3 days later. In the experimental setting of Jacob et al. [37] (C57BL/6×BALB/c)F_1 mice (*Igh*$^{b/a}$) were immunized with nitrophenyl (NP) coupled to chicken gamma globulin (NP–CG). In this system it is expected that the majority of anti-NP B cells bear the λ_1 L chain and that this is *Igh*^{b}-associated. Then the fractions of GCs with B cells expressing λ_1, binding CG, or both were determined.

A general problem with those two experimental settings is that the responses to each Ag are not comparable. This is due to likely differences in the number of Ag-specific precursor B cells, the affinities of the responding cells for their respective Ag, and Ag valence (Ox vs DNP in the first case, and NP vs CG epitopes in the second case). These are important unknown factors that affect estimates of the proportions of pre-response B cells specific for each Ag (assumed to be 1:1 in [37]), and the differential quality of activated Ag-specific B cells' response before the GC reaction starts. For instance, small fluctuations in the in situ proportions of B cells specific for each Ag may be highly amplified by higher affinity/avidity of B cell–Ag interactions leading mostly to extrafollicular plasma cell differentiation [38,39]. Due to all of the above, these settings are inadequate because they introduce too many uncertainties.

(2) *Two phenotypically different B-cell populations.* Historically this was the first approach, and the most cited original papers are those of Kroese and colleagues [40–42]. In their experiments they used rats with two populations of lymphocytes bearing different alleles of a given membrane protein. In the first paper [40] lethally X-irradiated rats of AO strain were reconstituted with different proportions of thoracic duct lymphocytes (TDL) from (AO×BN)F_1 rats (with MHC class II molecules of hybrid allotypes RT$1^{u/n}$) and AO→(AO×BN)F_1 chimeras (with MHC class II molecules of allotype RT1^u). The reconstituted rats were immunized with sheep erythrocytes (SRBC) and their spleens analyzed for GCs 5 days later.

As the authors acknowledged, a major drawback of that system is the difference in MHC between TDL rat donors and rat hosts, which would lead to a graft-vs-host reaction. This was circumvented by using TDL derived from bone marrow chimeras. However, these cells were still weakly reactive against host type cells [40].

The weak reactivity in the above system was avoided in a second paper by using PVG rats congenic for the RT7.2 allotype (CD45, a cell-membrane phosphatase). Lethally X-irradiated PVG rats (RT7.1) were reconstituted with a 1:1 mixture of TDL from both allotypes, immunized with SRBC and their spleens analyzed for GCs 7 or 10 days later.

A common source of error in those two papers was that, manipulating cells ex vivo and injecting particular ratios of them into X-irradiated hosts, meant that the real proportion of surviving, functional B cells of each allotype able to respond to the immunizing Ag becomes highly uncertain. In the 1988 paper [41], this is reflected in the widely different measured percentages of pure RT7.1 and pure RT7.2 GCs, 19% and 45%, respectively, in animals receiving a 1:1 ratio of RT7.1 and RT7.2 B cells. This strongly contrasts with the expected similar frequencies for both types of GCs. Finally, in both reports only one of the cell types was scored for presence/absence. All this makes the data [40,41] highly unreliable for estimating the number of GC founding B cells.

In contrast to the above two experimental systems, another animal model [42] used rat hemopoietic chimeras. Those were obtained by transferring congenic PVG RT7.2 embryonic liver cells into PVG (RT7.1) newborn rats. As in the previous papers, only the presence/absence of RT7.2 cells was analyzed. For our present purpose the best set of data in that paper corresponds to a chimera where the allelic B cell population fractions were $p = 0.21$ (population B) and $q = 0.79$ (population A). In this case, the fractions of pure GC sections were found to be $f_A^{ob} = 0.14$ for population A (allotype RT7.1), $f_B^{ob} = 0.06$ for population B (allotype RT7.2), while 80 % of GC sections showed a mixed population. Using this simple model, the number of founder cells is estimated by (12.3) to be $n_A \approx 8$ and by (12.4) to be $n_B \approx 2$.

This system has the advantage of using normal animals, without ex vivo manipulation of cells, which leads to an accurate determination of allelic B-cell population fractions. We therefore considered this work best from a methodological point of view and will focus on their data in the following analyses.

Consistency Test

We first asked if experimental findings and model assumptions are internally consistent. Does the data reported by Hermans et al. point to a binomial distribution of B cell population fractions in GCs? Table 12.1 shows the percentage of GCs with cells of types A, B, and mixed populations determined experimentally, as well as the expected values calculated from (12.2) to (12.4) using the estimation of n_A or n_B, along with the corresponding χ^2 values. In all cases, χ^2 is much larger than 10.827 (corresponding to a probability of 0.001 and 1 df). This indicates that the reported values are inconsistent with the model assumption.

A first source of error is that the authors of the above papers took sections as to represent whole GCs. However, sections are small samples of GCs, and as such do not always reflect the true GC composition. Consider, for instance, a GC seeded by k cells of type A and $n-k$ cells of type B, and totaling M cells. This GC will have $M\frac{k}{n}$ cells of type A. However, a section containing m cells (with $m \leq M(1-\frac{k}{n})$) will consist, with a probability $\binom{M(1-\frac{k}{n})}{m}/\binom{M}{m}$, exclusively of type B cells, feigning a single population GC. This shows that sampling errors lead to overestimating the number of single population GCs.

Additionally, errors of observation can lead to a further overestimation. For one thing, the experimenter sets a subjective and arbitrary cut-off due to T cells possibly being counted as B cells [42]. As a consequence, estimating the absence of one of the

Table 12.1 Analysis of internal consistency of data in [42] used to estimate n from chimera with $p=0.21$. X^2 of observed vs expected fractions f_A, f_B and f_{Mix} were calculated taking either observed f_A (f_A^{ob}) or observed f_B (f_B^{ob}) as the reference values

	$n_A = \log_{(1-p)}(f_A^{ob})$ (%)		$n_B = \log_p(f_B^{ob})$ (%)	
	Observed	Expected[a]	Observed	Expected[b]
f_A	14	–	14	65.4
f_B	6	2.2×10^{-4}	6	–
f_{Mix}	80	86.0	80	28.6
X^2	161996[c]		132.6[c]	

[a] Expected f_B was calculated as p^{n_A}
[b] Expected f_A was calculated as $(1-p)^{n_B}$
[c] χ^2(prob.: 0.001; df: 1) = 10.827

two B cell populations in sections may dismiss 10–30% of that population. Finally, if only one population is positively analyzed for presence/absence, and the other B cell population is simply inferred, as in [40–42], GC sections scored as 100% positive for the former population because all cells were apparently labelled could in fact contain cells of the other population. Real absence of the second population can only be determined by staining for that population.

We propose the following modified, refined model to take measurement and sampling errors into account. Our aim is to more accurately estimate the expected fractions of single and mixed GCs.

An Extension of the Basic Model

To take the above sources of errors into account, we introduced a parameter $l_T \leq m$ representing the maximum number of cells that can be overlooked, that is, sections with up to l_T B cells of one type are scored as pure sections of the other type.

Now taking into account both sampling and observation errors, we extend (12.1) and find that the probability of a mixed GC, seeded by k cells of type B ($1 \leq k < n$), yielding an apparently pure section of type A is:

$$\binom{n}{k} p^k q^{n-k} \sum_{l=max\left[0,m-M(1-\frac{k}{n})\right]}^{min\left[l_T,\frac{k}{n}M\right]} \frac{\binom{M-\frac{k}{n}M}{m-l}\binom{\frac{k}{n}M}{l}}{\binom{M}{m}} \tag{12.5}$$

Hence, the total fractions of sections expected to be scored as pure A, pure B and mixed are, respectively:

$$f_A(n) = \sum_{k=0}^{n-1} \binom{n}{k} p^k q^{n-k} \sum_{l=max\left[0,m-M(1-\frac{k}{n})\right]}^{min\left[l_T,\frac{k}{n}M\right]} \frac{\binom{M-\frac{k}{n}M}{m-l}\binom{\frac{k}{n}M}{l}}{\binom{M}{m}} \tag{12.6}$$

$$f_B(n) = \sum_{k=1}^{n} \binom{n}{k} p^k q^{n-k} \sum_{l=max\left[0,m-\frac{k}{n}M\right]}^{min\left[l_T,M(1-\frac{k}{n})\right]} \frac{\binom{M-\frac{k}{n}M}{m-l}\binom{\frac{k}{n}M}{l}}{\binom{M}{m}} \tag{12.7}$$

$$f_{Mix}(n) = 1 - f_A(n) - f_B(n) \tag{12.8}$$

A further extension of the basic model addresses its simplifying assumption that all GCs are seeded by the same number n of cells. In fact, it can be expected a priori that the number of cells seeding GCs covers a relatively wide, rather than a very narrow distribution. We therefore replaced this assumption by considering an exponential distribution with mean $\langle n \rangle = 1/\lambda$, such that the fraction of GCs seeded by n_i B cells is:

$$P_\lambda(n_i) = \int_{n_i-1}^{n_i} \lambda e^{-\lambda n} \mathrm{d}n \tag{12.9}$$

Then the fractions of sections expected to be scored as pure A, pure B and mixed from GCs seeded by n B cells are, respectively:

$$F_A(\lambda, n) = P_\lambda(n) f_A(n), \quad F_B(\lambda, n) = P_\lambda(n) f_B(n), \quad F_{Mix}(\lambda, n) = P_\lambda(n) f_{Mix}(n). \tag{12.10}$$

Finally, the *total* fractions of sections expected to be scored as pure A, pure B and mixed are given by the equations:

$$F_A(\lambda) = \sum_{n=1}^{\infty} F_A(\lambda, n), \quad F_B(\lambda) = \sum_{n=1}^{\infty} F_B(\lambda, n), \quad F_{Mix}(\lambda) = \sum_{n=1}^{\infty} F_{Mix}(\lambda, n). \tag{12.11}$$

This refined model can now be used to estimate the fractions of GC sections expected to be scored as type A, type B or mixed, for different values of the average number $\langle n \rangle$ of GC founder B cells, the size m of GC sections and scoring errors, l_{T_A} and l_{T_B}. The results of a systematic analysis of (12.6)–(12.11), taking $l_{T_A} = l_{T_B} = l_T$ are shown in Fig. 12.1.

Consistent with the observed data shown in Table 12.1, the results with the extended model indicate that the scoring error must have been $l_{T_A} \leq 10\%$ for counting "pure" GC sections of type A (i.e., truly pure + false positives). However, this error for cells of type B must have been much higher than 30%, since the estimated F_B

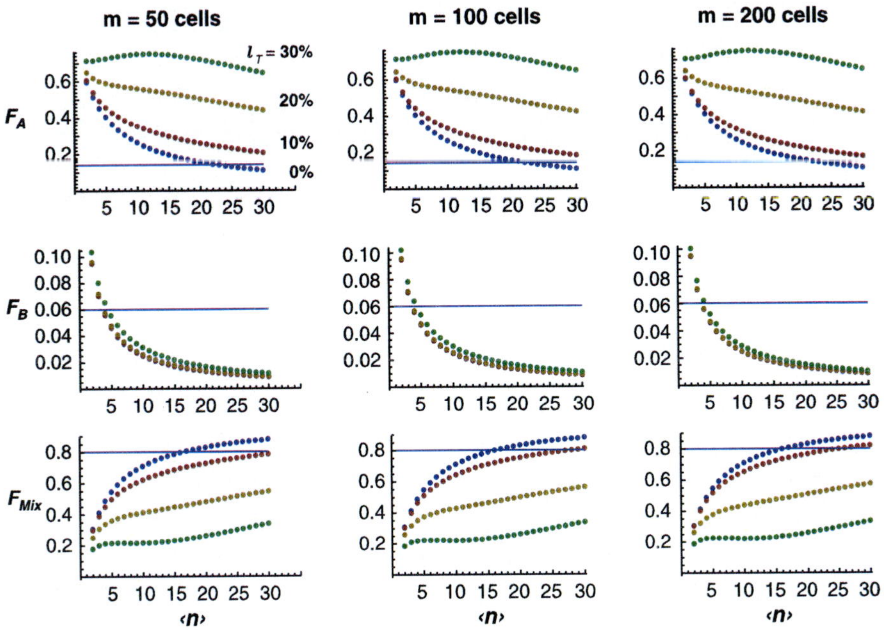

Fig. 12.1 Fractions of GC sections estimated as being scored type A (F_A), type B (F_B) or mixed (F_{Mix}), using the extended model (12.6)–(12.11). $\langle n \rangle$, Average number of GC founding B cells; l_T, scoring error; m, size of GC sections. *Horizontal lines* correspond to the observed fractions in [42]

should reach at least the observed value $f_B^{ob} = 0.06$. Hence, we need to assume that the observation errors are different for each of the two populations if the extended model is compatible with the observed data.

In addition, we can ask if the extended model and the data on pure and mixed GC sections in [42] are consistent. The values reported in the original paper were compared with the expected values according to the extended model (12.6)–(12.11). Figure 12.2 shows the corresponding X^2 for different average numbers of seeding B cells and different error thresholds l_T (once more taking $l_{T_A} = l_{T_B} = l_T$). Consistent with the previous result, in all cases the χ^2 test indicates that the reported values are too different from the expected ones to be due to random variations. We therefore re-estimated F_A for $l_{T_A} = 5\%$, 10% and 15%, and F_B for $l_{T_B} = 50\%$, 55% and 60%. The results clearly show that only scoring errors of $5 - 10\%$ for l_{T_A} and $55 - 60\%$ for l_{T_B} generate values of F_A, F_B and F_{Mix} compatible with the observed values and predict the same value for $\langle n \rangle$ (Fig. 12.3). Applying the χ^2 test for observation errors $l_{T_A} = 5\%$ and $l_{T_B} = 55\%$, we obtain a minimum value of

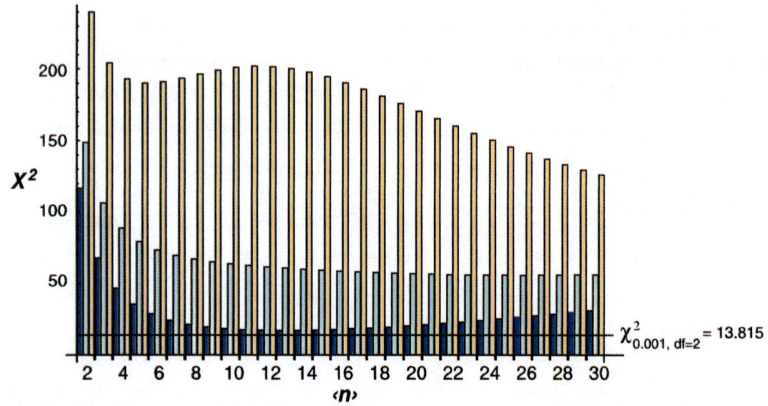

Fig. 12.2 χ^2 test of observed vs expected fractions of GC sections of type A (F_A), type B (F_B) or mixed (F_{Mix}), using the extended model (12.6)–(12.11). $\langle n \rangle$, average number of GC founder B cells. *Horizontal line*, χ^2 for a probability of 0.001 and 2 df. *Blue, green and yellow bars* are results obtained with $l_T = 10$, 20 and 30%, respectively. Calculations made assuming a GC size of $M = 5{,}000$ cells, and a GC section size of $m = 200$ B cells

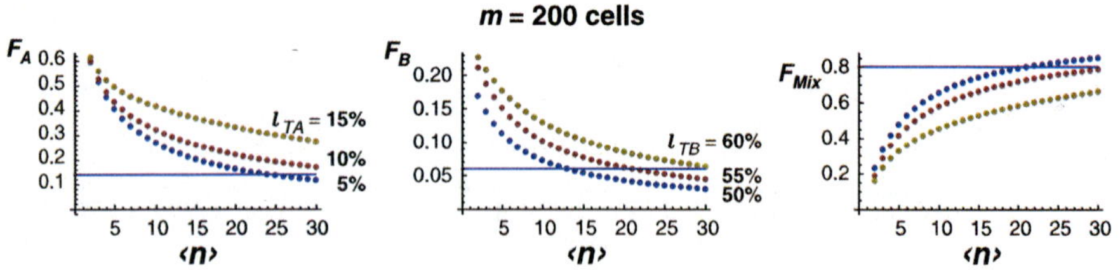

Fig. 12.3 Fractions of GC sections estimated as being scored type A (F_A), type B (F_B) or mixed (F_{Mix}) for the indicated values of l_{T_A} and $l_{T_s B}$, using the extended model (12.6)–(12.11). $\langle n \rangle$, average number of GC founder B cells. *Horizontal lines* correspond to the observed fractions in [42]. GC size, $M = 5{,}000$ cells

$X^2 = 0.064$ for $\langle n \rangle = 23$, while $l_{T_A} = 10\%$ and $l_{T_B} = 60\%$ gives a minimum value of $X^2 = 0.094$ for $\langle n \rangle = 37$. This indicates that the model is now sufficiently accurate to explain the data. Thus the predicted *average* number of GC founder B cells given by this refined model is $\langle n \rangle = 23 - 37$ cells.

Direct Experimental Evidence

Direct experimental observation of the B cell clonal diversity in GCs involves determining the number of distinct *Igh* and/or *Igl* sequences in individual GCs. Small clusters of B cells from single GC sections are isolated and then analyzed for the sequence diversity of rearranged V_H or V_L gene segments from particular V families. The data usually cited is that of *Igh* CDR3 region sequences obtained at different times in a primary IR [29, 43]. The experimental procedure for each GC analyzed involved the following steps: (a) sampling (sample A) of a cell cluster (~50 cells) from a GC section; (b) PCR amplification of sequences related to a particular reference V_H gene segment; (c) extremely small sampling (sample B) of the PCR products (the sample size was typically a fraction 10^{-12}-10^{-10} of the total sequences); (d) sequencing of sample B.

The published diversity within sample B from GC sections harvested at day 16 or earlier is one to nine different sequences, corresponding to one to nine different clones. The data on frequency distribution of the different sequences per GC section can be analyzed to estimate clonal diversity using methods borrowed from ecology, such as the abundant coverage estimator (ACE) [44] or the statistical model of Yule–Simon [45, 46]. Those methods have already been applied to estimate TCR diversity in murine splenic CD4$^+$ T cell populations [47, JF, unpublished]. Note that these methods underestimate true diversity, specially ACE which usually gives particularly low estimates [JF, unpublished]. The results of such analyses applied to V-region sequences from GCs are shown in Tables 12.2 and 12.3. Interestingly, the Yule estimator suggests a severe reduction of diversity after day 10 of a primary IR (Table 12.3), a time when selection likely starts to reduce, as well as bias the clonal composition of GCs [35]. Before such contraction (i.e., for GCs before day 14 of the response) the diversity is estimated to be in the range of 20–30 clones (Table 12.3; Yule estimator).

Note that again these diversity estimates are for GC sections, which are themselves small samples of GCs. Therefore the GC diversity can be expected to

Table 12.2 Clonal diversity estimation of GC B cells using the ACE and Yule methods. GCs are from day 10 of a primary IR. Data from [43]

	Diversity		
GC	Observed	Estimated (ACE)	Estimated (Yule)
B8	5	6	19
B12	4	13	13
B15	2	2	3

Table 12.3 Clonal diversity estimation of GC B cells using the ACE and Yule methods. GCs are from a primary IR. Data from [29]

		Diversity		
Day	GC	Observed	Estimated (ACE)	Estimated (Yule)
4	H4	3	4	7
6	J5	9	>45	29
8	M16	1	–	18
10	B8	6	9	32
14	D2	2	3	3
16	61AA02	2	3	4

be larger than the values estimated. Moreover, the experimental studies further underestimated polyclonality by restricting the analysis to particular V to J rearrangements. Thus, in contrast to the original interpretation that the data provided additional support for the pauciclonal/oligoclonal view of GC seeding, our diversity analysis of the little data available from individual GCs points to a clonal diversity before day 14 that is at least tenfold larger than the minimum values (2–5 clones) previously suggested.

Examples of Other Issues Worth Re-Assessing

As with the oligoclonal seeding of GCs, there are other issues that, being determinant elements in the GC scheme, lack strong experimental support (or any support at all) but are also generally presented as given facts.

Maximally efficient, affinity-dependent selection of GC B cells. This is at the heart of the recycling idea in most of its various versions. Initially proposed by MacLennan et al. [32], the recycling hypothesis gained great popularity after a rigorous analysis of the affinity maturation process based on SHM and selection in GCs [33]. Although not explicitly stated by Kepler and Perelson, the essence of their analysis can be summarized as follows:

- It assumes that GC B cells are subject to continuous selection for competitive Ag binding.
- It applies optimization theory to find the temporal mutational regime that, under the previous assumption, will maximize affinity increase within a relatively short time period.

The crux of the above analysis lies in the "hidden hypothesis" that *affinity maturation is maximized* in GCs. Despite its importance, this concept has never been put to test. Thus, this concept and the linked initial recycling idea, far from being established facts, are working hypotheses with many implications, and as such worth testing experimentally. Moreover, existing data are also consistent with the alternative hypothesis that selection is only moderately efficient, allowing for diversity generation in the memory B-cell pool in order to cope with antigenic variants of the original infectious Ag [48–50].

Affinity-based selection. This idea has been challenged by a few authors, based on very solid observations indicating that selection depends to a larger extent on k_{off} rates than on affinity [51, 52].

All-or-none behaviour of GCs. Available data on the distribution of GCs with respect to their relative content of B cells with key mutations (i.e., conferring higher affinity to an Ab) have been interpreted as indicating that once a key mutation is fixed in a GC population, it will rapidly take over all or most of the GC [35, 53]. These data, however, are still very limited [35, 53], and so do not allow one to draw major conclusions. Moreover, that data are yet to be investigated in the light of other statistical models, such as uniform distribution of the percentage of key-mutant B cells in GCs for values higher than 10%.

GC's lifespan. The global GC reaction in a primary IR to protein Ags lasts ~3 weeks. However, such kinetics of GC reactions are derived from analysis of different sets of GCs. In other words, they must be obtained from different animals at different time points after immunization. Therefore, these studies do not allow one to infer the average (and variance) of the lifespan of individual GCs [34]. Experimental work with current methods aimed at elucidating this question will likely require an elaborate experimental design supported by previous theoretical analyses.

Cell death in GCs. Mutant, low affinity B cells in GCs are assumed to undergo massive cell death "in situ" by apoptosis. However, little is known about migration of Ag-specific B cells in and out of GCs at different stages of the reaction. For instance, within 1–2 days after a primary immunization, the total number of GCs decrease ("GC dissociation") [11, 54–56, our unpublished observations]. The number of GCs start to increase again at day 2–3 postimmunization, reaching a maximum at about day 10. However, it has been proposed that some GC dissociation continues up to day 4 based on the fact that continuous trails of PNA^{hi} cells can be observed between the newly formed GCs and the medullary cords [57]. A related observation is that a second immunization with an unrelated protein Ag administered less than 9 days after the first immunization seems to prevent the full development of the previous GC reaction [19]. Also, T-independent Ags can induce GC formation, but at about day 5 postimmunization these GC reactions are aborted within 24 h [58–60]. Finally, blocking B-T cell interactions with an anti-CD40 Ab, aborts ongoing GC reactions within 24 h [61]. This was initially interpreted as being due to massive B-cell death caused by the lack of rescuing T cell derived signals. However, the same authors subsequently showed that blocking CD40 rescuing signals in this way in fact induced GC dissociation by B cell migration and not massive cell death [62]. In conclusion, the relative contribution of apoptosis vs migration to the rise-and-fall dynamics of GCs and to selection remains unclear.

Discussion

Here, we re-evaluated the methodology of assessing GC clonal diversity, putting special emphasis on the claim that GCs are oligoclonal. In an indirect approach, immunohistology of GC sections was used in conjunction with assuming a binomial distribution of the fractions of two phenotypically distinct B cell populations to determine the number of seeding B cells. This approach further assumed that GCs are closed to continuous B cell entry. Using such a model, Hermans et al. concluded from their data that GCs contain 2–8 clones [42]. Re-evaluation led to the present extended mathematical model. This model includes an estimation of errors created by data sampling, two new parameters that take into account possible mistakes in classification of single population GC sections, and the likely variability in the number of seeding B cells. Fitting this new model to experimental data resulted in an estimated mean number of $\langle n \rangle = 23 - 37$ seeder B cells and scoring errors of $l_{T_A} = 5 - 10\%$ and $l_{T_B} = 55 - 60\%$. A χ^2-test confirmed that under these conditions the extended model is consistent both internally and externally with the experimental data.

Furthermore, these results on the number of seeding B cells match well with the estimates obtained from directly observed numbers of clones. Although 1–9 clones were detected in GC sections of mice 4–10 days after immunization [29, 43], data analysis using the Yule estimator of diversity shows that these findings more likely reflect the fact that more than 20–30 clones are present in individual GC sections.

Previous reports indicated that very low affinity B cells can be recruited into the GC reaction, and then outcompeted later by higher affinity B cells unless they acquire higher affinity by mutation [63, 64]. Moreover, studies of the primary IR in BSA immunized chickens, following the direct method, found that GCs are seeded by more than 20 founder B cells [65]. Finally, modelling simulations of GC affinity maturation by Shlomchik and colleagues also suggested that GCs are seeded by 40–50 B cells [66]. The analysis described here is in agreement with these results and conclusions.

Remarkably, the present work is strictly based on classic experimental results often cited to support the oligoclonality idea as if it was an established fact. Here we reveal a number of drawbacks. Besides the problem that observations with GC sections were taken as representative of whole GCs, these experimental data must have high measurement errors. In addition, the analysis of the indirect evidence might include erroneous model assumptions, and B-cell receptor sequences of the direct evidence were poorly analyzed in terms of diversity. We conclude that although apparently pure sections are often observed, this is more likely a reflection of the fact that clonal sizes are unequally distributed and that a few clones are especially large, rather than diversity being restricted to a few clones.

The potential for changes in the basic model's assumptions is not exhausted with the present work. For instance, a model permitting B cells to constantly enter the GC would allow completely different interpretation of the data on GC B cell clonal diversity. Indeed, very recently, studies by Nussenzweig and colleagues [67] and Haberman and colleagues [68] on in vivo imaging of murine GCs suggest that these are open structures, receiving frequent visits from follicular B cells, and capable of recruiting high affinity Ag-specific B cells. This is in full agreement with another recent study of GC samples microdissected from three reactive human lymph nodes [69]. Here, analysis of IgM and IgG V_H transcripts detected single B cell clones in multiple GCs. One clone was even found in as many as 19 GCs, including IgM and IgG variants of the same clonal origin and the offspring of individual hypermutated IgG clones. Finally, when high affinity cells are deleted from mature GCs, a diverse lower affinity subpopulation re-emerges [62], indicating that either they were already there, or they continually enter GCs, or both.

The present analysis, in our opinion, is an example of what should be a widely used critical technique. We hope that it will stimulate and help in the design of more robust experimental settings able to provide clear-cut answers to immunological questions. It is our belief that the scientific community should remain aware of how a concept arose. If conclusions are based on direct observation of large enough data samples, then the concept can be considered established. However, a conclusion derived from a small data set and/or based on additional, as yet unverified assumptions, has to be constantly re-evaluated, specially in the light of new data and analysis techniques, as well as regarding consistency with other concepts.

Acknowledgments This work was supported by grants SAF2007-63152 (MICINN, Spain) and PIRSES-GA-2008-230665 (7th FP, EC) to JF and by the BMBF (grant 0315005B) and the VolkswagenStiftung to MOG. The authors wish to thank R. Ribeiro and V. Ganusov for critical reading of the manuscript.

References

1. Janeway C (2005) Immunobiology: the immune system in health and disease, 6th edn. Garland Science, New York
2. Flajnik MF (2002) Comparative analyses of immunoglobulin genes: surprises and portents. Nat Rev Immunol 2:688–698
3. Manser T (2004) Textbook germinal centers? J Immunol 172:3369–3375
4. Flemming W (1885) Studien ueber regeneration der gewebe. Arch Mikrosk Anat 24:50–97
5. Nieuwenhuis P, Opstelten D (1984) Functional anatomy of germinal centers. Am J Anat 170:421–435
6. Conway E (1937) Cyclic changes in lymphatic nodules. Anat Rec 69:487–513
7. Ehrich WE, Harris TN (1945) The site of antibody formation. Science 101:28–31
8. Hellman T, White G (1930) Das verhalten lymphatischen gewebes wihrend eines immunisierungsprozesses. Virchows Arch Path Anat 278:221–257
9. Ringertz N, Adamson CA (1950) The lymph-node response to various antigens; an experimental-morphological study. Acta Pathologica Et Microbiologica Scandinavica. Supplementum 86:1–69
10. Congdon CC, Makinodan T (1961) Splenic white pulp alteration after antigen injection: relation to time of serum antibody production. Am J Pathol 39:697–709
11. Congdon CC (1964) The early histologic effects of antigenic stimulation. Arch Pathol 78:83–96
12. Hanna MG (1964) An autoradiographic study of the germinal center in spleen white pulp during early intervals of the immune response. Lab Invest 13:95–104

13. Thorbecke GJ, Asofsky RM, Hochwald GM, Siskind GW (1962) Gamma globulin and antibody formation in vitro. III. induction of secondary response at different intervals after the primary; the role of secondary nodules in the preparation for the secondary response. J Exp Med 116:295–310
14. Hanna M, Netteshe P (1966) Studies of lymphatic tissue germinal centers during an immune reaction – localization of 125I-labeled heterologous and isologous proteins. J Reticuloendothel Soc 3:363
15. Hanna MG, Szakal AK (1968) Localization of 125I-labeled antigen in germinal centers of mouse spleen: histologic and ultrastructural autoradiographic studies of the secondary immune reaction. J Immunol 101:949–962
16. Humphrey JH, Askonas BA, Auzins I, Schechter I, Sela M (1967) The localization of antigen in lymph nodes and its relation to specific antibody-producing cells. II. comparison of iodine-125 and tritium labels. Immunology 13:71–86
17. Nossal GJ, Ada GL, Austin CM (1964) Antigens in immunity. IV. cellular localization of 125-I- and 131-I-labelled flagella in lymph nodes. Aust J Exp Biol Med Sci 42:311–330
18. Nossal GJ, Abbot A, Mitchell J, Lummus Z (1968) Antigens in immunity. XV. ultrastructural features of antigen capture in primary and secondary lymphoid follicles. J Exp Med 127: 277–290
19. White R (1963) Functional recognition of immunologically competent cells by means of the fluorescent antibody technique. In: Wolstenholme GEW, Knight J (eds) The immunologically competent cell. Churchill (Ciba Foundation Study Group, No. 16), London, pp 6–19
20. Tew JG, Mandel TE (1979) Prolonged antigen half-life in the lymphoid follicles of specifically immunized mice. Immunology 37:69–76
21. Tew JG, Phipps RP, Mandel TE (1980) The maintenance and regulation of the humoral immune response: persisting antigen and the role of follicular antigen-binding dendritic cells as accessory cells. Immunol Rev 53:175–201
22. Coico RF, Bhogal BS, Thorbecke GJ (1983) Relationship of germinal centers in lymphoid tissue to immunologic memory. VI. transfer of B cell memory with lymph node cells fractionated according to their receptors for peanut agglutinin. J Immunol 131:2254–2257
23. MacLennan IC, Gray D (1986) Antigen-driven selection of virgin and memory B cells. Immunol Rev 91:61–85
24. Rouse RV, Reichert RA, Gallatin WM, Weissman IL, Butcher EC (1984) Localization of lymphocyte subpopulations in peripheral lymphoid organs: directed lymphocyte migration and segregation into specific microenvironments. Am J Anat 170:391–405
25. Weigert MG, Cesari IM, Yonkovich SJ, Cohn M (1970) Variability in the lambda light chain sequences of mouse antibody. Nature 228:1045–1047
26. Eisen HN, Siskind GW (1964) Variations in affinities of antibodies during the immune response. Biochemistry 3:996–1008
27. Berek C, Berger A, Apel M (1991) Maturation of the immune response in germinal centers. Cell 67:1121–1129
28. Jacob J, Kelsoe G, Rajewsky K, Weiss U (1991) Intraclonal generation of antibody mutants in germinal centres. Nature 354:389–392
29. Jacob J, Przylepa J, Miller C, Kelsoe G (1993) In situ studies of the primary immune response to (4-hydroxy-3-nitrophenyl)acetyl. III. the kinetics of V region mutation and selection in germinal center B cells. J Exp Med 178:1293–1307
30. McHeyzer-Williams MG, McLean MJ, Lalor PA, Nossal GJ (1993) Antigen-driven B cell differentiation in vivo. J Exp Med 178:295–307
31. Ziegner M, Berek C (1994) Analysis of germinal centres in the immune response to oxazolone. Adv Exp Med Biol 355:201–205
32. MacLennan IC, Johnson GD, Liu YJ, Gordon J (1991) The heterogeneity of follicular reactions. Res Immunol 142:253–257
33. Kepler TB, Perelson AS (1993) Somatic hypermutation in B cells: an optimal control treatment. J Theor Biol 164:37–64
34. Or-Guil M, Wittenbrink N, Weiser AA, Schuchhardt J (2007) Recirculation of germinal center B cells: a multilevel selection strategy for antibody maturation. Immunol Rev 216:130–141

35. Radmacher MD, Kelsoe G, Kepler TB (1998) Predicted and inferred waiting times for key mutations in the germinal centre reaction: evidence for stochasticity in selection. Immunol Cell Biol 76:373–381
36. Liu YJ, Zhang J, Lane PJ, Chan EY, MacLennan IC (1991) Sites of specific B cell activation in primary and secondary responses to T cell-dependent and T cell-independent antigens. Eur J Immunol 21:2951–2962
37. Jacob J, Kassir R, Kelsoe G (1991) In situ studies of the primary immune response to (4-hydroxy-3-nitrophenyl)acetyl. i. the architecture and dynamics of responding cell populations. J Exp Med 173:1165–1175
38. Paus D, Phan TG, Chan TD, Gardam S, Basten A, Brink R (2006) Antigen recognition strength regulates the choice between extrafollicular plasma cell and germinal center B cell differentiation. J Exp Med 203:1081–1091
39. Phan TG, Paus D, Chan TD, Turner ML, Nutt SL, Basten A, Brink R (2006) High affinity germinal center B cells are actively selected into the plasma cell compartment. J Exp Med 203:2419–2424
40. Kroese FG, Wubbena AS, Seijen HG, Nieuwenhuis P (1987) Germinal centers develop oligoclonally. Eur J Immunol 17:1069–1072
41. Kroese FG, Wubbena AS, Seijen HG, Nieuwenhuis P (1988) The de novo generation of germinal centers is an oligoclonal process. Adv Exp Med Biol 237:245–250
42. Hermans MH, Wubbena A, Kroese FG, Hunt SV, Cowan R, Opstelten D (1992) The extent of clonal structure in different lymphoid organs. J Exp Med 175:1255–1269
43. Jacob J, Kelsoe G (1992) In situ studies of the primary immune response to (4-hydroxy-3-nitrophenyl)acetyl. II. a common clonal origin for periarteriolar lymphoid sheath-associated foci and germinal centers. J Exp Med 176:679–687
44. Lee SM, Chao A (1994) Estimating population size via sample coverage for closed capture-recapture models. Biometrics 50:88–97
45. Simon H (1955) On a class of skew distribution functions. Biometrika 42:425–440
46. Yule G (1925) A mathematical theory of evolution, based on the conclusions of Dr. J.C. Willis, FRS. Philos Trans R Soc Lond B 213:21–87
47. Hsieh C, Liang Y, Tyznik AJ, Self SG, Liggitt D, Rudensky AY (2004) Recognition of the peripheral self by naturally arising $CD25^{+}CD4^{+}$ T cell receptors. Immunity 21:267–277
48. Leanderson T, Kllberg E, Gray D (1992) Expansion, selection and mutation of antigen-specific B cells in germinal centers. Immunol Rev 126:47–61
49. Longo NS, Lipsky PE (2006) Why do B cells mutate their immunoglobulin receptors? Trends Immunol 27:374–380
50. Moreira JS, Faro J (2006) Re-evaluating the recycling hypothesis in the germinal centre. Immunol Cell Biol 84:404–410
51. Foote J, Milstein C (1991) Kinetic maturation of an immune response. Nature 352:530–532
52. Foote J, Eisen HN (1995) Kinetic and affinity limits on antibodies produced during immune responses. Proc Natl Acad Sci USA 92:1254–1256
53. Ziegner M, Steinhauser G, Berek C (1994) Development of antibody diversity in single germinal centers: selective expansion of high-affinity variants. Eur J Immunol 24:2393–2400
54. Hanna MG, Congdon CC, Wust CJ (1966) Effect of antigen dose on lymphatic tissue germinal center changes. Proc Soc Exp Biol Med (New York, N.Y.) 121:286–290
55. Szakal AK, Taylor JK, Smith JP, Kosco MH, Burton GF, Tew JJ (1990) Kinetics of germinal center development in lymph nodes of young and aging immune mice. Anat Rec 227:475–485
56. Tew JG, DiLosa RM, Burton GF, Kosco MH, Kupp LI, Masuda A, Szakal AK (1992) Germinal centers and antibody production in bone marrow. Immunol Rev 126:99–112
57. Kosco MH, Burton GF, Kapasi ZF, Szakal AK, Tew JG (1989) Antibody-forming cell induction during an early phase of germinal centre development and its delay with ageing. Immunology 68:312–318
58. de Vinuesa CG, Cook MC, Ball J, Drew M, Sunners Y, Cascalho M, Wabl M, Klaus GG, MacLennan IC (2000) Germinal centers without T cells. J Exp Med 191:485–494

59. Gaspal FMC, McConnell FM, Kim M, Gray D, Kosco-Vilbois MH, Raykundalia CR, Botto M, Lane PJL (2006) The generation of thymus-independent germinal centers depends on CD40 but not on CD154, the T cell-derived CD40-ligand. Eur J Immunol 36:1665–1673
60. Lentz VM, Manser T (2001) Cutting edge: germinal centers can be induced in the absence of T cells. J Immunol 167:15–20
61. Han S, Hathcock K, Zheng B, Kepler TB, Hodes R, Kelsoe G (1995) Cellular interaction in germinal centers. roles of CD40 ligand and B7-2 in established germinal centers. J Immunol 155:556–567
62. Han S, Zheng B, Porto JD, Kelsoe G (1995) In situ studies of the primary immune response to (4-hydroxy-3-nitrophenyl)acetyl. IV. affinity-dependent, antigen-driven B cell apoptosis in germinal centers as a mechanism for maintaining self-tolerance. J Exp Med 182:1635–1644
63. Porto JMD, Haberman AM, Shlomchik MJ, Kelsoe G (1998) Antigen drives very low affinity B cells to become plasmacytes and enter germinal centers. J Immunol 161:5373–5381
64. Porto JMD, Haberman AM, Kelsoe G, Shlomchik MJ (2002) Very low affinity B cells form germinal centers, become memory B cells, and participate in secondary immune responses when higher affinity competition is reduced. J Exp Med 195:1215–1221
65. Arakawa H, Kuma K, Yasuda M, Furusawa S, Ekino S, Yamagishi H (1998) Oligoclonal development of B cells bearing discrete Ig chains in chicken single germinal centers. J Immunol 160:4232–4241
66. Shlomchik MJ, Watts P, Weigert MG, Litwin S (1998) Clone: a Monte-Carlo computer simulation of B cell clonal expansion, somatic mutation, and antigen-driven selection. Curr Top Microbiol Immunol 229:173–197
67. Schwickert TA, Lindquist RL, Shakhar G, Livshits G, Skokos D, Kosco-Vilbois MH, Dustin ML, Nussenzweig MC (2007) In vivo imaging of germinal centres reveals a dynamic open structure. Nature 446:83–87
68. Hauser AE, Shlomchik MJ, Haberman AM (2007) In vivo imaging studies shed light on germinal-centre development. Nat Rev Immunol 7:499–504
69. Bende RJ, van Maldegem F, Triesscheijn M, Wormhoudt TAM, Guijt R, van Noesel CJM (2007) Germinal centers in human lymph nodes contain reactivated memory B cells. J Exp Med 204:2655–2665

Chapter 13
B Cell Strategies of Ag Recognition in a Stratified Immune System

Belen de Andrés, Ana R. Sánchez-Archidona, Isabel Cortegano, Natalia Serrano, Sharmili Jagtap, María-Luisa Gaspar, and Miguel-Angel Rodríguez Marcos

Abstract From the old good times in which pioneering immunologists were simply specialised in either T or B lymphocytes, to the current lymphoid complexities that powerful quantitative technologies have revealed, the adaptive immune system has become a highly complicated subject of study. Also, as a part of the hematopoietic system, the lymphoid tissue is constituted by cells that undergo great turnover rates, with continuous renewal during an individual's lifespan, extending from immature, bone marrow-derived precursors up to the final exhaustion of highly-differentiated cells. To exert their functions, B and T lymphocytes first need to recognise small molecular epitopes, through their highly diverse clonotypic receptors, with different degrees of specificity (the result of combinatorial gene rearrangements, with inexact joining ends). But lymphocytes also bear an evolutionary history that "informs" them about those more or less efficient receptor choices that previously influenced the survival of the species. Those cellular adjustments that were previously successful have remained preserved and are therefore preferentially exploited by the immune system (IS). Subsequently, immune strategies to "see" the antigenic world (self or foreign, harmless or dangerous) are balanced between an unrestricted openness to novelty (which has been named the promethean character of the IS [1]) and the evolutionary preservation of cellular activities and receptors that have already been tested and demonstrated their utility. In the following pages, we try to provide a few insights regarding when and where those functional polarisations are at work, with a special focus on B lymphocytes. We will leave out the differentiation steps and genetic programmes leading to the production of mature B lymphocytes, except in those selected cases when they may provide relevant information.

B. de Andrés (✉)
Centro Nacional de Microbiología, ISCIII, Majadahonda, 28220 Madrid, Spain
e-mail: bdandres@isciii.es

C. Molina-París and G. Lythe (eds.), *Mathematical Models and Immune Cell Biology*,
DOI 10.1007/978-1-4419-7725-0_13,

The Life and Death of a B Lymphocyte

More than other organs of the body, the lympho-hematopoietic system is characterised by the continuous regeneration of most of its cellular components (with half-lives extending from 12 h for neutrophils up to several months for macrophages or red blood cells). Mature lymphocytes are also constantly replenished by newly-emerging cells that result from the differentiation of self-renewing hematopoietic stem cells (HSCs) located in endosteal and, perhaps, in vascular-related niches from the adult Bone marrow (BM). Even intermediate cell-precursor stages cannot maintain themselves without the refuelling from HSCs. The adult B cell compartments are preserved in homeostatic conditions by means of recent cellular input, proliferation, survival, inter-clonal competition for "resources" and final differentiation to death; this lively cellular dynamics demands check-points to maintain the immune physiology. During an individual's lifetime, the steady state equilibrium of B cell compartments is also punctuated by vigorous antigenic encounters, which frequently occur in the context of innate inflammatory reactions, and trigger conventional immune responses (Fig. 13.1).

The persistence of functional B cell compartments requires the influx of immature precursors. If this input is blocked, B cells become exhausted, although to a variable cell subset-specific degree. The generation of new mature B cells ($10–20 \times 10^6$ day^{-1}) markedly overcomes the peripheral homeostatic demands: only one-tenth of maturing B cells are exported to the periphery and, from those, one-half are included in the functional B cell compartments [2]. This implies that B cell newcomers need to compete with both emerging and available B cell clones for limiting "resources" in order to successfully engraft in the periphery, expand and/or survive, and differentiate [3,4]. The competitors are clones of the same cell lineage, given that B, T or NK cells have specific requirements for survival (IL-7, IL-15, IL-2, BAFF, antigens, etc.), and, in particular, the emerging B cells only "sense" B cell deficiencies [5–8]. As we will discuss below, discrete cell subsets of the same B cell lineage show maintenance requirements that differ from those of other B cell populations. Various kinds of resources exist, which extend from antigens (Ags) that trigger B cell receptor (BCR) signalling up to innate Toll-like receptor (TLR) ligands, selected cytokines or restricted access to survival niches. The surface receptors involved in rescuing B cell subsets can be classified as BCRs, immune TLRs [9] and members of the TNF-receptor family, such as BCMA, TACI and BAFF-R [10]. Both BCR and TLR signals can be transmitted through Btk-related intracellular pathways (absent in xid mice), while BAFF-transducing signals are Btk-independent. Also, BCR and BAFF signal pathways may be coupled at the level of classical and non-classical NF-κB complexes [6]. Ag-specific inter-clonal competition has been demonstrated in immunoglobulin (Ig) transgenic and xid mouse models: in the former case, the endogenous, polyclonal Ig repertoire rapidly displaces the initially-predominant Ig-transgenic clone from functional roles in the IS; also, xid B cells are able to replenish B cell devoid hosts, except in the case when they are simultaneously reconstituted by normal B cells [11, 12]. It is important to remark that not all mature B cells and Ig repertoires are the result of

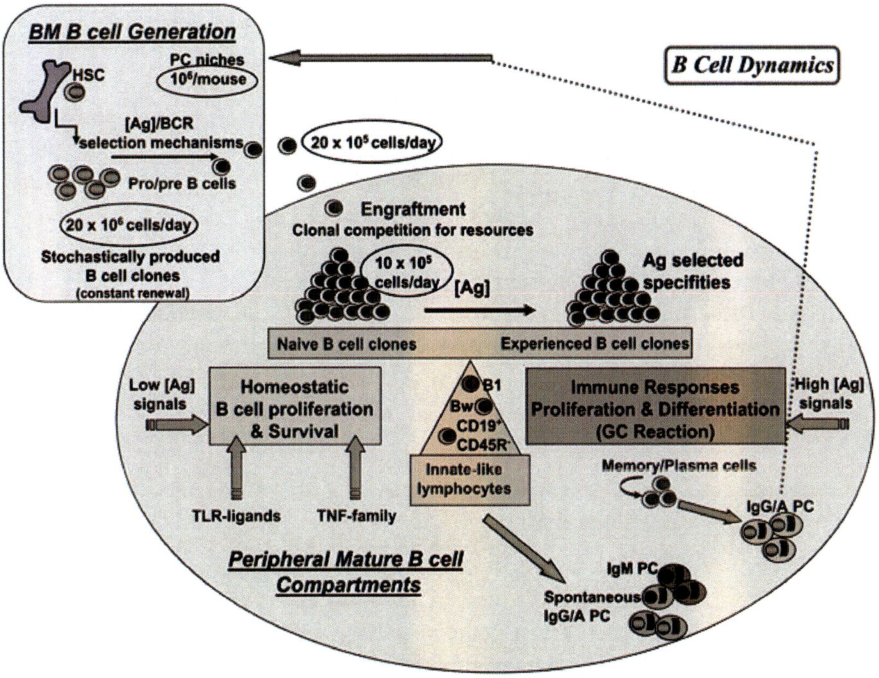

Fig. 13.1 Stratified B cell dynamics. Stochastically generated B cell clones are selected in the BM by Ag/BCR signals, and lead to the generation of naïve B cell clones. These naïve B cells migrate to the peripheral B cell compartments, and engraft there on distinctive niches after interclonal competition for resources that direct their proliferation, survival and/or maturation. Mature functional B cells respond to Ag encounters generating immune responses in the context of the GC, and generate memory and plasma cells. These plasma cells migrate again to the BM, where they engraft in limited niches to allow their long-term survival. Small numbers of innate-like lymphocytes are also present in peripheral organs, participating in different functions of the adaptive immune responses

Ag/BCR-specific selection steps but that, in selected scenarios or cell types, they also may be produced by Ag-independent, TLR- or BAFF-dependent pathways, for example B1 and marginal zone (MZ) B cells. Finally, superantigen-like ligands that recognise VH/Vβ common framework determinants may provide "tonic" signals that influence the establishment/survival of some mature lymphocytes [13, 14].

Classical experiments of both immuno-exhaustion by cytotoxic drugs and of adoptive cell transfers revealed two discrete B cell populations, with low- and high-cell turnover behaviour, respectively [15, 16]. In those protocols, around two-thirds of B cells were highly cycling and short-lived, while the remaining were more quiescent and persisted for long time periods. The first component reminded us of emerging, highly-proliferating, immature B cells (now defined as transitional), where new specificities are stochastically bred and submitted to selective processes, and the second one mimicked the already successful, established B cell clones in the immune periphery. The experimental replenishment of B cell compartments after adoptive B cell transfers has also been used as a recapitulation of normal B cell development and a surrogate of immune B cell dynamics in homeostasis. An important aspect revealed by later experiments is that the transfer of limited B cell numbers in a B cell-devoid host is initially biased to the reconstitution of normal amounts of IgM-secreting plasma cells and serum IgM levels. In a second phase, the subset of naturally activated B cells is predominantly reconstituted and, finally, significant, but not normal numbers of mature B cells (which would require continuous BM-derived cell input) become established [17, 18]. Alternatively, when B cell production is arrested in the BM, the most persistent populations are also Ig-secreting plasma cells and activated B cells [2, 19] (including B1 and MZ B cells; see below). The absolute numbers of these different B cell populations are self-regulated, such as, for instance, the replenishment of IgM-secreting plasma cells (or of B1 cells) in a first B cell transfer selectively blocks the inclusion of cells derived from a second transfer into the same compartment [18, 20].

Peripheral B Cell Dynamics

The fate of recently emerged BM B cells relies on BCR–Ag interactions (Fig. 13.2). Analyses made on different Ig transgenic mice have shown that recognition of high-affinity self-Ags mainly results in negative selection of the responding autoreactive B cell clones (through either apoptosis or functional anergy), subsequently eliminating pathogenic autoantibodies. Alternatively, the IgL chains from autoreactive BCRs

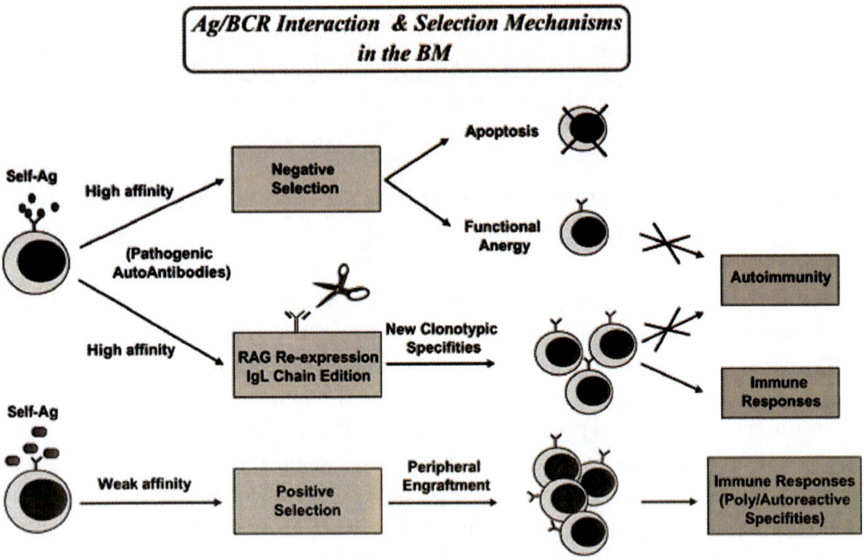

Fig. 13.2 Selection mechanisms induced by Ag/BCR interactions in the BM. During the generation of naïve B cells, those bearing high affinity BCR for self-Ags are eliminated by negative selection processes or by changing their BCR specificities by IgL edition. These B cells bearing weak affinity BCRs for self-Ag are positively selected and expanded

can be edited by re-expressing the recombination-activating gene (RAG) enzymatic machinery, and then give rise to new clonotypic specificities that escape to autoimmunity. However, BCR–self-Ag interactions of weak affinity are also essential to induce positive selection of the complementary B cell clones and to engraft them into the functional B cell periphery [21]. Actually, up to two-thirds of emerging B cells in humans bear poly- and autoreactive Ig specificities [22]. As also postulated for thymic T cell development (apoptosis by neglect), Ag-driven BCR signals are absolutely required to maintain the pools of mature B cells [23, 24].

Without attempting an exhaustive phenotypical dissection of all kinds of minor cell types, a progression of peripheral B cell subsets could be defined as follows: transitional B, mature B, Ig-secreting plasma and memory B cells [25]. The early transitional B cells are highly cycling, very sensitive to BCR interactions with low apoptosis thresholds, and they lack survival-promoting receptors such as CD21; late transitional B cells, by contrast, become progressively resistant to BCR-induced death and are expanded through low-affinity, self-Ag stimulation plus other environmental cues [26]. Among these later, BAFF-receptors are up-regulated and BAFF becomes the most critical B-specific survival signal from now on, being able even to overcome BCR negative signals [27]. Transitional B cells rapidly differentiate into more mature cellular stages, but do not include any long-lived cell, as shown by BrdU pulse-chase experiments. Mature B cells, predominantly of the follicular type, are established in response to Ag- and BAFF-dependent stimuli [28], and most of them are quiescent, although they also include a minor component of activated B cells (likely subjected to differential selection pressures). In the context of immune responses, plasma and memory B cells are generated inside the germinal centre (GC) reaction (Fig. 13.1). Early after Ag immunisation, Ag-cognate and non-cognate B cell clones proliferate and expand, differentiating into highly-proliferative centroblasts in the dark zone, a fraction of which will migrate and mature into quiescent centrocytes in the light zone, where they are in contact with T cells and immunocomplex-bearing follicular dendritic cells. An initial wave of cycling plasmablasts results into short-lived plasma cells that secrete a first antibody (Ab) barrier against the pathogen. Highly regulated processes of both inter- and intraclonal cell competition occur in the GC for selecting high-affinity, Ig-secreting cells [29, 30]. The early plasmablasts may be or not the source for Ag-selected, long-lived plasma cells that migrate towards chemokine gradients and home into a finite pool of BM survival niches; new plasmablasts only become established if they displace previous clones that are sitting in these BM niches. The longevity of BM plasma cells subsequently is not an intrinsic cell property, but the result of selective engrafting into the available survival niches that are calculated to be around 10^6/mouse; the signals involved in plasma cell survival are yet unclear [31]. A second branching of the GC-responding B cells leads to the formation of memory B cells, which display long half-lives that may extend up to the whole lifespan of an individual [32,33]. The differentiation pathways to both memory B and long-lived plasma cells require not only TLR signals, but also Ag binding and T cell help [34]. It seems that the generation of these two latter cell types are mutually independent: the selective depletion of memory B cells leaves untouched the compartment of long-lived BM plasma cells [35]. After immunisation with particulate Ag, discrete memory cell subsets have been recently identified both in GC-like structures and outside B cell follicles, bearing different effector functions for differentiation to plasmocytes and/or reinitiation of GC reactions [36]. It has been suggested that the expression ratios between the BCL-6 and Blimp-1 transcription factors inside the GC modulate the fates between memory and plasma cells [37].

Interestingly, the cell biology strategies that evolve inside the GC reaction do not substantially differ from those exploited in the establishment of naïve B cells and Ig repertoires:

- In both cases, Ig diversity is stimulated by either the process of RAG-dependent, Ig gene rearrangements in pre-B cells, or through the activation-induced deaminase (AID)-dependent mechanisms of Ig class switch and Ig somatic hypermutation in GC B cells, thus offering novel specificities to improve the chances of fitting the Ag (although some of them may need to be filtered out to preserve tolerance)
- Both transitional B cells and centroblasts display high rates of proliferation and, simultaneously, are very sensitive to apoptosis
- They both require BCR–Ag interactions for positive selection, establishment and longevity of the responding B cell clones

Mature B cell compartments are not only the consequence of cell proliferation but also the result of pro-survival signals. Transitional B cells are positively selected by low-affinity Ag interactions that lead them to the stage of BAFF responsiveness. This later is the crucial factor that maintains most naïve and peripheral B cells, and dictates their longevity in competition with other B cell clones. By contrast, it seems that the survival of memory B cells and natural Ab-secreting plasma cells is BAFF-independent [38].

The adult B cell immune compartment subsequently represents an heterogeneous mixture of clones (Fig. 13.1) derived from:

- Homeostatic signals, where the basal level of immune activity is likely maintained by weak interactions with Ag determinants that move in small concentration ranges [6] (somatic self-Ags, colonising microbes in epithelial layers, immune clonotypic epitopes, etc.), and from
- B cell clones that are secondary to encounters with high input of life-threatening pathogens, that have overcome the innate mechanisms of inflammation and generate Ag-specific, conventional immune responses [39].

In both situations, lymphoid cell proliferation includes an initial stage of exponential B cell growth and death (e.g., the early phase of B cell adoptive transfers into B-deficient hosts or the early wave of short-lived plasmablasts after Ag immunisation), and a second period of steady state, low proliferation rates (e.g., long-lived follicular B cells or BM-homed plasma cells). In general, homeostatic B cell proliferation (Table 13.1) is characterised by low cell division rates, T cell independency, preservation of the naïve cell phenotype (although with a transient up-regulation of activation markers), absence or low rates of Ig mutation, competition for BAFF

Table 13.1 Differential B cell proliferation

	Homeostasis	Immune responses
T cell dependence	−	+
Response to Ag	Low [Ag]	High [Ag]
Ag/BCR mediated	+	+ + +
BAFF-mediated	+ + +	+
Usage of Ig sequences	Germline	Diversified
AID-dependence	−	+ (CSR/SHM)
Cells generated	Naive phenotype	Memory and plasma cells

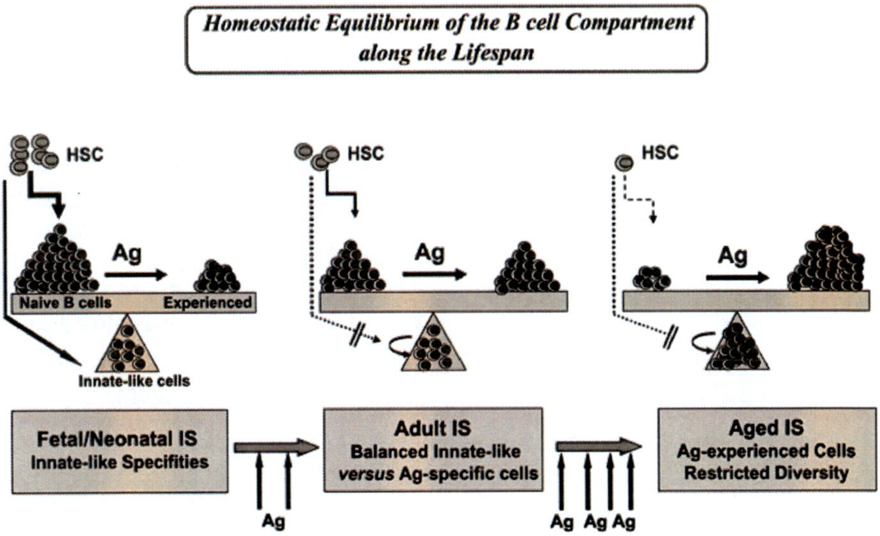

Fig. 13.3 The homeostatic equilibrium of peripheral B cell compartments. HSC differentiate into naïve B cells and replenish peripheral organs in the absence of Ag stimulation. These cellular inputs are high during the fetal/neonatal period, and most of the cells present in the periphery are naïve B cells. Innate-like cells, too, are mostly generated in this period (*left balance*). Exposure to Ag encounters induces the generation of experienced B cells in such a way that, in a stable adult IS, both newly arriving naïve cells and Ag experienced cells are present, as well as innate-like subpopulations that are self-maintained (*middle balance*). Aged individuals that have been exposed to continuous Ag-encounters accumulate experienced Ag-specific B cell clones. Also, a progressive exhaustion in the BM for the generation of new naïve B cells takes place, both facts producing a restricted repertoire of specificities (*right balance*)

and use of specific signal transduction pathways. By contrast, the proliferation rates accompanying immune responses tend to be greater, the process is predominantly T cell dependent, the responding clones become experienced B cells that mature to both memory and plasma cells, the AID-mediated process of Ig class switch recombination (CSR) and somatic hypermutation (SHM) becomes activated, and high-affinity BCR clonotypes are selected by binding to the specific Ag [6]. The stability and robustness of the immune behaviour (Fig. 13.3) likely has a lot to do with maintaining the equilibrium between those clones selected in homeostasis (naïve?) and those preserving the memory of dangerous Ag encounters (experienced?). With ageing, the production of polyclonal naïve B cells from HSCs becomes much smaller, and the mature B cell compartments mainly accumulate expanded clones of Ag-experienced cells. Both mechanisms lead to a progressive restriction in Ig diversity and to reduced potential of response to new Ag demands [40].

Evolutionary Forces Designing the Adaptive Immune System

The adaptive IS is not only a homogeneous cellular complex in which the exclusive driving force is the constant renewal of stochastically produced clones that are submitted to somatic Ag selection. It also includes a conservative component represented by discrete cellular subsets, so-called unconventional, innate-like or bridge lymphocytes (to remark their shared features with innate immune cells [41]), and cells with clonotypic specificities that are positively selected and highly expanded, in some cases, before any Ag encounter [42, 43]. The list of innate-like lymphocytes includes: $\gamma\delta^+$ T cells [44], B1 B, MZ B, NKT [45], some intestinal, intraepithelial CD8$\alpha\alpha^+\alpha\beta^+$T, Bw B [46], CD19$^+$ CD45R$^-$ pre-plasma B [47], extrafollicular B cells, etc. Without extending into minor, cell subset-specific details, we would like to underline a few relevant characteristics that are common to these lymphocytes (Table 13.2):

1. They are either oligoclonal and/or frequently include clonally dominant specificities, which might be (at least, in part) the result of Ag-independent selection

Table 13.2 Cells in the adaptive immune system

Innate-like lymphocytes	Conventional B cells
Embryo-fetal precursors: self-renewal in adult life	BM precursors: highly proliferating emerging B cells
Oligoclonal. Evolutionarily-tested dominant specificities (self-Ag/common pathogens/commensal bacteria)	Unrestricted, random generation of clonotypic diversity
Ag-independent selection	Ag-dependent selection. Affinity and specificity improvement. AID mediated CSR and SHM
Relevant in homeostatic conditions and initial phases of immune responses. Immunomodulatory roles	Involved in specific immune responses
Particular niches (gut epithelia, liver, coelomic cavities, extrafollicular splenic areas)	Classical secondary lymphoid organs: spleen, lymph nodes
Pre-activated state. Rapid, multi-specific responses, limitation of tissue damage	Quiescent status in peripheral organs. Slow and highly specific responses after Ag-stimulation

2. They are more functionally relevant in homeostatic conditions and in initial phases of the immune response, and display powerful immuno-modulatory roles over other lymphocyte subsets
3. They exploit niches different from those of conventional lymphocytes, such as gut epithelia, liver, coelomic cavities, extrafollicular splenic areas, etc.
4. In the cases studied, the signals involved in the proliferation/survival of innate-like lymphocytes differ from those of other lymphocytes
5. Frequently, innate-like lymphocytes are generated early in life from embryo-fetal precursors that are exhausted or highly reduced in the adult BM, leading to the suggestion that these mature lymphocytes should maintain their cell numbers by self-renewal mechanisms

To select just a few examples, immature multipotential progenitors from the early mouse ontogeny are highly efficient in the generation of B1, MZ B and the recently described $CD19^{+}$ $CD45R^{-}$ splenic B cells, although minor subsets of B1-specific pro-B cells may persist in the adult BM (yet, unexpectedly translating self-renewal behaviour to intermediate precursors [48]). When BM B cell input is arrested, these B cell subsets selectively persist for extended periods [2, 19]. Both B1 and MZ B cells (as well as IL-4 producing, adult $\gamma\delta^{+}$ T cells) include high fractions of N-negative V(D)J joints in their Ig repertoires, likely deriving from TdT-negative perinatal precursors [49, 50], although selection against N additions has also been suggested [51]. The clonal dominance of B1-enriched anti-phosphatidylcholine Abs (VH11-Vκ9) is pre-established before Ig expression [43]. An essential aspect of innate-like B cells is that they are highly biased to plasma cell differentiation and, more specifically, that they support most of the natural serum Abs, which likely have protective functions early after birth (together to mother-derived IgGs) and important immunoregulatory roles in the adult [52–54]. The first $\gamma\delta^{+}$ T cells emerge in developmentally-controlled, monoclonal waves that home to selected body areas (e.g., Vγ5Vδ1 to the epidermis, Vγ6Vδ1 to the uterine epithelium, Vγ1Vδ6 to the liver), although other specificities are also generated in the adult thymus [44]. The vast majority of NKT cells express the Vα14–Jα18 chain, preferentially bound to Vβ8, Vβ7 and Vβ2 chains and recognise glycolipids presented on CD1d, although they seem to emerge weeks after birth in the mice [45]. Even if the restricted clonotypic repertoires of innate-like cell populations are modulated by Ag selection, they are definitively constrained by developmental mechanisms and/or particular niche homing. This later aspect is highly influential in the formation of the gut-associated lymphoid tissue (GALT), which dominates the B cell system of avians, rabbits and sheep (but also cattle, pigs, dogs, etc.) In the GALT, gene conversion and SHM mechanisms play more relevant roles in establishing Ig diversity than gene rearrangements. Recently, the presence of discrete B cells associated to the intestine has also been reported in mice and humans, whose Ig repertoires diversify in Ag-independent manners [55, 56]. Compared to conventional B-lymphocytes, BCR signals are particularly deleterious to B1 and MZ B cells, leading them to death; by contrast, B1/MZ B cells are good responders to TLR-specific ligands that promote their survival and polyclonal cell expansion, through Btk-dependent intracellular signals [57]. While follicular B cells require both BCR and TLR signals to

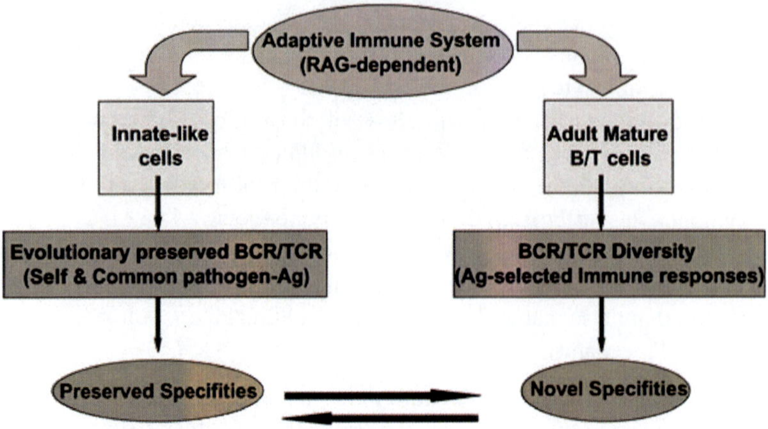

Fig. 13.4 Strategies of response used by the adaptive IS. Two complementary pathways are used to recognise Ags and to develop efficient immune responses. The Innate-like cells represent a rapid, albeit low efficient, way to cope with antigenic exposures. This ensures time for the conventional mature B- and T cells to develop slower but highly specific immune responses, capable to completely eliminate the antigens and to generate long term memory cells

differentiate into Ab-secreting cells [34], B1 and MZ B cells are able to differentiate even after single TLR stimuli [58]. Also, these two later B cell subsets rapidly respond to bacterial mitogens by modifying their adhesive properties and migrate from the coelomic cavities to the bloodstream [59] or from the marginal zone into the GC, to collaborate in the early aspects of immune responses [60].

We are now immersed into a scenario in which two complementary strategies of Ag recognition are present inside the adaptive IS (Fig. 13.4): firstly, an immune compartment that displays an unrestricted, random generation of clonotypic diversity on highly-proliferating emerging B cells, from which the Ag selects and expands those specificities displaying the highest complementarities. Further Ag encounter in the periphery refines the initial choice to further improve the affinity and specificity of naïve Abs. The whole process guarantees the deployment of specific immune responses, but it is somewhat slow to defeat highly-duplicating, invasive microorganisms. Alternatively and before the RAG evolutionary innovation, mechanisms of innate defence have demonstrated their survival value from insects up to jawless fishes, in different kinds of hemocytes (macrophage-like cells). They are also used by lymphocytes of the RAG-dependent, adaptive IS in upper vertebrates. In line with other authors, we suggest that a core of evolutionarily-tested cells and clonotypes, which are developmentally preserved to respond to either life-threatening pathogens present in early life and/or to developing self-somatic Ags (roles in tissue remodelling, scavenging apoptotic bodies, etc.), make use of mechanisms which are characteristic of the innate system. These innate-like lymphocytes exploit both the abilities of inflammation and innate immunity (speed of response, limitation of tissue damage, pre-activated state, etc.) and the focused clonotypic specificity of the adaptive system. This later feature is restricted to pre-defined receptors that recognise molecular patterns frequently present in common pathogens, in commensal bacteria and in self-Ags, perhaps as a result of biological convergence between the three life forms, which might be required for their mutual survival.

The Imprints of the Ontogeny in the Adult Immune System

The first mature sIgM$^+$ B cells appear at the end of mouse gestation. From then on and concerning B cell biology, two distinctive life periods are defined, that is, perinatal (up to 2–4 weeks of postnatal life in mice) and adult. The perinatal IS and, particularly, the Ig repertoires there expressed, are considered to be naïve or immature: they are little diversified, enriched in polyreactivities and in self-reactive, germline IgM with weak affinities for the Ag [61, 62] (Fig. 13.3). It is thought that newborns are mostly protected by mother-derived IgG Abs. By contrast, the adult IS deploys all the genetic mechanisms of clonotypic diversity that endow it with the ability to recognise and efficiently eliminate any kind of pathogen. However, the adult Ig clonotypic repertoires emerge over the "scaffold" of neonatal ones, and these later are preserved along the whole lifespan. But, how "immature" or "naïve" is the perinatal Ig repertoire? Alternatively, is it the result of selective pressures established during fetal development? In recent years, we have obtained some insight (Fig. 13.5): the lymphoid-specific programmes of differentiation and the first B-lineage precursors do emerge very early on in fetal ontogeny, at 10–12 gestational days, in both liver and extra-liver niches [47, 63–65]. The RAG [66] and non-homologous end-joining (NHEJ) enzymatic machineries are present at those early periods, and B-lineage specific genes are already expressed (Pax5,λ 5, VpreB, CD19, etc.). Also, the Ig gene rearrangement process starts and the first embry onic DJ$_H$ joints, displaying multiple mechanisms of diversity, are detected in the post-gastrulation mouse embryo [63,67]. An extensive analysis of these early, unselectable DJ$_H$ joints in normal mouse embryos and in heterozygous embryos derived from RAG2$^{-/-}$ mothers, has revealed unexpected features, such as the frequent addition of non-templated N nucleotides (in the absence of TdT), large coding joint deletions and a random usage of stretches of sequence micro-homology in the joining process of N$^-$ sequences [68]. These genetic patterns change around E14 when full VDJ$_H$ and VJ$_L$ chain gene rearrangement and transcription, and pre-BCRs, are detected [64]. The embryo N additions are due to the activity of the DNA polymerase μ(Pol μ), a member of the X family of DNA polymerases (where TdT is included [69]. The differences between random *versus* predominant micro-homology usage in both embryo and perinatal N$^-$ DJ$_H$ rearrangements, respectively, further suggested the existence of a process of positive selection of those micro-homology-bearing DJ$_H$s along the last third of gestation. We have reasoned that, similar to what happens with DNA repair mechanisms acting on other NHEJ processes, the joining enzymatic machinery (including Polμ) used in the early embryos may be preferentially focused onto the ligation of dangerous, free DNA ends introduced by RAG in this highly-proliferative cellular context of the ontogeny [70], although at the price of creating many "useless" products. Later on, the emergence of perinatal Ig repertoires might be due either (1) to an autonomous compartment of perinatal cell subsets, using different genetic machineries of gene rearrangement (unlikely), and/or (2) to the establishment of mechanisms both of positive selection of open-reading frame 1^+ sequence homology$^+$, germline V(D)J_Hs and/or negative selection of open-reading frame 2^+, N^+, and highly-deleted joints, perhaps related to selective pairing of emerging V(D)J_Hs to λ5/Vpre-B surrogate light chain, along the last third of mouse gestation (Fig. 13.5). On these bases, we are tempted to suggest that the perinatal IS is not simply immature and non-functional, but rather the result of selective pressures at work during late gestation acting on the random gene rearrangements that first emerge in the post-gastrulation mouse embryo. Many aspects of the perinatal clonotypic repertoire will become later fixed in

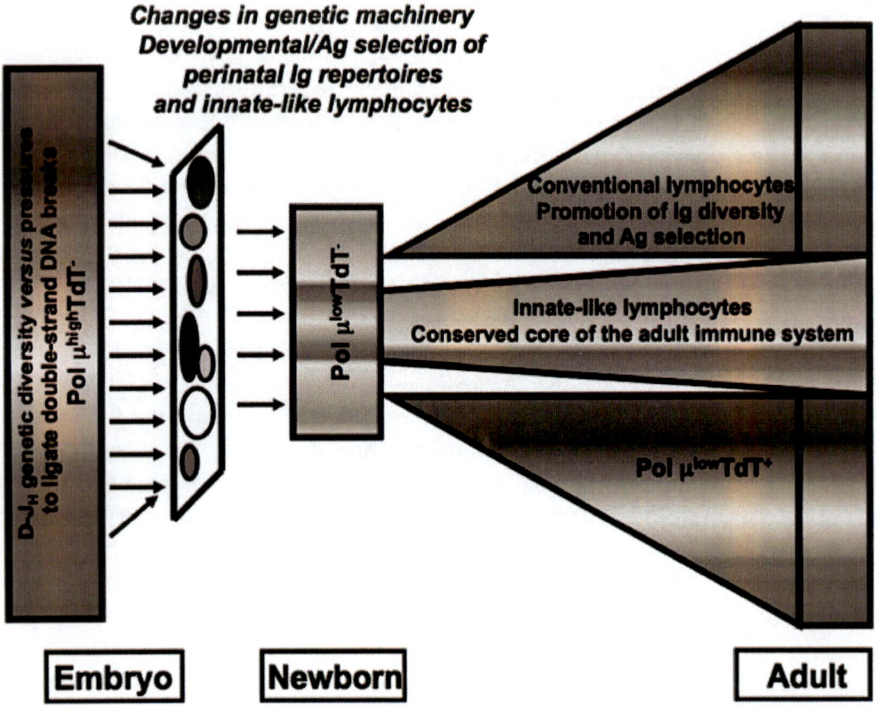

Fig. 13.5 *Left, multicolour rectangle* represents the stochastic DJ_H diversity displayed in the early post-gastrulation mouse embryo, as a consequence of the genetical attempts to ligate RAG-mediated double-strand DNA breaks in an ontogenical period when cells are highly proliferative. Most of these "useless" DJ_H joints are filtered out during the last third of gestation, and novel selective mechanisms are set at work, to generate the germline, restricted Ig repertoires present in perinatal life (*central rectangle*). Many of the perinatal Ig specificities are maintained inside the innate-like lymphocyte populations along the whole lifespan of the individual. In parallel, the mechanisms leading to the predominant, adult-type, immune diversity are up-regulated (TdT, Ig CSR, SHM, etc.) (*right, multicolour polygons*)

the innate-like lymphocytes, partly, because they derive from the same precursors in the early ontogeny and they persist for very long periods. Actually, selective experimental manipulations over specific clonotypes restricted to early life periods, remain established for the whole lifespan of the individual. In newborns, Ag is able to both positively and negatively select B cells, a phenomenon that becomes restricted to B1 B cells, these latter being able to rescue autoreactive B2 cells in the adult by yet unclear mechanisms. A developmental switch takes place around 4 weeks post-birth, when most of BM-derived B2 cells become submitted to Ag-driven negative selection [71], and these changes are likely on the basis of neonatal tolerance. Whatever the fine molecular mechanisms are involved in the process (and, probably, there are many, from somatic microenvironmental factors to differential thresholds of lymphocyte responses), the early generation/selection of long-lived cell populations of innate-like lymphocytes that preserve functional strategies of great survival value could contribute to the efficient protection of the individual against common pathogens, while limiting "collateral" damages to the normal tissues. Maybe these innate-like lymphocytes have also been evolutionarily exploited to maintain under control the unrestricted, molecular consequences (VDJ diversity) of the selfish RAG transposon "infection" that leaded to the generation of the adaptive IS.

Acknowledgements This work was supported by grants SAF2007-65265, ISCIII05/2007, SAL0304-2006, BFU2006-14390/BMC, PI 08/0685, SAF 2009:12596, Fundación Mutua Madrileña and CSD2007-00015. The Centro de Biología Molecular receives institutional funding from Fundación Ramón Areces.

References

1. Ohno S, Epplen J, Matsunaga T, Hozumi T (1981) The curse of Prometheus is laid upon the immune system. Prog Allergy 28:8–39
2. Hao Z, Rajewsky K (2001) Homeostasis of peripheral B cells in the absence of B cell influx from the bone marrow. J Exp Med 194:1151–1164
3. Gaudin E, Rosado M, Agenes F, Mclean A, Freitas A (2004) B-cell homeostasis, competition, resources, and positive selection by self-antigens. Immunol Rev 197:102–115
4. McLean A, Rosado M, Agenes F, Vasconcellos R, Freitas A (1997) Resource competition as a mechanism for B cell homeostasis. Proc Natl Acad Sci USA 94:5792
5. French J, Roark C, Born W, O'Brien R (2005) $\gamma\delta$ T cell homeostasis is established in competition with $\alpha\beta$ T cells and NK cells. Proc Natl Acad Sci USA 102:14741
6. Woodland R, Schmidt M (2005) Homeostatic proliferation of B cells. Semin Immunol 17: 209–217
7. Ranson T, Vosshenrich C, Corcuff E, Richard O, Muller W, Di Santo J (2003) IL-15 is an essential mediator of peripheral NK-cell homeostasis. Blood 101:4887–4893
8. Surh C, Sprent J (2008) Homeostasis of naive and memory T cells. Immunity 29:848–862
9. Pasare C, Medzhitov R (2005) Control of B-cell responses by Toll-like receptors. Nature 438:364–368
10. Miller J, Stadanlick J, Cancro M (2006) Space, selection, and surveillance: setting boundaries with BLyS. J Immunol 176:6405–6410
11. Freitas A, Rosado M, Viale A, Grandien A (1995) The role of cellular competition in B cell survival and selection of B cell repertoires. Eur J Immunol 25:1729–1738

12. Sprent J, Bruce J (1984) Physiology of B cells in mice with X-linked immunodeficiency (xid). III. Disappearance of xid B cells in double bone marrow chimeras. J Exp Med 160:711–723
13. Izcue A, Morales G, Minguet S, Sánchez-Movilla A, Morales P, Martínez-M J, Gaspar M, Marcos M (2001) Both B and $\gamma\delta$TCR$^+$ lymphocytes regulate $\alpha\beta$TCR$^+$ lymphocytes involved in superantigen specific responses. Eur J Immunol 31:2811–2817
14. Rhee K, Jasper P, Sethupathi P, Shanmugam M, Lanning D, Knight K (2005) Positive selection of the peripheral B cell repertoire in gut-associated lymphoid tissues. J Exp Med 201:55–62
15. Freitas A, Rocha B, Coutinho A (1986) Life span of B lymphocytes: the experimental basis for conflicting results. J Immunol 136:470–476
16. Freitas A, Rocha B, Coutinho A (1986) Two major classes of mitogen-reactive B lymphocytes defined by life span. J Immunol 136:466–469
17. Godin I, Garcia-Porrero J, Coutinho A, Dieterlen-Lièvre F, Marcos M (1993) Para-aortic splanchnopleura from early mouse embryos contains B1a cell progenitors. Nature 364:67–70
18. Agenes F, Freitas A (1999) Transfer of small resting B cells into immunodeficient hosts results in the selection of a self-renewing activated B cell population. J Exp Med 189:319–330
19. Carvalho T, Mota-Santos T, Cumano A, Demengeot J, Vieira P (2001) Arrested B lymphopoiesis and persistence of activated B cells in adult interleukin $7^{-/-}$mice. J Exp Med 194:1141–1150
20. Lalor P, Herzenberg L, Adams S, Stall A (1989) Feedback regulation of murine Ly-1 B cell development. Eur J Immunol 19:507–513
21. Gu H, Tarlinton D, Muller W, Rajewsky K, Forster I (1991) Most peripheral B cells in mice are ligand selected. J Exp Med 173:1357–1371
22. Wardemann H, Yurasov S, Schaefer A, Young J, Meffre E, Nussenzweig M (2003) Predominant autoantibody production by early human B cell precursors. Science 301:1374–1377
23. Kraus M, Alimzhanov M, Rajewsky N, Rajewsky K (2004) Survival of resting mature B lymphocytes depends on BCR signaling via the Igα/β heterodimer. Cell 117:787–800
24. Lam K, Kuhn R, Rajewsky K (1997) In vivo ablation of surface immunoglobulin on mature B cells by inducible gene targeting results in rapid cell death. Cell 90:1073–1084
25. Allman D, Pillai S (2008) Peripheral B cell subsets. Curr Opin Immunol 20:149–157
26. Meyer-Bahlburg A, Andrews S, Yu K, Porcelli S, Rawlings D (2008) Characterization of a late transitional B cell population highly sensitive to BAFF-mediated homeostatic proliferation. J Exp Med 205:155–168
27. Thien M, Phan T, Gardam S, Amesbury M, Basten A, Mackay F, Brink R (2004) Excess BAFF rescues self-reactive B cells from peripheral deletion and allows them to enter forbidden follicular and marginal zone niches. Immunity 20:785–798
28. Stadanlick J, Kaileh M, Karnell F, Scholz J, Miller J, Quinn III W, Brezski R, Treml L, Jordan K, Monroe J, Sen R, Cancro MP (2008) Tonic B cell antigen receptor signals supply an NF-κB substrate for prosurvival BLyS signaling. Nat Immunol 9:1379–1387
29. Le T, Kim T, Chaplin D (2008) Intraclonal competition inhibits the formation of high-affinity antibody-secreting cells. J Immunol 181:6027–6037
30. Phan T, Paus D, Chan T, Turner M, Nutt S, Basten A, Brink R (2006) High affinity germinal center B cells are actively selected into the plasma cell compartment. J Exp Med 203: 2419–2424
31. Radbruch A, Muehlinghaus G, Luger E, Inamine A, Smith K, Dorner T, Hiepe F (2006) Competence and competition: the challenge of becoming a long-lived plasma cell. Nat Rev Immunol 6:741–750
32. Crotty S, Felgner P, Davies H, Glidewell J, Villarreal L, Ahmed R (2003) Cutting edge: long-term B cell memory in humans after smallpox vaccination. J Immunol 171:4969–4973
33. Hammarlund E, Lewis M, Hansen S, Strelow L, Nelson J, Sexto G, Hanifin J, Slifka M (2003) Duration of antiviral immunity after smallpox vaccination. Nat Med 9:1131–1137
34. Richard K, Pierce S, Song W (2008) The agonists of TLR4 and 9 are sufficient to activate memory B cells to differentiate into plasma cells in vitro but not in vivo. J Immunol 181: 1746–1752
35. Ahuja A, Anderson S, Khalil A, Shlomchik M (2008) Maintenance of the plasma cell pool is independent of memory B cells. Proc Natl Acad Sci USA 105:4802–4807

36. Dogan I, Bertocci B, Vilmont V, Delbos F, Mégret J, Storck S, Reynaud C, Weill J (2009) Multiple layers of B cell memory with different effector functions. Nat Immunol 10: 1292–1299
37. Tarlinton D, Radbruch A, Hiepe F, Dörner T (2008) Plasma cell differentiation and survival. Curr Opin Immunol 20:162–169
38. Scholz J, Crowley J, Tomayko M, Steinel N, O'Neill P, Quinn W, Goenka R, Miller J, Cho Y, Long V, Ward C, Migone T, Shlomchik MJ, Cancro MP (2008) BLyS inhibition eliminates primary B cells but leaves natural and acquired humoral immunity intact. Proc Natl Acad Sci USA 105:15517–15522
39. van Zelm M, Szczepanski T, van der Burg M, van Dongen J (2007) Replication history of B lymphocytes reveals homeostatic proliferation and extensive antigen-induced B cell expansion. J Exp Med 204:645–655
40. Guerrettaz L, Johnson S, Cambier J (2008) Acquired hematopoietic stem cell defects determine B-cell repertoire changes associated with aging. Proc Natl Acad Sci USA 105:11898–11902
41. Bendelac A, Bonneville M, Kearney J (2001) Autoreactivity by design: innate B and T lymphocytes. Nat Rev Immunol 1:177–186
42. Hardy R, Wei C, Hayakawa K (2004) Selection during development of VH11 B cells: a model for natural autoantibody-producing CD5 B cells. Immunol Rev 197:60–74
43. Yoshikawa S, Kawano Y, Minegishi Y, Karasuyama H (2009) The skewed heavy-chain repertoire in peritoneal B-1 cells is predetermined by the selection via pre-B cell receptor during B cell ontogeny in the fetal liver. Int Immunol 21:43–52
44. Hayday A, Pennington D (2007) Key factors in the organized chaos of early T cell development. Nat Immunol 8:137–144
45. Matsuda J, Gapin L, Sidobre S, Kieper W, Tan J, Ceredig R, Surh C, Kronenberg M (2002) Homeostasis of Vα14i NKT cells. Nat Immunol 3:966–974
46. Thiriot A, Drapier A, Vieira P, Fitting C, Cavaillon J, Cazenave P, Rueff-Juy D (2007) The Bw cells, a novel B cell population conserved in the whole genus Mus. J Immunol 179:6568–6578
47. de Andres B, Cortegano I, Serrano N, del Rio B, Martin P, Gonzalo P, Marcos M, Gaspar M (2007) A population of $CD19^{high}CD45R^{low}CD21^{low}$ B lymphocytes poised for spontaneous secretion of IgG and IgA antibodies. J Immunol 179:5326–5334
48. Montecino-Rodriguez E, Leathers H, Dorshkind K (2006) Identification of a B-1 B cell–specified progenitor. Nat Immunol 7:293–301
49. Carey J, Moffatt-Blue C, Watson L, Gavin A, Feeney A (2008) Repertoire-based selection into the marginal zone compartment during B cell development. J Exp Med 205:2043–2052
50. Grigoriadou K, Boucontet L, Pereira P (2003) Most IL-4-producing $\gamma\delta$ thymocytes of adult mice originate from fetal precursors. J Immunol 171:2413–2420
51. Cassady-Cain R, Kaushik A (2006) Increased negative selection impairs neonatal B cell repertoire but does not directly lead to generation of disease-associated IgM auto-antibodies. Int Immunol 18:661–669
52. Baumgarth N, Herman O, Jager G, Brown L, Herzenberg L, Chen J (2000) B-1 and B-2 cell-derived immunoglobulin M antibodies are nonredundant components of the protective response to influenza virus infection. J Exp Med 192:271–280
53. Ehrenstein M, Cook H, Neuberger M (2000) Deficiency in serum immunoglobulin (Ig) M predisposes to development of IgG autoantibodies. J Exp Med 191:1253–1258
54. Kumazaki K, Tirosh B, Maehr R, Boes M, Honjo T, Ploegh H (2007) $AID^{-/-}micro s^{-/-}$mice are agammaglobulinemic and fail to maintain B220-$CD138^{+}$ plasma cells. J Immunol 178:2192–2203
55. Shimomura Y, Ogawa A, Kawada M, Sugimoto K, Mizoguchi E, Shi H, Pillai S, Bhan A, Mizoguchi A (2008) A unique B2 B cell subset in the intestine. J Exp Med 205:1343–1355
56. Weller S, Mamani-Matsuda M, Picard C, Cordier C, Lecoeuche D, Gauthier F, Weill J, Reynaud C (2008) Somatic diversification in the absence of antigen-driven responses is the hallmark of the IgM^{+} IgD^{+} $CD27^{+}$ B cell repertoire in infants. J Exp Med 205:1331–1342
57. Choi Y, Baumgarth N (2008) Dual role for B-1a cells in immunity to influenza virus infection. J Exp Med 205:3053–3064

58. Genestier L, Taillardet M, Mondiere P, Gheit H, Bella C, Defrance T (2007) TLR agonists selectively promote terminal plasma cell differentiation of B cell subsets specialized in thymus-independent responses. J Immunol 178:7779–7786
59. Ha S, Tsuji M, Suzuki K, Meek B, Yasuda N, Kaisho T, Fagarasan S (2006) Regulation of B1 cell migration by signals through toll-like receptors. J Exp Med 203:2541–2550
60. Cinamon G, Zachariah M, Lam O, Foss Jr F, Cyster J (2008) Follicular shuttling of marginal zone B cells facilitates antigen transport. Nat Immunol 9:54–62
61. Holmberg D, Freitas A, Portnoi D, Jacquemart F, Avrameas S, Coutinho A (1986) Antibody repertoires of normal BALB/c mice: B lymphocyte populations defined by state of activation. Immunol Rev 93:147–169
62. Kearney J, Vakil M (1986) Idiotype-directed interactions during ontogeny play a major role in the establishment of the adult B cell repertoire. Immunol Rev 94:39–50
63. Marcos M, Morales-Alcelay S, Godin I, Dieterlen-Lievre F, Copin S, Gaspar M (1997) Antigenic phenotype and gene expression pattern of lymphohemopoietic progenitors during early mouse ontogeny. J Immunol 158:2627–2637
64. de Andres B, Gonzalo P, Minguet S, Martinez-Marin J, Soro P, Marcos M, Gaspar M (2002) The first 3 days of B-cell development in the mouse embryo. Blood 100:4074–4081
65. Martinez M, Minguet S, Gonzalo P, Soro P, de Andres B, Izcue A, Marcos M, Gaspar M (2001) Long-lived polyclonal B-cell lines derived from midgestation mouse embryo lymphohematopoietic progenitors reconstitute adult immunodeficient mice. Blood 98:1862–1871
66. Yokota T, Huang J, Tavian M, Nagai Y, Hirose J, Zuniga-Pflucker J, Peault B, Kincade P (2006) Tracing the first waves of lymphopoiesis in mice. Development 133:2041–2051
67. Chang Y, Paige C, Wu G (1992) Enumeration and characterization of DJH structures in mouse fetal liver. EMBO J 11:1891–1899
68. Gozalbo-Lopez B, Andrade P, Terrados G, de Andres B, Serrano N, Cortegano I, Palacios B, Bernad A, Blanco L, Marcos M, Gaspar ML (2009) A role for DNA polymerase micro in the emerging DJH rearrangements of the postgastrulation mouse embryo. Mol Cell Biol 29: 1266–1275
69. Moon A, Garcia-Diaz M, Batra V, Beard W, Bebenek K, Kunkel T, Wilson S, Pedersen L (2007) The X family portrait: structural insights into biological functions of X family polymerases. DNA Repair 6:1709–1725
70. Heyer B, MacAuley A, Behrendtsen O, Werb Z (2000) Hypersensitivity to DNA damage leads to increased apoptosis during early mouse development. Genes Dev 14:2072–2084
71. Ferry H, Crockford T, Leung J, Cornall R (2006) Signals from a self-antigen induce positive selection in early B cell ontogeny but are tolerogenic in adults. J Immunol 176:7402–7411

Chapter 14
Dynamics of Peripheral Regulatory and Effector T Cells Competing for Antigen Presenting Cells

Nuno Sepúlveda and Jorge Carneiro

Abstract A healthy immune system requires a balance between effector T (T_E) cells that mount immune responses, and regulatory T (T_R) cells that prevent them. Understanding this balance requires knowing how the repertoires of T_E and T_R cells in the periphery are patterned and related to each other. The present work addresses this issue in the framework of the Crossregulation model, which was formulated to contemplate the dynamics of a large number of T cell clones recognizing a non-exclusive set of antigen-presenting cells (APCs). Assuming a continuous thymic export, three distinct patterns for the peripheral repertoire emerge from the simulations. These patterns are distinguished by the composition of the resident population, i.e., the subset of clones that can persist in the periphery. On one extreme, there is a repertoire pattern characterised by a predominance of T_E cell clones and a too small T_R-cell pool only maintained in the periphery by the continuous thymus export. Thus, this repertoire pattern cannot be the basis of a healthy immune system, because peripheral tolerance to self antigens is not ensured. On the other extreme, there is a repertoire pattern showing a predominance of T_R cells maintained by their T_E counterparts coming from the thymus. In spite of ensuring peripheral tolerance, this repertoire pattern does not appear efficient to fight infections occurring throughout life. In the middle of these two extremes lies a third repertoire pattern exhibiting a partition of the peripheral compartment similar to the one suggested in a previous work. In this repertoire structure, a small subset of highly crossreactive T_R-cell clones prevents expansion of T_E-cell clones driven by body antigens, and a more diverse subset of T_E-cell clones remains uncontrolled by T_R cells, which allows the setup of immune responses against harmful pathogens. This repertoire pattern seems then the most adequate to represent a healthy immune system. For each emergent repertoire pattern, clear and testable predictions are given for different properties related to the diversity of T_E and T_R cells. For the most adequate repertoire pattern, a higher diversity of T_E cells than of T_R cells is expected as well

J. Carneiro (✉)
Instituto Gulbenkian de Ciência, Oeiras, Portugal
e-mail: nunosep@igc.gulbenkian.pt

C. Molina-París and G. Lythe (eds.), *Mathematical Models and Immune Cell Biology*, DOI 10.1007/978-1-4419-7725-0_14, © Springer Science+Business Media, LLC 2011 as a negative correlation between clonal size distributions for clones belonging to the overlapping repertoire. These predictions can then be compared to available data on T cell receptor diversity.

Introduction

A complete characterization of the peripheral T cell repertoire is fundamental to understand how the immune system avoids deleterious autoimmune responses during the lifespan of the organism and, at the same time, reacts efficiently to the harmful pathogens it faces. The interest on T cell repertoires has recently increased with the discovery of opposing roles of the so-called regulatory $CD4^+CD25^+Foxp3^+$ T (T_R) cells in the immune system [1, 2]. On the one hand, T_R cells are beneficial to the host owing to the fact that they control autoimmunity, graft-vs-host rejection, and inflammation [3]. On the other hand, they can also be detrimental to the host by suppressing protective immune responses against potentially harmful pathogens or cancer [4, 5].

Several experimental studies attempted to quantify the main properties of T cell receptor (TCR) repertoire of T_R and conventional effector (T_E) T cells from different mouse models [6–10]. All these studies showed that TCR samples of T_R cells were more diverse than those of T_E cells, both in the thymus and periphery, suggesting a higher TCR diversity of the sampled T_R-cell populations. Moreover, samples of both cell types show some similarity in terms of shared TCR variants and their respective sample abundances. Yet, this similarity considerably varies from one study to another and, therefore, it is difficult to ascertain the exact extension of the overlap between the respective sampled repertoires.

Besides the basic knowledge of the TCR repertoire, it is also important to determine which antigens are recognized by T_R cells. Two lines of evidence support the notion that these cells preferentially react to body antigens. First, T_R cells can prevent autoimmune pathologies when transferred to animals lacking these cells [11, 12]. Second, thymic T_R-cell differentiation seems to be driven by high TCR affinity/avidity towards self antigens [13, 14]. However, T_R cells also retain some reactivity to exogenous antigens since only heterologous and not autologous antigen-presenting cells (APCs) could activate T cell hybridomas loaded with T_R-driven TCRs [9]. Therefore, it is still unclear what are in fact the main targets of T_R cells, either body or foreign antigens.

Crossreactivity of T_R cells is yet another important property to be characterized when studying the role of these cells in the immune system. However, due to the lack of accurate experimental techniques that can simultaneously deal with a large diversity of T cells and antigens, this aspect of T_R-cell physiology is just a matter of speculation. In theory, all T cells should be highly crossreactive because their number is dramatically low in comparison with the number of antigens [15]. The same argument can be applied to T_R cells but how their crossreactivity relates to that of T_E cells remains unclear.

Since powerful experimental assays are still lacking to properly characterise structural and functional properties of T cell repertoires at the same time, mathematical modelling provides an alternative tool to perform such task. This was recently illustrated with the usage of the Crossregulation model in the study of the peripheral T_R-cell repertoire [16]. The Crossregulation model essentially describes a basic interaction network between APCs, T_R, and T_E cells, in which T cell proliferation is promoted upon conjugation with APCs [17–21]; a distinct interaction network between these three cellular key players was recently proposed, providing though similar qualitative behaviours [22]. To study the peripheral T cell repertoire, the Crossregulation model was previously formulated under the assumption that every T cell clone in the periphery could only recognize an exclusive set of APCs [16]. In this simple scenario, several predictions for the repertoire structure were put forward. Peripheral T cell repertoire should then be partitioned into three subsets. The first one is composed of few clones with short life-span that interact with few APCs and, thus, cannot be sustained in the periphery. The second subset refers to a diverse set of barely autoreactive T_E cell clones, whose expansion is limited only by APC availability. The third subset is composed of a less diverse set of highly autoreactive T cell clones, exhibiting both T_R and T_E cells that regulate each other's growth. Since the majority of T_R cells belong to this third subset, the peripheral T_R pool is expected to be less diverse than its T_E counterpart, in clear contrast to what observed in experimental TCR samples. Moreover, since the amount of APCs recognized by each T cell clone can be related to crossreactivity, it is expected that T_R-cell clones would be highly crossreactive and targeted to self antigens. However, these predictions were obtained under the strong assumption of no interclonal competition for APCs. It is then important to know whether they would hold in a more general scenario, in which distinct T cell clones could share some APC specificity with each other. This is the goal of the present work.

The Crossregulation Model

The Crossregulation model describes the peripheral T cell population dynamics taking into account three mutually interacting cell types: APCs displaying membrane MHC-peptide complexes; T_E cells that can potentially induce autoimmunity or mount immune responses against pathogens depending on their specificity; and T_R cells with the same antigen specificity that may prevent clonal expansion of T_E cells.

The model satisfies a set of postulates concerning the life cycle of the above-mentioned three cell types and the interactions they make with each other. The postulates are the following:

1. There are $\mathcal{A}$ distinct APC subpopulations in the periphery, distinguished by the set of peptides they present. Each subpopulation presents a unique but non-exclusive set of peptides, as also assumed in a stochastic model describing the

dynamics of the naive T cell repertoire [23]. Every APC of a particular subpopulation presents the same set of peptides, being equivalent as far as recognition by and conjugation with T cells is concerned.

2. Each APC subpopulation is continuously renewed from precursors, and we assume that its cell density is stationary. Thus, the density of each APC subpopulation is a fixed parameter in the model.
3. T_E and T_R cells are also classified into a subpopulations according to their cognate APCs, defined as the set of different APC subpopulations they recognize. To study T cell repertoire dynamics, it is useful to aggregate T cells into subpopulations, which are equivalent with respect to the interactions they make with their cognate APCs. For short, we will refer to these equivalent classes of T cells as 'clones', although it should be noticed that these 'clones' may not be identical to real 'clones' defined by sequence of the TCR.
4. The thymus exports clones to periphery at a constant rate s_c.
5. Free T_E and T_R cells die with a constant rate δ in the periphery, independently of their antigen specificity.
6. Activation of T_E and T_R cells requires productive conjugation with cognate APCs, and depends on interactions these T cells make with each other in multicellular conjugates.
7. T_E and T_R cells interact indirectly by competition for cognate APCs and directly by molecular processes that require the co-localization on the same multicellular conjugate with APCs.
8. Proliferation of specific T_E cell populations is promoted by productive conjugation with cognate APCs, and is inhibited by T_R cells on the same multicellular conjugates.
9. T_R-cell proliferation depends on interactions with cognate APCs and T_E cells on the same multicellular conjugates.

A more detailed discussion of these postulates can be found elsewhere [16].

Mathematical Formulation

Let $E_{i,t-t_i^*}$ and $R_{i,t-t_i^*}$ be the densities of T_E and T_R cells in clone i, respectively, at time t given that this clone had entered the periphery at time t_i^*. Let also $E^*_{i,t-t_i^*}$ and $R^*_{i,t-t_i^*}$ be their activated cell densities. The basic dynamics of each clone is described by the following set ordinary differential equations

$$\begin{cases} \dfrac{dE_{i,t-t_i^*}}{dt} = \pi_e E^*_{i,t-t_i^*} - \delta E_{i,t-t_i^*} \\ \dfrac{dR_{i,t-t_i^*}}{dt} = \pi_r R^*_{i,t-t_i^*} - \delta R_{i,t-t_i^*}, \end{cases} \quad i = 1, \ldots, \mathcal{D}_t, \tag{14.1}$$

where $\mathcal{D}_t$ is the number of different T cell clones in the periphery at time t (i.e., T cell diversity), δ is the T cell death rate, π_e and π_r are the proliferation rates of activated T_E and T_R cells, respectively.

The densities of activated T_E and T_R cells of each clone are calculated in a stepwise manner [17, 24]. Let the i-th T cell clone recognize the j-th APC subpopulation. Let $C_{ij,t}$ be the total density of conjugates formed by T cells from clone i and APCs from subpopulation j at time t. The dynamics of these conjugates is generically described by the following system of equations

$$\frac{dC_{ij,t}}{dt} = \gamma_c \left(A_j - \sum_{i=1}^{\mathcal{D}_t} C_{ij} \right) \left(T_{i,t-t_i^*} - \sum_{j=1}^{\mathcal{A}} C_{ij} \right) - \gamma_d C_{ij}, \tag{14.2}$$

where γ_c and γ_d are the conjugation and deconjugation rates between APCs and T cells, respectively. These rates are assumed to be identical for any T cell clone i and cognate APC j. In the above equation, new conjugates are formed by the free sites on APCs of population j with free T cells of clone i at rate γ_c, while existing conjugates deconjugate at rate γ_d. Assuming that the conjugation and deconjugation of T cells from the APCs is a fast process with respect to the overall T cell dynamics, we solve in each time step the steady state of the above equation system by the Euler method, which is slow but with guaranteed convergence.

Given the conjugate density $C_{ij,t}$, we next calculate the density of conjugated T_E and T_R cells as being proportional to the relative frequency of T_E and T_R in the clone, i.e.,

$$e_{ij,t} = C_{ij,t} \times \frac{E_{i,t-t_i^*}}{E_{i,t-t_i^*} + R_{i,t-t_i^*}} \quad \text{and} \quad r_{ij,t} = C_{ij,t} \times \frac{R_{i,t-t_i^*}}{E_{i,t-t_i^*} + R_{i,t-t_i^*}}. \tag{14.3}$$

To calculate the amount of conjugated T cells becoming activated, we assume that an APC has s independent conjugation sites, each one with the possibility of being occupied by a T cell [18]. With this assumption, we can determine, respectively, the fractions of sites per APC occupied by T_E and T_R cells

$$\tilde{e}_{ij,t} = \frac{e_{ij,t}}{A_j s} \quad \text{and} \quad \tilde{r}_{ij,t} = \frac{r_{ij,t}}{A_j s}. \tag{14.4}$$

The next step is to compute the fraction of activated T_E and T_R cells upon conjugation with the APCs. For the sake of simplicity, we consider only 2 sites per APC. The fraction of activate T_E cells is equivalent to the fraction of T_E cells conjugated alone or with other T_E cell. In turn the fraction of activated T_R cells corresponds to the fraction of T_R conjugated together with a T_R cell. We approximate these fractions based on multinomial distribution with probabilities given by the fractions of APC sites occupied by T_E and T_R cells, and introduce factors that increase the efficiency of suppression. These extra factors allow the simplistic scenario $s = 2$ to capture the higher efficiency of suppression obtained with larger more realistic

s values, when assuming that a single T_R cell per APC is sufficient to suppress all the T_E cells (one or more) on the same multicellular conjugate. Thus, we calculate the fraction of conjugated T_E and T_R cells that will be activated by the following expressions, respectively:

$$e^*_{ij,t} = 1 - \frac{2\tilde{r}_{ij,t}}{2 - \tilde{e}_{ij,t}} \quad \text{and} \quad r^*_{ij,t} = \frac{2\tilde{e}_{ij,t}}{2 - \tilde{r}_{ij,t}}. \tag{14.5}$$

Finally, we calculate the densities of activated T_E and T_R in the clone i at time t as:

$$E^*_{i,t} = \sum_{j=1}^{A} \alpha_{ij} e_{ij,t} \times e^*_{ij,t} \quad \text{and} \quad R^*_{i,t} = \sum_{j=1}^{A} \alpha_{ij} r_{ij,t} \times r^*_{ij,t}, \tag{14.6}$$

where α_{ij} is a dummy variable indicating whether the i-th T cell clone recognizes or not the APC population j.

Relevant Properties of the Basic Crossregulation Model

Before we describe how the crossreactivity distribution is generated in the model, it is better to recall some of the basic dynamical properties of the model when APCs, T_E, and T_R cells are assumed to be homogeneous populations. This is useful to understand how APC-related parameters were specified in the simulations. A fundamental property of this model is the dependence of T_R- and T_E-cell steady states on the APC density [16]. Bifurcation analysis of this parameter reveals three qualitatively distinct regimes for the dynamical system (Fig. 14.2a). The first regime refers to the trivial situation where the APC density is too low to sustain either T_E or T_R cell populations in the system (that is, T_R- and T_E-cell densities are equal to 0 in steady state). When the APC density reaches a critical point (a_E in Fig. 14.2a), the system leads to the competitive exclusion of T_R cells by their T_E counterparts. When APC density is high enough (a_R in Fig. 14.2a), the system enters into a bistable regime in which, depending on the initial condition for T_R and T_E cell densities, one can obtain either the coexistence of both cell types or the competitive exclusion of T_R cells by their T_E counterparts.

For the sake of simplicity, we performed all simulations by generating every T cell clone under the same initial conditions for T_E and T_R cells (that is, $E_{i,t_i^*} = E$ and $R_{i,t_i^*} = R$, $\forall i$). To understand these complex simulations, it is important to know the behaviour of the simplified system in the bistable parameter regime, as illustrated in Fig. 14.2b. When specifying a particular initial condition for T_R and T_E cells, the parameter region where T_R cells cannot be sustained in the system is extended to the interval between a_E to $\tilde{a}_R$ in Fig. 14.2b. When the APC density reaches $\tilde{a}_R$, the system falls into a steady state where both cell types coexist. Finally, when the APC density is too high, the system falls again into a state in which T_E cells outcompete their T_R counterparts (not shown in Fig. 14.2b; [16]).

APC Subpopulations and Their Density

In the model, the peripheral APC pool is divided into different subpopulations, each one presenting a distinct yet overlapping set of peptides. In this scenario, the abundance of each APC subpopulation is somehow dependent on the availability of the respective set of peptides in the periphery. Moreover, by postulate 2, the abundance of each APC subpopulation in the periphery is assumed to be stationary. Therefore, every APC subpopulation and its respective density in the periphery is generated in the beginning of the simulations. For the sake of simplicity, we assume that most APC subpopulations are at low densities reflecting the presentation of diverse and rare antigens in the body, and few APC subsets at high densities (that present ubiquitous antigens or abundant antigens). With this rationale in mind, we assumed that the density of each APC subpopulation in the periphery follows an Exponential distribution with mean value λ_a, i.e.,

$$A_j|\lambda_a \rightsquigarrow Exp(\lambda_a), j = 1, \ldots, \mathcal{A}. \tag{14.7}$$

Connecting T Cell Clones with APC Subpopulations

When new T cell clones enter the periphery, we need to define their respective sets of cognate APCs (Fig. 14.3a). Since any T cell must pass through positive and negative selection in the thymus, it is reasonable to assume that every generated T cell clone would recognize a limited set of APC subpopulations in the periphery. On the one hand, positive selection seems to ensure that any T cell clone coming out of the thymus recognizes at least some APC subpopulations in the periphery. On the other hand, negative selection purges from the repertoire clones that recognize too many APC subpopulations. With this in mind, we constructed an appropriate Markov chain model with discrete steps to generate a narrow connectivity distribution around its mean. The states of this Markov chain are the number of APC subpopulations recognized by a clone (number of connections, for short), ranging from 0 to $\mathcal{A}$. The initial state of a clone is zero connections. In each step of the chain, we test whether an APC subpopulation will be recognized or not by the newly-generated T cell clone. Therefore, there are as many chain steps as APC subpopulations. Moreover, the number of APC subpopulations recognized by a given T cell clone can only be maintained or increase along the steps of the chain. We assume that the transition matrix $(\mathcal{A}+1) \times (\mathcal{A}+1)$ is given by

$$p_{i,j} = \begin{cases} 1 - \theta e^{-\lambda_c \times j}, & j = i \\ \theta e^{-\lambda_c \times j}, & j = i+1 \ , i, j = 0, \ldots, \mathcal{A} \\ 0, & \text{otherwise} \end{cases} \tag{14.8}$$

where θ is the initial probability of generating a connection with a newly-generated clone, and λ_c is the exponential decay by which the probability of making a new connection decreases (Fig. 14.3b). By standard Markov chain theory, the connectivity distribution is then given by the vector that indicates the process started with zero connections multiplied by transition matrix after $\mathcal{A}$ steps. Figure 14.3c gives an example of the connectivity distribution produced by this stochastic process. In practical terms, we connect the different APC subpopulations to the new clone as follows: (1) sample successively without replacement each APC subpopulation; (2) connect randomly each one to the new clone with the probability of increasing connectivity included in (14.8).

In all simulations, the parameter θ was setup at 0.90 to ensure that every newly-generated clone recognizes with a high probability at least one APC subpopulation, in agreement with thymic positive selection. Furthermore, the parameter λ_c can be determined to generate a connectivity distribution with a given mean value. Since this distribution cannot be calculated analytically, except for the case of few APC populations, we used numerical equation-solving methods through the software `Mathematica`.

As discussed above, the persistence of a T cell clone in the periphery is critically dependent on the total density of its cognate APCs, denoted as $A_{T,i}$. Therefore, it is of interest to determine the probability distribution of $A_{T,i}$ in order to control how many T cell clones would be in the different regimes of the system, as illustrated in Fig. 14.3d. To this end, we first note that the density of each APC subpopulation comes from an Exponential distribution. Thus, the total APC density given the number of connections k_i is a sum of independent and identical Exponential distributions, which leads to a Gamma distribution with shape parameter k_i and scale parameter λ_a, that is,

$$A_{T,i}|K_i = k_i \rightsquigarrow Ga\left(k_i, \lambda_a\right), k_i = 1, \ldots, \mathcal{A}. \quad (14.9)$$

where K_i is the random variable indicating the number of connections with clone i.

We then apply the well-known total probability theorem to obtain the probability density distribution of $A_{T,i}$ given that a clone recognizes at least one APC subpopulation

$$f_{A_{T,i}}(a) = \sum_{k_i=1}^{\mathcal{A}} f_{A_{T,i}}(a|K_i = k_i) \frac{P\left[K_i = k_i|\theta, \lambda_c\right]}{1 - P\left[K_i = 0|\theta, \lambda_c\right]}, \quad (14.10)$$

where $f_{A_{T,i}}(a|K_i = k_i)$ is the probability density function of the Gamma distribution given in equation (14.9), and $P\left[K_i = i|\theta, \lambda_c\right]$ is the probability of having a clone that recognizes k_i distinct APC subpopulations. The above probability distribution is then a mixture of appropriate Gamma distributions (Fig. 14.3d). In the simulations, we control the parameter λ_a to obtain a probability distribution of $A_{T,i}$ that implies, for example, that 10% of the clones generated lie in the bistable regime (Fig. 14.3d). To this end, we again used the numerical methods implemented in the software `Mathematica` to determine λ_a that satisfy the above condition.

The model has many parameters, summarized in Table 14.1. Some of the parameters were chosen according to previous work [16]. The remaining ones are set in

Table 14.1 Parameters and variables of the T cell dynamics model. The values adopted here for some parameters were taken from Carneiro et al. [16]. The remaining parameters were determined in order to satisfy specific conditions on the crossreactivity distribution, as explained in the main text

Parameter/variable	Values	Description
Basic T cell dynamics		
$\mathcal{D}_t$	—	Number of T cell clones at time t (T cell diversity)
t_i^*	—	Time of entrance of the i-th T cell clone in the periphery
E_{i,t_i^*}, R_{i,t_i^*}	—	T_R- and T_E-cell densities from T cell clone i when entering the periphery
$E_{i,t-t_i^*}$, $R_{i,t-t_i^*}$	—	T_E and T_R cell densities from T cell clone i at time t given that the clone entered in the periphery at time t_i^*
$E_{i,t}^*$, $R_{i,t}^*$	—	Activated T_E- and T_R-cell densities from i-th clone at time t
π_e	1.00	T_E-cell proliferation rate upon stimulation by APCs
π_r	1.10	T_R-cell proliferation rate upon stimulation by APCs
δ	0.02	T cell death rate
s_c	—	Number of T cell clones entering the periphery per unit of time
Characterization of APC population		
$\mathcal{A}$	—	Number of distinct APC subpopulations in the periphery (APC diversity)
A_j	—	APC density of subpopulation j
s	2	Number of conjugation sites per APC
λ_a	—	Mean of the exponential deviate of APC subpopulation density
Interaction between T cells and APCs		
$C_{ij,t}$	—	Density of APC-T cell conjugates formed by i-th T cell clone and j-th APC population
γ_c	1.00	Rate at which T cells conjugate with cognate APCs
γ_d	1.00	Rate at which T cells deconjugate from cognate APCs
k_i	—	Number of APC subpopulations recognized by the i-th T cell clone (crossreactivity of the i-th T cell clone)
θ	0.90	Initial probability of connecting an APC population with a new clone
λ_r	—	Exponential decay in the probability of increasing crossreactivity of a clone given that it recognizes a certain number of APC subpopulations
$A_{T,i}$	—	Total APC density recognized by the i-th T cell clone

order to specify the proportion of clones falling in the bistable regime at the time of their entrance in the periphery (Fig. 14.3d). For simplicity in the simulations, we generate all T cell clones with similar initial conditions, $E_{i,t_i^*} = E$ and $R_{i,t_i^*} = R$, $\forall i$. Clonal extinction is assumed when the overall density of a clone drops below a certain threshold (e.g., 10^{-3}). The corresponding equations are then removed from the system, as previously done [19, 25].

Results

Dynamics and Structure of the Peripheral T Cell Repertoire in the Absence of Thymic Export

We first study the simplest situation where the whole T cell repertoire is generated at the beginning of the simulations. This situation is conceptually related to an adoptive transfer experiment, in which a pool of T cells is transferred into an animal lacking T cells, such as nude or $rag^{-/-}$ mouse [11, 26].

Less Crossreactive T Cell Clones are Outcompeted from the Repertoire

It has been suggested that T cell competition is one of the major forces shaping the T cell repertoire [27]. T cell competition was studied before but not accounting for the presence of T_R cells in the repertoire [28, 29] and the heterogeneity of the APC pool [19]. In the present model, T cells should form multicellular conjugates with their cognate APCs in order to survive and proliferate (Fig. 14.1). Thus, the model features competition among T cells within the same clone and also between different clones partaking the same APC. Since we assume that the pairwise conjugation rate has the same value for all clones and APC subpopulations, the competitiveness of a clone depends on its cognate APC density relatively to those of the remaining clones. In general, it is expected that T cell clones that can conjugate with more APCs will outcompete the ones with lower cognate APCs. Yet, this expectation may not hold true owing to the fact that T cell clones share different sets of APC subpopulations, forming a complex network of interactions between these two cell types (Fig. 14.3a). In this scenario, the persistence of a given T cell clone in the periphery is dependent not only on the APC density but also on the different APC subsets shared with the remaining clones. It is then reasonable to conceive a situation where a T cell clone recognizing a large density of APCs can be outcompeted if it has to partake these APCs with many different T cell clones.

It is clear from the simulations that the mean APC density per T cell clone increases in time until reaching a plateau (Fig. 14.4a). This result suggests that clones that can form more multicellular conjugates tend to persist longer in the simulations. A similar result can be extracted from the temporal evolution of the mean crossreactivity of T cell clones (Fig. 14.4b). Since less crossreactive T cell clones are purged from the repertoire, T cell diversity decreases in time (Fig. 14.4c). This decrease is more pronounced in the beginning of the simulation, where the difference between T cell clones with low and high level of crossreactivity is larger, and less pronounced in the end of the simulation, where this difference is smaller. In terms of total T cell density, the respective steady-state is reached very early in the simulation (Fig. 14.4d), maintaining it even when T cell diversity is still decreasing and far from its final value (Fig. 14.4c).

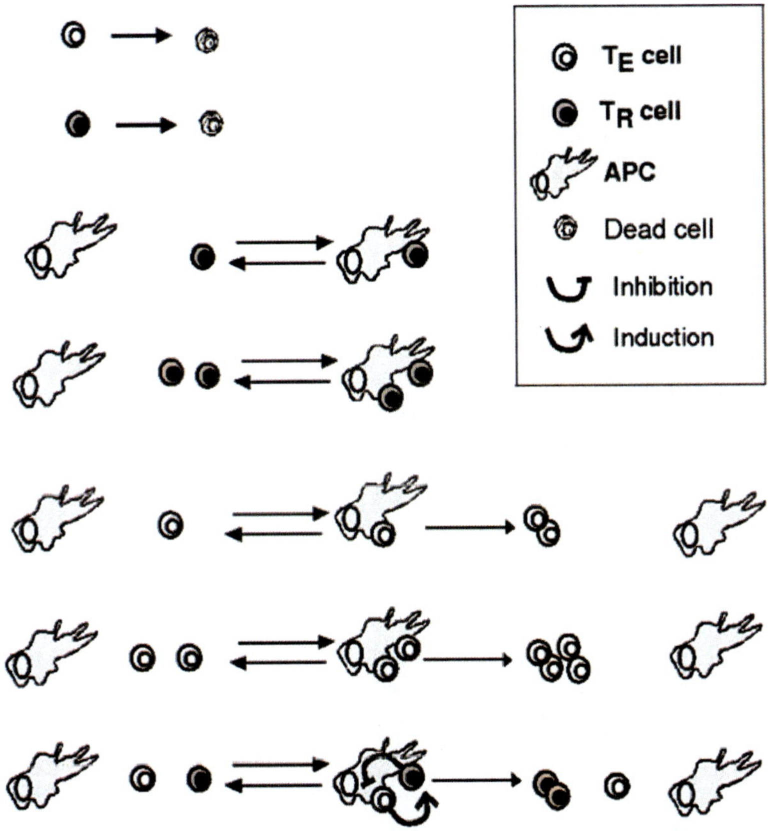

Fig. 14.1 The Crossregulation Model. The reaction diagram indicates the events and interactions underlying the dynamics of APCs, T_E cells and T_R cells as assumed in the model. In this simple scenario the APC can only form conjugates with a maximum of two T cells, which can belong to any of the clones that recognize this specific APC subpopulation

Partition of the T Cell Repertoire

We previously studied the T cell repertoire assuming that T cells recognize independent mutually exclusive APC subsets [16]. In this particular case, if all clones are generated with similar initial composition, the structure of the repertoire is straightforwardly predicted from the bifurcation diagram shown in Fig. 14.2b. This shaping of the repertoire leads to a partition of the peripheral compartment in three distinct T cell clone subsets. The first subset contains T cell clones that eventually go extinct in the periphery, because they can conjugate with just a few APCs (at densities lower than a_E). The second subset is made of T cell clones that can conjugate with APCs at densities between a_E and $\tilde{a}_R$. At equilibrium, these clones are composed of T_E cells only, because T_R-cell populations cannot be sustained. Therefore, this

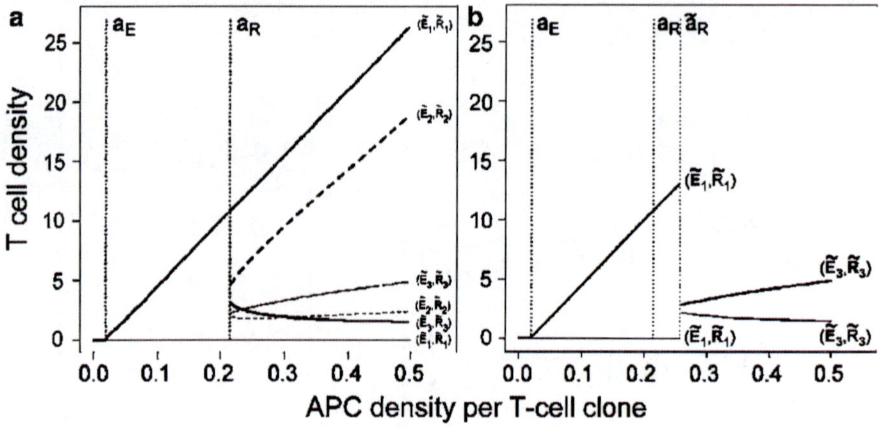

Fig. 14.2 Steady state analysis of the Crossregulation model as a function of APC density per T cell clone. (**a**) Bifurcation diagram where stable (*solid lines*) and unstable (*dashed lines*) steady states for T_E (*thick lines*) and T_R (*thin lines*) cells are represented. The system qualitatively changes its dynamic behaviour in the critical points a_E and a_R, defining three distinct parameter regimes. First, for APC density less than a_E, the stable steady state is $(\tilde{E}_0, \tilde{R}_0) = (0, 0)$ (not labelled). Second, for APC density between a_E and a_R, the stable steady state is given by $(\tilde{E}_1, \tilde{R}_1)$ with $\tilde{R}_1 = 0$. Third, for APC density higher than a_R, the system enters in a bistable regime that, depending on the initial conditions for T_R and T_E cells, the system either goes to $(\tilde{E}_1, \tilde{R}_1)$ or $(\tilde{E}_3, \tilde{R}_3)$. These two separated by the saddle point $(\tilde{E}_2, \tilde{R}_2)$. (**b**) Stable steady states of T_E and T_R cells for a given initial condition: $E(0) = R(0) = 0.50$. Parameters values are given in Table 14.1

subset represents an exclusive T_E-cell repertoire. The third subset includes clones that recognize cognate APCs in densities higher than $\tilde{a}_R$. Because of the specified initial condition, the T_R-cell populations can be sustained in this parameter regime, coexisting with their T_E counterparts. Therefore, this subset comprises the overlap repertoire between T_E and T_R cells, characterized by a higher average T_R-cell density than that of T_E cells and a negative correlation between the respective T_R and T_E-cell densities (Fig. 14.2b). Since there is no parameter regime in which T_R cells can be sustained alone, the previously-studied model provides no means of forming an exclusive T_R-cell repertoire.

With these previous results, we ask whether they are robust under interclonal competition for APCs. Can a similar partition of the repertoire be obtained in these conditions? Can T cell competition generate an exclusive T_R-cell repertoire? Representative results of the simulation of peripheral repertoire dynamics contemplating interclonal competition in the absence of thymic export are illustrated in Fig. 14.5.

It is clear that, at equilibrium, a partition of the peripheral repertoire is also obtained in the presence of interclonal competition (Fig. 14.5a). Clones whose cognate APCs are in densities less than a_E vanish. Moreover, since highly crossreactive T cell clones can outcompete less crossreactive ones (Fig. 14.4a and b), the effective

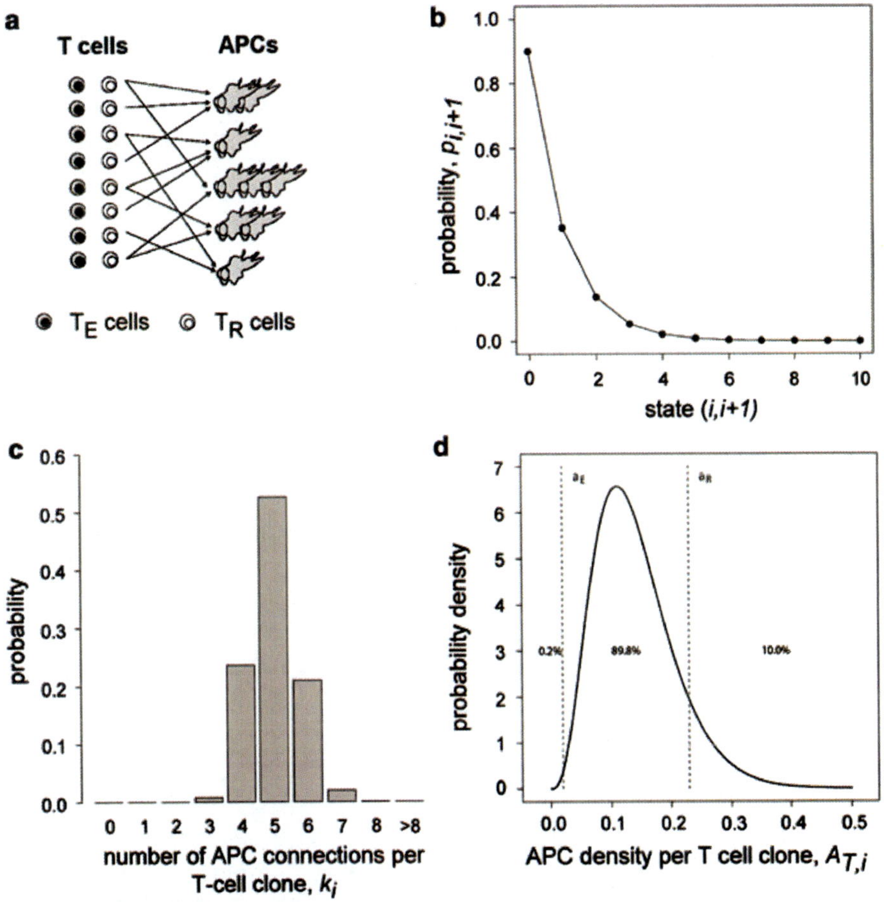

Fig. 14.3 Network between T cell clones and APC subpopulations. **(a)** T cell clones recognize a different subset of APC subpopulations; **(b)** Probability of generating a new connection ($i + 1$) given a clone has already i connections (14.8); **(c)** Probability distribution of the number of connections per T cell clone; **(d)** Probability distribution of APC density recognized per T cell clone. Parameters were specified in order to generate 0.2% T cell clones in an extinction regime ($A_t < a_E$), 89.8% T cell clones in immunity regime ($a_E \leq A_t \leq a_R$) and 10% T cell clones in the bistable regime ($A_t > a_R$). Parameter values of plots B, C and D: $\mathcal{A} = 100$, $\lambda_a = 2.8 \times 10^{-3}$, $\theta = 0.90$, and $\lambda_c = 1.07$

limit for T cell survival is higher than a_E, as shown by the respective APC density associated to the clone that recognizes the smallest APC density. The exact value for this threshold cannot be computed because it is dependent on the actual connectivity structure between APC subpopulations and T cell clones, which is randomly generated in each simulation. A similar result is obtained for the threshold that allows T_R cells to be sustained within a clone. In this case, clones recognizing APCs at densities higher than $\tilde{a}_R$ will help to maintain T_R cells in less crossreactive clones. As a consequence, the threshold for T_R-cell persistence decreases, being closer to a_R,

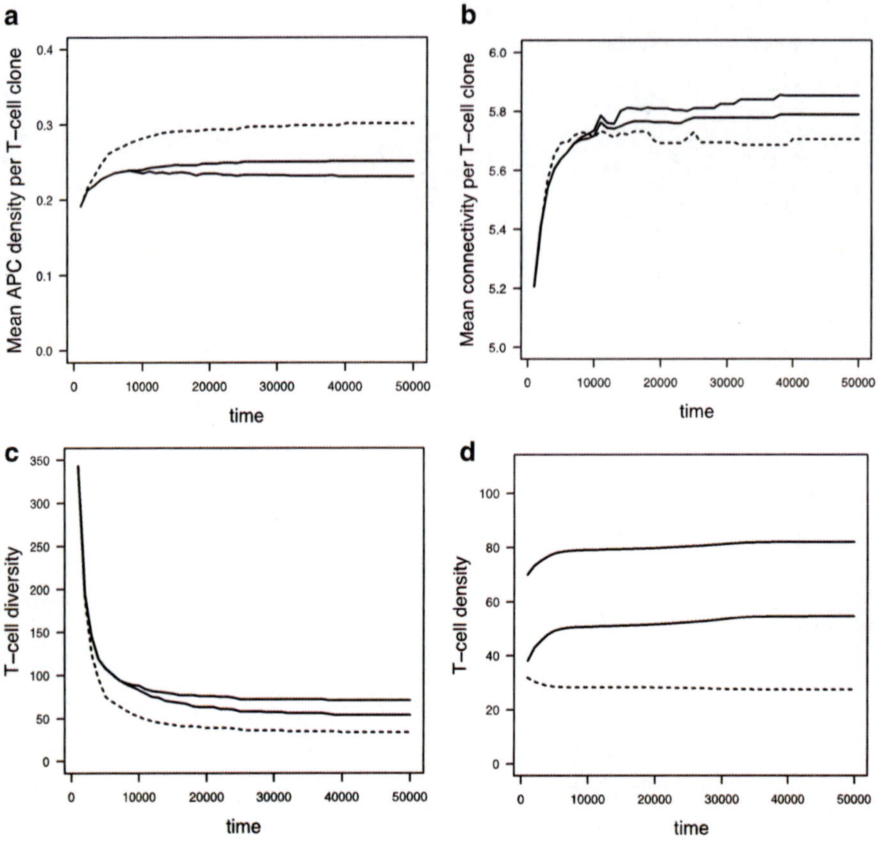

Fig. 14.4 Peripheral T cell repertoire dynamics without thymic export. The *panel* shows the time course of several quantities during a representative simulation in which 500 distinct clones were generated at time $t = 0$, with similar initial conditions, competing for $\mathcal{A} = 100$ APC subpopulations (*thick line* - T cells; *solid line* - T_E cells; *dashed line* - T_R cells; Initial condition for each clone: $t_i^* = 0$, $E_{i,0} = R_{i,0} = 0.50, \forall i = 1, \ldots, 500$). The parameters λ_a and λ_c were specified in order to generate T cell clones with a mean crossreactivity of five cognate APCs and 10% probability of recognizing APC densities necessary for a bistable dynamics, as illustrated in Fig. 14.3 ($\lambda_a = 2.8 \times 10^{-3}$ and $\lambda_c = 1.07$). The remaining parameters are given in Table 14.1

the lower APC density limit that defines the bistable regime for the homogeneous population case (Fig. 14.2a). It is worth noting that, in this scenario, the equilibrium T cell density of each clone driven by a given APC density is lower than in the scenario without interclonal competition for APCs.

Since a similar partition of the peripheral repertoire is observed in the context of interclonal competition, previous suggestions for the diversity and related properties of T_R and T_E cells also hold [16]. Therefore, T_R cells seem to be less diverse than their T_E counterparts (Fig. 14.5b). As a corollary, the diversity of the T_E-cell exclusive repertoire has the largest contribution to the total diversity (Fig. 14.5c).

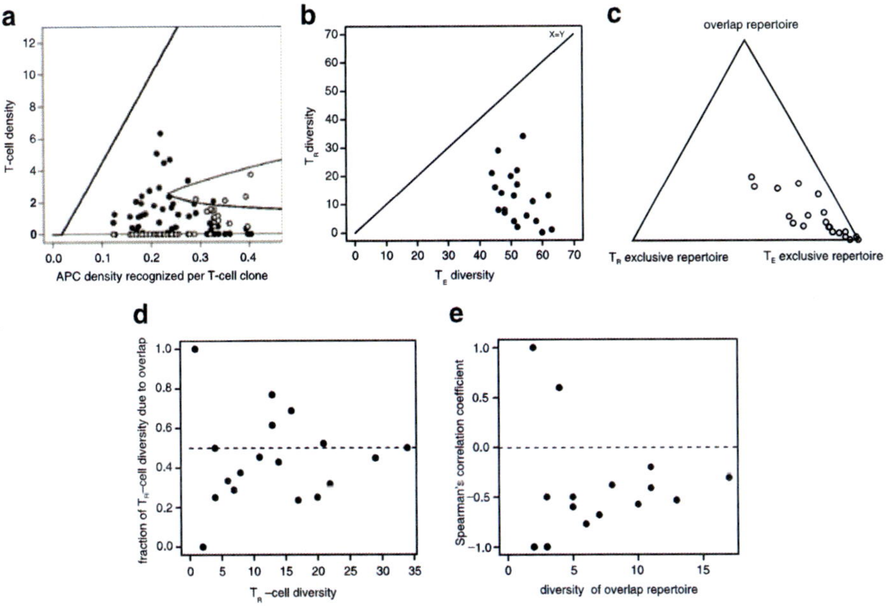

Fig. 14.5 Partition of the peripheral T cell repertoire in simulations without thymic export. (**a**) A representative snapshot of a T cell repertoire at equilibrium ($t = 50000$), where *solid and empty circles* represent T_E and T_R clones, respectively. (**b**) Diversity of T_R and T_E cells, where the *diagonal line* corresponds to equal diversity of both cell types. (**c**) *Ternary diagram* representing the contribution of the diversities of T_R-exclusive, T_E-exclusive, and overlap repertoires to the total T cell repertoire diversity. (**d**) Diversity of the T_R/T_E overlap repertoire plotted as a fraction of total T_R-cell diversity, where the *horizontal line* corresponds to the case in which T_R-cell exclusive and overlap repertoires are equally diverse. (**e**) Spearman's correlation coefficient calculated between T_R and T_E-cell densities in clones of the overlap repertoire. Each *dot* in plots B, C, D and E represents a distinct simulation of the T cell repertoire (20 simulations in total). The parameters of these simulations are given in Fig. 14.4 and Table 14.1

Interestingly, the T_R-cell repertoire can now be divided into an exclusive subset and a subset overlapping with the repertoire of T_E-cells (Fig. 14.5c). Therefore, in contrast to the previous model [16], competition between T cell clones with different specificities can give rise to an exclusive T_R-cell repertoire. The contribution of this repertoire to the overall T_R-cell diversity varies across simulations. The diversity of the exclusive T_R-cell repertoire can be lower, equal or higher than the diversity of overlap between T_E and T_R repertoire (Fig. 14.5d). Finally, as predicted by Fig. 14.2b, T_R and T_E-cell densities of clones belonging to the intersection repertoire tend to be negatively correlated according to Spearman's coefficient (Fig. 14.5e).

Dynamics and Structure of the T Cell Repertoire in the Presence of a Constant Thymic Export

We now present the results of simulations with a continuous thymic export. These simulations are more complex than the previous ones and, as we will see below, it is important to distinguish two pools within the total peripheral T cell repertoire: the pool of recent thymic emigrants and the pool of resident T-clones. The repertoire of resident clones can exhibit a structure similar to that obtained without thymic export, while the repertoire of recent thymic emigrant clones will exhibit a structure related to the initial conditions imposed for the generation of each clone.

Three Distinct Classes of T Cell Repertoire Patterns Arise in the Presence of a Continuous Thymic Export

The peripheral T cell repertoires emerging in simulations with interclonal competition and constant thymic export can be classified into three classes according to their overall structure and the partitioning of clones made exclusively of T_R or T_E cells, or containing both cell types (Fig. 14.6a, e and i). These three classes for repertoire structures can be obtained even for the same parameter values.

The first class of repertoire structure is characterised by clones containing only T_E cells, even if these clones recognize cognate APCs at densities within the bistability regime (Fig. 14.6a). Newly generated T_R-cell clones persist transiently in the periphery until their extinction, and the T_R-cell repertoire is maintained only by thymic influx. All T_R-cell clones are in the subset regarding the newly-generated T cells clones (i.e., recent thymic emigrant population). Nonetheless, T_R-cell mean connectivity, density, and diversity increase in the initial phase of the simulations until a plateau is reached (Fig. 14.6b, c and d). This is explained by the fact that, even if all T_R cells eventually become extinct, T_R-cell clones that recognize few APCs vanish faster than the ones recognizing more APCs, resulting in an apparent increase of the above properties.

The second class of repertoire structure is associated with a partition similar to that shown in Fig. 14.5a. The total repertoire is partitioned into three subsets, even though recent thymic emigrants are present in all of them (Fig. 14.6e). The first subset comprises a set of transient T cell clones that recognize cognate APCs at low densities and, thus, eventually go extinct after entering periphery. This subset is maintained by thymic influx. The second subset refers to T cell clones that recognize intermediate densities of APCs and form the T_E-cell exclusive repertoire. The third subset encompasses T cell clones that recognize APCs at sufficiently high amounts. In most cases, these clones are regulated by T_R cells. Occasionally, this third subset might include some expanded clones containing exclusively T_E cells. As for the previous repertoire class, the mean connectivity increases in time (Fig. 14.6f), but now due to a highly crossreactive T_R and T_E-cell resident population. Therefore, the mean connectivity of both cell types increases not only due to the initial purging

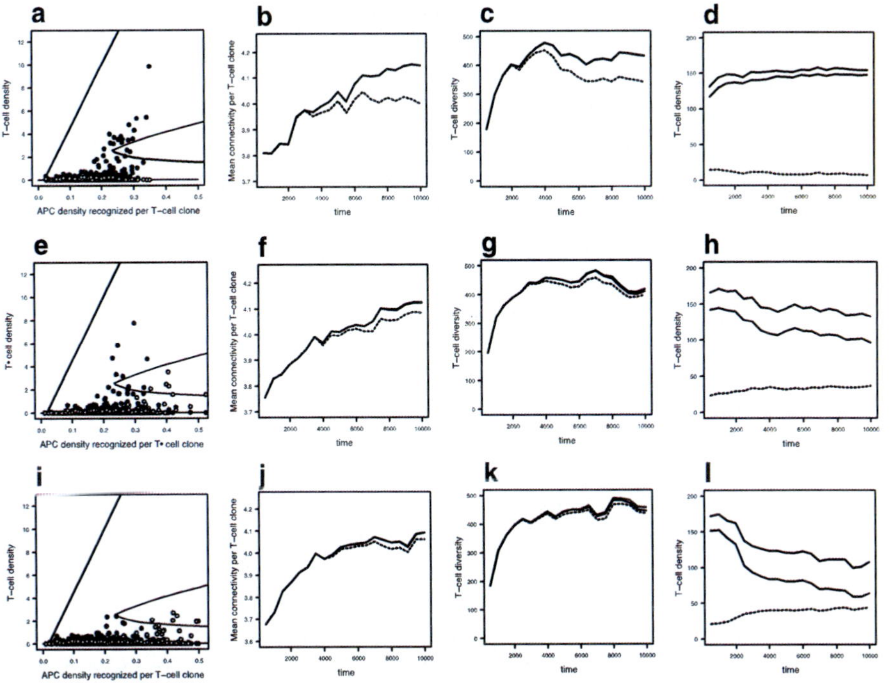

Fig. 14.6 Three distinct classes of T cell repertoire patterns arise in the simulations with thymic export (*thick line* - T cells; *solid line* - T_E cells; *dashed line* - T_R cells; $s_c = 0.3$, initial condition for each clone: $E_{i,t_i^*} = 0.75$, $R_{i,t_i^*} = 0.25, \forall i$). Plots in *first row* represent total repertoire structures where T_E-cell clones predominate within repertoire (repertoire structure of class I). In this scenario, T_R cells are mostly maintained by continuous thymic export. Plots in *second row* represent total repertoires showing a partition similar to the one shown in Fig. 14.5a (repertoire structure of class II). Plots in *third row* represent total repertoire structures in which T_R cell clones predominate (repertoire structure of class III). In this case, T_E cells are mostly maintained by the constant thymic export. Parameters λ_a and λ_c were setup in order to generate T cell clones with a average connectivity of 3.75 APC species and a 10% probability of recognizing APC densities in the bistable regime ($\lambda_a = 3.6 \times 10^{-3}$ and $\lambda_c = 0.72$). The remaining parameters are given in Table 14.1

of more specific clones from the repertoire, but also to the formation and maturation of a T_R and T_E resident pool with increasingly high crossreactivity. Crossreactivity increases slowly in this resident pool, as old resident clones are replaced by newly arrived clones, which happen to be more crossreactive. As in Fig. 14.4c, T_E-cell diversity is higher than that of T_R cells (Fig. 14.6g). Interestingly, the total T cell density reaches its peak very early in the simulation but after that, with the maturation of the resident T_R-cell pool, decreases slowly in time (Fig. 14.6h), eventually reaching a steady state determined by an overall regulation of all resident clones by T_R cells. Thus, while the total T cell density decreases the total T_R-cell density increases.

The third class of repertoire structure refers to a predominance of T_R cells within the clones early on in simulations (Fig. 14.6i and l). The repertoire is composed of a recent thymic emigrant population and a resident population made almost exclusive of clones with T_R cells and without T_E cells (Fig. 14.6i). Few T_E-cell clones are maintained in the intermediate APC density region. However, the populations of T_R cells, which require T_E cells for growth, can only persist in the periphery due to the source of T_E cells coming from the thymus. Since T cell clones are exported to the periphery with both cell types, T_R-cell diversity tends to be close to that of T_E cells (Fig. 14.6k). In some cases, T_R-cell diversity can be even higher (simulations not shown). Finally, as in the previous class, the evolution of total T cell density has a biphasic behaviour: it reaches a peak early in the simulations, followed by an abrupt drop. This drop correlates with the overtaking of the repertoire by T_R-cell clones (Fig. 14.6l). Therefore, these repertoires can be interpreted as a situation where all the immune responses are globally regulated by T_R cells.

At this point, it is worth noting that the first and third classes of repertoire patterns have deleterious functional consequences for the organism. The organism's phenotype corresponding to the first class of repertoire structures, in which T_E-cell predominate in all the clones and T_R-cells cannot expand and persist in the periphery, is that of a massive systemic autoimmune pathology, akin to what is observed in animals or humans deficient in the *foxp3* gene. [30]. Conversely, in the third class of repertoire structures, in which T_E-cells are under strong control of highly cross-reactive T_R-cell clones, the phenotype of the organism would be effectively that of an immunocompromised individual unable to mount immune responses. Therefore, the repertoire structure of class II seems to be the most adequate to describe the structure and function of the peripheral T cell repertoire in a healthy individual, as suggested in a previous work [16]. Yet, there is a price to pay in this class. In these simulations, some occasional clonal expansions containing exclusively T_E cells are observed, which might be interpreted as organ-specific autoimmune diseases.

The Three Repertoire Classes are Better Distinguished by the Structure of Their Resident Populations

As mentioned above, any peripheral T cell repertoire is composed of a recent thymic emigrant population and a resident one. The latter population is represented by a small number of clones that expand and persist at T_R and/or T_E densities above those initially set (as can be observed in Fig. 14.6a, e and i). Nevertheless, it is difficult to define exactly which clones effectively form the resident population. For the sake of simplicity, the 25 oldest clones in the simulations are considered representative of the resident population. In contrast, recent thymic emigrant clones represent the majority of the repertoire by the large number of clones with densities equal or lower than the initial ones (as also observed in Fig. 14.6a, e and i). It is worth noting that most of these recent thymic emigrant clones will eventually go extinct.

Due to the predominance of recent thymic emigrant clones, all three repertoire classes show similar structures when the whole T cell repertoire is analyzed (Fig. 14.7a, b and c). That is, T_E-cell diversity tends to be higher than that of T_R

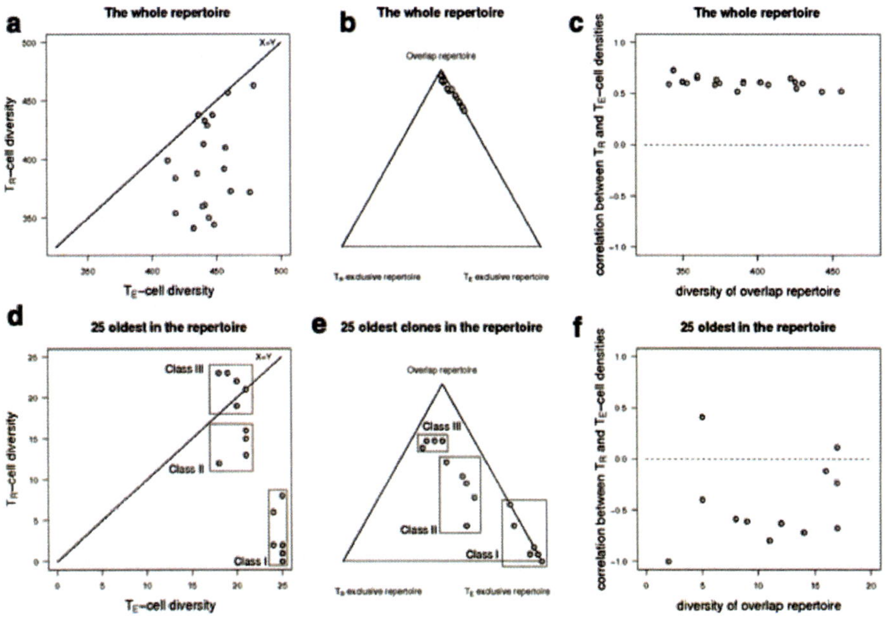

Fig. 14.7 The three T cell repertoire classes are better distinguished within the structure resident cell populations. *Plots* A, B, and C refer to the analysis of the whole T cell repertoire, while *plots* D, E, and F correspond to the analysis of the resident T cell population, represented by the 25 oldest clones in the simulations. Each *dot* in the plot represents a distinct simulation (20 simulations in total). See Fig. 14.6 for further details

cells but with no clear distinction between these three classes of repertoire structures (Fig. 14.7a). The same happens when analyzing the contribution of the exclusive repertoires of T_R and T_E cells, and their overlapping repertoire, to the total T cell diversity (Fig. 14.7b). As expected, the overlap between the repertoires of T_R and T_E recent thymic emigrants represents the larger fraction of the total repertoire. This implies in turn a positive correlation between T_R and T_E-cell clones belonging to the overlapping repertoire (Fig. 14.7c).

The main differences between the three repertoire classes are found within the minor pool of resident T cell clones. In fact, the first class of repertoires has a resident population with a large difference between T_R and T_E-cell diversities (Fig. 14.7d). In this class, the resident repertoire is mainly composed of T_E-cell exclusive repertoire. The second class of repertoires has an intermediate difference between T_R and T_E-cell diversities. Using a ternary diagram representation we plot simultaneously the diversities of the T_R and T_E exclusive repertoires as well as that of the overlap. The points representing the resident population in the second class of repertoires lay almost in the central region of the triangle, which means that exclusive T_R-cell and T_E-cell diversities and the shared diversity contribute almost equally to the overall T cell diversity (Fig. 14.7e). Since the third class of repertoires is characterized by a strong predominance of T_R cells, the diversity of these cells within the resident population is equal or higher than that of the T_E cells

(Fig. 14.7d). The repertoire of the resident population is then mostly composed of T cell clones in the overlapping repertoire and few clones from the exclusive T_R-cell repertoire (Fig. 14.7e). Notwithstanding the differences between the three classes of repertoire, they all tend to show a negative correlation between T_R and T_E-cell densities within the clones that represent the overlap. A positive correlation can also be obtained when some of the clones in the resident population are replaced by newly arising T cell clones.

Parameter Dependence of the Peripheral T Cell Repertoire Structure

We now ask how the structure of the peripheral T cell repertoire changes with some of the parameters in the simulations, such as the average clonal connectivity coming from the thymus, number of APC subpopulations (APC diversity), and the number of clones exported from the thymus (Fig. 14.8). We pay special attention to the probability of obtaining each of the three classes of T cell repertoire structures and the decomposition of T cell diversity into their three components. To calculate the proportion of each repertoire structure for a given parameter set, we visually classified the repertoires according to the diversity composition exhibited by their 25 oldest clones at the end of the simulation ($t = 10{,}000$), as illustrated in Fig. 14.7e.

In the simulations, for fixed values of APC diversity and thymic export, increasing the average connectivity per clone increases the chance of obtaining repertoire structures of class III (Fig. 14.8a). Because repertoires in this class are characterized by a predominance of T_R cells, the overall T cell diversity tends to decrease as connectivity increases (Fig. 14.8b), and the repertoire of the resident population becomes increasingly dominated by the T_R-cell exclusive repertoire (Fig. 14.8c). In fact, with a higher average connectivity for clones coming from the thymus, the ratio of T_R cells per APC increases, promoting in this way peripheral tolerance.

Fixing the parameters controlling connectivity and thymic export, a higher APC diversity increases the overall T cell diversity (Fig. 14.8e). This is just a consequence of a higher number of combinations of different APCs that can be recognized by the T cell clones. In theory, this parameter when increased should transiently decrease interclonal competition, unless compensated by an increase of the thymic export. T_E cells would then benefit of this diminished interclonal competition in the beginning of the simulation to proliferate and competitively exclude T_R-cell within the same clone. Therefore, individuals with a higher APC diversity but with similar thymic output tend to exhibit repertoires of class I (Fig. 14.8d). For this reason, the diversity of the T_E-cell exclusive repertoire (Fig. 14.8e) and its relative contribution to the diversity of the resident population (Fig. 14.8f) tend to increase as a function of the APC diversity.

The overall T cell diversity is also positively correlated with the thymic export (Fig. 14.8h), as expected from the fact that there is more clones from the thymus. Since clones enter the periphery with both T_R and T_E cells, diversity of the overlap

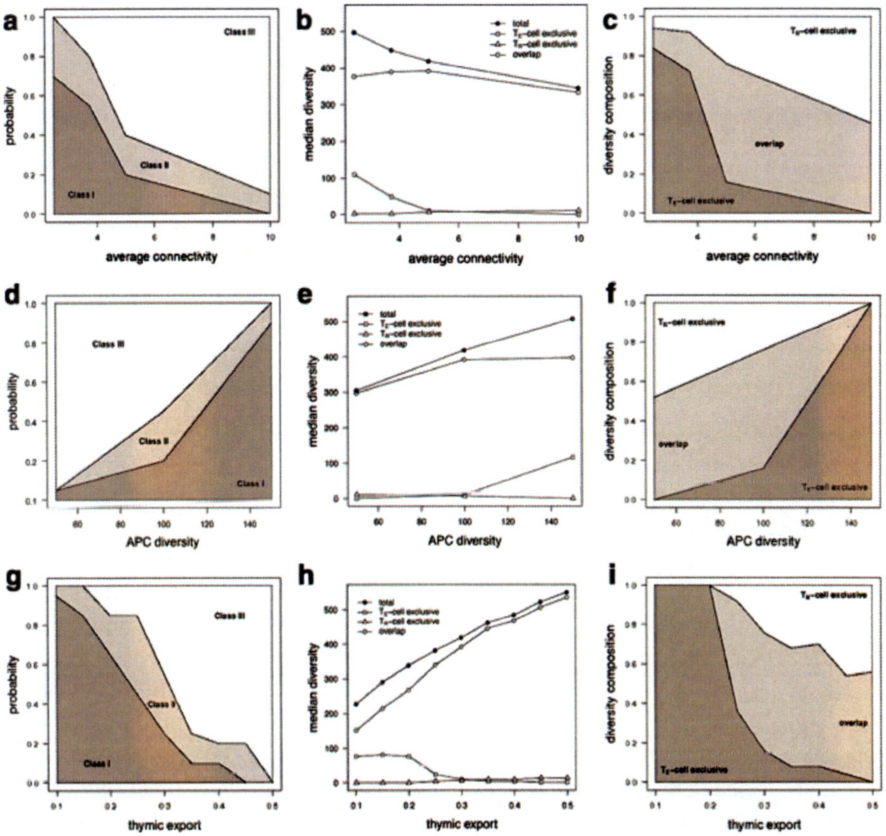

Fig. 14.8 Effect of some parameters on the structure of the repertoire. *First, second and third rows* refer to the effect of mean connectivity coming from the thymus, APC diversity, and thymic export, respectively. *First, second, and third columns* refer to the probability of generating each repertoire structure (classes I, II, and III), the diversity of each T cell repertoire, and diversity composition for the 25 oldest clones at the end of the simulations ($t = 10{,}000$). Parameters: $s_c = 0.30$ and $\mathcal{A} = 100$ (plots **a**, **b**, and **c**); $s_c = 0.30$, $\lambda_a = 2.08 \times 10^{-3}$ and $\lambda_c = 1.07$ (plots **d**, **e**, and **f**; thymic average connectivity of five APC populations); $\mathcal{A} = 100$, $\lambda_a = 2.08 \times 10^{-3}$ and $\lambda_c = 1.07$ (plots **g**, **h**, and **i**; thymic average connectivity of five APC populations). The parameter λ_a was calculated in order to generate 10% of clones in the bistable regime. For each parameter set, we performed 20 simulations

repertoire also increases with this parameter (Fig. 14.8h). A priori, a higher thymic export is expected to exhibit two major effects. On the one hand, it increases the chance of generating highly crossreactive clones. On the other hand, as it was the case for increased connectivity, it promotes the control by T_R cells by favoring the ratio of T_R cells per APC and, at the same time, increasing interclonal competition, which slows down the taking over of APC by T_E cells. Because of these effects, high values of thymic export favor the emergence of repertoires of class III (Fig. 14.8g), leading to a high contribution of T_R-cell diversity in the resident population (Fig. 14.8i).

Discussion

Recently, there is a growing interest in the structure of the T cell repertoire, in particular in its decomposition in subsets that are exclusive of either T_R- or T_E-cell populations, and the subset corresponding to the overlap between the two populations. Here, we tried to understand how this decomposition comes about based on the simulations of the Crossregulation model [16], describing the peripheral dynamics of T_R- or T_E-cell populations.

We previously studied a scenario in which the peripheral T cell repertoire was, by construction, composed of independent clones that recognize mutually exclusive sets of cognate APCs [16]. This model precluded the possibility of generating a T_R-cell exclusive repertoire, because the persistence of T_R-cell populations requires their T_E counterparts in the same clone. To overcome this limitation of the model, each T cell clone in the repertoire was interpreted as an independent functional class of T cells exhibiting similar TCR specificity but that may contain distinct T_E and T_R-cells with different specificities. According to this possibility, a T_R-cell exclusive repertoire was implicitly included, but the model could not be used to make predictions about the fractions of exclusive or overlapping repertoires.

We now show that shared specificity between different clones is enough to originate a T_R-cell exclusive repertoire. Shared specificity is then a sufficient but not a necessary condition for the generation of a T_R-cell exclusive repertoire, since the mechanism of differentiation in the thymus of T_E and T_R cells might produce in the first place exclusive repertoires of each cell type. Studies on the fate of individual thymocytes bearing a specific transgenic TCR demonstrated the differentiation of cells with the same specificity to both T_R and T_E phenotypes [31]. However, a small overlap was observed between thymic T_R and T_E TCR samples in recent studies on animal models with limited TCR diversity [7, 8]. At first glance, this observation suggests that the thymus is actually able to seed the periphery with a T_R-cell exclusive repertoire. Yet, a small overlap in these small samples might not imply a similar interpretation for the whole repertoire of an individual, as discussed in depth in Sepúlveda [32]. Small or no overlap between pairs of samples can arise by chance due to small or different sample sizes. Venturi [33] alluded to this problem when studying the evolution of TCR repertoire of an individual during an influenza infection, and provided a method to standardize the calculations. A small overlap was observed in small peripheral T_R and T_E TCR samples, which was also interpreted as evidence for the existence of peripheral T_R-cell exclusive repertoire [6, 8]. Yet, the cautionary note on the interpretation of these observations based on small samples also hold here. We then conclude that current statistical analysis of available data is not sufficiently accurate to determine whether or not an exclusive T_R-cell repertoire exists in the thymus or periphery.

Notwithstanding this conclusion, our simulations of peripheral T cell clone dynamics with continuous thymic export suggest that a small T_R-cell exclusive repertoire may arise in the periphery, irrespective of how the repertoire is structured (Fig. 14.7e). This result can be further extended to the more general situation in which the thymus exports T_R-cell exclusive specificities. In fact, we demonstrate that peripheral repertoire selection would purge those specificities that are less crossreactive, maintaining only the highly crossreactive ones. At this point, it is worth to recall that current models and observations of thymic T_R-cell development suggest that these cells are educated or selected on high affinity/avidity interactions with thymic APC [13]. Therefore, it is tempting to speculate that thymic T_R-cell differentiation process has been 'optimized' in order to generate T_R-cell clones that, in theory, would have a higher chance to survive peripheral selection.

Three distinct structures of the peripheral T cell repertoire emerged in our simulations. On the one extreme, repertoire structures of class I are characterized by a dominance of T_E cell and exclusion of T_R cells in most of the resident clones. This scenario can be interpreted as the devastating systemic autoimmune pathology observed in *foxp3*-deficient individuals. On the other extreme, there are repertoires patterned in such a way (class III repertoires) that, in the vast majority of resident clones, T_R cells predominate over their T_E counterparts, which often vanish. A repertoire patterned in this way is expected to avoid autoimmunity but will be mostly likely inappropriate to fight infections. From previous studies of the Cross-regulation model in the presence of growing antigen [20, 21], one expects this repertoire pattern would benefit mild and moderate growing infections that should become chronic due the inhibition of specific immune responses by (specific) T_R cells [20]. In contrast, strong acute infections that elicit an overwhelming increase in specific APC subpopulations, might break the global regulation state of the repertoire, as demonstrated in Leon et al. [20]. The last repertoire structure arising in simulations lies in the middle of the above two extremes, showing a similar structure to that discussed in Carneiro et al. [16] and obtained here in simulations without thymic export (class II repertoire). The repertoire is partitioned in three subsets of clones: (1) a small resident subset avoiding autoimmune responses against recurrent body antigens by the presence of T_R cells; (2) a relatively large subset lacking T_R cells that allows T_E cells to respond to non-recurrent foreign antigens; (3) a third subset solely maintained by the thymus that would otherwise go extinct.

As argued above, repertoire structures of class II seem the most adequate to describe a healthy immune system. Assuming that the parameters used in the simulations are more or less realistic, this pattern for peripheral T cell repertoire has a small chance of being generated in the context of a continuous thymic export, as shown in Fig. 14.8. In fact, this repertoire structure should be better understood as quasi-stationary transient rather than a true steady state. Since T_R cells are present in the resident population of this repertoire structure, a continuous thymic export would help the system to reach asymptotically the above-mentioned global regulation repertoire structure. This is evident by the sustained increase in time of the fraction of T_R cells during the simulations in which the repertoire were characterized as class II. As a consequence, we expect that aged animals would be in general more susceptible to infections due to a higher fraction of T_R-dominated clones in the peripheral repertoire. Interestingly, T_R cells seem to be at a higher fraction in older individuals than in young ones both in humans and mice (refs. [34, 35]; R. Paiva and colleagues, unpublished results). For example, Nishioka et al. [35] reported that aged C57BL/6 mice show a fraction of T_R cells in the $CD4^+$ pool around 30%, which is statistically higher than the 14% observed in young mice. Moreover, these T_R cells in older mice seem to maintain their suppressive function. In agreement with our above interpretation, it was suggested that the increased fraction of T_R cells in the peripheral repertoire might be an explanation for the increased incidence of infectious diseases in the elderly [34, 35].

One of the simplifications introduced in our simulations was to assume that thymic export is constant. However, thymic output could be controlled in order to maintain the repertoire as much time as possible in its healthy structure. Since it was more easy to obtain such repertoire structures in simulations without thymic export (data not shown), one can raise the hypothesis that in healthy organisms the thymic export should be divided into two distinct phases. First, during early ontogeny of the organism, the thymic export should be strong (perhaps enriched with multireactive clones) to fill up the periphery and seed it with sustainable T_R-cell clones. Then, thymic export should decrease to avoid the filling of the periphery with further T_R-cell clones that, "helped" by early resident T_R-cell clones, ultimately push the system to a global regulation repertoire structure. This biphasic thymic export is a well established fact both in humans and mice, the so-called thymic involution. Yet, this phenomenon is generally viewed as being deleterious, as remarked by Dowling and Hodgkin [36], since it is usually associated with a decline in immune function in the elderly. In the vein of the above discussion, we suggest that thymic involution might be crucial to seed and maintain the repertoire in a healthy structure. Dowling and Hodgkin [36] also argue that thymic involution is an important process shaping the T cell repertoire in young individuals. In their view, T cell clones exhibiting TCR specificities close to positive and negative thresholds have a shorter lifespan. In this scenario, the thymus involutes to allow to peripherally select T cell clones exhibiting TCR with affinities lying between positive and negative thresholds and, thus, have longer lifespans and are more useful for immunity.

Interestingly, a biphasic thymus export has been previously proposed but using a different argument [13]. It was suggested that T_R cells against ubiquitious peptides are produced in the thymus during perinatal period in order to cover every available tissue-specific antigen. Then, it was assumed that the thymus stops producing T_R cells, exporting only naive T_E cells, which undergo a process of education upon encountering antigen that converts them to the T_R pool in the periphery. In this line of thought, this process of peripheral education assumes a key role in the maintenance of the T_R pool in adulthood. Although peripheral education is now known to occur in the periphery [26, 37–39], its exact contribution to the structure of the peripheral repertoire is still unclear, since current estimates for the fraction of peripherally-converted T_R cells are disparate across different experimental settings, and furthermore, T_R cells are known to be continuously produced in the thymus (reviewed in [40]).

The long-term outcome of our simulations was somehow dependent on the early history of repertoire formation. In this regard, if T_E-cell clones take over the repertoire in the first place, it would be difficult to maintain later-generated T_R-cell clones in the periphery. This would lead to a repertoire structure of class I with its deleterious consequences to the host. In contrast, if early-generated T_R-cell clones could be sustained in the periphery, they would help later-generated ones to survive in the repertoire, giving rise to repertoire structures of class III. This dependence on the early history of repertoire formation seems in agreement with previous studies on neonatally thymectomized animals [41], where thymus ablation between 2 and 4 days of life but not later can generate an autoimmune condition in adults. This short time window for the consequences of thymectomy suggests that very early events of repertoire formation are crucial to determine the respective long-term outcome. This is in line with the functional significance of thymic involution alluded above, since a large thymic output in early life would favor the emergence of a repertoire with a healthy pattern.

It is well known that autoimmune diseases are affected by multiple genetic and environmental factors, which lead to a complex inheritance pattern [42]. Complexity in these diseases arises due to the incomplete penetrance phenomenon, that is, only a proportion of individuals with risk-associated genotypes will manifest disease. Traditionally, one invokes environmental factors to explain this observation, such as the occurrence of specific infections. Interestingly, twins maintained in the same environment do not show a 100% concordance rate for disease status. Moreover, different autoimmune-disease animal models also show incomplete penetrance, even under strict environmental conditions. This is particularly evident in the NOD mouse strain that spontaneously develops type I diabetes with incomplete penetrance in both males and females [43]. To explain this phenomenon, we and others have previously argued that a fraction of incomplete penetrance could be attributed to an internal stochastic component governing the expression of the phenotype, such as a stochastic allelic expression [44–47]. In the case of autoimmune diseases, we invoked the stochasticity in T cell repertoire generation as a putative explanation for this internal component of penetrance [45], following the observations in monozygotic twins when studying different autoimmune diseases [48, 49]. In this line of reasoning, the simulations reported here showed that the three above-mentioned repertoire patterns can arise in simulations with the same parameter set (Fig. 14.8). If one interprets these simulations as different individuals with the same genetics, they indicate that the stochasticity in the T cell repertoire formation can indeed be an important factor in the understanding of incomplete penetrance in the context of complex autoimmune diseases. Recently, a lower T_R-cell TCR diversity in NOD mice than in non-diabetic C57BL6 mice was reported, suggesting important differences between the T cell repertoires of these mouse models [50]. However, the comparison of TCR diversity between unaffected and affected NOD mice was not performed, which would allow to assess the putative role of stochastic T cell repertoire formation in incomplete penetrance. In this scenario, one would expect to observe clear differences in the T cell repertoire between unaffected and affected NOD mice.

Our simulations provide simple predictions for some structural properties of the peripheral repertoire. These predictions are distinct for each repertoire class, namely, in what concerns to the resident T cell population. For instance, repertoire structures of class I exhibit a higher diversity of resident T_E cells than that of T_R cells, and a small overlap between the respective resident repertoires. Class II repertoires also predict a higher diversity of resident T_E cells but the difference between T_E and T_R-cell diversity should be smaller than in class I. Repertoire structures of class III would in turn show comparable diversities of T_R and T_E cells. Besides these predictions, our simulations suggest that, when there is an overlap between resident T_R and T_E-cell specificities, the respective clonal size distributions tend to be negatively correlated.

Since the three classes repertoire patterns are fairly distinguished by their resident population, a way to discriminate them experimentally is to study the evolution of the T cell repertoire in time. In theory, one can detect the resident population by following the TCR specificities that consistently appear in different time points. However, current peripheral T_R-cell repertoire studies just take a snapshot of the repertoire at a given time point [6, 8–10, 51], which might obscure the underlying complexities [52]. These studies show that TCR samples of T_R cells are more diverse than those from their T_E counterparts, which prompted the authors to extend this conclusion to the whole repertoire of an individual [6, 8–10]. However, we demonstrate that this conclusion might not be true [32, 53]. In fact, when re-analyzing data from Hsieh et al. [6, 7], we could not reject the hypothesis of equal diversity between peripheral T_R and T_E cells at the level of the whole peripheral repertoire. Since there is no information available on the resident T cell population, the comparable diversity of peripheral T_R and T_E cells might be due to a situation where the thymus preferentially exports clones with both cell types, as in simulations performed here. If this holds true, little can be said about the underlying repertoire structure, as illustrated in Fig. 14.7a and b. In fact, stochastic simulations demonstrate that the available data of Hsieh et al. [6, 7] could not rule out a complete overlap between thymic T_R and T_E-cell repertoires and, thus, little information about the underlying structure can be extracted by a snapshot of the repertoire [32]. In line with these results, we conclude that current T_R-cell studies need to be redesigned in order to gain more insight on peripheral T_R-cell repertoire structure.

Acknowledgements This work was supported by Fundação para a Ciência e a Tecnologia (grants PTDC/SAU-MII/71402/2006) and a fellowship to NS (SFRH/BD/19810/2004). The authors would like to thank Prof. Ramit Mehr (Bar-Ilan University) for many comments on an early version of this paper.

References

1. Hori S, Nomura T, Sakaguchi S (2003) Control of regulatory T cell development by the transcription factor Foxp3. Science 299:1057–1061
2. Fontenot JD, Gavin MA, Rudensky AY (2003) Foxp3 programs the development and function of $CD4^+CD25^+$ regulatory T cells. Nat Immunol 4:330–336
3. Sakaguchi S, Sakaguchi N, Shimizu J, Yamazaki S, Sakihama T, Itoh M, Kuniyasu Y, Nomura T, Toda M, Takahashi T (2001) Immunologic tolerance maintained by $CD25^+$ $CD4^+$ regulatory T cells: their common role in controlling autoimmunity, tumor immunity, and transplantation tolerance. Immunol Rev 182:18–32

4. Mills KHG (2004) Regulatory T cells: friend or foe in immunity to infection? Nat Rev Immunol 4:841–855
5. Belkaid Y (2007) Regulatory T cells and infection: a dangerous necessity. Nat Rev Immunol 7:875–888
6. Hsieh CS, Liang Y, Tyznik AJ, Self SG, Liggitt D, Rudensky AY (2004) Recognition of the peripheral self by naturally arising CD25$^+$ CD4$^+$ T cell receptors. Immunity 21:267–277
7. Hsieh CS, Zheng Y, Liang Y, Fontenot JD, Rudensky AY (2006) An intersection between the self-reactive regulatory and nonregulatory T cell receptor repertoires. Nat Immunol 7:401–410
8. Pacholczyk R, Ignatowicz H, Kraj P, Ignatowicz L (2006) Origin and T cell receptor diversity of Foxp3$^+$CD4$^+$CD25$^+$ T cells. Immunity 25:249–259
9. Pacholczyk R, Kern J, Singh N, Iwashima M, Kraj P, Ignatowicz L (2007) Nonself-antigens are the cognate specificities of Foxp3$^+$ regulatory T cells. Immunity 27:493–504
10. Wong J, Obst R, Correia-Neves M, Losyev G, Mathis D, Benoist C (2007) Adaptation of TCR repertoires to self-peptides in regulatory and nonregulatory CD4$^+$ T cells. J Immunol 178:7032–7041
11. Sakaguchi S, Sakaguchi N, Asano M, Itoh M, Toda M (1995) Immunologic self-tolerance maintained by activated T cells expressing IL-2 receptor alpha-chains (CD25). Breakdown of a single mechanism of self-tolerance causes various autoimmune diseases. J Immunol 155: 1151–1164
12. Sakaguchi S, Wing K, Miyara M (2007) Regulatory T cells - a brief history and perspective. Eur J Immunol 37:S116–S123
13. Modigliani Y, Bandeira A, Coutinho A (1996) A model for developmentally acquired thymus-dependent tolerance to central and peripheral antigens. Immunol Rev 149:155–200
14. van Santen HM, Benoist C, Mathis D (2004) Number of Treg cells that differentiate does not increase upon encounter of agonist ligand on thymic epithelial cells. J Exp Med 200:1221–1230
15. Mason D (1998) A very high level of crossreactivity is an essential feature of the T-cell receptor. Immunol Today 19:395–404
16. Carneiro J, León K, Caramalho I, van den Dool C, Gardner R, Oliveira V, Bergman ML, Sepúlveda N, Paixao T, Faro J, Demengeot J (2007) When three is not a crowd: a Crossregulation model of the dynamics and repertoire selection of regulatory CD4$^+$ T cells. Immunol Rev 216:48–68
17. León K, Peréz R, Lage A, Carneiro J (2000) Modelling T-cell-mediated suppression dependent on interactions in multicellular conjugates. J Theor Biol 207:231–254
18. León K, Peréz R, Lage A, Carneiro J (2001) Three-cell interactions in T cell-mediated suppression? A mathematical analysis of its quantitative implications. J Immunol 166:5356–5365
19. León K, Lage A, Carneiro J (2003) Tolerance and immunity in a mathematical model of T-cell mediated suppression. J Theor Biol 225:107–126
20. León K, Faro J, Lage A, Carneiro J (2004) Inverse correlation between the incidences of autoimmune disease and infection predicted by a model of T cell mediated tolerance. J Autoimmun 22:31–42
21. León K, Garcia K, Carneiro J, Lage A (2007) How regulatory CD25$^+$CD4$^+$ T cells impinge on tumor immunobiology: the differential response of tumors to therapies. J Immunol 179:5659–5668
22. Fouchet D, Regoes R (2008) A population dynamics analysis of the interaction between adaptive regulatory T cells and antigen presenting cells. PLoS ONE 3:e2306
23. Stirk ER, Molina-París C, van den Berg HA (2008) Stochastic niche structure and diversity maintenance in the T cell repertoire. J Theor Biol 255:237–249
24. Carneiro J, Paixao T, Milutinovic D, Sousa J, León K, Gardner R, Faro J (2005) Immunological self-tolerance: lessons from mathematical modeling. J Comput Appl Math 184:77–100
25. Carneiro J, Coutinho A, Faro J, Stewart J (1996) A model of the immune network with B-T cell cooperation I - prototypical structures and dynamics. J Theor Biol 182:513–529
26. Zelenay S, Lopes-Carvalho T, Caramalho I, Moraes-Fontes MF, Rebelo M, Demengeot J (2005) Foxp3$^+$ CD25- CD4 T cells constitute a reservoir of committed regulatory cells that regain CD25 expression upon homeostatic expansion. Proc Natl Acad Sci USA 102:4091–4096

27. Freitas AA, Rocha B (2000) Population biology of lymphocytes: the flight for survival. Annu Rev Immunol 18:83–111
28. de Boer RJ, Perelson AS (1994) T cell repertoires and competitive exclusion. J Theor Biol 169:375–390
29. de Boer RJ, Perelson AS (1997) Competitive control of the self-renewing T cell repertoire. Int Immunol 9:779–790
30. Brunkow ME, Jeffery EW, Hjerrild KA, Paeper B, Clark LB, Yasayko SA, Wilkinson JE, Galas D, Ziegler SF, Ramsdell F (2001) Disruption of a new forkhead/winged-helix protein, scurfin, results in the fatal lymphoproliferative disorder of the scurfy mouse. Nat Genet 27:68–73
31. Bensinger SJ, Bandeira A, Jordan MS, Caton AJ, Laufer TM (2001) Major histocompatibility complex class ii-positive cortical eptihelium mediates the selection of $CD4^{+}25^{+}$ immunoregulatory T cells. J Exp Med 194:427–438
32. Sepúlveda N (2009) How is the T-cell repertoire shaped? PhD thesis, University of Porto
33. Venturi V, Kedzierska K, Tanaka MM, Turner SJ, Doherty PC, Davenport MP (2008) Method for assessing the similarity between subsets of the T cell receptor repertoire. J Immunol Methods 329:67–80
34. Dejaco C, Duftner C, Schirmer M (2006) Are regulatory T cells linked with aging? Exp Gerontol 41:339–345
35. Nishioka T, Shimizu J, Yamazaki S, Sakaguchi S (2006) $CD4^{+}CD25^{+}Foxp3^{+}$ T cells and $CD4^{+}CD25^{-}Foxp3^{+}$ T cells in aged mice. J Immunol 176:6586–6593
36. Dowling MR, Hodgkin PD (2009) Why does the thymus involute? A selection-based hypothesis. Trends Immunol 30:295–300
37. Thorstenson KM, Khoruts A (2001) Generation of anergic and potentially immunoregulatory $CD25^{+}CD4$ T cells in vivo after induction of peripheral tolerance with intravenous or oral antigen. J Immunol 167:188–195
38. Apostolou I, von Boehmer H (2004) In vivo instruction of suppressor commitment in naive T cells. J Exp Med 199:1401–1408
39. de Lafaille MAC, Lino AC, Kutchukhidze N, Lafaille JJ (2004) CD25- T cells generate $CD25^{+}Foxp3^{+}$ regulatory T cells by peripheral expansion. J Immunol 173:7259–7268
40. Annacker O, Pimenta-Araujo R, Burlen-Defranoux O, Bandeira A (2001) On the ontogeny and physiology of regulatory T cells. Immunol Rev 182:5–17
41. Asano M, Toda M, Sakaguchi N, Sakaguchi S (1996) Autoimmune disease as a consequence of developmental abnormality of a T cell subpopulation. J Exp Med 184:387–396
42. Gregersen PK, Behrens T (2006) Genetics of autoimmune diseases – disorders of immune homeostasis. Nat Rev Genet 7:917–928
43. Anderson M, Bluestone J (2005) The nod mouse: a model of immune dysregulation. Annu Rev Immunol 23:447–485
44. Sepúlveda N, Paulino CD, Carneiro J, Penha-Gonçalves C (2007) Allelic penetrance approach as a tool to model two-locus interaction in complex binary traits. Heredity 99:173–184
45. Sepúlveda N, Paulino C, Penha-Gonçalves C (2009) Bayesian analysis of allelic penetrance models for complex binary traits. Comput Stat Data Anal 53:1271–1283
46. Alper CA, Awdeh Z (2000) Incomplete penetrance of MHC susceptibility genes. Tissue Antigens 56:199–206
47. Alper CA, Husain Z, Larsen CE, Dubey DP, Stein R, Day C, Baker A, Beyan H, Hawa M, Ola TO, Leslie RD (2006) Incomplete penetrance of susceptibility genes for MHC-determined immunoglobulin deficiencies in monozygotic twins discordant for type diabetes. J Autoimmun 27:89–95
48. Hohler T, Hug R, Schneider PM, Krummenauer F, Gripenberg-Lerche C, Granfors K, Merker-Hermann E (1999) Ankylosing spondylitis in monozygotic twins: studies on immunological parameters. Ann Rheum Dis 58:435–440
49. Haegert DG, Galutira D, Murray TJ, O'Connor P, Gadag V (2003) Identical twins discordant for multiple sclerosis have a shift in their T-cell receptor repertoire. Clin Exp Immunol 134:532–537

50. Ferreira C, Singha Y, Furmanskia AL, Wong FS, Gardena OA, Dysona J (2009) Non-obese diabetic mice select a low-diversity repertoire of natural regulatory T cells. Proc Natl Acad Sci USA 106:8320–8325
51. Lathrop SK, Santacruz NA, Pham D, Luo J, Hsieh CS (2008) Antigen-specific peripheral shaping of the natural regulatory T cell population. J Exp Med 205:3105–3117
52. Magurran AE, Henderson PA (2003) Explaining the excess of rare species in natural species abundance distributions. Nature 422:714–716
53. Sepúlveda N, Paulino CD, Carneiro J (2008) Diversity of regulatory and effector T lymphocytes (in portuguese). In: Hill MM, Ferreira MA, Dias JG, Salgueiro MF, Carvalho H, Vicente P, Braumann CA (eds) Estatística: da Teoria à Prática. SPE editions, pp 513–523

Chapter 15
Mathematical Models of the Role of IL-2 in the Interactions Between Helper and Regulatory CD4^{+} T Cells

Kalet León and Karina García-Martínez

Abstract Mathematical models for the role of IL-2 in the dynamic interplay between CD4^{+} helper and regulatory T cells are studied. These models are extensions of the crossregulation model of CD4^{+} T cell dynamics including IL-2 molecules. The goal is to understand how the immune system is dynamically organized, structured by this interaction with self antigens, and how such organization might determine its overall function. We consider two model variants. In the first, regulatory T cells suppress helper T cells by competing with them for IL-2 in the lymph node. The second variant adds a direct inhibition of helper T cell activation which requires their co-localized activation on the APCs. We use the models to study the impact of treatments that either sequester or inject IL-2 in the immune system response. We show that treatment sequestering IL-2 could be used in particular conditions, both to render tolerant a preexistent immune/autoimmune system or to break a preexistent tolerant state, inducing an immune response. However, IL-2 injections will always reinforce the preexistent state, further expanding either the regulatory or helper T cells for a preexistent tolerant or immune state.

Interleukin-2 (IL-2) was the first T cell growth factor to be identified and molecularly cloned. Since its discovery, it was shown to promote T cell proliferation and survival in vitro [1], and to enhance T cell immunity in vivo in particular instance of viral infection [2] and vaccination [3–5]. However, this classical role of IL-2 in promoting T cell immunity has been recently challenged by experiments using the IL-2 or IL-2R KO mice [6], which have shown that in vivo IL-2 is an essential homeostatic growth factor for the CD4^{+}CD25^{+}FoxP3^{+} (regulatory) T cells, supporting an additional role for this cytokine in the maintenance of natural and induced tolerance.

CD4^{+}CD25^{-}FoxP3^{-} (helper) T cells have been identified as the principal source of IL-2 in vivo, while CD4^{+}CD25^{+}Foxp3^{+} regulatory T cells are unable to produce this cytokine [7], suggesting that they have to sequester the IL-2 produced by

K. León (✉)
Center of Molecular Immunology, Havana, Cuba
e-mail: kalet@cim.sld.cu

C. Molina-París and G. Lythe (eds.), *Mathematical Models and Immune Cell Biology*,
DOI 10.1007/978-1-4419-7725-0_15,

the helper cells in order to proliferate and survive. Moreover, in vitro and in vivo experiments have shown that regulatory T cells inhibit the production of IL-2 in the responder's helper cells [8–10], limiting also in this way their own source of this cytokine. Altogether, the latter facts reveal a complex dynamic relationship between IL-2 and CD4$^+$ T cells. On one hand, IL-2 promotes the proliferation of the helper T cells, which may drive effective immunity and foster IL-2 production. But, on the other hand, it promotes the expansion of regulatory T cells, which may turn off the immune reaction, as well as the IL-2 production on its own. The dynamic balance between these opposing forces might result in a complex response to IL-2 modulation treatments, such as those ones being investigated in clinical [11, 12] and preclinical [2, 13] studies.

In this chapter, mathematical models describing the regulatory role of IL-2 in CD4$^+$ T cell dynamics are studied. We review modelling of the dynamical aspects of interleukin-2 interaction with T cells. Then, recently developed mathematical models, which were designed to study the interplay of IL-2 with helper and regulatory CD4$^+$ T cells, are described. These models are used to compare different hypotheses regarding the specific role of IL-2 in the mechanism by which regulatory T cells suppress helper T cells, and to explore the effect of IL-2 modulating therapies on the dynamics. Overall, this chapter will illustrate, through these particular examples, how mathematical models can be used to analyze a specific biological problem, explaining known experimental facts and deriving predictions that might guide new experimentation.

Brief Review of Mathematical Models of IL-2 Dynamics

Several mathematical models have addressed molecular and cellular aspects of IL-2 interaction with its specific receptor at the T cell surface. IL-2 was one of the first biological molecules discovered to signal through interactions with a receptor composed of multiple protein chains, the α, the β and the γ chains [14]. Debate continues [15] on the detailed dynamics of IL-2 molecular interaction with these receptor chains. Some evidence suggests, for instance, the existence of a preassembled form of the α and β subunit of IL-2 receptor; while other evidence suggests that the latter dimer is dynamically formed upon IL-2 ligations. Mathematical models like the one developed by Goldstein et al. [16], call attention to the potential pitfalls of using classical Scatchard plot technique to study the binding properties of dynamically assembled receptors. Later models [17] directly analyzed experimental data of the IL-2/IL-2 receptor interaction.

Mathematical models from the group of Lauffenberger [18] have significantly contributed to elucidating the dynamics of IL-2 and IL-2 receptor internalization and degradation by T cells upon signalling. These models help to understand and quantify the dynamics and have practical implications, predicting the effect of particular IL-2 mutants, which, by altering the latter dynamics, exhibits super-agonist effects [19, 20].

Svirshchevskaya et al. [21] developed a mathematical model of IL-2 dependent proliferation for the CTLL2 T cell line. This model can be used to quantify IL-2 or IL-2-inhibitor activity in an in vitro proliferation assay. Borisova et al. [22] integrate different aspect of IL-2 and IL-2 receptor dynamics, in a model for the cell-cycle of T cells. This model was used to analyze experimental data [23], predicting the effect of different concentrations of IL-2 in the cultures, both for the expression of IL-2/IL-2 receptor complexes at T cell surface and the overall dynamics of cell cycle progression. Morel et al. [24] also developed a mathematical model of lymphocyte proliferation, including the effect of both IL-2 and IL-4 cytokines. They used the model to study synergistic and antagonistic effects of these cytokines, which might derive from the competition of these molecules at the cell surface for the common γ chain that integrate their receptors. A more recent model, developed by De Boer et al. [25], analyzes also experimental data of T cell proliferation in vitro in the presence of different levels of IL-2. However they use data from a more informative technique to score T cell proliferation. This technique uses a fluorescent dye, known as CFSE, to allow a dissection of the generational structure of proliferating T cell populations. The analysis of available data, with the model, suggests that IL-2 levels significantly influence T cell death rate in the culture, but as a function of the division history of individual T cells (i.e. the number of division cycles each T cell has undergone).

Published modelling results, either on the regulatory role of IL-2 in the dynamics of helper and regulatory CD4$^+$ T cells or on the biological impact of therapies based on modulating IL-2 activity, are rare. Burroughs et al. [26], proposed a model to study the dynamic interaction of helper and regulatory T cells, including IL-2 as a homeostatic factor shared by these T cell populations. This model predicts an interesting quorum threshold effect, which regulates helper T cell activation and is dynamically controlled by the presence of regulatory T cells in different situations. However, the latter model relies on some biologically unrealistic assumptions, which might limit its applications (it assumes the existence of an unknown homeostatic cytokine which is exclusively used by regulatory T cells in the absence of IL-2, contradicting existing experimental observations). Our group has recently developed a mathematical model for helper and regulatory CD4$^+$ T cell dynamics [29], accounting for many aspects of IL-2 dynamics. This model is an extension of the crossregulation model of immunity [27,28], used to investigate the dynamics of the interactions between regulatory and helper CD4$^+$ T cells. The following section in this chapter will review the mathematical formulation and properties of this model.

IL-2 in Helper and Regulatory CD4 T Cell Dynamics

The models introduced in this section are a natural extension of the crossregulation model [29]. They model the interaction between helper and regulatory T cell populations inside a lymph node, which contain a constant amount of their cognate antigen presenting cells (APCs) and homeostatic cytokines. The processes of IL-2 production, degradation and function, which are relevant to CD4$^+$ T cell dynamics, are explicitly included. Two model variants, with different mechanisms for the role of IL-2 in T cell mediated suppression, are studied. In the first model variant, regulatory T cells suppress helper T cell proliferation by effectively reducing IL-2 concentration, producing competition for this cytokine [8–10, 30, 31]. The regulatory T cells exploit, therefore, their natural over-expression of the high-affinity IL-2 receptor to dominate the system. The second model variant includes, in addition to competition for IL-2, that the regulatory T cells inhibit the activation of helper cells on the APCs, perhaps by temporally conditioning the APCs [32–34], as postulated in the original crossregulation model.

Postulates of the Models

The models rely on eight postulates which capture current knowledge regarding the relationship between IL-2 and CD4$^+$ T cells. Bellow we enunciate each postulate; provide a brief biological justification with relevant references, and explain how each postulate is implemented in the models.

1. ***Helper and regulatory cells go through at least three different functional states during their life cycle.***
   The cell cycle for T cells has been extensively studied [35]. It is well established that T cells require signalling through their T cell receptor (TCR) in order to leave the G0 phase and enter the G1 phase of cell cycle. However TCR signalling is not enough to proceed into the cell cycle. These cells are arrested at the G1 checkpoint, if insufficient levels of relevant cytokines are encountered. IL-2, IL-4, IL-7 and other γ chain family of cytokines are potent growth factors for T cells.

   The models here assume that both helper (E) and regulatory (R) T cells pass through the same three functional states during their life cycles. These functional states are:

   (a) Resting T cells which have not received signal through TCR recently (since its last division or its de novo generation)
   (b) Activated T cells which are in the functional state derived from the interaction of resting T cells and cognate APCs
   (c) Cycling T cells, irreversibly committed to cell division, derived from the activated T cells that receive enough cytokine-related signal

2. ***Helper cells produce IL-2, upon activation with APCs, being the major source for this cytokine in vivo.***
   In vitro studies have shown that CD4$^+$CD25$^-$ T cells, cultured with APC and anti-CD3, produce large amounts of IL-2 [36]. Moreover, in vivo experiments [37] have shown that CD4$^+$CD25$^-$FoxP3$^-$ (helper) T cells capable of producing IL-2 are mandatory to produce the IL-2, which sustain a functional

$CD4^+CD25^+FoxP3^+$ (regulatory) T cell population. In the models, only helper T cells produce IL-2. This IL-2 production is explicitly modelled here as a burst associated with the transition of helper T cells, from the resting state to the activated state (although qualitatively similar results are obtained in the model if IL-2 is assumed to be produced continuously by helper T cells while in the activated state). No other source of IL-2 is considered, since we are interested on the interplay between helper and regulatory T cells.

3. ***Helper T cells proliferate and survive, upon TCR ligation on APCs and signalling received from IL-2 or other homeostatic cytokines.***
   In vitro experiments have shown that IL-2 is a growth/survival factor for activated $CD4^+$ T cells [1]. However, normal T cells response can be obtained in vivo in IL-2 deficient mice [38,39], suggesting that helper T cells can use other cytokines to proliferate/survive in vivo additionally to IL-2 (perhaps IL-7 [40–42], IL-15 [42, 43] or IL-21 [44]). In the model, resting helper T cells are activated after conjugation with APCs, which provide MHC-peptide and other costimulatory signals. Once activated, these cells can proceed into the cell cycle if enough cytokine signal is received. This required signal comes, for this cell type, either from available IL-2 or from an alternative cytokine, referred as IL-α, which is present at a constant homeostatic level in the lymph node. The amount of IL-α is a control parameter of the model, whose value is assumed to be set by cellular subsets not included in the model.
4. ***Regulatory T cells do not produce IL-2, but their proliferation and survival is strictly dependent on both TCR ligation and IL-2-derived signalling.***
   Regulatory T cells do not express IL-2 mRNA in vitro [36] or in vivo [7], even after activation. In vitro, they proliferate when stimulated simultaneously with APCs and external IL-2. The required IL-2 mediated signal can be recovered by co-culturing regulatory cells with helper cells capable of producing IL-2 [36]. Moreover, in vivo experiments have shown that the absence of $CD4^+CD25^-FoxP3^-$ T cells capable of producing IL-2 leads to the absence of regulatory T cell population and furthermore to the development of autoimmunity [36]. Thus, in the model regulatory T cells do not produce IL-2. Like helper T cells, they are activated after conjugation with cognate APCs. However, to proceed into the cell cycle, they need a cytokine related signal provided exclusively by available IL-2.
5. ***Regulatory T cells inhibit IL-2 production by helper T cells upon their co-localized activation on the APCs. This interaction might also inhibit other processes on helper T cell proliferation.***
   Regulatory T cells have been shown to inhibit the proliferation of helper T cells both in vitro [8,10,45,46] and in vivo [6,47,48]. In vitro, this interaction requires cell to cell contact between Regulatory T cells, Helper T cells and APCs [8, 10, 46, 49]. This interaction reduces IL-2 production and mRNA expression by $CD4^+CD25^-$ T cells [8, 10]. However, regulatory T cells have been shown to control autoimmunity [6, 47] and to inhibit in vivo expansion of $CD4^+CD25^-$ T cells from IL-2 or IL-2R KO mice [48], suggesting that inhibition is more than just suppression of IL-2 production.

Accounting for the suppressive interaction exerted by regulatory T cells over the helper T cell is the core of the crossregulation model [28, 50]. In that model suppression was considered as dependent on multicellular conjugate formation. That is, helper T cells activation is assumed effective only when these cells are conjugated to APCs that are not simultaneously conjugated to regulatory T cells. However, as discussed in [27], such a modelling strategy accounts for many detailed mechanisms of suppression, which assume that suppression occurs upon the colocalized activation of helper and regulatory T cells. The two model variants in this chapter use the latter formalism to treat suppression. The first model variant assumes that suppression exerted by regulatory cells does not affect the rate of helper T cell activation (transitions from resting to activated state), but does affect the burst of IL-2 production associated with it. The second model variant assumes, however, that suppression does reduce the activation rate of helper T cells, also affecting IL-2 production.

6. ***IL-2 signal upon ligation to specific receptors which are differentially expressed on helper and regulatory T cells.***
In the model, a single class of IL-2 receptor is included, resembling the high affinity form of the receptor, which is the most relevant for $CD4^+$ T cell biology. The numbers of IL-2 receptors on helper and regulatory T cells in different activation states (resting, activated and cycling) are controlled by independent parameters, which may be chosen to fulfil desired biological constraints. Signalling through IL-2 receptors is taken to be equivalent to its binding to IL-2. The model equation computes the mean number of receptors bound per cell, for each cell type at a given time. The fraction of particular cells taking action, depending on the IL-2 derived signal, is computed with a sigmoid function (a Hill function) of the mean level of bound IL-2 receptors per cell. The sigmoid type of function was selected following in vitro studies [51], which have shown this type of dose response curve for activated T cell proliferation in response to external IL-2.
The signalling machineries of helper and regulatory T cells are assumed to be similar. Thus we use the same sigmoid type of function for these cells: i.e with a similar value of Hill coefficient h. Differences between IL-2 signalling on these cell types are assumed to rely mainly on their differential expression of IL-2 receptor and in the value of the control parameters S_E and S_R which set the sensitivity threshold of each specific cell type to the signal. In the case of helper T cells, IL-α is assumed to provide an additional signal, which is equivalent to the IL-2 mediated signal. Therefore, we modelled the presence of a constant level of this cytokine in the system as equivalent to the signalling through a fixed number of IL-2 receptors.
7. ***IL-2 is internalized and degraded upon ligation to its receptor.***
In vitro kinetic experiments have shown internalization of IL-2 by T cells [18]. A fraction of the internalized IL-2 is recycled back, bound to an α chain of the receptor, while another fraction is degraded along with the β and γ chain of the receptor [19, 52]. Mathematical models [18] and experiments have estimated rates of internalization and degradation of $0.04\,\text{min}^{-1}$ [19] and $0.035\,\text{min}^{-1}$ [53]. Consequently, the model includes, in the equations for IL-2 dynamics,

a degradation term proportional to the total number of IL-2 molecules bounded to the receptors. This term sums the net effect of internalizing and recycling IL-2 at the cell surface. Different cell types may have different IL-2 degradation rates.

8. ***Processes, such as diffusion inside the Lymph Node, IL-2 association and dissociation to receptors, or T cell conjugation and dissociation from APCs, are assumed to be quasi steady-state equilibrium.***
Characteristic free movement of T cells inside the lymph node has a typical observed velocity of 10 μm/min [54]. Thus if the diameter of a Lymph Node is around 1 mm, a lymphocyte can go through it in less than 2 h. IL-2 association and dissociation to its receptor is a process with a time scale of seconds, while T cell conjugation to APCs can last for around 1 h [54]. These processes are, in any case, faster than those cellular processes in the model, which alter T cell population sizes, for instance cell proliferation (8–12 h) [55–58] and cell life span (at least 7 days). Quasi steady-state assumptions are common in mathematical biology [59]. They rely on identifying fast and slow processes in the same dynamics. The approximation allow then to uncouple the fast and slow dynamics, essentially by following the slow dynamic with ordinary differential equations, where the variables of the fast dynamics are computed at equilibrium as function of the slow dynamical variables for each given time. Quasi steady state assumptions were already included in the crossregulation model [28], taking T cells movement and conjugation to APCs as the fast dynamics process. In the models here, this assumption is extended by assuming processes of IL-2 diffusion and its binding to IL-2 receptor also as fast processes.

Equations for the Dynamics of T Cell Population in the Lymph Node

The model equations for the dynamics of CD4$^+$ helper (E) and regulatory (R) T cells are:

$$\frac{\mathrm{d}E_N}{\mathrm{d}t} = \Gamma_E - K_A^E E_N^B \left(1 - \frac{R_T^B}{sA}\right)^{\eta(s-1)} + 2K_P^E E_C + \alpha K_S^E \frac{(S_E)^h}{(S_E)^h + \left(IL2_B^{E_A}\right)^h + (I l\alpha)^h} E_A - K_d^E E_N^F \quad (15.1)$$

$$\frac{\mathrm{d}E_A}{\mathrm{d}t} = K_A^E E_N^B \left(1 - \frac{R_T^B}{sA}\right)^{\eta(s-1)} - K_S^E E_A \quad (15.2)$$

$$\frac{\mathrm{d}E_C}{\mathrm{d}t} = K_S^E \frac{\left(IL2_B^{E_A}\right)^h + (I l\alpha)^h}{(S_E)^h + \left(IL2_B^{E_A}\right)^h + (I l\alpha)^h} E_A - K_P^E E_C \quad (15.3)$$

$$\frac{\mathrm{d}R_N}{\mathrm{d}t} = \Gamma_R - K_A^R R_N^B + 2K_p^R R_C + \alpha K_S^R \frac{(S_R)^h}{(S_R)^h + \left(IL2_B^{R_A}\right)^h} R_A - K_d^R R_N^F \tag{15.4}$$

$$\frac{\mathrm{d}R_A}{\mathrm{d}t} = K_A^R R_N^B - K_S^R R_A \tag{15.5}$$

$$\frac{\mathrm{d}R_C}{\mathrm{d}t} = K_S^R \frac{\left(IL2_B^{R_A}\right)^h}{(S_R)^h + \left(IL2_B^{R_A}\right)^h} R_A - K_P^R R_C. \tag{15.6}$$

Variables and parameters are defined in Tables 15.1 and 15.2. Note that, two model variants are encoded in these equations, being its realization dependent of the value of switching parameter η.

Equations (15.1)–(15.6) represent the following processes and interactions which implement models postulates (see Fig. 15.1):

(a) Resting *E* and *R* cells are produced by an external source term, the thymus [first term of (15.1) and (15.4)]; They die at a constant rate [last term of (15.1) and

Table 15.1 Description of model variables and intermediate quantities

Item	Description
Independent variables	
E_N, E_A, E_C	Total number of resting, activated and cycling E Cells
R_N, R_A, R_C	Total number of resting, activated and cycling R cells
IL-2	Total number of IL-2 molecules in the Lymph Node
Intermediate quantities	
E_N^B, E_A^B, E_C^B	Total number of resting, activated and cycling E Cells conjugated to APCs
E_T^B	$= E_N^B + E_A^B + E_C^B$
E_N^F, E_A^F, E_C^F	Number of resting, activated and cycling E cells not conjugated to the APCs (note that, free cells = total cells-conjugated cells)
R_N^B, R_A^B, R_C^B	Total number of resting, activated and cycling R Cells conjugated to APCs
R_T^B	$= R_N^B + R_A^B + R_C^B$
R_N^F, R_A^F, R_C^F	Number of resting, activated and cycling R cells not conjugated to the APCs (note that, free cells = total cells-conjugated cells)
$IL2_B^{E_N}, IL2_B^{E_A}, IL2_B^{E_C}$	Mean number of IL-2 molecules bound per cell, respectively for the resting, activated or cycling E cells
$IL2_B^{R_N}, IL2_B^{R_A}, IL2_B^{R_C}$	Mean number of IL-2 molecules bound per cell, respectively for the resting, activated or cycling R cells
$IL2_B^T$	Total IL-2 bound to receptors ($= IL2_B^{E_N} + IL2_B^{E_A} + IL2_B^{E_C} + IL2_B^{R_N} + IL2_B^{R_A} + IL2_B^{R_C}$)
$IL2_F$	Number of IL-2 molecules not conjugated to the receptors ($= IL2 - IL2_B^T$)

Table 15.2 Description of parameters

Parameters	Description	Value
Γ_E, Γ_R	Influx rate of new resting E and R cells from the thymus	$10^{-2}, 10^{-2}$
A, s	Number of antigen presenting cells (APCs) and conjugations sites for T cells on them	$10^5, 6$
K_A^E, K_A^R	Activation rate for resting E and R cells conjugated to APCs	$0.69\,h^{-1}, 0.09\,h^{-1}$
K_P^E, K_P^R	Division rate for cycling E and R cells	$0.17\,h^{-1}$
K_S^E, K_S^R	Deactivation rate for activated E and R cells	$0.69\,h^{-1}$
K_d^E, K_d^R	Death rate for free resting E and R cells	$0.004\,h^{-1}$
α	Fraction of activated E and R cells that return to the resting state without receiving successfully the IL-2 related signal	0.95
Ilα	Level of homeostatic cytokine, additional to IL-2, that the activated E cells, but not the R cells, can use to become cycling cells	250
S_E, S_R	Sensitivities thresholds for E and R cells to cytokines signal	500
η	Switch parameter that determines the existence ($\eta = 1$) or not ($\eta = 0$) of direct regulation by the R cells of the activation of E cells	0, 1
Γ_i	External influx of IL-2, typically used to simulate treatments	0.002
K_P^i	Rate of IL-2 production by helper cells upon activation	$45.38\,h^{-1}$
K_d^i	Degradation rate of IL-2 not conjugated to its receptor	$0.138\,h^{-1}$
K_{in}^E, K_{in}^R	Internalization/degradation rate of IL-2 by E and R cells	$610^{-10} min^{-1}$
$RI^{E_N}, RI^{E_A}, RI^{E_C}$	Number of IL-2 receptors in each resting, activated or cycling E cells	$0, 10^3, 10^3$
$RI^{R_N}, RI^{R_A}, RI^{R_C}$	Number of IL-2 receptors in each resting, activated or cycling R cells	$10^3, 10^4, 10^3$

(15.4)]; and they become activated following conjugation to APCs [second term in (15.1) and (15.4), and their complementary terms in (15.2) and (15.5)]. The activation process can be inhibited for a helper T cell (model case $\eta = 1$) if a regulatory T cell is conjugated on the same APC. Note the factor inversely proportional to the total number of R cells bound to APCs (R_T^B) which is included on the activation terms for the helper cells. The mathematical form of this factor follows the formalism developed in [60].

(b) The activated E and R cells require enough cytokine-derived signals to become cycling cells [first terms on (15.3) and (15.6)]. The activated R cells receive this signal only from IL-2, while the E cells could also use an unknown homeostatic cytokines referred as IL-α. In the absence of enough cytokine derived signal a fraction α of the activated cells revert to the resting state [see the fourth terms

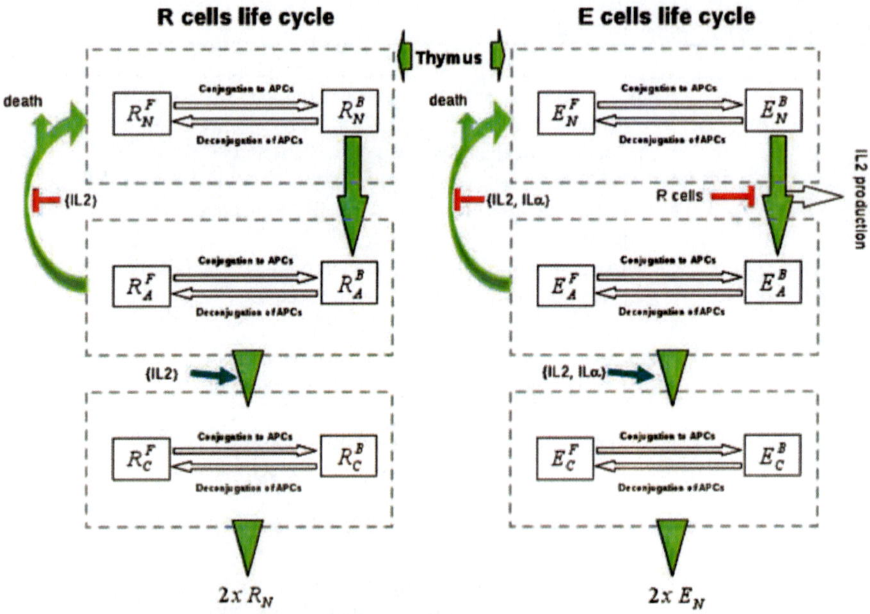

Fig. 15.1 Diagram of helper (*E*) and regulatory (*R*) T cells life cycle in the model. New resting *E* (E_N) and *R* (R_N) cells are constantly generated by the thymus. These resting T cells are activated by interaction with their cognate APCs (upon conjugation to APCs). During activation the *E* cells produce IL-2, although the whole process can be inhibited by the presence of co-localized *R* cells. Activated *E* (E_A) and *R* (R_A) enter cell cycle (becoming cycling cells) when receiving enough signal from IL-2, or another external cytokine (IL-α) in the case of *E* cells. In the absence of enough cytokines, activated T cells become deactivated, a fraction α return back to the resting state an the fraction (1-α) simply die. Cycling *E* (E_C) and *R* (R_C) cells divide at a constant rate generating two new resting *E* or *R* cell respectively. The process related to T cell conjugation and deconjugation to APCs (processes inside *dashed squares*) are assumed to be fast and are modelled in quasi-steady state equilibrium

of (15.1) and (15.4)] and the remaining fraction (1-α) simply die. The fraction of activated T cells getting enough cytokine derived signal to proceed with the cell cycle at any given time is computed with a sigmoid function of the mean number of bound IL-2 receptors per activated cell. This sigmoid function contains parameters S_E and S_R to set different sensitivities thresholds, respectively for *E* and *R* cells, to the delivered signal. The cycling *E* and *R* cells divide producing two new resting cells with constant rate [last terms of (15.3) and (15.6) and third terms in (15.1) and (15.4)].

The numbers of helper and regulatory T cells conjugated to APCs (E_N^B, E_A^B, E_C^B, R_N^B, R_A^B, R_C^B) are computed in quasi-steady state equilibrium following postulate 8. Briefly [28] equations for the conjugation and deconjugation of T cells to APCs (see Fig. 15.1) are formulated, assuming these individual processes obey first-order kinetics events with characteristic association and dissociation rates. Then, in virtue of the fast time scale of these processes, the equations are taken to equilibrium, for a fixed number of T cells of each given type at time t, obtaining the following set of equations:

$$\frac{K^E}{V_{LN}} = \frac{E_l^B}{E_l^F F} \qquad \frac{K^R}{V_{LN}} = \frac{R_l^B}{R_l^F F} \qquad F = sA - \sum_l E_l^B - \sum_l R_l^B,$$

where the label l takes values denoting the functional states of the T cells: $l = N$ resting state, $l = A$ activated state and $l = C$ cycling state; K^E, K^R are the equilibrium conjugation constant to APC sites for helper and regulatory T cells ($K^E = K^R = 10^{-9} L \cdot \text{cell}^{-1}$); V_{LN} is the volume of the lymph node. This set of algebraic equations is solved numerically to obtain the number of conjugated T cells of each type as a function of model parameters and independent variables.

Model Equations for the Dynamics of IL-2 in the Lymph Node

The dynamics for IL-2 molecules is given by the following equation:

$$\begin{aligned}\frac{\mathrm{d}IL2}{\mathrm{d}t} &= \Gamma_i + K_p^i E_N^B \left(1 - \frac{R_T^B}{sA}\right)^{(s-1)} - K_d^i IL2 \\ &- K_{in}^E \left(IL2_B^{E_A} E_A + IL2_B^{E_C} E_C\right) \\ &- K_{in}^R \left(IL2_B^{R_N} R_N + IL2_B^{R_A} R_A + IL2_B^{R_C} R_C\right);\end{aligned}$$

Variables and parameters are defined in Tables 15.1 and 15.2. This equation considers that (see Fig. 15.2):

1. IL-2 is produced by helper T cells upon activation (second term). This IL-2 production can be inhibited by the co-localized activation of regulatory T cells (note: the part of this term, which is inversely proportional to the total number of regulatory cells bound to APCs in the system, R_T^B);
2. That IL-2 is degraded spontaneously most likely because of unspecific consumption by non-T cells or because of migration out of the lymph node and renal elimination (third term in the equation);
3. IL-2 is also degraded by E and R cells, after being internalized in the form of complexes with its receptor (fourth and fifth terms).

The mean number of IL-2 molecules bound to the receptors per cell in each cell type is computed in quasi steady-state equilibrium. For this, the parameter RI^{X_l} is defined as the number of IL-2 receptors a cell of type X in the state l has on

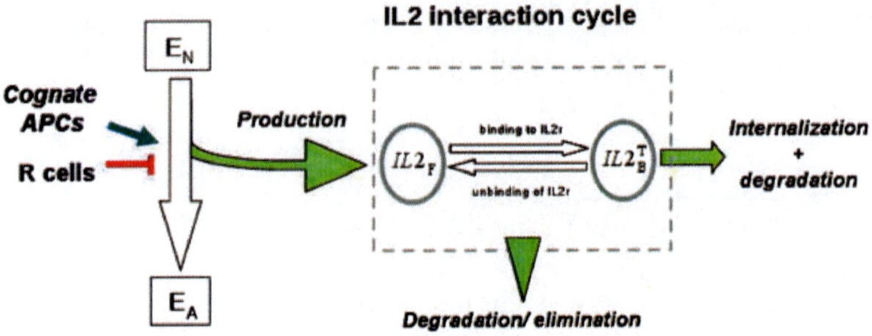

Fig. 15.2 IL-2 interaction cycle in the model. IL-2 is produced as a burst upon *E* cells activation on cognate APCs. This production of IL-2 can be inhibited by the presence of co-localized *R* cells. IL-2 is conjugated to the receptor (IL2r) at the *E* or *R* cell's surface, being subsequently internalized and degraded. IL-2 is further degraded in model because of renal elimination or because of internalization/degradations by non-T cells. The processes of IL-2 association and dissociation from IL-2 receptors (processes inside *dashed squares*) are assumed to be fast and are modelled in quasi-steady state equilibrium

its surface. Then the following equations are written down, for the formation and dissociation of IL-2/IL-2-receptor complexes:

$$\frac{dIL2_F}{dt} = -k_{on}^I \frac{IL2_F\left(RI^T - IL2_B^T\right)}{(V_{LN})^2} + k_{off}^I \frac{IL2_B^T}{V_{LN}}$$

$$\frac{dIL2_B^T}{dt} = k_{on}^I \frac{IL2_F\left(RI^T - IL2_B^T\right)}{(V_{LN})^2} - k_{off}^I \frac{IL2_B^T}{V_{LN}}$$

and

$$RI^T = RI^{E_A}E_A + RI^{E_C}E_C + RI^{R_N}R_N + RI^{R_A}R_A + RI^{R_C}R_C.$$

Where, imposing the condition of equilibrium, the total amount of bound IL-2 ($IL2_B^T$) is computed as:

$$IL2_B^T = \frac{K^I(IL2 + RI^T) + V_{LN}}{} \times \frac{-\sqrt{\left(K^I(IL2 + RI^T) + V_{LN}\right)^2 - 4(K^I)^2 IL2 RI^T}}{2K^I},$$

where $K^I = k_{on}^I / k_{off}^I$ is the affinity of IL-2 for its receptor ($K^I = 10^{11} M^{-1}$) and V_{LN} is the volume of the lymph node. Finally, the mean number of IL-2 receptors bound to IL-2 per cell for each cell type is obtained as:

$$IL2_B^{E_l} = \frac{RI^{E_l}}{RI^T} IL2_B^T \qquad IL2_B^{R_l} = \frac{RI^{R_l}}{RI^T} IL2_B^T \qquad \forall l \in \{N, A, C\}.$$

Parameter Values Used in Simulations

The analyses performed in this chapter are mostly qualitative and therefore describe properties of the models which do not rely on particular parameter values. However in order to minimally test the realism of our model behaviour, and to provide figures with values for the different variables with have a clear biological meaning, we have used the parameter set provided in Table 15.2. This particular set of parameter values was estimated using the experimental data and biologically reasonable assumptions as explained on [29]. All simulations were performed using software Mathematica 4.1 from Wolfram Research Inc.

Dynamical Properties of the Models

The models in this chapter are primarily used to study the dynamics of polyclonal CD4$^+$CD25$^+$FoxP3$^+$ (Regulatory) and CD4$^+$CD25$^-$FoxP3$^-$ (Helper) T cells populations which interact, in the lymph node, where multiple self-antigens are presented to the T cells, by a constant population of antigen presenting cells (APCs). They study the dynamics of the immune system in the absence of invading pathogens, when the CD4$^+$ T cells interact with normal body components (self-antigens). The goal is to understand how the immune system is dynamically organized, structured by this interaction with self antigens, and how such organization might determine its overall function. We focus on CD4$^+$ T cell dynamics assuming, on the one hand, that the expansion of helper CD4 T cells implies an increase of immune effectors functions, which are mediated by the CD4 T cells themselves as well as B cells and CD8 T cells that cooperate with them. The expansion of regulatory T cells, on the other hand, is assumed to induce tolerance with no expansion of immune effectors.

In the following subsection, the main dynamical properties of the formulated models are analyzed. The biological implications, of these properties, are discussed. In particular, we pinpoint, how these models conserve the most relevant dynamical properties of the crossregulation model [28], further gaining new interesting features, which derives from the explicit introduction of IL-2 dynamics.

Steady State Analysis: Searching for Bi-Stability

A key dynamical property of the models is the existence of a parameter regime of bi-stable behaviour, with two stable steady states of clear biological interpretation (Fig. 15.3). That is:

- a steady state, which is interpreted as an immune/autoimmune state (Fig. 15.3a), label *IS*), because the helper T cells (E) are substantially expanded out-competing the regulatory T cells (R) from the system; and

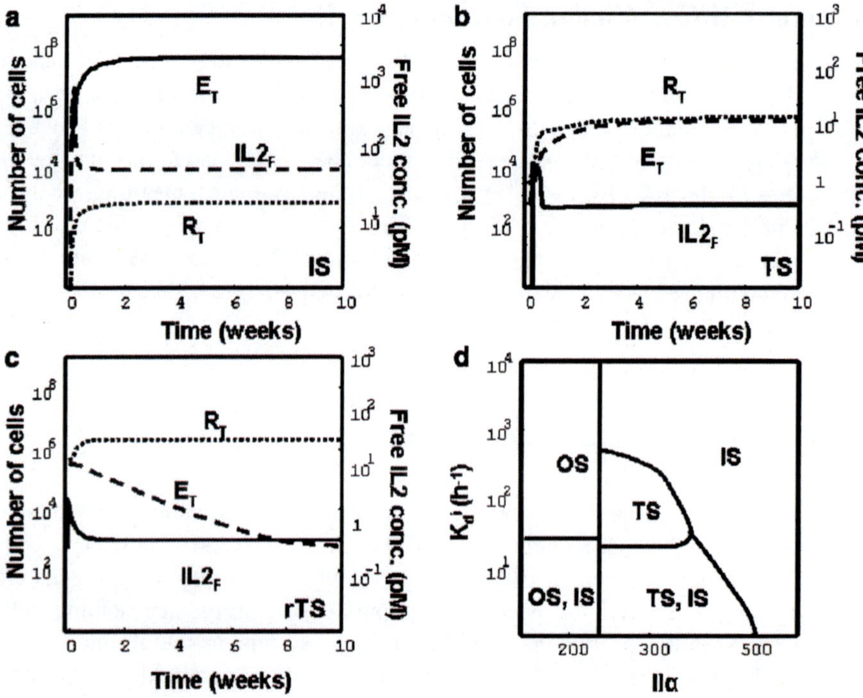

Fig. 15.3 Steady states obtained in numerical simulations of the model and their dependence on nondimensional parameters. The graphs (**a**)–(**c**) show examples of kinetic evolution of the total number of effectors T cells ($E_T = E_N + E_A + E_C$), the total number of regulatory T cells ($R_T = R_N + R_A + R_C$) and the free IL-2 concentration ($IL2_F$), where the system evolves into: (**a**) the immune/autoimmune steady state (*IS*), (**b**) the tolerant steady state (*TS*), and (**c**) the reinforced tolerant steady state (*rTS*). The graph in (**d**) show five regions delimiting parameter values of K_d^i and IL-α, at which the system can evolve into the indicated types of steady states, depending on the initial condition chosen in the simulation. The most relevant parameter region is the bistable region (*lower right region*) where both the tolerant (*TS*) and the immune (*IS*) states coexist in the system. The particular simulations in (**a**)–(**c**) used values of K_d^i and IL-α inside this bistable region ($K_d^i = 0.138\ h^{-1}$, IL-$\alpha = 250$), but differ in the initial number of resting regulatory T cells and the size of the external source of IL-2 ((**a**) $R_N(0) = 50, \Gamma_i = 10^{-6}$; (**b**) $R_N(0) = 2500, \Gamma_i = 10^{-6}$; (**c**) $R_N(0) = 2500, \Gamma_i = 10^5$). Other parameter values were taken from Table 15.2. The remaining initial conditions were taken as $E_N(0) = 500, E_A(0) = 0, E_C(0) = 0, R_A(0) = 0, R_C(0) = 0, IL2(0) = 80pM$

– a second steady state, which is interpreted as tolerance (Fig. 15.3b), label *TS*), because the helper and regulatory T cells coexist at low amounts, being the expansion of the helper T cells actively controlled by the interaction with the regulatory cells.

The existence of such a bistable parameter regime is one of the properties inherited from the original crossregulation model [27, 28]. Two other types of stable steady states are possible in the models, but in other parameter regimes. These states are: a trivial steady state (label *OS*), where both helper and regulatory population collapse to very small values, which are proportional to the size of the external source term of new resting T cells (the size of thymic output); and a steady state interpreted as reinforced tolerance (Fig. 15.3c), label *rTS*), because the helper T cells collapse to very small values dictated by thymic output, while the regulatory T cells are expanded and sustained in the system by the influence of an external source of IL-2.

The existence of the bistable regime of parameters referred above is essential, in these models, to properly explain the phenomenology of dominant tolerance, particularly the results of in vivo adoptive transfer experiments. In these experiments [61] the adoptive transference of different proportions of helper and regulatory T cells, extracted from the lymph node of normal healthy mice, into immune deficient mice (those primarily lacking T cells, i.e. Rag$^{-/-}$ or Nu$^{-/-}$), results either in the reconstitution of a normal immune system (tolerant to the self-body components), containing both helper and regulatory T cells, or in the development of autoimmune diseases which are mediated by the uncontrolled expansion of the transferred helper T cells and the absence of the counteracting regulatory populations. Note that to simulate adoptive transfer experiment in the models, one will have to start the system with no T cells (neither E nor R), but with a fixed population of APCs. Then different proportions of the E and R populations have to be introduced and the outcome of the system has to be simulated numerically. In other to obtain as results that the system could either go to the autoimmune steady state or to the tolerant steady state, such states have to exist and to be stable for the same set of parameter values. That is the system have to be bistable.

Parameter Constraints for Biologically Reasonable Model Behaviour

Only those parameter values leading to the bistable behaviour (lower right region in Fig. 15.3d), will be compatible with the results of adoptive transfer experiments. This argument can be then used to derive parameter constraints with interesting biological meaning. Three of these parameter constraints, analytically derived in [29] are:

$$\frac{S_E}{RI^{E_A}} > \frac{S_R}{RI^{R_A}} \qquad Il\alpha > 0 \qquad \text{and} \qquad K^E\frac{K_A^E}{K_d^E} > K^R\frac{K_A^R}{K_d^R}.$$

The first of these parameter constraints means that regulatory T cells have to be more efficient using IL-2 at low concentrations than helper T cells. This can be either because the regulatory cells have a larger sensitivity to IL-2 signal (low value of S_R) or a higher number of IL-2 receptors (large RI^{R_A}). This constraint is strictly required in the first model variant ($\eta = 0$), but it is dispensable in the second one ($\eta = 1$). However, current experimental data would most likely support its validity. The main argument favoring that indeed regulatory T cells use IL-2 more efficiently than helper T cells, is that they naturally over express the alpha chain of the high affinity receptor for this cytokine [62]. Interestingly, upon activation the helper cells also up regulate this alpha chain of the receptor, but always to a lessser extent than their activated regulatory counterparts [62]. Moreover, in vitro studies have shown that regulatory T cells can capture, internalizes and degrades IL-2 more efficiently than helper T cells [63], supporting that indeed this over expression of the receptor can result in a more efficient capture of the IL-2.

The second parameter constraint expresses the need for a cytokine, alternative to IL-2, that can promote helper T cell proliferation and survival. This constraint is strictly required for both the first model variant ($\eta = 0$) and the second model variant ($\eta = 1$), if the regulatory T cells are assumed equal or more sensitive to IL-2 than the helper T cells. Experimental support for the existence of such alternative cytokines comes, indirectly, from the observation both in vivo (IL-$2^{-/-}$ [38, 39]; IL-2R$\alpha^{-/-}$ [64]; IL-2R$\beta^{-/-}$ [65]) and in vitro ([66–69]) of significant proliferation and survival of IL-$2^{-/-}$ or IL-2R$\alpha^{-/-}$ or IL-2R$\beta^{-/-}$ helper CD4$^+$ T cells. This alternative cytokine has been proposed to belong to the γ chain family of cytokine [70], with candidates including IL-7, IL-15 and IL-21. In the models, however, the absence of this cytokine prevents the existence of a tolerant equilibrium (lower left region Fig. 15.3d). That is, for low values of IL-α only the immune or the trivial immune state could exist, predicting a severe effect of knocking out this cytokine in the peripheral CD4 T cell compartment. Actually the KO animal is expected to be either autoimmune (falling into the autoimmune steady state) or to have few T cells in the peripheral lymph nodes (falling into the trivial immune state). Of the candidate cytokines, only IL-7 has been shown to have an important role in peripheral CD4$^+$ T cell dynamics. IL-7 knockout mice have no peripheral T cells [41, 42], and moreover prevent the homeostatic proliferation and accumulations of freshly-transferred naïve CD4$^+$ T cells [42]. Therefore, the models strongly suggest IL-7 as the likely γ chain family cytokine that supports the proliferation and survival of peripheral helper CD4$^+$ T cells in conditions of IL-2 limitation. Other γ chain related cytokines might contribute, as well, to helper T cell proliferation, but perhaps in more specific situations or in more specific CD4$^+$ T cells subsets.

The third and last parameter constraint means, on biological grounds, that helper T cells must expand faster than their regulatory counterpart on conditions of IL-2 excess, either because helper T cells have a faster activation rate (large K_A^E); or a larger capacity to E conjugate to the APCs (large K_E), or a larger live spans (low K_d^E). This parameter condition is strictly required by the first model variant ($\eta = 0$) and is most likely required in the second model variant ($\eta = 1$), if the regulatory T cells are assumed equal or more sensitive to IL-2 than the helper T cells. To our knowledge there are no direct experimental observations supporting or invalidating this parameter constraint. An indirect observation suggesting, perhaps, that helper T cells might indeed expands faster in vivo than regulatory T cells is that, in typical adoptive transfer experiments, regulatory and helper T cells are mixed in 50 : 50 ratio to recover a final tolerant animal (an autoimmune disease free animal). The ratio of this mixture is heavily biased in favor of regulatory T cells, compared to the 10 : 90 ratio (R : E) in the normal or reconstituted tolerant mice. A simple explanation for the necessity of such a bias is that one has to give advantage to the regulatory T cells to compensate the larger expansion capacity of the helper T cells if autoimmunity is to be prevented. Nevertheless, a simple direct experiment can be designed to test some aspects of this constraint. For instance, the kinetics of proliferation of mouse $CD4^{+}CD25^{-}FoxP3^{-}$ and $CD4^{+}CD25^{+}FoxP3^{+}$ T cells can be compared, stimulating them in vitro with anti-$CD3^{+}$APCs and supplementing different concentrations of human IL-2, while blocking endogenous mouse IL-2 with commercial mAbs S4B6.

Model Dependence on the Level of Cognate APC Stimulation

Other key properties of the models in this chapter, inherited from the crossregulation model [27], rely on their peculiar dependencies on the level of cognate APC stimulation in the system, via the number of cognate APC in the lymph node, model parameter A.

The first important component on this dependence is that the bifurcation diagram of model steady states, with the value of parameter A, shows three qualitatively different regions (Fig. 15.4a):

- There is an initial region of low values of parameter A (region I), where neither the immune nor the tolerant state can exist
- There is an intermediate range of values of A (region II) where the immune but not the tolerant state can exist
- There is a final range for higher number of APCs (region III) where the system is bistable and both the tolerant and immune state can exist

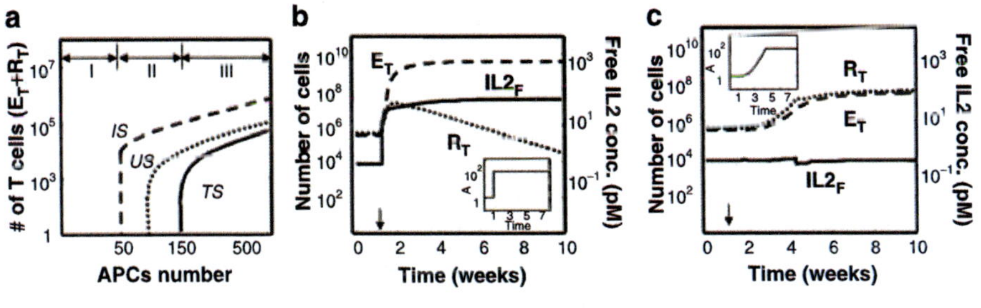

Fig. 15.4 Dependence on APC numbers. The graph (**a**) shows bifurcations diagram of model steady states with the number of APCs in the system. *Dashed line* indicates the total number of T cells ($E_T + R_T$) in the immune state (*IS*); the *solid line* indicate the total T cell number in the stable tolerant state *TS*; and the *dotted line* give the total T cell number in an unstable saddle node state *US*. The regions I, II and II indicated in the graph exhibit different model dynamics as discussed in the main text. The graphs (**b**) and (**c**) show kinetic evolution of the total number of effectors (E_T) and regulatory (R_T) cells and the free IL-2 concentration ($IL2_F$), when starting in the tolerant steady state reached in Fig. 15.3b. The number of APCs is increased 100 times. (**b**) Illustrate the effect of an abrupt increase that break the pre-existent tolerance and take the system into an immune steady state. (**c**) Illustrate the effect of a slow increase where T cells are expanded but tolerance is always preserved. *Vertical arrows* indicate the starting-point of the APCs increase in the simulation and the inserted graphs show the particular kinetics of the APCs number

Such bifurcation structure is critical to argue an impact of these models [27] in dynamically partitioning T cell repertoires based on the interaction with different self-antigens, which might reach different levels of presentations by the APCs. Briefly, the models, like the crossregulation model, proposes that the composition of peripheral T cell repertoire, in terms of the proportions of regulatory and helper T cells, for each particular T cell clone (TCR specificities, recognizing the same antigens), is determined both by the initial proportion of regulatory to helper T cells from the thymus and the peripheral selection mediated by their interactions with cognate APCs (those ones presenting self-body components). Only for those T cell clones recognizing a sufficiently large number of cognate APCs, a tolerant steady state can be attained in the periphery, sustaining a large number of regulatory T cells. Therefore the peripheral T cell repertoire will be dynamically partitioned [27] by these interactions, into at least two subsets of clones: Those T cell clones recognizing high level of self-antigen in the APCs, which will contain many regulatory T cells and therefore enforce tolerance to self-body-components, and those clones recognizing lower levels of self-antigens in the APCs, which will be enriched of helper T cells and could them be used for mounting effective immune reactions to newly incoming antigens (those from pathogens).

The second important aspect of the dependence of the models with parameter A, is that, in these models, an abrupt (Fig. 15.4b), but not a slow (Fig. 15.4c), increase in the number of APCs, in simulations which have being parameterized inside the bistable parameter regime and have been initialized in the tolerant steady state, can drive the system into the immune steady state. This property is relevant to argue [27, 71] a cooperative dynamic interplay between the innate and the adaptive immune system components in the T cell response to incoming pathogens or well-adjuvated vaccines. Briefly, innate immune system is composed of several subsets of immune cells (non T cells) that have receptors detecting specific pathogens components and triggering the fast maturation of precursor of the antigen presenting cell, which process and present the specific antigens of the pathogens to the T cells. Therefore, in the models, as in the crossregulation model, the fast increase in the number of APCs induced by the activation of innate immune system (inflammation) facilitate the induction of the immune response by helper T cells, since it weaken the activity of any possible pre-existent regulatory T cells. Interestingly, this property of the models, also explains a phenomena known as adjuvant induced autoimmunity [71], where in some animal models the injection of adjuvant or the infection with some specific pathogens, which contain many substance able to activate innate immune system, causes the induction of autoimmune disorders as arthritis. In contrast, other natural variations in the level of antigen presentation (driving slow variations on APCs numbers), like those induced by developmental chains in self component concentrations (i.e. the expression of some hormones in the puberty), which do not involve inflammation and are gradual (slow kinetics) will not facilitate immunity.

Local Versus Global Regulation

Perhaps the most relevant theoretical aspect of the models presented in this chapter is that the introduction of IL-2 in T cell dynamics breaks the local/specific character of interaction in the original crossregulation model. In the cross regulation model all the interactions required co-localizations of helper T cells, regulatory T cells and APCs. The locality of these interactions implies at least partial specificity, because T cell clones recognizing APCs that present different sets of antigens will never interfere with each other. However in the models analyzed here, since IL-2 is shared inside the lymph node, different T cell clones in the same lymph node will use it in common and therefore will interfere with each other's dynamics. To study the implication of the mixture of local/specific and non-local/unspecific interactions in the models, this section explores the situation of two set of APCs ($A1$ and $A2$) recognized by two different clones containing both helper and regulatory T cells ($E1$, $R1$ and $E2$, $R2$) which share the same lymph node. That is sharing the same pool of IL-2 in the lymph node. In particular, we simulate a system were $R1$ and $R2$ cells are set to initially control the $E1$ and $E2$ populations (in their respective tolerant states), but the number of APCs of type 1 ($A1$) is abruptly increased. We analyze them, how responses induced by a perturbation directed towards antigens on APCs of type 1 might influence the response to antigens on APCs of types 2.

In the first model variant, where regulation is mediated only by competition for IL-2 (case $\eta = 0$), the response to APCs of type 1 and of type 2 are fully coupled to each other. Therefore it is impossible to break tolerance expanding an immune response to APCs of type 1, without breaking tolerance in response to APCs of type 2. An illustration of such coupling is given on Fig. 15.5a, where an abrupt increase in $A1$ values is shown to induce the expansion of both $E1$ and $E2$ cells, with a consequent elimination of both $R1$ and $R2$ cells. The absolute character of this dynamical coupling was formally proved on [29]. On biological grounds this dynamical coupling will mean that all clones inside a given lymph node will have to be in the same dynamical state. If one particular clone expands the helper T cells, reaching the immune state, then all the other clones in the same lymph node will also have to reach the immune state. Moreover, to keep one clone in the tolerant state, dominated by regulatory T cells, every clone in the same lymph node will have to be in the same state. Therefore, this sort of global coupling of the response of different clonal specificities is obviously unrealistic, since tolerance and immunity do coexist toward different antigens in the real immune system. An argument to overcome this problem is to claim further locality on the use of IL-2 inside the lymph nodes. Cytokines does not diffuse instantaneously and therefore might be preferentially used close to its initial production site. However, it is difficult to defend such argument, in this model variant, because competing for the paracrine IL-2 is precisely what the regulatory T cells inside a given clone do to control their helper T cell counterpart. Thus unless one find a way to argue a sorting in small local clusters inside the lymph node of the APCs presenting different sets of antigens, the latter argument would be unrealistic.

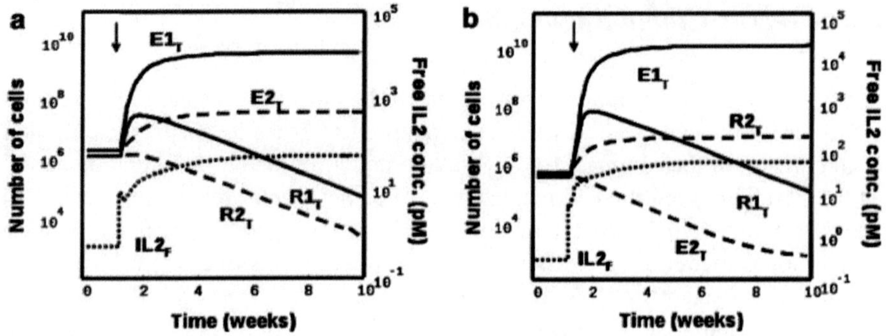

Fig. 15.5 Simulations of lymph nodes containing two sets of APCs (*A1* and *A2*) recognized by two different T cell clones ($\{E1, R1\}$ and $\{E2, R2\}$). Figures shows the characteristic kinetic outcomes obtained from simulations were *R1* and *R2* cells are set to initially control the *E1* and *E2* cell populations (i.e. in their respective tolerant states), and the number of APCs of type 1 (*A1*) is abruptly increased. (**a**) Result of simulation for model variant $\eta = 0$, where the breakdown of tolerance for clone 1 is always followed by a breakdown of tolerance for clone 2. (**b**) Result of simulation for model variant $\eta = 1$, where tolerance can be broken for clone 1 keeping, and even reinforcing, tolerance in clone 2. In these particular simulations the number of APCs of type 1 increase from it basal value 5.10^4 to 5.10^6 (**a**) and 10^7 (**b**), while the number of APCs of type 2 is maintained constant in 5.10^4. In both graphs, the kinetics of free IL-2 is indicated by a *dotted curve*, and the number of T cells of type 1 and 2 with *continuous and dashed curves* respectively. *Vertical arrow* inserted in (**a**) and (**b**) indicates the starting-point of the increase of APCs of type 1

In the second model variant, which include, additionally to competition for IL-2, a direct regulation of helper T cell activation by regulatory T cells (case $\eta = 1$), the response to APCs of type 1 and of type 2 also influence each other. However in this case one can easily get mixed responses, such as the one illustrated in Fig. 15.5b, where an abrupt increase of $A1$ values is shown to induce the expansion of $E1$ cells with a consequence elimination of $R1$ cells, but retaining an expanded dominant population of $R2$ cells. Interestingly, as can be also observed in Fig. 15.5b, the expansion of $E1$ cells reinforce the tolerance to antigens in APCs of type 2, since it leads to a net increase of $R2$ cells, with a concomitant reduction of the $E2$ cells (this second T cell clone falls into a reinforced tolerant state such as the one showed in Fig. 15.3c). Therefore, on biological grounds, for this second model variant T cell clones inside the same lymph node interfere with each others, but mixed immune and tolerant responses can be achieved. This property solves the problem posed to the first model variant, suggesting the requirement of some local/specific suppressive interaction between helper and regulatory T cells for a biologically reasonable model behaviour. Moreover, in this model variant the existence of an expanded immune response of a given clone could reinforce the tolerant state of other T cell clones in the same lymph node. This property does not exist in the original crossregulation model and it is a direct consequence of the mixture of local/specific suppressive interactions with non-local/unspecific interaction in this model variant. This property is, in our view, biologically relevant since it might help to focus an immune reaction in the immune system to some undesirable foreign antigen, avoiding damage to some concomitant self-antigens that needs to be tolerated. Nevertheless a complete study of the implication of this new property in the theoretical framework of this model will require a more complete study, including in the model perhaps several T cell clones, allowing some crossreactivity on their recognition of different APCs sets and taking into account the variability of lymph node volume with the T cell content. Future studies might address this issue, which could be intuitively expected to explain phenomena like epitope dominance in immunity, antigen spreading in some cases of autoimmunity, which are all phenomena that can not be explained by the original crossregulation model.

Overall the latter analysis suggests that suppression exerted by regulatory T cells over the helper T cells can not be reduced to competition for IL-2 (case $\eta = 0$). It further requires that the regulatory T cells inhibit helper T cell activation by a mechanism involving some locality and therefore some level of specificity (case $\eta = 1$). However, this conclusion does not mean that competition for IL-2 is irrelevant. It is quite probable that in some experimental system, the observed suppression can be explained by such competition for IL-2. This notion might reconcile some in vitro studies [7, 8] which have claimed an unspecific suppression by regulatory T cells, with the bulk of data supporting the specificity of T cells mediated suppression [72]. Interestingly, the analysis in this work further suggests that the combination of competition for IL-2 with some local/specific inhibition of helper T cell activation might confer some biologically relevant properties to the modelled system.

The Effect of IL-2 Modulating Therapies

In this section, the effects of IL-2 modulating therapies are studied in the second model variant ($\eta = 1$), which according to the analysis in previous sections exhibits a more realistic behaviour. Two classes of therapies are studied:

1. Therapies where IL-2 activity is increased, by injecting an excess of this cytokine into the system. Such type of therapy is simulated by increasing the external source of IL-2 molecules in the model (increasing the value of model parameter Γ_i).
2. Therapies that limit IL-2 activity, by effectively reducing the availability of this cytokine in the system, for instance, by using drugs that sequester IL-2 in circulation or simply block its binding to the IL-2 receptor. This latter type of therapy is simulated by increasing the degradation rate of IL-2 in the model (increasing the value of model parameter K_d).

In the following subsections, the effects of these therapies are explored in two particular treatment scenarios: Long term continuous treatments and acute (transient) treatments. In each case, the impact of the dose of the applied treatment is determined, and the implications of the obtained results are discussed.

The Effect of Long-Lasting Modulation of IL-2

The effect of long term (continuous) treatments, which modulate IL-2 activity, are simulated as a permanent change in the parameter values, Γ_i or K^i_d, for the two different types of therapies studied here. We explore whether changes in these parameters compromise the existence and/or the stability of model steady states equilibrium. In particular, those ones interpreted, in previous sections, as the tolerant (Fig. 15.3b) and the immune/autoimmune (Fig. 15.3a) states of the modelled system. To this end, the bifurcation diagrams of model steady states with the parameter Γ_i and K^i_d, were computed in a broad, but reasonable, range of values for the remaining model parameters. The typical (robust) structure found on these bifurcation diagrams is shown in Fig. 15.6. There, different regions of the parameter space, defined according to the subset of model stable steady states that exist in the system, are identified and labelled. The region labelled *B* corresponds to parameter values leading to a bistable regime, with a stable tolerant state and a stable immune/autoimmune state. This parameter regime, following the discussion

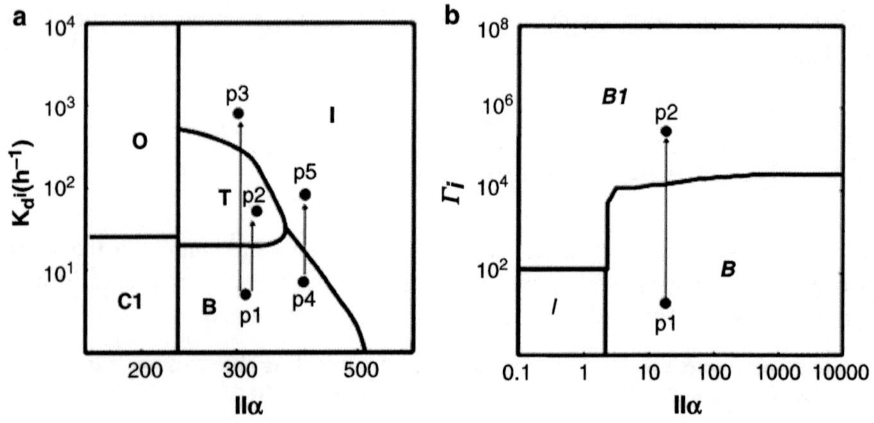

Fig. 15.6 Two dimensional bifurcation diagrams for model steady states, with the value of treatment-related parameters K^i_d (**a**) and Γ_i (**b**) versus the parameter IL-α. *Continuous lines* in the graphs demark regions of parameter values, which differ on the set of stable steady states existing in the modelled system. Region label *B* is a parameter region of bistable behaviour where the tolerant (*TS*) and the immune (*IS*) steady states exists; Region label *I* is a parameter region where only the immune (*IS*) steady state exists; Region label *T* is a parameter region where only the tolerant (*TS*) steady state exists; Region label *B1* is a parameter region of bistable behaviour, where the reinforce tolerant (*rTS*) and the immune (*IS*) steady states exists; Region label *C1* is a parameter region of bistable behaviour, where the trivial (*OS*) and the immune (*IS*) steady states exists; and Region label *O* is a parameter region, where only the trivial (*OS*) state exist. The specific graphs given in (**a**) and (**b**) were computed, using the default set of parameter provided in Table 15.2, except for the parameters indicated in the axes. They show the typical form/structure observed in these bifurcation diagrams for many other parameter values computed. The *small arrows* in the graphs indicate transitions in parameter space induced by specific treatment as discussed in the main text

in previous sections, is the expected situation in normal immune systems in the absence of perturbation, treatment or diseases. Thus, the model system is assumed to be parameterized inside this region before the application of any treatment. Other parameter regions in Fig. 15.6 could be only reached in the model by perturbations (like the one induced by long lasting treatments) which modify particular parameter values, conditioning certain types of responses as result in the system. In particular, interesting cases could be, (1) Perturbations taking the system into parameter region label *T*, which lead to reinforcement or reestablishment of tolerance, since inside this parameter region only the tolerant state or the reinforced tolerant state exist and/or are stable; and (2) Perturbations taking the system into parameter region label *I*, which unavoidably lead to immune reaction, since in this region the immune/autoimmune state is the only stable steady state.

Based on the diagrams of Fig. 15.6, the effect of long-lasting treatments modulating IL-2 activity can be analyzed. It can be seen, that IL-2 depletion treatments have a strong, dose-dependent impact on the system dynamics, by changing the structure of possible steady states. This treatment could lead, at intermediate doses (the dose being assumed proportional to the magnitude of the increment in K_d^i value), to the unavoidable reinforcement or reestablishment of tolerance, by taking the system into parameter region *T* (see transition p1–p2 in Fig. 15.6a). However, at higher doses (see transition p1–p3 in Fig. 15.6a) this treatment leads to immunity/autoimmunity, by taking the system into parameter region *I*. The latter dependence, is the case for low values of IL-α (transitions p1–p2–p3 in Fig. 15.6a), because for larger values of IL-α the study treatment could only induce immunity, with a direct transition between parameter regions *B* and *I*, without ever passing through region *T* (transition p4–p5 in Fig. 15.6a). In sharp contrast to the latter case, treatments increasing IL-2 activity have a negligible impact on the structure of steady states of the system. They could at most take the system, at high dose, into region *B*1 (see in Fig. 15.6b transition p1–p2), where the system remains bistable, with a stable immune state (as in region *B*) and a stable state of reinforced tolerance (very similar to the tolerant state of region *B*, see Fig. 15.3c). This result suggests, in practical terms, that this latter type of treatment would neither break tolerance inducing immunity nor reestablish tolerance in ongoing immune/autoimmune response. It must likely further expand the preexistent class of response in the system either tolerance or immunity.

The bifurcation analysis carried out in this section provides intuition on the potential effect of different treatments in the modelled system. It predicts a strong impact of IL-2 depletion treatments, which could forcedly take the system into the tolerant or the immune/autoimmune steady state, depending on the treatment dose and exact model parameterization (values of IL-α). It predicts, in contrast, a discrete, rather negligible, effect of IL-2 addition treatments, which most likely left the system in their starting steady state, either the tolerant or the immune/autoimmune state. Despite its general value, bifurcation analysis, only give a rough approximation of the effect of desired treatments. They could hide kinetics effects of treatments applications and, they only approximate long term effect of treatments that induce permanent or long lasting change on model parameters. Real treatments, however, typically induce a transient effect over the system. They transiently affect the value of some relevant parameter, which sooner or later returns back to its original value. In the next sections, these limitations are overcome by simulating the effect of treatment as transient perturbations in model parameters.

The Effect of Transient IL-2 Depletion Treatments

The effects of treatments that transiently reduce IL-2 activity are simulated here, by a sudden (instantaneous) increase in the value of model parameter K_d^i (to take the values $K_d^i + DosS$), which is followed by its exponential decay, towards its initial value K_d^i. Such exponential decay, on treatment effect, is assumed to account for the clearance of the used drugs and it is controlled in the simulations by parameter K_a (which set the rate of this decay). For instance, let's say the drug is an antibody, which sequesters IL-2 on circulation, then once injected into the system this antibody will be cleared from it, with a typical clearance rate, K_a. The latter treatment is applied in simulation, where the model has been parameterized inside the parameter regime B of Fig. 15.6 and has been initialized either in the tolerant or the immune/autoimmune steady state. The overall goal in these simulations is to explore, if the applied treatment can trigger a switch of system steady state: a switch from tolerant state to immune/autoimmune state or vice versa. The results are illustrated on Fig. 15.7.

IL-2 depletion treatment applied in the context of an ongoing immune/ autoimmune reaction (model initialized in the immune state) could reestablish tolerance (Fig. 15.7a), if the dose of the treatment (value of $DosS$) is large enough and the rate of drug clearance (K_a) is low enough (Fig. 15.7b). This result can be qualitatively explained using the bifurcation analysis of the previous section. Note that, this treatment could take the system (by increasing parameter K_d^i) transiently into region T (Fig. 15.6a), where tolerant state is the only steady state of the system. Thus, if the clearance rate K_a is small enough, the perturbed system, will spend enough time in region T, as to reach tolerant steady state. Then by the time the drug is fully cleared, the system will be already locked into tolerant equilibrium. In consonance with the latter analysis, the tolerating effect of IL-2 depletion is observed, only, for low values of IL-α.

IL-2 depletion treatments applied in the context of immune tolerance (model set in the tolerant state), can breakdown this preexistent tolerance, inducing an immune/autoimmune reaction (Fig. 15.7c). This effect is observed in a broad range of parameter IL-α values, but with different dependencies of treatment efficacy with the drug clearance rate K_a (Fig. 15.7d). For high values of parameter IL-α, the treatment requires a sufficiently slow clearance rate K_a to be effective (dashed line Fig. 15.7d), but for lower values of IL-α, its effectiveness is further restricted to a window of intermediate K a values (solid line Fig. 15.7d). These treatment effects and parameter dependencies can be explained from the bifurcation analysis in previous sections. Note that, an increase in K_d^i values can take the system into the region label I (where only the immune steady state exists), for a broad range

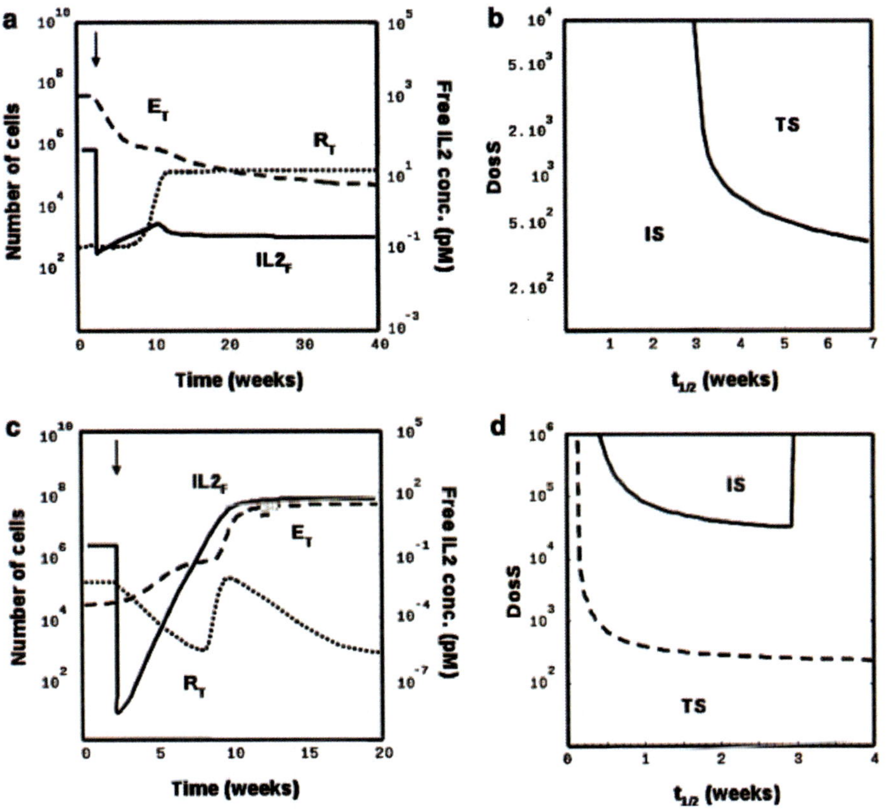

Fig. 15.7 Result of numerical simulations of the effect of transient IL-2 depletion treatments. The panels (**a**, **c**) show examples of kinetic evolution of the number of *E* cells, the number of *R* cells and the concentration of free IL-2 ($IL2_F$) in model simulations, where the system is parameterized inside region *B* of Fig. 15.6 and initialized either in the immune steady state (**a**) or the tolerant steady state (**c**), being then perturbed by transient increase of the value of model parameter K_d^i (see main text for details). Panel (**a**) illustrate the case in which the applied treatment ($DosS = 10^3, t_{1/2} = 7.6$ *weeks*), induce a switch from the immune steady state to the tolerant steady state. i.e. it reverts a pre-existence immune/autoimmune state, re-establishing tolerance. Panel (**c**) illustrates the case, in which the applied treatment ($DosS = 2.10^6, t_{1/2} = 5$ *days*), induces a switch from the tolerant steady state to the immune steady state. i.e. it induces an immune/autoimmune response in an initially tolerant system. The panels (**b**) and (**d**) show the dependence of treatment effects with the parameter *DosS* (proportional to the dose of the applied drug) and $t_{1/2}$ (proportional to drug clearance rate, $K_a = ln2/t_{1/2}$), which control treatment application in the simulations. These panels shows parameter conditions leading, upon treatment application, either to immunity or to tolerance in a system initialized in the immune state (panel **b**) or the tolerant state (panel **d**). In panel (**d**) the *solid line* divide parameter condition leading to immunity or tolerance for a system with a low value of IL-α (=250) and the *dashed line* dose so for a system with a higher value of IL-α (=400)

of IL-α values (Fig. 15.6a). However, for large values of K_a, the applied drug is cleared so fast, that it does not affect T cell population dynamics, setting up a general upper limit for the K_a values, which are compatible with effective treatment.

Interestingly, for low values of IL-α treatment takes the system into parameter region I (Fig. 15.6a transition p1–p3), but, it always pass through region T while the drug is cleared (transition p3 to p2 in Fig. 15.6a). Therefore, if drug clearance is slow enough, tolerance state is always reached inside regions T, setting up an additional lower limit for K_a in this parameter case. Other relevant parameter dependence for the predicted efficacy of IL-2 depletion treatments to induced immune/autoimmune responses is obtained with the parameters Γ_E,Γ_R, which sets the size of thymic output in the model (Fig. 15.8). As thymic output is increased (keeping Γ_E,Γ_R constants) the effectiveness of the treatment is dramatically reduced (Fig. 15.8). Actually, if thymic output became large enough, treatment effect is completely lost. Such parameter dependence derives, qualitatively, from the dependence with Γ_E,Γ_R values of the amount of regulatory T cells, remaining in the system in the absence of IL-2 induced by depletion therapy. The higher the size of thymic output, the higher the number of R cells remaining in the system, and therefore the higher their capacity to prevent a driving of the system towards the immune state, once the acute effect of IL-2 depletion has stopped.

Summarizing, the results here confirm those of the previous section, predicting that IL-2 depletion treatments could be used in appropriate conditions both to break a preexistent tolerant state, inducing immune/autoimmune reactions and to

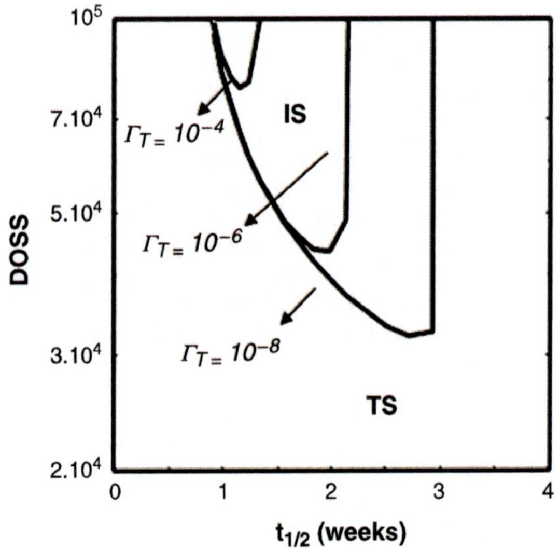

Fig. 15.8 The effect of thymus, on preventing the induction of an immune/autoimmune response in an initially tolerant system, by IL-2 depletion treatment. The panel shows parameter conditions (value of *DosS* and K_a) leading, upon IL-2 depletion treatment, either to immunity or to tolerance in a system initialized in the tolerant steady state. The *solid lines* in the graph corresponds to different interfaces between the immune and the tolerant outcome regions, for simulation which differ on the indicated value of parameter $\Gamma_T = \Gamma_E = \Gamma_R$. This parameter set the size of thymic output in the model simulations. Thus, the graph shows, that the larger the size of thymic output in the modelled system the smaller the region where immunity/autoimmunity could be induced by treatment application

render tolerant a preexistent immune/autoimmune system. These results, apparently counter-intuitive and of potential practical implication, derive from the highly non-linear relationship of IL-2 with CD4$^+$ T cell dynamics in the immune system. Interestingly, these model predictions are indeed compatible with existent experimental observations. On the one hand, the predicted capacity of treatments blocking IL-2 activity to promote immunity, explains observations where monoclonal antibodies, against IL-2, have been shown to promote effective immune responses to tumors [13] and to induce autoimmune disease in mice [73]. In both cases, the model explains the observed effects as being associated to the treatment capacity to weaken regulatory cell activity, as argued by their original authors. Moreover, the model explains the requirement of thymectomy in the induction of autoimmunity as reported in [73], suggesting that the apparent tolerating effect of thymic output is rooted in system dynamics, and might not be easily overcome by increasing the dose of anti-IL-2 MAbs. Relevantly the model, further, predicts an ideal IL-2 depletion scheme to enhance/optimize this treatment immune stimulatory capacity. This ideal scheme will require sustaining a high constant concentration of the IL-2 depleting drug in the system, for an initial long enough period of time, clearing it fast from the system afterwards (as fast as possible).

On the other hand, the model predicted capacity of IL-2 blocking therapies to reestablish tolerance in the context of ongoing immune/autoimmune reactions, is less well documented in the literature. This prediction might explains experimental data, supporting the use for the therapy [74, 75] of autoimmune diseases and some instance of allograft rejection of non-depleting antibodies against CD25 (the α chain of IL-2 receptor), which do block IL-2 signalling in vivo. Moreover, this prediction could also explain some experimental data, which show a positive effect of anti-IL-2 antibodies in the treatment of atherosclerosis (a disease of elderly people involving inflammation in the arteries). However, one must be aware, that in none of the latter observations a direct participation of regulatory T cells, as proposed by the model here, has been proved. In any case this second model prediction is very interesting from the practical perspectives for the treatment of autoimmune diseases. In particular, the fact the predicted treatment effect only occurs, in the model, for high drugs doses and low degradation rates, provides interesting guidelines for its practical implementation.

Simulating the Effect of IL-2 Injection Treatments

Treatments that transiently increase IL-2 levels in the system are simulated here as a sudden, but transient increase of the value of model parameter Γ_i, which stands for an external source of IL-2. Two parameters are used to control treatment application. The parameter *DosI*, which sets the total amount of IL-2 injected and the parameter T_i, which sets the length of the time period in which the sets total amount of IL-2 is administered. The study of this treatment, in model simulations, confirms the result of previous sections, predicting that it is unable to induce a switch of preexistent

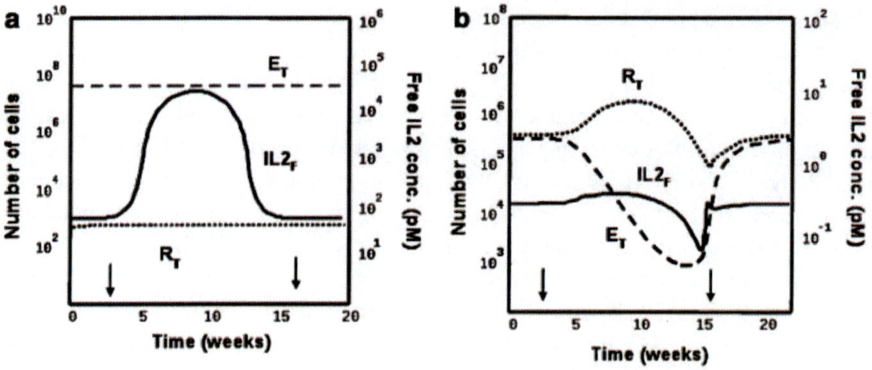

Fig. 15.9 Result of numerical simulations of the effect of transient IL-2 addition treatments. The panels (**a**, **b**) show examples of kinetic evolution of the number of E cells, the number of R cells and the free IL-2 concentration ($IL2_F$) in model simulations, where the system is parameterized inside region B of Fig. 15.6 and initialized either in the immune steady state (**a**) or the tolerant steady state (**b**), being then perturbed by transient increase of the value of model parameter Γ_i (see main text for details). Panel (**a**) illustrates how the applied treatment ($DosS = 10^4$, $td = 20$ *weeks*), slightly reinforce the pre-existent immune steady state, being unable to switch the system into the tolerant steady state. Panel (**b**) illustrate, how the applied treatment ($DosS = 10^3$, $td = 20$ *weeks*), reinforces the pre-existent tolerant steady state, being unable to switch the system into the tolerant steady state

steady state at any dose (*DosI*) or timing T_i. When this treatment was applied in a system with a preexistent immune reaction (initialized in the immune state), it was unable, to revert the system to the tolerant state. Furthermore it promotes a transient slight expansion of helper T cells (Fig. 15.9a). Moreover, when the treatment was applied to a previously tolerant system (initialized in the tolerant state) it transiently, but significantly, reinforces the preexistent tolerant state, further expanding the regulatory T cell population (Fig. 15.9b). The length of the transient effect induced by the treatment relates directly to the timing of treatment application (value of T_i) as soon as the treatment is stopped the reinforcement effect is rapidly lost.

The latter results are somehow counter-intuitive, since, in this model where IL-2 is the key molecule in immune regulation, its external addition to the system appears with little, or almost irrelevant, dynamical effect. However, the specific model predictions for this type of treatment are indeed compatible with existing experimental observations and further provide a guide for its further practical application. On one hand, the first model prediction, the reinforcement of ongoing immune reactions by IL-2 addition, explains classical observations on in vivo animal models, where IL-2 has been shown to augment immune reactions to viral infection [2] and to well-adjuvated vaccines [3–5]. In these systems the immune response, induced to the involved foreign antigens, which are controlled weakly if at all by regulatory T cell activity, is further promoted by the injected IL-2. Furthermore, the observed enhancement of immunity, in these experimental systems, might not rely only on the model predicted expansion of helper CD4$^+$ T cells. It might also involve important direct effects of IL-2 on CD8$^+$ T cell and NK cells, which are known to be relevant in many of these particular systems. In any case, the models here will further predict that optimal application of IL-2 for the purpose of enhancing immunity, will be obtained when providing IL-2 after the immune reaction has already started and not before, because some remnant of immune regulation might still exist. On the other hand the second model prediction, the capacity of IL-2 addition to reinforce natural tolerance mediated by regulatory T cells, also explains several experimental observations. In particular, it explains clinical data stating that regulatory T cell populations are significantly expanded, both in cancer [11, 76] or HIV [12] patients, treated with IL-2. Such effect might be related to the poor efficacy observed in these clinical applications of IL-2. In the case of cancer, less than 20% of the treated patients show some anti-tumor effect, perhaps, according to the model here, because a small fraction of the patients, happen to have a naturally preexistent immune response against tumor antigens, which could be further enhanced by the added IL-2. In the case of HIV patients, IL-2 based therapy have led to the recovery of CD4$^+$T cell counts, but the patients do not seem to recover their capacity to fight general infections, perhaps, according to the model here, because this treatment is simply reinforcing tolerance mediated by regulatory T cell activity. Furthermore, this second model prediction also explains results in preclinical animal models. It explains, for instance, that IL-2 injections can prevent allograph rejection [77] and attenuate the induction of Experimental Autoimmune Encephalomyelitis (EAE) [77]. Interestingly, as clearly expected from the model here, the latter effects are observed for schemes of IL-2 applications where this cytokine is used in the system before implanting the halogenic tissue or before vaccinating to induce EAE. This is before there will be any ongoing immune reaction in the system, thus when preexistent natural tolerance could be reinforced by the applied treatment.

Finally we would like to quote here the current accumulation of preclinical data [78, 79] regarding the biological effects of immunocomplexes of IL-2 (Antibody+IL-2). This novel therapeutic strategy has shown very interest results, especially on the prevention of autoimmune diseases. This therapeutic strategy seems to be more effective than therapies with IL-2 or antibody anti-IL-2 alone and it has the extra dimension of the relevant role conferred to the immunocomplexes by the particular site of the IL-2 recognized by the antibody. The model studied here is not appropriate to study the effects of such relevant therapies, thus further model development is currently on the way to address this issue.

References

1. Smith K (1988) A. Interleukin-2: inception, impact, and implications. Science 240:69–76
2. Blattman J, Grayson J, Wherry E, Kaech S, Smith K, Ahmed R (2003) Therapeutic use of IL-2 to enhance antiviral T-cell responses in vivo. Nat Med 9:540–547
3. Fishman M, Hunter T, Soliman H, Thompson P, Dunn M, Smilee R, Farmelo M, Noyes D, Mahany J, Lee J, et al (2008) Phase II trial of B7-1 (CD-86) transduced, cultured autologous tumor cell vaccine plus subcutaneous interleukin-2 for treatment of stage IV renal cell carcinoma. J Immunother 31:72

4. Kudo-Saito C, Garnett C, Wansley E, Schlom J, Hodge J (2007) Intratumoral delivery of vector mediated IL-2 in combination with vaccine results in enhanced T cell avidity and anti-tumor activity. Cancer Immunol Immunother 56:1897–1910
5. Lin C, Tsai Y, He L, Yeh C, Chang T, Soong Y, Monie A, Hung C, Lai C (2007) DNA vaccines encoding IL-2 linked to HPV-16 E7 antigen generate enhanced E7-specific CTL responses and antitumor activity. Immunol Lett 114:86–93
6. Almeida A, Legrand N, Papiernik M, Freitas A (2002) Homeostasis of peripheral $CD4^+$ T cells: IL-2R {alpha} and IL-2 shape a population of regulatory cells that controls $CD4^+$ T cell numbers. J Immunol 169:4850
7. Papiernik M, do Carmo L, Pontoux C, Joret A, Rocha B, Penit C, Dy M (1997) T cell deletion induced by chronic infection with mouse mammary tumor virus spares a CD25-positive, IL-10-producing T cell population with infectious capacity. J Immunol 158:4642
8. Thornton A, Shevach E (1998) $CD4^+CD25^+$ immunoregulatory T cells suppress polyclonal T cell activation in vitro by inhibiting interleukin 2 production. J Exp Med 188:287
9. Thornton A, Donovan E, Piccirillo C, Shevach E (2004) Cutting edge: IL-2 is critically required for the in vitro activation of $CD4^+CD25^+$ T cell suppressor function. J Immunol 172:6519
10. Takahashi T, Kuniyasu Y, Toda M, Sakaguchi N, Itoh M, Iwata M, Shimizu J, Sakaguchi S (1998) Immunologic self-tolerance maintained by $CD25^+CD4^+$ naturally anergic and suppressive T cells: induction of autoimmune disease by breaking their anergic/suppressive state. Int Immunol 10:1969
11. Ahmadzadeh M, Rosenberg S (2006) IL-2 administration increases $CD4^+CD25^{hi}Foxp3^+$ regulatory T cells in cancer patients. Blood 107:2409
12. Sereti I, Imamichi H, Natarajan V, Imamichi T, Ramchandani M, Badralmaa Y, Berg S, Metcalf J, Hahn B, Shen J, Powers A, Davey R, Kovacs J, Shevach E, Lane H (2005) In vivo expansion of $CD4CD45RO^-CD25^+$ T cells expressing foxP3 in IL-2-treated HIV-infected patients. J Clin Investig 115:1839–1847
13. Kamimura D, Sawa Y, Sato M, Agung E, Hirano T, Murakami M (2006) IL-2 in vivo activities and antitumor efficacy enhanced by an anti-IL-2 mAb. J Immunol 177:306
14. Smith K (2006) The structure of IL-2 bound to the three chains of the IL-2 receptor and how signaling occurs. Med Immunol 5:3
15. Bodnár A, Nizsalóczki E, Mocsár G, Szalóki N, Waldmann T, Damjanovich S, Vámosi, G (2008) A biophysical approach to IL-2 and IL-15 receptor function: Localization, conformation and interactions. Immunol Lett
16. Wofsy C, Goldstein B (1992) Interpretation of Scatchard plots for aggregating receptor systems. Math Biosci 112:115
17. Goldstein B, Jones D, Kevrekidis I, Perelson A (1992) Evidence for p55 p75 heterodimers in the absence of IL-2 from Scatchard plot analysis. Int Immunol 4:23
18. Fallon E, Lauffenburger D (2000) Computational model for effects of ligand/receptor binding properties on interleukin-2 trafficking dynamics and T cell proliferation response. Biotechnol Prog 16:905–916
19. Fallon E, Liparoto S, Lee K, Ciardelli T, Lauffenburger D (2000) Increased endosomal sorting of ligand to recycling enhances potency of an interleukin-2 analog. J Biol Chem 275:6790
20. Rao B, Driver I, Lauffenburger D, Wittrup K (2004) Interleukin 2 (IL-2) Variants engineered for increased IL-2 receptor {alpha}-subunit affinity exhibit increased potency arising from a cell surface ligand reservoir effect. Mol Pharmacol 66:864
21. Svirshchevskaya E, Sidorov I, Viskova N, Dozmorov I (1993) Quantitative analysis of interleukin-2-induced proliferation in the presence of inhibitors using a mathematical model. J Immunol Methods 159:17
22. Borisova L, Kuznetsov V (1997) A mathematical model of T lymphocyte proliferation controlled by interleukin-2 internalization. Membrane Cell Biol 11:259
23. Borisova L, Andreev S, Kuznetsov V Kinetics of T cell proliferation: a mathematical model and data analysis. Membrane Cell Biol 12:111
24. Morel B, Burke M, Kalagnanam J, McCarthy S, Tweardy D, Morel P (1996) Making sense of the combined effect of interleukin-2 and interleukin-4 on lymphocytes using a mathematical model. Bull Math Biol 58:569–594

25. De Boer R, Ganusov V, Milutinović D, Hodgkin P, Perelson A (2006) Estimating lymphocyte division and death rates from CFSE data. Bull Math Biol 68:1011–1031
26. Burroughs N, Miguel Paz Mendes de Oliveira B, Adrego Pinto A (2006) Regulatory T cell adjustment of quorum growth thresholds and the control of local immune responses. J Theor Biol 241:134–141
27. Carneiro J, Leon K, Caramalho Í, van den Dool C, Gardner R, Oliveira V, Bergman M, Sepúlveda N, Paixão T, Faro J, et al (2007) When three is not a crowd: a crossregulation model of the dynamics and repertoire selection of regulatory CD4 T cells. Immunol Rev 216:48–68
28. Leon K, Perez R, Lage A, Carneiro J (2000) Modelling T-cell-mediated suppression dependent on interactions in multicellular conjugates. J Theor Biol 207:231–254
29. García-Martínez K, León K (2009) Modeling the role of IL-2 in the interplay between CD4$^+$ helper and regulatory T cells: assessing general dynamical properties. J Theor Biol
30. de la Rosa M, Rutz S, Dorninger H, Scheffold A (2004) Interleukin-2 is essential for CD4$^+$CD25$^+$ regulatory T cell function. Eur J Immunol 34:2480–2488
31. Scheffold A, Huhn J, Hofer T (2005) Regulation of CD4$^+$CD25$^+$ regulatory T cell activity: it takes (IL-) two to tango. Eur J Immunol 35:1336–1341
32. Taams L, Van Rensen A, Poelen M, Van Els C, Besseling A, Wagenaar J, van Eden W, Wauben M (1998) Anergic T cells actively suppress T cell responses via the antigen-presenting cell. Eur J Immunol 28:2902–2912
33. Cederbom L, Hall H, Ivars F (2000) CD4$^+$CD25$^+$ regulatory T cells down-regulate co-stimulatory molecules on antigen-presenting cells. Eur J Immunol 30:1538–1543
34. Tang Q, Adams J, Tooley A, Bi M, Fife B, Serra P, Santamaria P, Locksley R, Krummel M, Bluestone J (2006) Visualizing regulatory T cell control of autoimmune responses in nonobese diabetic mice. Nat Immunol 7:83–92
35. Smith K (2005) Determining to divide: how do cells decide? J Biol Phys 31:261–272
36. Barthlott T, Moncrieffe H, Veldhoen M, Atkins C, Christensen J, O'Garra A, Stockinger B (2005) CD25$^+$CD4$^+$ T cells compete with naive CD4$^+$ T cells for IL-2 and exploit it for the induction of IL-10 production. Int Immunol 17:279
37. Almeida A, Zaragoza B, Freitas A (2006) Indexation as a novel mechanism of lymphocyte homeostasis: the number of CD4$^+$CD25$^+$ regulatory T cells is indexed to the number of IL-2-producing cells. J Immunol 177:192
38. Horak I (1995) Immunodeficiency in IL-2-knockout mice. Clin Immunol Immunopathol 76:172–173
39. Sadlack B, Merz H, Schorle H, Schimpl A, Feller A, Horak I (1993) Ulcerative colitis-like disease in mice with a disrupted interleukin-2 gene. Cell 75:253–261
40. Peschon J, Morrissey P, Grabstein K, Ramsdell F, Maraskovsky E, Gliniak B, Park L, Ziegler S, Williams D, Ware C (1994) Early lymphocyte expansion is severely impaired in interleukin 7 receptor-deficient mice. J Exp Med 180:1955
41. von Freeden-Jeffry U, Vieira P, Lucian L, McNeil T, Burdach S, Murray R (1995) Lymphopenia in interleukin (IL)-7 gene-deleted mice identifies IL-7 as a nonredundant cytokine. J Exp Med 181:1519
42. Purton J, Tan J, Rubinstein M, Kim D, Sprent J, Surh C (2007) Antiviral CD4$^+$ memory T cells are IL-15 dependent. J Exp Med
43. Lodolce J, Boone D, Chai S, Swain R, Dassopoulos T, Trettin S, Ma A (1998) IL-15 receptor maintains lymphoid homeostasis by supporting lymphocyte homing and proliferation. Immunity 9:669–676
44. King C, Ilic A, Koelsch K, Sarvetnick N (2004) Homeostatic expansion of T cells during immune insufficiency generates autoimmunity. Cell 117:265–277
45. Read S, Mauze S, Asseman C, Bean A, Coffman R, Powrie F (1998) CD38 CD45RB low CD4 T cells: a population of T cells with immune regulatory activities in vitro. Eur J Immunol 28:3435–3447
46. Lombardi G, Sidhu S, Batchelor R, Lechler R (1994) Anergic T cells as suppressor cells in vitro. Science 264:1587

47. Malek T, Yu A, Vincek V, Scibelli P, Kong L (2002) CD4 Regulatory T cells prevent lethal autoimmunity in IL-2Rβ-deficient mice implications for the nonredundant function of IL-2. Immunity 17:167–178
48. Wolf M, Schimpl A, Hunig T (2001) Control of T cell hyperactivation in IL-2-deficient mice by CD4 CD25 and CD4 CD25 T cells: evidence for two distinct regulatory mechanisms. Eur J Immunol 31:1637–1645
49. Frasca L, Carmichael P, Lechler R, Lombardi G (1997) Anergic T cells effect linked suppression. Eur J Immunol 27:3191–3197
50. León K, Peréz R, Lage A, Carneiro J (2001) Three-cell interactions in T cell-mediated suppression? A mathematical analysis of its quantitative implications. J Immunol 166:5356
51. Smith K (2004) The quantal theory of how the immune system discriminates between "self and non-self". Med Immunol 3:3
52. Hemar A, Subtil A, Lieb M, Morelon E, Hellio R, Dautry-Varsat A (1995) Endocytosis of interleukin 2 receptors in human T lymphocytes: distinct intracellular localization and fate of the receptor alpha, beta, and gamma chains. J Cell Biol 129:55
53. Duprez V, Dautry-Varsat A (1986) Receptor-mediated endocytosis of interleukin 2 in a human tumor T cell line. Degradation of interleukin 2 and evidence for the absence of recycling of interleukin receptors. J Biol Chem 261:15450
54. Miller M, Safrina O, Parker I, Cahalan M (2004) Imaging the single cell dynamics of $CD4^+$ T cell activation by dendritic cells in lymph nodes. J Exp Med 200:847
55. Murali-Krishna K, Altman J, Suresh M, Sourdive D, Zajac A, Ahmed R (1998) In vivo dynamics of anti-viral CD8 T cell responses to different epitopes. An evaluation of bystander activation in primary and secondary responses to viral infection. Adv Exp Med Biol 452: 123
56. León K, Faro J, Carneiro J (2004) A general mathematical framework to model generation structure in a population of asynchronously dividing cells. J Theor Biol 229:455–476
57. Jelley-Gibbs D, Lepak N, Yen M, Swain S (2000) Two distinct stages in the transition from naive CD4 T cells to effectors, early antigen-dependent and late cytokine-driven expansion and differentiation. J Immunol 165:5017
58. Usherwood E, Crowther G, Woodland D (1999) Apoptotic cells are generated at every division of in vitro cultured T cell lines. Cell Immunol 196:131–137
59. Borghans J, de Boer R, Segel L (1996) Extending the quasi-steady state approximation by changing variables. Bull Math Biol 58:43–63
60. Leon K, Garcia K, Carneiro J, Lage A (2007) How regulatory $CD25^+CD4^+$ T cells impinge on tumor immunobiology? On the existence of two alternative dynamical classes of tumors. J Theor Biol 247:122–137
61. Sakaguchi S, Ono M, Setoguchi R, Yagi H, Hori S, Fehervari Z, Shimizu J, Takahashi T, Nomura T (2006) $Foxp3^+CD25^+CD4^+$ natural regulatory T cells in dominant self-tolerance and autoimmune disease. Immunol Rev 212:8–27
62. Kuniyasu Y, Takahashi T, Itoh M, Shimizu J, Toda G, Sakaguchi S (2000) Naturally anergic and suppressive $CD25^+CD4^+$ T cells as a functionally and phenotypically distinct immunoregulatory T cell subpopulation. Int Immunol 12:1145
63. Pandiyan P, Zheng L, Ishihara S, Reed J, Lenardo M (2007) $CD4^+CD25^+Foxp3^+$ regulatory T cells induce cytokine deprivation-mediated apoptosis of effector $CD4^+$ T cells. Nat Immunol 8:1353–1362
64. Willerford D, Chen J, Ferry J, Davidson L, Ma A, Alt F (1995) Interleukin-2 receptor alpha chain regulates the size and content of the peripheral lymphoid compartment. Immunity 3:521
65. Suzuki H, Kundig T, Furlonger C, Wakeham A, Timms E Matsuyama T, Schmits R, Simard J, Ohashi P, Griesser H, et al (1995) Deregulated T cell activation and autoimmunity in mice lacking interleukin-2 receptor beta. Science 268:1472
66. Schorle H, Holtschke T, Hünig T, Schimpl A, Horak I (1991) Development and function of T cells in mice rendered interleukin-2 deficient by gene targeting. Nature 352:621–624
67. Razi-Wolf Z, Höllander G, Reiser H (1996) Activation of $CD4^+$ T lymphocytes form interleukin 2-deficient mice by costimulatory B7 molecules. Proc Natl Acad Sci USA 93:2903

68. Van Parijs L, Biuckians A, Ibragimov A, Alt F, Willerford D, Abbas A (1997) Functional responses and apoptosis of CD25 (IL-2Rα)-deficient T cells expressing a transgenic antigen receptor. J Immunol 158:3738
69. Malek T, Yu A, Scibelli P, Lichtenheld M, Codias E (2001) Broad programming by IL-2 receptor signaling for extended growth to multiple cytokines and functional maturation of antigen-activated T cells. J Immunol 166:1675
70. Malek T, Bayer A (2004) Tolerance, not immunity, crucially depends on IL-2. Nat Rev Immunol 4:665–674
71. León K, Faro J, Lage A, Carneiro J (2004) Inverse correlation between the incidences of autoimmune disease and infection predicted by a model of T cell mediated tolerance. J Autoimmunity 22:31–42
72. Miyara M, Sakaguchi S (2007) Natural regulatory T cells: mechanisms of suppression. Trends Mol Med 13:108–116
73. Setoguchi R, Hori S, Takahashi T, Sakaguchi S (2005) Homeostatic maintenance of natural Foxp3$^+$CD25$^+$CD4$^+$ regulatory T cells by interleukin (IL)-2 and induction of autoimmune disease by IL-2 neutralization. J Exp Med 201:723
74. Schippling D, Martin R (2008) Spotlight on anti-CD25: daclizumab in MS. In: International MS journal/MS Forum. vol 15:94
75. Waldmann T (2007) Anti-Tac (daclizumab, Zenapax) in the treatment of leukemia, autoimmune diseases, and in the prevention of allograft rejection: a 25-year personal odyssey. J Clin Immunol 27:1–18
76. Cesana G, DeRaffele G, Cohen S, Moroziewicz D, Mitcham J, Stoutenburg J, Cheung K, Hesdorffer C, Kim-Schulze S, Kaufman H (2006) Characterization of CD4$^+$CD25$^+$ regulatory T cells in patients treated with high-dose interleukin-2 for metastatic melanoma or renal cell carcinoma. J Clin Oncol 24:1169
77. Montero E, Alonso L, Perez R, Lage A (2007) Interleukin-2 mastering regulation in cancer and autoimmunity. Clin Immunol 123:125–125
78. Boyman O, Surh C, Sprent J (2006) Potential use of IL-2/anti-IL-2 antibody immune complexes for the treatment of cancer and autoimmune disease. Expert Opin Biol Therapy 6:1323–1331
79. Wilson M, Pesce J, Ramalingam T, Thompson R, Cheever A, Wynn T (2008) Suppression of murine allergic airway disease by IL-2: anti-IL-2 monoclonal antibody-induced regulatory T cells. J Immunol 181:6942

Chapter 16
A Physicist's Approach to Immunology

Mario Castro

Abstract In this chapter some heuristic techniques borrowed from Physics are used to formulate some hypotheses and validate them through available experimental data. The main goal is to present a quantitative approach to motility and antigen presentation from simple energetic and statistical arguments.

Introduction

How Biology Works

Biology is an experimental science: observations are made and put in the fundamental framework of some basic underlying principle. Most of the current knowledge in the field has been achieved in the last 50 years (catalyzed by the discovery of the DNA double helix structure). Traditionally, Biology problems have been approached from two independent points of view: geneticists tried to unveil the fundamental mechanisms of the cell by expressing or inhibiting this or that gene and analyzing what changes they produced. Alternatively, biochemists focus primarily on proteins, their structure, or their interactions, to cite a few.

With the advent of DNA recombination techniques, both fields merged (in which we now call Molecular Biology) and the synergy of both fields has been the responsible of the explosion of knowledge since the mid-seventies. A similar pattern has occurred in the last two decades with Theoretical Biology.

On the other hand, Physics and its practitioners (physicists) were comfortable with the so-called reductionism: living bodies are made of organs, which are made of tissues, which are made of cells, which are made of molecules, which are made of atoms, who obey quantum mechanics. So, Physics explains everything! On the other

M. Castro (✉)
Grupo Interdisciplinar de Sistemas Complejos (GISC), Grupo de Dinámica No Lineal, Universidad Pontificia Comillas, Madrid E-28015, Spain
e-mail: mariocastro73@gmail.com

C. Molina-París and G. Lythe (eds.), *Mathematical Models and Immune Cell Biology*, DOI 10.1007/978-1-4419-7725-0_16,

hand, biologists claimed that the laws of Physics have not been able to explain even the simplest living organism. These antagonistic points of view are, naturally, far being completely true, and Theoretical Biology (in which biologists, mathematicians and physicists have found a niche) has reached some successful milestones in last years.

One of the aims of the book you are reading is to try to show that Theoretical Biology, in general, and Theoretical Immunology, in particular, can be a crucial ingredient (perhaps a third leg in the study of Biology) in the field of Immunology and, hopefully, it may be able to shed some light on what directions the experimentation should go.

How Physics Works: How Physicists Work

An interesting question is if fundamentals laws in Biology can be put in a quantitative mathematical predictable language so we can make Biology a science which is a little more like Physics or Chemistry where we can actually analyze mathematically a given problem and make predictions. The great success of Physics in the last century was that it was able to explain, in a *quantitative* way, all events known to occur in the inanimate world. In contrast, as I mentioned above, it cannot explain or predict any behaviour in even the simplest living organisms *using those fundamental laws*. The reason is that atoms or galaxies are intrinsically predictable, and experiments are essentially reproducible. Thus, Physics formalizes the relationships between those inanimate objects in the form of mathematical equations. These equations (or *laws*) are obtained combining experimental evidence and some intellectual abstractions such as conservation laws or symmetries.

The reason why living systems are essentially unpredictable is that their interactions with the environment are governed by nonlinear dynamical laws. Moreover, *cells* (unlike atoms) are closer in essence to engineering devices: they accomplish a specific function which cannot be predicted from its constituents (like the flight of an airplane cannot be inferred from the knowledge of every screw, engine, or metal sheet it has). In summary, a biology understood in terms of universal reproducible laws simply doesn't work.

But what about physicists? Well, behind the rigorous, precise and exact physical theories there are physicists which do not consider reductionism too seriously. Thus, their main aim is to find the relationships between physical observables, paying attention to their relative scales, orders of magnitude and simplicity principles. So, the structure of the nucleus is irrelevant to understand how a fish can swim or knowing the amount of energy released by ATP does not really help to understand the time needed for a mammal to mount an effective immune response. So the whole is more than the parts. Prominent examples of this way of thinking can be found in Physics. For instance, a fluid vortex or a phase-transition belong to this realm of science where microscopic knowledge is not as important as collective behaviour.

Theoretical Biology shares this viewpoint: the way in which it can be effective is through the concept of hierarchy of biological entities: different scales need different models and tools. Essentially, all the levels of description must be related, but this *separation of scales* provides accurate descriptions of the phenomena and, more importantly, with predictive power. In Biology perhaps the best we can do is to obtain a coarse-grained description of the phenomena and put that description in the form of a law. As an example, if we were able to write the number of T cells, N_T in a (mature) mammal as function of its mass, M, [1] in the form

$$N_T = aM^b, \tag{16.1}$$

where a and b are positive constants, we would gain some insight about the differences between the immune system in mice and humans.

So, I want to emphasize that my aim in this chapter is to illustrate, *naively*, the way in which Physics, or more precisely, physical tools can be applied as a first approach to certain immunology problems. By *naively* I mean that detailed models (based on, for instance, stochastic methods or differential equations) are still mandatory to really understand the underlying biological processes but, however, those traditional models can be supplemented with some physically meaningful reasoning. Fortunately for the mathematician, those standard models will be presented in subsequent chapters in companion with novel experimental results (fortunately for the biologist).

A Naive Approach to Quantitative Immunology: Divide and Conquer

The immune system is one of the most complex system in Biology. Globally, it works with precision identifying a potential menace, preparing an adequate response and eliminating it. Of course, a model of the whole system cannot be easily formulated. Hence, we need to divide the problem in well-defined subproblems and, at the same time, be able to draw a general picture in order to not lose perspective.

Following this programme, here I summarize some evidence that can help us to make that division:

1. The immune system is formed by a hierarchy of biological entities
   (a) At small scales, those entities are heterogeneous and posses a high degree of specialisation (lymphocytes, leukocytes, ...)
   (b) At large scales, those entities can be described as (statistical) populations
2. Stability
   (a) The cells and the whole body are open systems (energy and entropy are exchanged with the environment)
   (b) The systems self-regulate (homeostasis)
   (c) After an immune response it returns to a steady state

3. Adaptation
   (a) Adaptive amplification (populations)
   (b) Antigen recognition (and presentation)
   (c) Information (learning, memory, ...)
4. Replication
   (a) Evolution
   (b) Adapted diversity
   (c) Natural selection

This list is far from being exhaustive (or rigorous) but one can recognize different problems in Theoretical Immunology (such as clonal selection, lymphocyte homeostasis, ...) but more importantly it stresses the fact that multiple scales are involved in the process. Again, this hierarchy will help me to identify some problems in which something quantitative can be said in very simple terms.

In this chapter I will focus on three different aspects (operating at different scales of energy, length and time) of one of those problems: antigen presentation. The sequence of events that I am interested in proceeds as follows:

- First, the lymphocyte and the antigen or the dendritic cell (DC) must meet.
- Subsequently, at least in the case of the dendritic cell, its surface needs to be scanned by the T cell.
- Finally, the ligand (peptide-Major Histocompatibility Complex, pMHC) and the T cell receptor (TCR) must *fit*.

These problems will be addressed with the use of some tools borrowed from Physics: order of magnitude estimation, dimensional analysis, statistical partition function,....

Case Study: Motility and Antigen Presentation

Immunity depends crucially on how antigens are recognized by the cells of the immune system. In order to do this efficient and accurately three conflicting conditions have to be accomplished [2]: the number of presented peptides must be large; the specificity of recognition must be large and, finally, the frequency of naive T cells that can recognize a foreign peptide must also be sufficiently large.

Consider presentation: It occurs whenever an antigen (both in free or bound to a presentation cell) meets a lymphocyte and a ligand on the former binds to a receptor on the latter. Following our divide-and-conquer strategy, we recognize two independent events: transport (meeting) and binding (chemical attraction). In chemical notation, the reaction

$$X + Y \leftrightarrow Z, \tag{16.2}$$

where X is the antigen or dendritic cell, for instance, Y the immune cell and Z the complex $X + Y$, can be split into

$$X + Y \leftrightarrow XY \leftrightarrow Z, \tag{16.3}$$

where the first *reaction* contains the transport information and the second one the binding information. Let us consider both mechanisms separately.

Cell Motility

Cells (in general) are small bodies with sizes ranging from a few to a few tens of microns (μm), namely $10^{-6} - 10^{5}$ m. The transport may be persistent (directed) or diffusive (random). Let us explore the differences.

In the first case, the cells are directed towards larger gradients of some substances (generically referred to as chemokines). The higher the gradient the higher the *force* experimented by the cell.

On the contrary, if cells (or large molecules) move diffusively, they may be locally in thermal equilibrium with their environment. Thus the erratic motion (fluctuation) is closely related to the resistance of the environment to motion (dissipation). Then, one can determine the diffusivity (assuming that the cell is an sphere of radius R) as [3]:

$$D = \frac{kT}{6\pi\mu R}. \tag{16.4}$$

Thus, the diffusion coefficient or motility (D) emerges as the capacity to extract thermal energy from the environment and dissipate it producing motion. As mass scales as $M \sim R^3$ (assuming constant density), one would expect that diffusivity scales as:

$$d \sim M^{-1/3}. \tag{16.5}$$

In Table 16.1, I collect some values of D for large molecules and cells taken from [4]. In Fig. 16.1, I show that the agreement with (16.5) is remarkable (at least in the shaded region).

However, for a dendritic cell with a typical size of 20 μm, (16.4) gives $D_{DC} \simeq 1\,\mu\text{m}^2\,\text{min}^{-1}$ and for a T cell, $D_T \simeq 4\,\mu\text{m}^2\,\text{min}^{-1}$. Those values are not close to the values reported from the experiments [5], where $D_T \simeq 50 - 100\,\mu\text{m}^2\,\text{min}^{-1}$.

Indeed this discrepancy must have a biological explanation. Two choices appear naturally at those scales: the T cell is either *advected* (chemotaxis) to specific targets or the diffusion theory must be refined. For instance, many bacteria diffuse randomly but with some persistence in their motion.

If T cells respond to external signalling (e.g., chemokine gradients), the motion would be less random and more directed. In addition, one would expect higher cell velocities and, as a side effect, collective motion towards the source of chemokines

Table 16.1 Diffusivities of some molecules and cells. Adapted from [4]

M ($g\,mole^{-1}$)	$D(\mu m^2\,min^{-1})$	Name	M ($g\,mole^{-1}$)	$D(\mu m^2\,min^{-1})$	Name
18	139,200	H_20	43,500	4,860	Ovalbumin
32	120,000	O_2	68,000	4,380	Hemogoblin
32	90,000	Methanol	68,430	3,900	Serum
36.5	216,000	HCl	74,000	3,720	Transferrin
44	114,000	CO_2	98,630	2,820	Gonadotropin
58.5	90,000	NaCl	109,000	2,700	Collagenase
60	78,000	Urea	130,000	3,180	Actin
75	60,000	Glycine	140,000	1,860	Plasminogen
89	57,000	α-alanine	143,000	3,000	Ceruloplasmin
89	58,200	β-alanine	153,100	2,520	γ-globulin
92	52,800	Glycerol	158,500	2,520	immunoglobulin
111	72,000	$CaCl_2$	190,000	2,160	Glucose
180	42,600	Glucose	339,700	1,260	Fibrinogen
182	42,600	Mannitol	482,700	2,220	Urease
192	41,400	Citric	529,800	2,280	Cytochrome
342	32,400	Sucrose	820,000	1,500	α-macroglobulin
6,660	9,000	Milk	2,663,000	1,080	β-lipoprotein
13,683	7,800	Ribonuclease	4,200,000	780	ribosome
24,430	4,620	Insulin	31,340,000	336	Tobacco
27,100	5,640	Somatotropin	~8e + 11	24.6	1 μm nanodevice
30,640	6,600	Carbonic	~4e + 12	9.6	Platelet
44,070	3,240	Plasma	~6e + 13	4.1	Red blood cell

(because the signalling would be shared by many similar cells). Let us revise recent evidence that supports this idea. In [5, 6], the authors showed that B and T cells migrate as autonomous agents and there is no trace of collective motion that support the chemokine hypothesis. Paradoxically, they also report travelling persistence times of 1–2 min (see also [7, 8]).

What about the second choice? If pure diffusion is considered, persistent motion is sustained with velocities of the order of D/R. For a T cell (using the experimental $D_T \sim 75\,\mu m^2\,min^{-1}$ and $R \sim 3.5\,\mu m$) $v \sim 20\,\mu m\,min^{-1}$. This velocity means that the persistence time is on the order of 1–2 s. So we need to consider persistence in the description of the random motion. One simple approach is to assume that the changes in the velocity of the cell are at an angle, θ, randomly chose from a given distribution, $P(\theta)$. In this case the diffusivity is corrected by:

$$\tilde{D}_T = \frac{D_T}{1-\alpha}, \tag{16.6}$$

where α is mean value of the cosine of the angle between successive runs [9]. If the distribution of successive angles is completely random $\alpha = 0$. If it is peaked at 0 (ballistic motion instead of diffusive motion), $\alpha \to 1$.

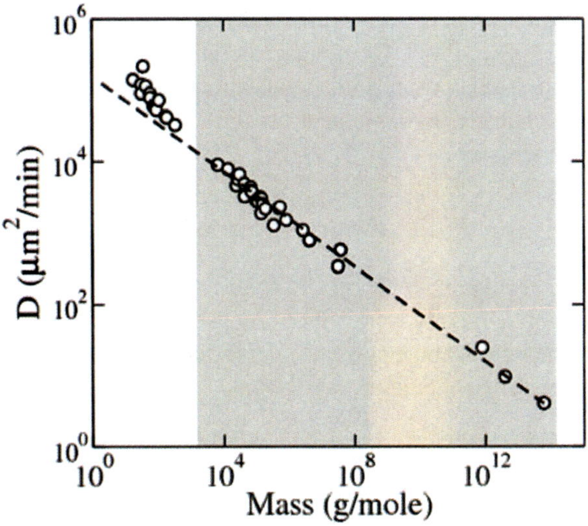

Fig. 16.1 Log–log plot of the diffusivity and the mass (data taken from Table 16.1). The *dashed straight line* is a fit to $D \sim M^{-1/3}$. The *shaded region* is a visual help to identify where the scaling law (16.5) is valid

At this point, we need to support or discard the hypothesis made through the experiments. In summary,

1. There is a discrepancy between the theoretical 4 $\mu m^2\,min^{-1}$ and the experimental 50–100 $\mu m^2\,min^{-1}$ diffusivities.
2. Persistence may be an essential ingredient behind pure diffusion.

Again, in [5] it is reported that the distribution of successive angles is not at all random and that the cells can reach velocities about 25 $\mu m\,min^{-1}$, in good agreement with our rough arguments. Moreover, agent based modelling [8] of cell diffusion with persistence produce excellent agreement with the experiments.

Biologically, this means that, in the short term, the cells are somehow directed but at long times (distances) they seem to diffuse randomly. The quantitative separation between these scales deserves further experimental development. For instance, chemokines could be introduced artificially in lymph nodes to change the variability of, for instance, the parameter α. The heterogeneity if the lymph node needs also to be taken into account as it can influence locally the value of the effective D.

Moreover, as shown in [1] the average time, τ, until first contact between the lymphocyte and the antigen (or the dendritic cell), initially a distance R apart, is given by:

$$\tau = \frac{R^3}{3D_T(R_T + R_a)}, \tag{16.7}$$

where R_a is the radius of the antigen. Thus, controlling D_T can be an strategy to control τ and consequently the mean response time of the adaptive immune system.

Another biological implication of this analysis is that T cells do not spend all their available energy finding DCs or antigens. Thus, diffusion is an effective way to explore (instead of self-propulsion, for instance). A simple energetic calculation can illustrate this. The energy dissipated per unit time (power) by friction with its surroundings is given by:

$$P = F_{friction} \times V = 6\pi \mu R V^2 \text{ (W)}, \tag{16.8}$$

where V is the mean velocity of the cell, R its size and μ the viscosity of the medium (let us assume that it is mainly water). The order of magnitude of the latter quantity is much smaller than the average power available per cell. This power can be roughly estimated from the human metabolic rate (about 100 W) and its total number of cells (about 10^{14}) [10].

T Cell and Dendritic Cell Encounters

When a dendritic cell and a T cell meet, the latter scans the surface of the former in the so called *selection phase*. Cells, unlike engineering devices, are not programmed (in the common meaning of the expression) to localize a *target*. Instead, the process is somehow driven by fluctuations (that's why stochastic modelling is a promising approach at this level of description, see Chap. 17). This scanning time can last from a few minutes to hours [11]. This variability is due to the dynamics over the dendritic cell surface.

Let us try to gain some insight in this number. If we assume that the T cell diffusivity is $D_T = 75\,\mu\text{m}^2\,\text{min}^{-1}$, the average time for scanning without any kind of binding is on the order of 1 h (assuming that the radius of a dendritic cell is $R_{DC} = 20\,\mu\text{m}$ [5] and that its surface is $4\pi R_{DC}^2$). If there were any kind of binding between the T cell receptor and the pMHC complex (peptide-Major Histocompatibility Complex) this time would be even larger. This suggest that the T cell do no scan completely the surface of the dendritic cell or, alternatively, that the diffusivity of the T cell surrounding a dendritic cell is larger (an order of magnitude) than moving freely. Notwithstanding, this second choice is less probable.

Then the conclusion is that the T cell do no scan completely the dendritic cell in each encounter but, rather, it must do it in independent events. This speculation seem to be supported by the experiments (see [12]).

Ligand–Receptor Dynamics

Ligand–receptor dynamics takes place at the level of the protein. As in other fields in Biology, the structure and function of T cells depend crucially on the physical interaction at the atomic level. Typical distances are about 2 Å and the energies depend on the type of bond. For instance, covalent bonds are strong (on the order of 80 kcal mol^{-1}). This strength guarantees that the proteins preserve their primary structure. On the contrary, hydrogen bonds are about 1/20th of the energy of the covalent ones. Thus, translations, rotations, twisting and deformations are statistically probable. The factor 1/20 is important because, at temperatures that are relevant for life, a *population* of proteins will have enough energy to break that bond even in the absence of a tensile forces.

Being more specific, I am interested in the binding process which, schematically, can be written as

$$XY \leftrightarrow Z$$

The left and right *reactions* have independent rates k_{on} and k_{off}. Traditionally, the most important factor is the *affinity*

$$K = \frac{k_{on}}{k_{off}} \ (\mathrm{M}^{-1}). \tag{16.9}$$

Thermodynamically, it depends on temperature and the so-called *free-energy* of association through the relation

$$K = e^{-\Delta F/RT}, \tag{16.10}$$

where $R = 8.31\,\mathrm{J\,K^{-1}\,mol^{-1}}$ is the perfect gas constant.

The free energy ΔF has both energetic and entropic contributions. The former are related with the binding energies and the latter with the number of possible configurations of the rotation, translation, vibration and conformation degrees of freedom.

In addition,

$$\Delta F = \Delta E - T\Delta S \tag{16.11}$$

(if we consider that process is at constant pressure, $F \rightarrow G = H - TS$, where $H = E + PV$ is the enthalpy).

Equation (16.10) is quite useful since it allows to calculate microscopic properties (ΔE and ΔS) from kinetic experiments and, in addition, it emphasized that bond rupture is a stochastic event (16.10) can be seen as a probability of an event to happen).

From the point of view adopted here the entropic term makes the estimation rather complicated. Notwithstanding I shall adopt the same divide-and-conquer strategy as before. The following steps are more involved so I will anticipate that strategy.

1. The free energy difference ΔF has two contributions. Binding energetic are on the order of magnitude of 5 kcal mol^{-1} (typical of hydrogen bonds), so it is reasonable to expect that energies are that of 1 or 2 bonds (typical of TCR-pMHC binding). Thus ΔE is the range $[-10, 0]$ kcal mol^{-1} (the negative sign is because binding energies are negative).

2. Statistical Mechanics allows to calculate ΔF in some ideal cases. Those cases will help me to set upper bounds to entropic calculations under certain restrictive assumptions.
3. Finally, using the latter estimations, I will provide some orders of magnitude of variation of the affinity and, more importantly, emphasize the way in which this particular problem should be addressed.

From an Statistical Mechanics perspective, when the pMHC and the TCR bind, there are, apparently a loss of degrees of freedom related to the lack of relative motion between them. Thus, a possible approach to estimate the free energy change is consider what happens when the, for instance, pMHC goes from free to bound. For simplicity, I will consider that it is a rigid molecule (and discuss the implications of relaxing this assumption below).

All the statistical information is contained in the so-called partition function, Z. The partition function is related to the free energy as:

$$F = -k_B T \log Z, \tag{16.12}$$

where k_B is the Boltzmann's constant and T the temperature in Kelvin.

Thus, following any basic textbook in Statistical Mechanics we find that the partition function of the translational degrees of freedom.

$$Z = \frac{1}{N!h^{3N}} \int \mathrm{d}\mathbf{p}^N \exp\left[-\frac{|\mathbf{p}_i|^2}{2mk_BT}\right] \tag{16.13}$$

So it can deduced that the variation of the free energy is:

$$\Delta F = -k_B T \log\left[\frac{(2\pi m k_B T/h^2)^{3/2}}{V_p}\right] - k_B T \tag{16.14}$$

The molecular weight of the TCR-pMHC can be on the order of 10^6 and the typical length scales around 3 nm [13] so, at $T = 310\,\mathrm{K}$, $\Delta F \simeq 6\,\mathrm{kcal\,mol^{-1}}$. This is the contribution of the translational degrees of freedom.

The rotational degrees of freedom can be computed provided that the principal moments of inertia of the ligand are known. From our hand-waving perspective, we can assume that each rotational mode carries the same change in the free energy than the translational (because the moment of inertial approximately mass$\times$distance^2 and the calculation of the partition function are the same).

So, I have found that, if the ligand–receptor complex is rigid, the total change in the free energy is about

$$\Delta F \simeq 6 + 3 \times 6 = 24\ \mathrm{kcal\,mol^{-1}}.$$

Finally, from that value, ΔS is in the range $[-110, -80]$ cal mol^{-1} K^{-1}. The value of ΔF estimated from Statistical Mechanics is large compared to ΔE so, at the light of the experiments, we will be able to determine if its really so large or if it is an overestimate. In the latter case it would mean that, despite the loss of translational and rotational degrees of freedom, new configurational degrees of freedom are relevant (presumably vibrational). If that is the case, ΔS could even take positive values, thus supporting the idea of the relevance of other degrees of freedom that compensate the loss of translational and rotational mobility.

At this point the heuristic approach cannot be stressed further so experimental data is need to test the assumptions made. In [14] some values of ΔS are reported and ΔS ranges, typically, between -90 to $40\,\mathrm{cal\,mol^{-1}\,K^{-1}}$.

In the most negative case, ($\Delta S = -90$), the value calculated from the partition function is in good agreement, but note that the agreement seems to break down in general. As I mentioned above, this would indicate that the flexibility (as opposite to rigidity) and the vibrational modes are key ingredients in the theory. In [14] some other mechanisms are discussed but, again, the heuristic method of reasoning has provided some hypotheses that are worth studying experimentally.

Conclusion

In this introductory chapter, the main message that I want to convey is that, far from the useless reductionism of traditional Physics or the quest for universal laws in Biology, Physics often address the problems by combining different tools: separation of scales (divide and conquer) and coarse-grained modelling. Here, I have only illustrated the surface of this fruitful strategy with simple arguments.

The choice of problems in this chapter has been made to emphasize that the underlying physical principles are valid at many different scales. Other simple arguments such as scaling can be used to determine the connection between experiments in mice and humans (see, for instance [1] and [15]).

So, Theoretical Biology models can be enriched with this knowledge: finding mathematical relationships between biological quantities (for instance, differential equations) is the main task of Theoretical Biology but, in many situations, those relationships may be full of physical meaning.

Mathematics and Physics are already part of other disciplines of Biology, such as Ecology or Neurology. In those fields, it is customary to include mathematical concepts even in textbooks. Thus, Lotka–Volterra or Hodgkin–Huxley models coexist with biological descriptions. In the case of Immunology, it appears that this symbiosis between fields is emerging. It is our hope in this book to establish roots that could help the field evolve in that direction. Hopefully, in a few years we could find some of the ideas, experiments and theories collected in the present book in general textbooks of Immunology.

The naive approach presented here tries to exemplify the formulation of simple hypotheses and some basic methods to grant those hypotheses useful physical meaning.

References

1. Wiegel FW, Perelson AS (2004) Some scaling principles for the immune system. Immunol Cell Biol 82:127–131
2. Mason D (1998) A very high level of crossreactivity is an essential feature of the T-cell receptor. Immunol Today 19:395–404
3. Einstein A (1956) Investigations on the theory of the Brownian movement. Dover, Mineola
4. Freitas RA (1999) Nanomedicine, vol I. Basic capabilities. Landes Bioscience
5. Miller MJ, Wei S, Cahalan M, Parker I (2003) Autonomous T cell trafficking examined in vivo with intravital two-photon microscopy. Proc Natl Acad Sci USA 100:2604–2609
6. Miller MJ, Wei S, Parker I, Cahalan M (2002) Two-photon imaging of lymphocyte motility and antigen response in intact lymph node. Science 296:1869
7. Beltman JB, Marée AFM, Lynch JN, Miller MJ, de Boer RJ (2007) Lymph node topology dictates T cell migration behavior. J Exp Med 204(4):771–780
8. Beltman JB, Marée AFM, de Boer RJ (2007) Spatial modelling of brief and long interactions between T cells and dendritic cells. Immunol Cell Biol 85:306–314
9. Berg HC (1993) Random walks in biology. Princeton University Press, Princeton
10. Purcell E (1977) Life at low Reynolds number. Am J Phys 45:11
11. Gunzer M, Schfer A, Borgmann S, Grabbe S, Znker KS, Brcker E, Kmpgen E, Friedl P (2000) Antigen presentation in extracellular matrix: interactions of T cells with dendritic cells are dynamic, short lived, and sequential. Immunity 13:323–332
12. Stoll S, Delon J, Brotz TM, Germain RN (2002) Dynamic imaging of T cell-dendritic cell interactions in lymph nodes. Science 296:1873–1876
13. Murphy KM, Travers P, Walport M (2007) Janeway's immunobiology (immunobiology: the immune system (Janeway)), 7th edn. Garland Science
14. Sturtevant J (1977) Heat capacity and entropy changes in processes involving proteins. Proc Natl Acad Sci USA 74:2236–2240
15. Borghans JA, Noest AJ, de Boer RJ (1999) How specific should immunological memory be? J Immunol 163:569–575

Chapter 17
Timescales of the Adaptive Immune Response

Mark Day and Grant Lythe

Abstract We discuss continuum (lattice-free) and inherently stochastic models of immune cellular interactions, using the simplest hypothesis, that cells follow Brownian paths. The timescale for a cell to explore a volume such as a lymph node is L^2/D, where L is the radius of the region and D the diffusivity of a cell. The average time a cell spends in a volume with an exit is proportional to L^3/aD, where a is the radius of the exit. The mean time before a cell encounters a zone of attraction with radius b around, for example, an antigen presenting cell, is proportional to L^3/bD.

Introduction

Interactions between T lymphocytes and antigen-presenting cells (APCs) in lymph nodes are an essential step in the initiation of cell-mediated immune responses. T cells need to come into physical contact with APCs [1, 2]. In fact, the antigen-specific T cell can become activated only if it encounters an APC bearing cognate antigen in the midst of millions of unspecific lymphoid cells [3].

In the twenty-first century, observation of cell–cell contacts and cell migration in lymphoid tissues has become possible in vivo, with the technique of intravital two-photon microscopy [4–10]. Pioneering in vivo studies have raised a new set of questions, relating to cell–cell dynamics in secondary lymphoid organs and to the data sets produced in imaging experiments [11, 12]. What happens during the day or days that a T cell spends in a lymph node? What is the role of the reticular network in the movement of T cells? How do T cells collect and integrate signals from cognate antigen? Is a long (hours) contact with a Dendritic Cell (DC) necessary for activation, or is a series of short-lived (minutes) encounters sufficient [13]? To what extent do molecular mechanisms (amount and quality of peptide–MHC ligands) govern the length of time spent in contact? What are the roles of adhesion

G. Lythe (✉)
Department of Applied Mathematics, University of Leeds, LS29JT, UK
e-mail: grant@maths.leeds.ac.uk

C. Molina-París and G. Lythe (eds.), *Mathematical Models and Immune Cell Biology*,
DOI 10.1007/978-1-4419-7725-0_17,

molecules, co-stimulatory molecules and strength of affinity? Does clustering of T cells play a role in mounting an immune response? What controls differentiation into effector and memory cells? In the case of infection, what strategies can be used to maximise the time that lymphocytes spend in the particular lymph node draining the currently infected site? What is the role of inflammation [14] and expansion of the vascular input?

We will think of a lymphocyte as a particle undergoing Brownian motion inside a lymph node. Brownian motion is imagined here as the result of many constantly-changing influences that jostle the cell, without any consistently preferred direction. It can be expected, from the central limit theorem, to be a quantitatively correct description of T cell movement on some timescales, and most analyses of in vivo cell movement data seem to support this [4]. However, there are reasons to suspect that other descriptions may be needed, on short timescales (less than a minute in a lymph node) and perhaps on longer timescales, if phenomena such as microstreaming [15] and a preference for moving along the stromal network [16] are widespread. Even so, the end result will still appear to be Brownian motion if the stromal network or microstreams are themselves randomly-oriented [17].

The purpose of this chapter is to calculate timescales from the simple hypothesis that T cells move, and may encounter APCs, along Brownian paths. By not attempting to explicitly model the crowded environment of a lymph node, timescales are expressed in terms of very few parameters: the diffusivity of a cell, the size of the volume that the cell explores, and the radii of the efferent vessel (or other exit) and of the zone of attraction of an APC.

Brownian Motion

Each of the Cartesian components of a Brownian motion in three dimensions is independent. We use x to represent a position in three-dimensional space, and r its distance from the point $(0,0,0)$. If the position at time t of a cell undergoing Brownian motion is denoted by $\mathbf{B}_t$, and if there is no preferred direction and there are no boundaries or barriers, the probability density function of $\mathbf{B}_t$ is

$$P^{(n)}(x,t) = (4\pi Dt)^{-3/2}\exp\left(-r^2/4Dt\right).$$

The parameter D is the diffusivity.

The total distance travelled in a time t is $\mathbf{X}_t = |\mathbf{B}_t|$ (Fig. 17.1). This is a random variable with probability density, $\rho(r) = \frac{\mathrm{d}}{\mathrm{d}r}\mathcal{P}[\mathbf{X}_t < r]$, given by

$$\rho(r) = \frac{r^2}{2\sqrt{\pi}(Dt)^{3/2}}\mathrm{e}^{-r^2/4Dt}.$$

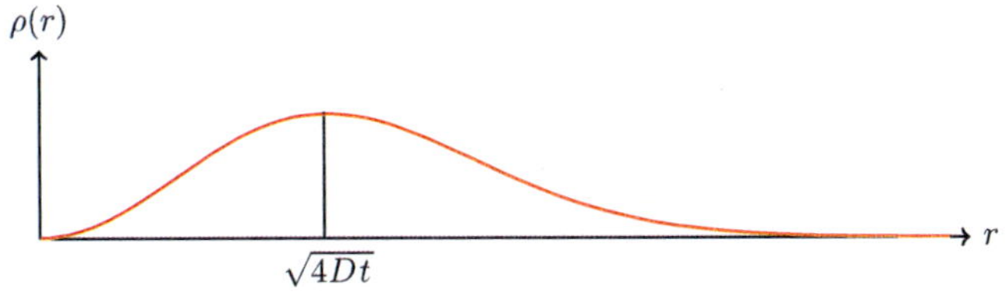

Fig. 17.1 The probability density of $\mathbf{X}_t$, the distance travelled in time t by a Brownian motion in three dimensions with diffusivity D

The mean distance travelled in time interval t is

$$\mathbb{E}(\mathbf{X}_t) = \left(\frac{16Dt}{\pi}\right)^{\frac{1}{2}};$$

the variance of $\mathbf{X}_t$ is $\mathbb{E}(\mathbf{X}_t{}^2) - (\mathbb{E}(\mathbf{X}_t))^2 = (3 - \frac{8}{\pi})2Dt$. Note that $\mathbb{E}(\mathbf{X}_t)$ is not proportional to t as $t \to 0$, so no well-defined velocity exists. Under the assumption of Brownian motion, the appropriate quantity to report is the diffusivity D, obtained by plotting mean displacement versus the square root of the time between measurements [4, 18].

First Passage to a Sphere

We'll suppose, in the spirit of the infamous spherical cow, that the lymph node, or its T cell zone, is spherical with radius R. How long does it take a T cell, executing Brownian motion in three dimensions with diffusivity D, to reach the edge of the sphere (Fig. 17.2)? In general, the mean time T to hit the boundary of a region is a function of the starting point, x_0, that satisfies the partial differential equation (PDE) known as Poisson's equation [19]:

$$D\nabla^2 T(x_0) = -1. \tag{17.1}$$

The spherical approximation means that, instead of solving (17.1) by the general method of separation of variables, we can look for a function of $r_0 = |x_0|$ only:

$$T(x_0) = F(r_0),$$

which satisfies

$$D\frac{1}{r^2}\frac{\partial}{\partial r}r^2\frac{\partial}{\partial r}F(r) = -1. \tag{17.2}$$

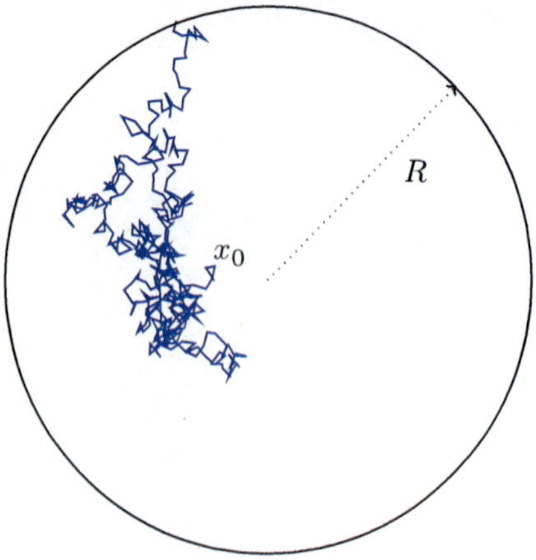

Fig. 17.2 Brownian motion inside a sphere of radius R: illustration of a sample path, starting at x_0

The general solution is

$$F(r) = -\frac{1}{6D}r^2 - \frac{c_1}{r} + c_2, \tag{17.3}$$

where c_1 and c_2 are constants. Requiring that the mean time to reach the surface starting from the centre of the sphere, $F(0)$, is not infinite, means that that $c_1 = 0$. We also know that $F(R) = 0$ because, if the starting point is already on the surface, then the time to reach the surface is certainly zero. The mean time to hit the surface of the sphere, starting a distance r_0 from its centre, is thus

$$T(x_0) = F(r_0) = \frac{1}{6D}(R^2 - r_0^2). \tag{17.4}$$

To derive a characteristic mean time that doesn't depend on initial conditions, we average (17.4) over all possible starting points in the sphere, by integrating over the sphere, and dividing by the volume of the sphere:

$$\tau = \frac{1}{\frac{4}{3}\pi R^3} \int_0^R \frac{1}{6D}(R^2 - r^2)4\pi r^2 \mathrm{d}r = \frac{1}{15D}R^2.$$

Before moving on, we also consider the slightly more difficult question of *where* the particle first hits the sphere. If the particle starts in the centre, then all parts are equally likely to be hit; if not, the probability density is maximised at the point on the surface closest to the initial condition.

Suppose the surface of the lymph node, S, contains a part that is special (an exit for example). The probability that our Brownian particle, starting at position x, hits a part, H, of the surface is a function, $p(x)$, that satisfies Laplace's equation [19,20]

$$\nabla^2 p(x) = 0, \tag{17.5}$$

with boundary conditions

$$p(x) = \begin{cases} 1 & x \in H \\ 0 & \text{elsewhere on } S. \end{cases} \tag{17.6}$$

This applies whatever the shape of the lymph node, but we will continue to make life easier for ourselves by considering it to be spherical.

A useful method for constructing solutions of equations like (17.1) and (17.5) is as follows. The solution of

$$\begin{cases} D\nabla^2 \Phi(x) = -\delta(x - x_0), & |x| < R \\ \Phi(x) = 0 & |x| = R \end{cases} \tag{17.7}$$

is

$$\Phi(x) = \frac{1}{4\pi D}\left(\frac{1}{|x - x_0|} - \frac{R/r_0}{|\frac{R^2}{r_0^2}x_0 - x|}\right). \tag{17.8}$$

The solution $\Phi(x)$ is called a Green function and may be written $G(x, x_0)$ to emphasise that it is a function of the initial condition x_0 as well as of x. (We have written it $\Phi(x)$ because it is identical to the electric potential at x, due to a point charge at x_0, inside a grounded sphere.)

Once we have $\Phi(x)$, we can calculate two quantities of interest. The mean exit time (the mean time until a particle following Brownian motion started at x_0 hits the sphere) is the integral

$$\mathbb{E}(\tau) = \int \Phi(x)\mathrm{d}x, \tag{17.9}$$

over the interior of the sphere. Thus $\Phi(x)$ has the nice interpretation of an *occupation density*: to find the average amount of time spent in some region of the interior of the sphere, before the sphere is hit, we integrate $\Phi(x)$ over that region. To find the total mean time, we integrate over the whole sphere, which is what (17.9) does.

The probability density of *where* on the sphere the first exit occurs is obtained by differentiating $\Phi(x)$ in the outward radial direction:

$$h(y) = -D\frac{\partial}{\partial r}\Phi(x)|_{r=R} = \frac{1}{4\pi R}\frac{R^2 - r_0^2}{|x - y|^3}, \tag{17.10}$$

where y is a position on the surface of the sphere. To obtain the probability that the exit occurs in some region on the surface, we integrate the function $h(y)$ over that region. Suppose that θ is the angle between the lines from the origin through x_0 and x, $r_0 = |x_0|$ and $r = |x|$. Then, using $|x - x_0|^2 = r^2 + r_0^2 - 2rr_0\cos\theta$,

$$\Phi(x) = \frac{1}{4\pi D}\left(\frac{1}{\left(r^2 + r_0^2 - 2rr_0\cos\theta\right)^{\frac{1}{2}}} - \frac{R/r_0}{\left(R^4/r_0^2 + r^2 - 2R^2\frac{r}{r_0}\cos\theta\right)^{\frac{1}{2}}}\right).$$

For example, we can evaluate the probability that the first part of the sphere that is hit is in the polar region with $\theta < \theta_e$ [19]:

$$\begin{aligned}\mathcal{P}[\text{hit polar region}] &= \frac{1}{4\pi R}\int_0^{\theta_e} \frac{R^2 - r_0^2}{(r^2 + r_0^2 - 2rr_0\cos\theta)^{3/2}} 2\pi R^2 \sin\theta d\theta \\ &= \frac{R^2 - r_0^2}{2r_0}\left(\frac{1}{R - r_0} - \frac{1}{(R^2 + r_0^2 - 2Rr_0\cos\theta_e)^{1/2}}\right).\end{aligned}$$

Escape Through a Small Hole

We can expand our model by adding an exit to the sphere. If a particle hits the exit, it will leave; if it hits the surface anywhere else it will be reflected back into the sphere (Fig. 17.3). This changes the boundary conditions in PDEs such as (17.7), and means that there is no longer symmetry in one of the angles. We now need $T(x_0) = f(r, \theta)$ such that

$$\nabla^2 f = \frac{1}{r^2}\frac{\partial}{\partial r}\left(r^2\frac{\partial f}{\partial r}\right) + \frac{1}{r^2 \sin\theta}\frac{\partial}{\partial\theta}\left(\sin\theta\frac{\partial f}{\partial\theta}\right) = -\frac{1}{D}, \tag{17.11}$$

with boundary conditions

$$\begin{aligned} f(R,\theta) = 0 &\quad \text{if} \quad 0 \le \theta < \theta_e \\ \frac{\partial f}{\partial r}(R,\theta) = 0 &\quad \text{if} \quad \theta_e \le \theta \le \pi, \end{aligned}$$

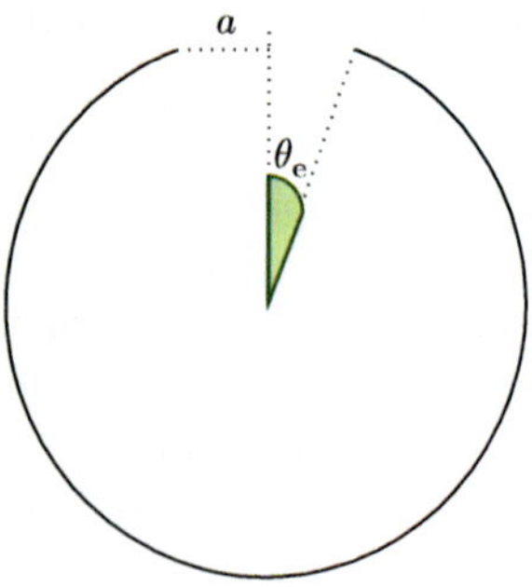

Fig. 17.3 A simple model: a sphere of radius R, whose surface is reflecting except in a polar region

where θ_e is the angle of the exit, given by $\theta_e = \arcsin \frac{a}{R}$ (Fig. 17.3). We can construct a solution using the Legendre polynomials:

$$f(r,\theta) = \frac{1}{6D}(R^2 - r^2) + \sum_{n=0}^{\infty} A_n r^n P_n(\cos\theta),$$

provided we can find coefficients A_n such that

$$\sum_{n=0}^{\infty} A_n R^n P_n(\cos\theta) = 0 \qquad 0 \leq \theta < \theta_e \tag{17.12}$$

$$\sum_{n=0}^{\infty} n A_n R^{n-1} P_n(\cos\theta) = \frac{1}{3}\frac{R}{D} \qquad \theta_e \leq \theta < \pi. \tag{17.13}$$

For example, taking $\theta = 0$ gives $\sum_n A_n = 0$. $\theta = \pi$ gives $\sum_n (-1)^n n A_n = \frac{2}{3}\frac{R}{D}$.

However, we will use the solution of a simpler problem. An exact solution exists for the steady-state probability density in the case of a flat reflecting surface with a circular absorbing hole, with radius a, using oblate spheroidal coordinates [21]. The circular hole is $z = 0$, $x^2 + y^2 < a^2$ and the reflecting surface is the remainder of the plane $z = 0$. The probability density is [22]

$$\rho(x, y, z) = \frac{2}{\pi}\rho_\infty \operatorname{atan} \zeta,$$

which is constant along ellipsoids

$$\frac{x^2 + y^2}{1 + \zeta^2} + \frac{z^2}{\zeta^2} = a^2.$$

On the absorbing circle, $\zeta = 0$. Elsewhere on $z = 0$, $a^2(\zeta^2 + 1) = x^2 + y^2$ and, because the ellipsoids intersect the plane at right angles, $\frac{\partial \rho}{\partial z} = 0$. On the absorbing circle $\frac{\partial \rho}{\partial z} = \frac{2}{\pi}\rho_\infty\left(x^2 + y^2\right)^{-\frac{1}{2}}$. Thus, the probability current [23] out of the hole, obtained by integrating $D\frac{\partial \rho}{\partial z}$ over the disc $x^2 + y^2 < a^2$, $z = 0$, is [21]

$$J = 4D\rho_\infty a. \tag{17.14}$$

If $a/R \ll 1$ then (17.14) will be a good approximation to the mean exit time from a volume $V = \frac{4}{3}\pi R^3$. A single particle, started at a random position inside the lymph node, corresponds to a probability density equal to $1/V$ [24]. The probability per unit time, or rate, of escape is thus

$$k = \frac{4Da}{V} = D\frac{3}{\pi}\frac{a}{R^3}.$$

The next term, in the small parameter a/R, can also be calculated [25].

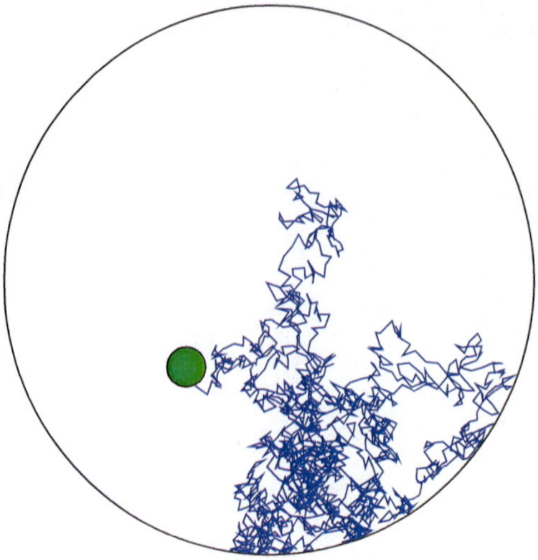

Fig. 17.4 Sample path of Brownian motion, reflected on the surface of the sphere, run until hitting a spherical zone of attraction *(green)*

Cell–Cell Encounters

If the surface of a lymph node is reflecting, how long does it take a T cell to find an APC? A simple approximation can be obtained by assuming that the T cell is a point particle, moving with diffusivity D, and the APC has radius b, is stationary and situated at the centre of a spherical lymph node with radius R (Fig. 17.4). Then the radially-symmetric version of the Poisson equation (17.2) and its general solution (17.3), yields the mean time, but with the conditions

$$F(b) = 0 \quad \text{and} \quad F'(R) = 0.$$

The mean time to hit the stationary central cell with radius b, starting a distance r from the centre, is

$$F(r) = \frac{R^3}{3D}\left(\frac{1}{b} - \frac{1}{r}\right) - \frac{1}{6D}(r^2 - b^2).$$

We obtain the average over all starting positions by taking the integral, and expanding in powers of b/R:

$$\tau = \frac{1}{\frac{4}{3}\pi(R^3 - b^3)} \int_b^R 4\pi r^2 F(r)\mathrm{d}r = \frac{R^3}{3Db} - \frac{3}{5}\frac{R^2}{D} + \frac{2}{3}\frac{b^2}{D} + \ldots. \qquad (17.15)$$

Modelling Interactions

While Brownian motion will undoubtedly be only an approximate description of lymphocyte motion, it has the advantage of simplicity. The only parameter is the diffusivity, which has been estimated from analysis of imaging data in vivo as

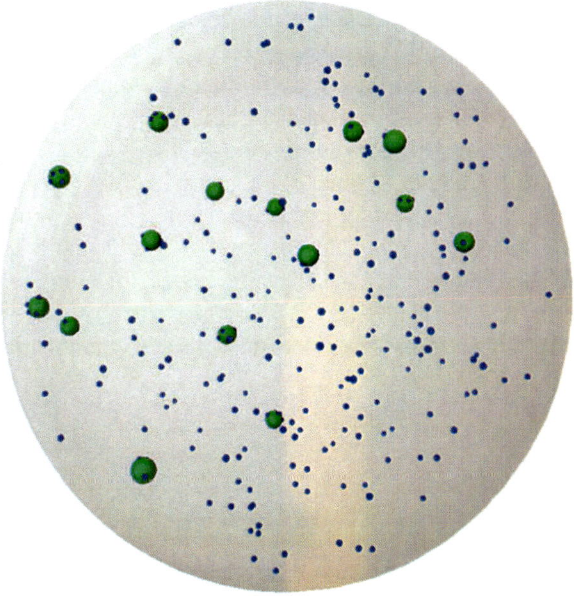

Fig. 17.5 Snapshot of a spatial model of cellular dynamics in a lymph node. The *green spheres* represent the zone of attraction of an APC. The *smaller blue spheres* represent T cells

50–100 $\mu m^2 min^{-1}$ [26–28]. It has also been reported that dendritic cells diffuse less rapidly [4, 27]. The assumption of Brownian motion is convenient for the construction of multi-cell lymph node models (Fig. 17.5).

In our computational model, T cells and DCs diffuse independently, unless they collide, inside a sphere. In a timestep Δt, the position of each cell is moved independently, with a displacement in each of the three space dimensions that is drawn from the zero-mean Gaussian distribution with variance $2D\Delta t$. Physical contacts between cells are deemed to occur when two or more cells are closer than a defined radius of attraction, that takes the sweeping effect of dendrites and local effect of chemokines into account. When a T cell and a DC meet, they remain together for a time that is an exponentially-distributed random variable: $\Pr(\tau_{\text{contact}} > t) = \exp(-\lambda t)$. The parameter λ is the inverse of the mean contact time. The choice of exponential distribution corresponds to a constant probability per unit time of the contact, if still formed, being broken.

With computational models it is possible to explore "sampling strategies" of T cells in the lymph nodes [29–36], for example numerous, short, contacts versus longer contacts [13] and strategies that maximise the time that T cells spend in the particular lymph node draining an infected site. It has been suggested that T cell priming by DCs follows three sequential stages [26]. The first, activation, phase is characterised by transient serial T cell–DC contacts, the second phase involves longer contacts and cytokine production and the last phase is characterised by high T cell motility and rapid proliferation. Three hypotheses have been proposed to explain how T cell–DC contacts change from being transient to long-lived [10]. In the first, DCs mature over time, increasing their expression of adhesion and costimulatory molecules, and so gain the ability to establish long-lived interactions with T cells. In the second, it is T cells that gain the ability to establish long-lived interactions, by summing signals received during short contacts with DCs until a threshold is reached. The third hypothesis is that, with a certain probability, each T cell–DC encounter can become a stable, long-lived interaction.

We intend to extend the computational model of T cells and DCs and their contacts to more realistic representations of lymph nodes, including reticular network and internal partitions. We hope that stochastic modelling of this type will help to identify relevant parameters that characterise T cells and their interactions with DCs in the lymph nodes.

References

1. Celli S, Garcia Z, Bousso P (2005) CD4 T cells integrate signals delivered during successive DC encounters in vivo. J Exp Med 202:1271–1278
2. Dustin M (2008) T-cell activation through immunological synapses and kinapses. Immunol Rev 221:77–89
3. Bajenoff M, Egen J, Qi H, Huang A, Castellino F, Germain R (2007) Highways, byways and breadcrumbs: directing lymphocyte traffic in the lymph node. Trends Immunol 28:346–352
4. Miller M, Wei S, Parker I, Cahalan M (2002) Two-photon imaging of lymphocyte motility and antigen response in intact lymph node. Science 296:1869–1873
5. Zinselmeyer B, Dempster J, Gurney A, Wokosin D, Miller M, Ho H, Millington O, Smith K, Rush C, Parker I, Brewer MCJ, Garside P (2005) In situ characterization of $CD4^+$ T cell behavior in mucosal and systemic lymphoid tissues during the induction of oral priming and tolerance. J Exp Med 201:1815–1823
6. Millington O, Zinselmeyer B, Brewer J, Garside P, Rush C (2007) Lymphocyte tracking and interactions in secondary lymphoid organs. Inflamm Res 56:391–401
7. Hauser AE, Junt T, Mempel TR, Sneddon MW, Keinstein SH, Henrickson SE, von Andrian UH, Shlomchik MJ, Haberman AM (2007) Definition of germinal-center B cell migration in vivo reveals predominant intrazonal circulation patterns. Immunity 26:655–667
8. Allen CD, Okada T, Tang HL, Cyster JG (2007) Imaging of germinal center selection events during affinity maturation. Science 315:528–531
9. Garside P, Brewer J (2008) Real-time imaging of the cellular interactions underlying tolerance, priming, and responses to infection. Immunol Rev 221:130
10. Bousso P (2008) T-cell activation by dendritic cells in the lymph node: lessons from the movies. Nat Rev Immunol 8:675–684
11. Beltman J, Henrickson S, von Andrian U, de Boer R, Marée A (2009) Towards estimating the true duration of dendritic cell interactions with T cells. J Immunol Methods 347:54–69
12. Beltman J, Marée A, de Boer R (2009) Analysing immune cell migration. Nat Rev Immunol 9:789–798
13. Beltman JB, Marée AFM, de Boer RJ (2007) Spatial modelling of brief and long interactions between T cells and dendritic cells. Immunol Cell Biol 85:306–314
14. Maroof A, Beattie L, Kirby A, Coles M, Kaye P (2009) Dendritic cells matured by inflammation induce CD86-dependent priming of naive CD8+ T cells in the absence of their cognate peptide antigen. J Immunol 183:7095
15. Beltman JB, Maree AF, Lynch JN, Miller MJ, de Boer RJ (2007) Lymph node topology dictates T cell migration behavior. J Exp Med 204:771–780

16. Bajénoff M, Egen J, Koo L, Laugier J, Brau F, Glaichenhaus N, Germain R (2006) Stromal cell networks regulate lymphocyte entry, migration, and territoriality in lymph nodes. Immunity 25:989–1001
17. Beauchemin C, Dixit N, Perelson A (2007) Characterizing T cell movement within lymph nodes in the absence of antigen. J Immunol 178:5505
18. Worbs T, Mempel T, Bolter J, von Andrian U, Forster R (2007) CCR7 ligands stimulate the intranodal motility of T lymphocytes in vivo. J Exp Med 204:489
19. Stirzaker D (2005) Stochastic processes and models. Oxford University Press, Oxford
20. Ito K, McKean Jr H (1974) Diffusion processes and their sample paths. Springer, New York
21. Berg H, Purcell E (1977) Physics of chemoreception. Biophys J 20:193–219
22. Hill T (1975) Effect of rotation on the diffusion-controlled rate of ligand-protein association. Proc Natl Acad Sci USA 72:4918–4922
23. Gardiner CW (2004) Handbook of stochastic methods for physics, chemistry and the natural sciences, 3rd edn. Springer, New York
24. Grigoriev I, Makhnovskii Y, Berezhkovskii A, Zitserman V (2002) Kinetics of escape through a small hole. J Chem Phys 116:9574
25. Singer A, Schuss Z, Holcman D (2008) Narrow escape and leakage of Brownian particles. Phys Rev E 78:51111
26. Mempel T, Henrickson S, Von Andrian U (2004) T-cell priming by dendritic cells in lymph nodes occurs in three distinct phases. Nature 427:154–159
27. Miller M, Wei S, Cahalan M, Parker I (2003) Autonomous T cell trafficking examined in vivo with intravital two-photon microscopy. Proc Natl Acad Sci USA 100:2604–2609
28. Miller M, Safrina O, Parker I, Cahalan M (2004) Imaging the single cell dynamics of CD4+ T cell activation by dendritic cells in lymph nodes I. Parker and MD Cahalan. J Exp Med 200:847–856
29. Catron D, Itano A, Pape K, Mueller D, Jenkins M (2004) Visualizing the first 50 Hr of the primary immune response to a soluble antigen. Immunity 21:341–347
30. Textor J, Westermann J (2007) Modeling migration, compartmentalization and exit of naive T cells in lymph nodes without chemotaxis. Lect Notes Comput Sci 4628:228
31. Riggs T, Walts A, Perry N, Bickle L, Lynch J, Myers A, Flynn J, Linderman J, Miller M, Kirschner D (2008) A comparison of random vs. chemotaxis-driven contacts of T cells with dendritic cells during repertoire scanning. J Theor Biol 250:732–751
32. Chavali A, Gianchandani E, Tung K, Lawrence M, Peirce S, Papin J (2008) Characterizing emergent properties of immunological systems with multi-cellular rule-based computational modeling. Trends Immunol 29:589–599
33. Bogle G, Dunbar P (2008) Simulating T-cell motility in the lymph node paracortex with a packed lattice geometry. Immunol Cell Biol 86:676–687
34. Bogle G, Dunbar P (2010) Agent-based simulation of T-cell activation and proliferation within a lymph node. Immunol Cell Biol 88:172–179
35. Baldazzi V, Paci P, Bernaschi M, Castiglione F (2009) Modeling lymphocyte homing and encounters in lymph nodes. BMC Bioinformatics 10:387
36. Linderman J, Riggs T, Pande M, Miller M, Marino S, Kirschner D (2010) Characterizing the dynamics of CD4+ T cell priming within a lymph node. J Immunol 184:2873

Chapter 18
Using Mathematical Models to Explore the Role of Cytotoxic T Lymphocytes in HIV Infection

Helen Fryer and Angela McLean

Abstract The combination of mathematical modelling and data analysis to understand the within-host dynamics of human immunodeficiency virus (HIV) infections has been one of the most informative uses of mathematical biology in the last decade. Simple models of viral dynamics together with viral load measurements provided an early estimate that the turnover of HIV infected cells is very rapid: most do not survive beyond 1 day. Although this estimate was initially a surprise to the field, further corroborating evidence has made it widely accepted. More recently, within-host models have been used to investigate the efficacy of cytotoxic T-lymphocytes (CTLs) in controlling HIV infection. Though there is clear experimental evidence that they play some role, the magnitude of this role remains contentious. Models have offered three insights on this topic. Firstly, in chronically infected humans fewer than 20% of HIV-infected cell death is attributed to killing by CTLs. Secondly, CTLs are more efficient in acute infection than chronic infection, but not dramatically so, and thirdly, CTLs are markedly more efficient in simian than human immunodeficiency viral infection. Although based on simple models and repeatable data, the main prediction of this work that CTL vaccines might work in macaques but not in humans is yet to gain recognition. This is despite the fact that this prediction was borne out by the failure of STEP vaccine. We contend that in time this assertion too will become more widely accepted.

With an estimated 33 million people affected by HIV/AIDS across the world [1], a huge research effort is now focussed on understanding the immune response to Human Immunodeficiency Virus (HIV) in a bid to creating an effective vaccine. Early HIV vaccines aimed to elicit antibodies, but quickly proved to be completely ineffective [2]. The reason for this is now clear: even during natural infection, HIV is able to evade antibody neutralisation, primarily by obstructing antibody binding sites. Over the last 15 years, several pieces of evidence have demonstrated that cytotoxic T lymphocytes contribute to the control of HIV infection in vivo. These

H. Fryer (✉)
The Institute for Emerging Infections, The Oxford Martin School,
Department of Zoology, University of Oxford, Oxford OX13PS, UK
e-mail: helen.fryer@zoo.ox.ac.uk

C. Molina-París and G. Lythe (eds.), *Mathematical Models and Immune Cell Biology*,
DOI 10.1007/978-1-4419-7725-0_18,

include a temporal correlation between the reduction of viremia in acute infection and the appearance of HIV-specific CTLs [3,4]; the selection of viral mutants that evade killing by CTLs [5,6]; and associations between particular HLA class I alleles and life expectancy [7–9]. Though the magnitude of CTLs' contribution to viral regulation is contentious, these findings have encouraged the vaccine field to direct research towards developing a vaccine that elicits CTLs. The aim of such a vaccine would be to dampen viral replication in hosts who become infected.

In 2004 the first large-scale human efficacy trial (the STEP Study) of a CTL-based vaccine candidate began. Considerable hope was pinned on this vaccine, but in 2007 the trial was halted when interim results showed the vaccine to be entirely ineffective both at preventing infection and reducing viral load in those who become infected [10, 11] (Table 18.1 and Fig. 18.1). This result is regarded as a

Table 18.1 Results of the STEP study. The number of HIV infections according to baseline Ad5 antibody titre. The Merck vaccine trial was halted early when it became apparent that the infection rate in individuals with high immunity to the Ad5 vector was higher in those who were vaccinated than in those who received the placebo. Even in individuals with low Ad5 immunity, the vaccine provided no protection [15]

	Number of HIV infections	
Ad5 antibody titre prior to innoculation	Vaccine	Placebo
<18	20/382	20/394
$18 < \text{Ad5} \leq 200$	8/140	4/142
$200 < \text{Ad5} \leq 1000$	14/229	7/229
$\text{Ad5} > 1000$	7/163	2/157
Total	49/914	33/922

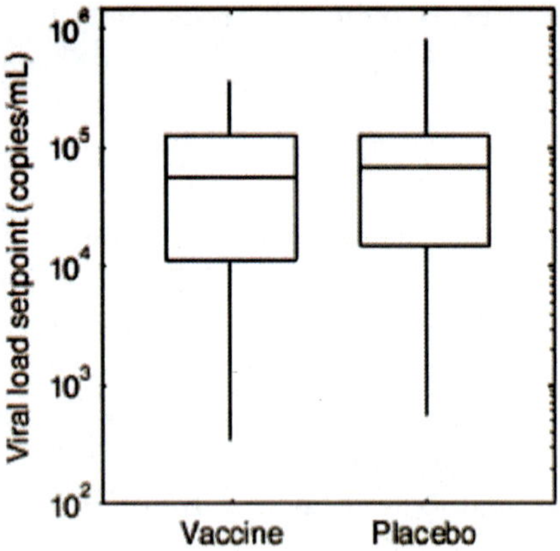

Fig. 18.1 Boxplots of the viral setpoints measured in individuals who became infected in the STEP study. There was no difference in the viral setpoints measured in the vaccine ($N = 46$) and placebo recipients ($N = 30$). When this analysis was restricted to patients who had low Ad5 immunity, the same result was obtained (data not shown) [15]

major setback for the HIV vaccine field, not least because the candidate (the Merck HIV DNA/adenovirus complex) had shown the most promise of several proposed CTL-based vaccines that were tested in monkeys challenged with simian-human immunodeficiency virus (SHIV)[12]. The failure of this vaccine has reignited doubt that CTLs play a pivotal role in controlling HIV replication and has thus led the field to question whether CTLs will ever prove a useful tool for an HIV vaccine. In addition, the failure has highlighted the inconsistency between CTL efficacy in monkeys and humans. In light of these issues, this chapter reviews how mathematical models have been used together with experimental results to address the question: how effective are CTLs are in controlling HIV-1 infection?

The chapter begins with a discussion of experimental and observational results that provide insight into the role of CTLs in controlling HIV infection. This is followed by a review of models of viral dynamics that have been used to interpret patient viral load data in relation to this question. Finally, a model of the infection of CD4 T cells with either the wildtype viral strain or a CTL escape mutant strain [13] is discussed. We show how this model has been used to analyse CTL escape mutant outgrowth data to quantify the magnitude of the CTL response. Further, we show how it has been used to compare the strength of the HIV-specific CTL response in humans to that of the SIV/SHIV-specific response in macaques [14].

Experimental and Observational Results that Provide Insight into the Role of CTLs in Controlling HIV Infection

Today there is considerable evidence to suggest that CTLs contribute to the control of HIV infection in vivo. What is not clear is exactly how important their contribution is at each stage of infection. Measuring their contribution directly is not possible and there is some evidence against a significant role for CTLs; notably certain correlations predicted on the basis of a strong CTL response have not been observed. Data indicating an important role for CTLs in controlling HIV in infected individuals are summarised in Fig. 18.2.

One of the early observations in favour of a role for CTLs came in 1994 when a temporal correlation was established between the reduction in viremia in acute infection and the appearance of HIV-specific CTLs [3, 4]. The appearance of the CTLs preceded the appearance of antibodies, indicating that antibodies were not responsible for the observed effect (Fig. 18.2(a)). Other evidence is provided by the selection of CTL escape mutations in vivo [5, 6]; by in vitro experiments that have revealed CTLs capable of exerting potent antiviral effects towards HIV [16–19] (Fig. 18.2(b)); and by the high levels of anti-HIV-1 CTLs seen in established HIV-1 infection. These can often be detected directly from the peripheral blood without in vitro stimulation. This confirms that the CTLs are ‘at work’, rather than simply waiting to be stimulated into action.

Probably the best evidence in favour of a *significant* contribution by CTLs, however, comes from clear associations between progression to AIDS and different

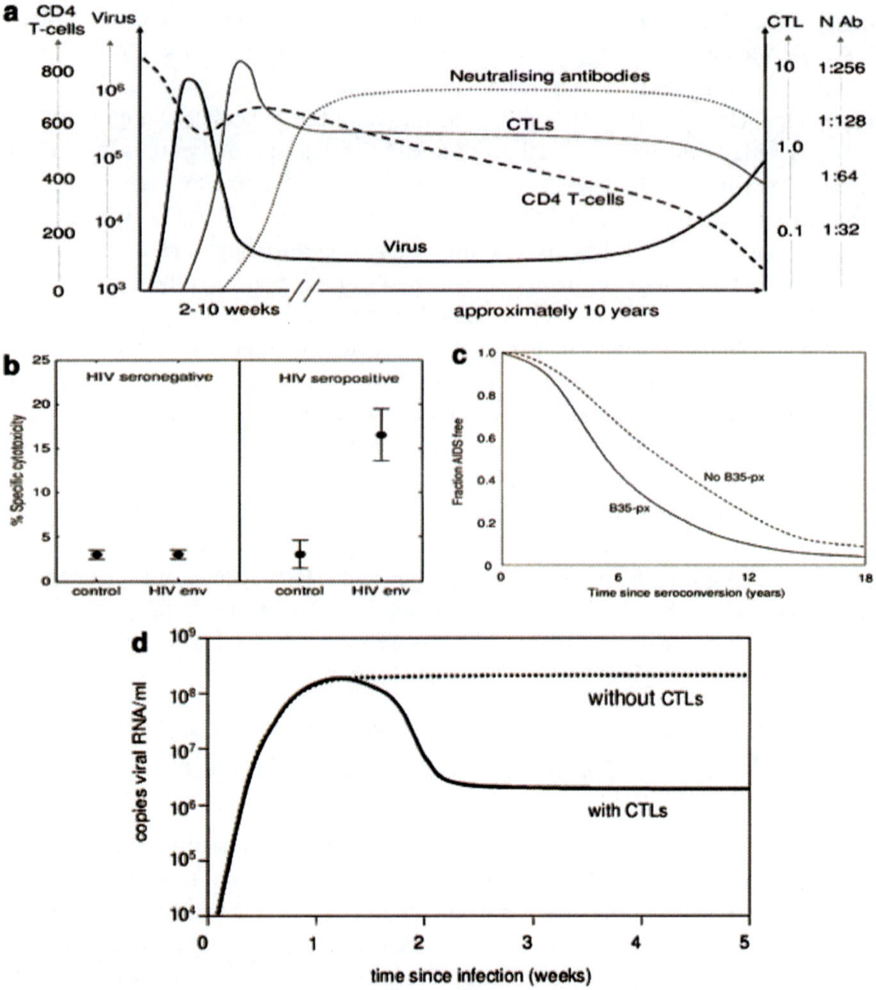

Fig. 18.2 (**a**) The time course of natural HIV infection in a typical individual. A temporal correlation exists between the containment of viremia during acute infection and the appearance of HIV-specific CTLs. The measurements shown are the CD4$^+$ T cell count (cells/μl), the plasma viral load (RNA copies/ml), the CTL response (% effector CTL in PBMC) and the neutralising antibody titre [31]. (**b**) Evidence that CTLs target HIV-infected cells in vitro. The mean percent specific CTL cytotoxicity and corresponding standard error is presented for four seropositive and five seronegative individuals. The percent specific cytotoxicity is shown for the effector:target ratio 25:1. Cytotoxicity is measured in vitro against two vaccinia recombinants: one containing the *env* region of the HIV genome and one containing the bacterial lacZ gene used as a control. The HIV-specific response to the env/Vac complex was significantly elevated in seropositive individuals compared to the response to the lacZ/Vac (control) complex in the same hosts and compared to the env/Vac complex in seronegative individuals. Note that some background cytotoxicity directed against the Vaccinia virus is expected in all hosts [16]. (**c**) The association between the HLA B35Px allele and progression to AIDS. Infected individuals who carry at least one copy of the HLA B35Px allele progress to AIDS more rapidly than those who contain no copies of HLA B35px [20, 22]. (**d**) The effect of depletion of CTLs during primary SIV infection in macaques. The early control of virus fails when CTLs are depleted [20, 22]

HLA alleles. These alleles encode for the HLA molecules involved in the process that results in CTL targeting. Notably, HLA types B57, B51 and B27 are associated with slow progression and HLA B35Px is associated with rapid progression (Fig. 18.2(c)). Patients who are homozygous at any of the three HLA class I loci also progress substantially more rapidly to disease [7–9]. More recently, Kiepiela and colleagues have also shown that the part of the genome targeted by CTLs is an important determinant of disease progression. They found that an increased breadth of gag-specific responses is associated with lower levels of viremia, whereas an increased breadth of env-specific and accessory/regulatory-specific responses is associated with increased levels of viremia. The reason for these protein-specific differences is unclear.

HIV research that cannot be conducted in humans is instead tested in non-human primates infected either with simian immunodeficiency virus (SIV) or the hybrid virus simian-human immunodeficiency virus (SHIV). The primates used for these tests are typically rhesus macaques or pigtail macaques, chosen because upon infection they reproduce many of the key elements of HIV infection in humans. Not only do they experience $CD4^+$ T cell depletion leading eventually to AIDS-like symptoms, but they also have similar immune systems to humans. Like humans, macaques have been shown to elicit CTL responses against immunodeficiency virus infection. Furthermore, when antibodies specific for CTLs were used to experimentally deplete CTLs during acute SIV/SHIV infection, the early control of virus failed [20–22] (Fig. 18.2(d)). When the antibodies were instead given in the chronic phase, the virus level rose until the effects of the antibody wore off. Depletion of B-cells had no effect. These results favour a role for CTLs in controlling the set point in natural HIV infection, though should be viewed in light of the unknown extent to which results from macaques can be extrapolated to humans.

Evidence of CTL involvement in controlling HIV infection is therefore clear. Exactly how important they are in this process, however, is put into question by the absence of certain results. If CTLs were crucial in containing HIV, one might expect that counts of HIV-specific CTLs would correlate with viral load. Despite early reports to this effect [23–25], it is now accepted that no such simple correlation exists [26, 27]. In an examination of CTL responses directed against the entire HIV genome in 57 patients, no correlation was found between either the breadth or the magnitude of the CTL responses and viral load [27]. A second study into the effects of drug therapy interruptions also showed no correlation between CTL count and either viral rebound rates at the start of the interruption, or viral clearance rates at the restart of therapy [26]. Furthermore, neither measures of absolute numbers nor changes in numbers of CTLs were able to predict viral load in this experiment. At the very least, these results indicate that different CTLs have different efficacies, a finding supported by the study by Kiepiela et al. [28] and by two recent studies that suggest that a correlation may exist between the number of polyfunctional CTL (as measured by five different immunological markers) and plasma viral load [29, 30].

Modelling HIV-1 Viral Dynamics to Understand What Patient Viral Load Data Tells Us About the Importance of CTLs in Controlling HIV

One way in which the importance of CTLs in controlling HIV infection can begin to be addressed is by investigating how well observed viral load data can be explained by mathematical models of viral dynamics. Viral dynamics during HIV infection have been widely analysed and it is now accepted that there are three main phases to the dynamics in natural infection [31] (Fig. 18.2(a)). During the first few weeks (acute infection), viral load rises to a peak before declining and settling to an equilibrium called the viral set point. The start of this decline is typically associated with the appearance of CTLs specific for HIV. Once at the set point, viral load remains at that level for a period of several years (the chronic phase), during which there is both rapid viral replication and rapid viral loss. The set point differs between hosts and is predictive of life expectancy – higher set points correlate with shorter life expectancies. Eventually this balance is disrupted and viral load gradually increases in concordance with a decline in CD4 count that eventually results in AIDS. Information about viral dynamics can also be attained by disrupting the natural course of infection. This can be achieved when antiretroviral therapy is started. These drugs disrupt the viral life cycle and when given during chronic infection, lead to viral load decaying in an exponential manner.

A Simple Model of Virus Infection

A simple model that has been proposed to explain some of the observed HIV viral dynamics is shown in Fig. 18.3 [32]. The model consists of three variables: the

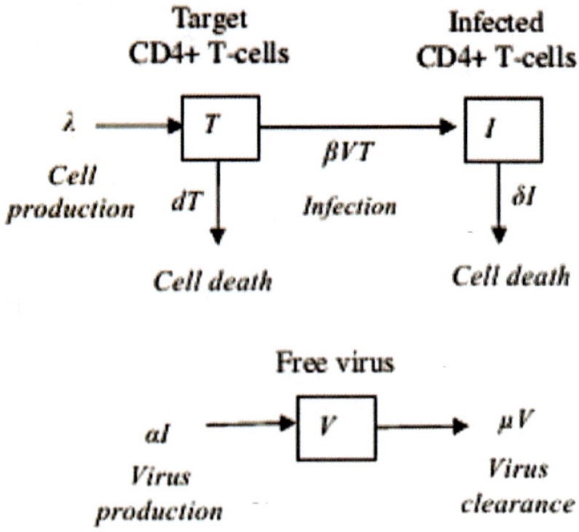

Fig. 18.3 A simple within-host model of viral infection. T, I and V represent the number of target cells, infected cells and free virions, respectively. Free virions are produced at a rate, α, per infected cell and infect target cells at a rate proportional to the abundance of free virions, βV. In addition, target cells are produced at rate λ and die at a rate d per cell; free virions are cleared at rate μ per virion; and infected cells die at rate δ per cell

population size of target cells (i.e. uninfected $CD4^+$ T cells), T, infected cells, I, and free virions, V. Free virions are produced at a rate, α, per infected cell and infect target cells at a rate proportional to the abundance of free virus, βV. In addition, target cells are produced at rate λ and die at a rate d per cell; free virions are cleared at rate μ per virion and infected cells die at rate δ per cell. The model is prescribed by the following set of coupled ordinary differential equations.

$$\begin{aligned}\dot{T} &= \lambda - dT - \beta VT \\ \dot{I} &= \beta VT - \delta I \\ \dot{V} &= \alpha I - \mu V.\end{aligned} \tag{18.1}$$

Fitting the Simple Target-Cell-Limited Model to Viral Load Data from Hosts with Natural HIV Infection

In the simplest form of this model, the death rate of infected cells, δ, can be assumed to be constant through time. As such, the model includes no explicit CTL response. Despite this, the model is nevertheless able to reproduce many of the viral dynamics observed in HIV-infected individuals; namely it predicts that virus will grow rapidly up to a peak before declining and settling at a stable equilibrium at which there is continuous production and loss of virus. It is perhaps surprising that a model that includes no explicit introduction of a CTL response during acute infection is able to reproduce the early containment of virus so often attributed to CTLs. In fact, in this model it is saturation of the target cell population that is the cause of viral suppression. The simple model of infection of target cells therefore puts into question the hypothesis that acute viral containment is due to CTLs and suggests that the observed temporal correlation may be coincidental. In support of this assertion, Phillips [33] has cited three studies in which infected individuals suppressed their primary viremia without mounting any detectable specific anti-viral immune responses.

Although saturation can explain the drop in viral load qualitatively, how well it can explain the drop quantitatively has been questioned. Stafford et al. [34] fitted viral load data from ten patients in acute infection. They showed that, by assuming only small variations in different model parameters across patients, the model could accurately explain both the increase in viremia up to the peak and the set point that was eventually reached, but could not explain the seemingly overly-rapid rates of viral decay seen in some individuals. These fast decay rates could be explained by the appearance of CTLs during acute infection. The fact that viral set points can be explained without inclusion of a CTL response, however, implies that even if the CTL response is important during acute infection, it is not explicitly important during chronic infection.

Does Progression to AIDS Happen Because of a Failing CTL Response?

It has often been suggested that it is the failure of the CTL response that eventually disrupts the equilibrium observed during chronic phase and leads to increasing viremia and eventual progression to AIDS. If this were true, it would favour the theory that CTLs are crucial in controlling HIV during chronic infection, as has been shown to be true in SIV/SHIV infected in macaques. In these monkeys, depletion of CTLs during chronic infection leads to an increase in viremia.

It has been proposed that, if progression to AIDS results from a failing CTL response, then individuals with more advanced disease would kill infected cells more slowly. Conversely if AIDS results from the virus evolving to infect cells more quickly, infected cells would be cleared more quickly in individuals with more advanced disease. To investigate this issue, curves showing the decay of viremia following treatment with a protease inhibitor (PI) – a type of antiretroviral – in chronically infected humans have been compared to decay curves predicted by a version of the simple target-cell limited model that is adapted to include PI therapy during chronic infection. Protease inhibitors do not affect the survival or rate of virion production of infected cells, but do prevent the production of *infectious* virions. To model the impact of PI therapy, infectious virions V_{I} and non-infectious virions V_{NI} are regarded separately and the production rate of infectious virions is set to zero. Furthermore, for the short period over which viral decay has been measured (approximately 1 week) the number of target cells is assumed to remain constant at the steady state value attained prior to therapy, T_0. Equations representing the adapted model are given by:

$$\begin{aligned} \dot{I} &= \beta V_{\mathrm{I}} T_0 - \delta I \\ \dot{V}_{\mathrm{I}} &= -\mu V_{\mathrm{I}} \\ \dot{V}_{\mathrm{NI}} &= \alpha I - \mu V_{\mathrm{NI}}. \end{aligned} \tag{18.2}$$

During the steady state achieved prior to therapy, the abundance of target cells, productively infected cells and free virus can be regarded as constant. Thus by setting $T = I = V = 0$ in the original model (18.1), expressions can be found for α and I_0 ($\alpha = \delta\mu/\beta T_0$ and $I_0 = \beta V_0 T_0/\delta$) which enable (18.2) to be solved to give an expression for the decay of virus ($V = V_{\mathrm{NI}} + V_{\mathrm{I}}$) upon commencement of therapy:

$$V(t) = V_0 \mathrm{e}^{-\mu t} + \frac{\mu V_0}{\mu - \delta}\left(\frac{\mu}{\mu - \delta}(\mathrm{e}^{-\delta t} - \mathrm{e}^{-\mu t}) - \delta t \mathrm{e}^{-\mu t}\right) \tag{18.3}$$

This function fits patient data in which plasma viral load after PI treatment is seen to decay in an exponential fashion following an initial lag [35] (Fig.18.4(a)). Importantly, this equation shows that the exponential viral decay rate will be faster if clearance of infected cells (rate δ) is faster. It is noteworthy that this correlation would hold true irrespective of whether clearance is CTL mediated. The hypothesis

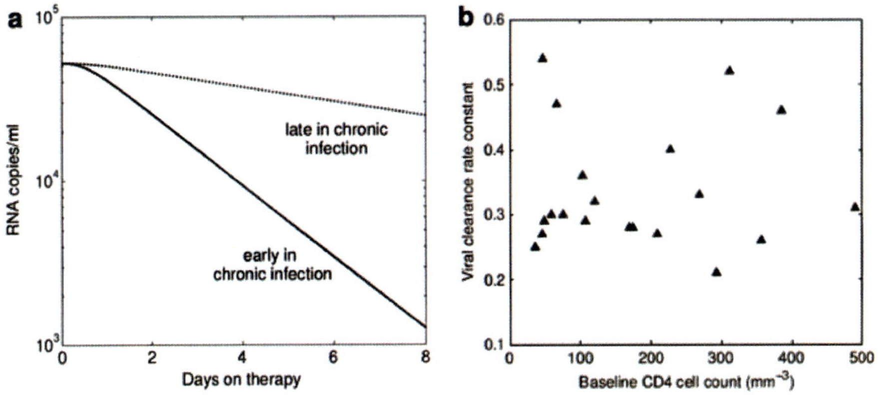

Fig. 18.4 The decline of viremia upon commencement of treatment with a protease inhibitor in individuals with chronic HIV infection. (**a**) An example of the predicted decline in viral load following therapy in patients late in chronic infection and in patients early in chronic infection under the assumption that patients with further progressed disease clear infected cells more slowly (see (18.3)). The rate of viral decline would be faster in patients early on during chronic infection. (**b**) A comparison of pre-therapy CD4 counts and viral clearance rates in 20 individuals. No relationship exists between these two variables, implying that progression to AIDS does not result from a failing CTL response [36]

that individuals with more advanced disease clear infected cells more slowly can therefore be tested by comparing therapy-induced viral decay rates to pre-therapy CD4 T cell counts. Interestingly, a comparison of such data shows that not only is there no relationship between pre-therapy CD4 T cell counts and infected cell clearance rates, but that infected cell clearance rates are remarkably consistent across different individuals [36]. These results imply that people do not progress to AIDS because clearance of infected cells diminishes. They therefore also provide no evidence that CTLs are important in controlling HIV during chronic infection.

In summary, a simple within-host mathematical model of viral dynamics can be used to understand what observed viral dynamics reveal about the importance of the CTL response in HIV infection. This analysis shows that much of the observed viral dynamics seen in natural infection can be explained *qualitatively* by a model that includes no explicit CTL response. This includes containment of virus during acute infection through saturation of the target cells. When the model and the data are compared *quantitatively*, there remains no explicit evidence that CTLs are important during chronic infection; however, in some individuals, rates of viral containment during acute infection are faster than predicted by the simple model, suggesting that CTLs may indeed be functionally important during acute infection, as has been shown to be to be the case for macaques using CTL depletion experiments. To investigate further whether CTLs are important during chronic infection, viral clearance rates have been compared to pre-therapy CD4 T cell counts. No relationship has been found between these two variables, suggesting that progression to AIDS does not occur because of a failing CTL response. This comparison therefore also provides no evidence in support of CTLs being functionally important during chronic HIV infection. This is in disagreement with results from macaques in which depletion of CTLs during chronic SIV/SHIV infection leads to significantly increased levels of viremia.

Modelling Within-Host Dynamics of CTL Escape and Reversion to Estimate the Proportion of Cell Lysis Attributable to CTLs

There is good evidence to show that the average life expectancy of productively infected $CD4^+$ T cells is considerably less than that of $CD4^+$ T cells present in uninfected individuals [35–40]. What is not known is what proportion of this cell death is attributable to CTLs and what proportion is attributable to other factors–notably viral cytotopathicity and by-stander activation.

Longitudinal data relating to the outgrowth of CTL escape mutants within individuals is able to provide us with insight into this issue (Fig. 18.5(a)). For a start, it is clear from the fact that selection of CTL escape mutants occurs, that CTLs are responsible for some proportion of infected cell death. Secondly, it is intuitive that the rate of outgrowth of a particular CTL escape mutant should reflect three main quantities: the strength of the CTL response that is being evaded, the fitness constraints associated with the mutant and the rate of cell turnover. Fitness costs cannot be measured directly. However, fitness costs are believed to drive the reversion of

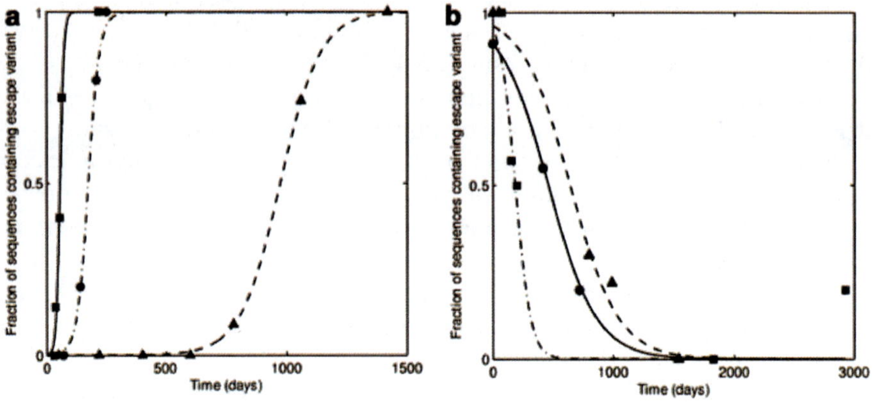

Fig. 18.5 Escape and reversion data and theoretical fits. (**a**) Three examples of the selection of an escape mutant in an individual who makes an immune response to the epitope of interest. The function $1/(ge^{-kt}+1)$ is fitted to each of the escape curves to estimate the value of the escape rate, k. The estimated escape rates for these three events are, 0.012, 0.119 and 0.41 day^{-1}. (**b**) Three examples of the reversion of an escape mutant in an individual who does not make an immune response to the epitope of interest. The function $1/(ge^{\phi t}+1)$ is fitted to each the reversion curves to estimate the value of the fitness cost (or reversion rate), ϕ. The estimated fitness costs for these three events are, 0.016, 0.005 and 0.005 day^{-1}

escape mutants back to the wildtype form in hosts who do not mount an immune response to the epitope [41, 42] (Fig. 19.5b). This would be most probable when an escape mutant selected in one host is transmitted to a different host who does not carry the selecting HLA and so cannot recognise the epitope. Longitudinal data on the reversion of escape mutants is therefore able to provide information on fitness costs, given that it is again intuitive that the rate at which reversion occurs in this setting is an indication of the fitness cost of the mutant–faster reversion implies higher cost.

Asquith et al. [13] recently formulised these ideas into a mathematical model. The model enables a quantitative estimate to be made from longitudinal CTL escape and reversion data on the proportion of infected cell lysis that can be attributed to CTLs. Here we describe the model and show how it can be used to derive this important result.

A Within-Host Model of Outgrowth of Escape Mutants

The model encapsulates the dynamics of outgrowth of an escape mutant at a single CTL epitope and includes two different types of productively infected $CD4^+$ T cells – those that are infected with a wildtype viral strain, y, and those that are infected with a viral strain that has an escape mutation in the epitope, x. Cells infected with the wildtype strain replicate at rate a, (the net contribution of all factors except CTL-mediated death), are killed by CTLs specific for epitopes other that the epitope of interest at rate b, and are killed by CTLs specific for the epitope of interest at rate c. For hosts who target the epitope c has a positive value ($c > 0$), whereas for hosts who do not target the epitope, c equals zero ($c = 0$). Cells infected with the mutant strain replicate at rate a' and are killed by CTLs specific for epitopes other than the epitope of interest at rate b. See Fig. 18.6. The pair of first-order ordinary differential equations that define the model are

$$\begin{aligned}\dot{y} &= ay - by - cy \\ \dot{x} &= a'x - bx\end{aligned} \tag{18.4}$$

It is therefore assumed that in any given host, the number of CD4 cells infected with the wildtype strain changes in an exponential manner at rate $a - b - c$ ($y = y(0)\mathrm{e}^{(a-b-c)t}$) and that the number of CD4 cells infected with the mutant strain change in an exponential manner at rate $a' - b$ ($x = x(0)\mathrm{e}^{(a'-b)t}$); note that these growth rates can be negative. In order to understand how effective CTLs are, the first aim is to estimate an average value for c, the rate of CTL-mediated lysis directed against an epitope. This quantity cannot be measured directly, but can be estimated indirectly from longitudinal escape and reversion data using two expressions derived from the model. These are the rate of escape of a mutant (given by k) and the fitness cost of a mutant (given by ϕ). To estimate these values, the model must first be manipulated to extract an expression for the quantity that can be measured in hosts, namely the proportion of virus that has the escape mutant, $p(t) = x(t)/(x(t) + y(t))$.

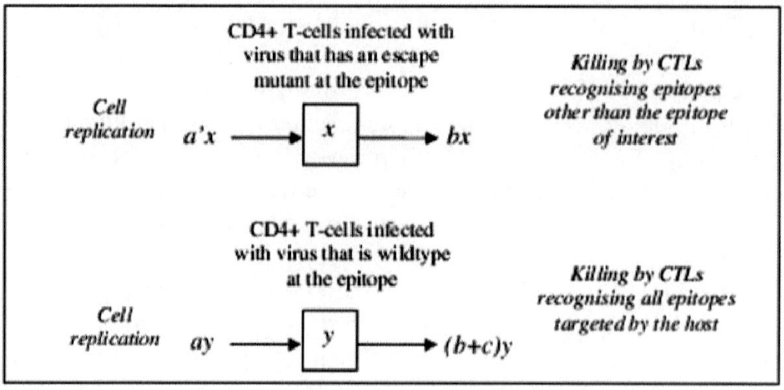

Fig. 18.6 A model of the dynamics of productively infected CD4$^+$ T cells. Cells can either be infected with virus that has an escape mutant at the epitope of interest, x, or with virus that is wildtype at the epitope, y. Cells infected with an escape mutant replicate at rate a' per cell and are killed at rate b per cell. Cells infected with wildtype virus replicate at rate a per cell and are killed at a faster rate, $b + c$ per cell

The Rate of Escape

In hosts who mount an immune response to the epitope ($c > 0$), $p(t)$ is able to increase over time as the mutant strain outgrows the wildtype strain. In the context of the model this can be expressed as

$$p(t) = \frac{1}{ge^{-kt} + 1} \tag{18.5}$$

where $g = y(0)/x(0)$ and $k = c - a + a'$ (the rate of escape) is the difference between the growth rate of the escape variant, $a' - b$, and growth rate of the wild-type variant, $a - b - c$. The rate of escape, k, can thus be estimated by fitting the longitudinal escape mutant outgrowth data to the function $1/(ge^{-kt} + 1)$ (Fig. 19.5).

Using this method Asquith et al. [14] derived estimates of the rate of escape, k, for 21 reported CTL escape variants. The rates were consistently slow across the different escape variants: in 20 of the 21 datasets the rate of escape was less than 1 day^{-1} and the median rate was 0.01 day^{-1}.

The Fitness Cost of an Escape Mutant (ϕ)

In hosts who are infected with an escape mutant strain but do not mount a CTL response to the epitope, the proportion of virus that is mutant, $p(t)$, can decline over time as the fitness-compromised mutant strain is replaced by the wildtype strain. Assuming $c = 0$ (no CTL directed against epitope), $p(t)$ can be expressed as

$$p(t) = \frac{1}{ge^{\phi t} + 1} \tag{18.6}$$

where $\phi = a - a'$ (the fitness cost or reversion rate of the escape mutant) is the difference in the replication rate between the wildtype strain, a, and the escape mutant strain, a'. The fitness cost ϕ can thus be estimated by fitting the longitudinal reversion data to the function $1/(ge^{\phi t} + 1)$ (Fig. 19.5b).

To see if the slow escape rates could be explained by high fitness costs, Asquith et al. estimated the fitness costs from seven reversion datasets using this method. Their analysis revealed that the fitness cost of an average CTL escape mutation is very low – approximately 0.005 day^{-1}. This slow rate is consistent both with data from in vitro competition assays [43] and with the observation that many escape mutants accumulate in the population over time [44–46].

The Rate of Killing of Productively Infected Cells by a CTL Response Against a Single CTL Epitope (c)

Since $k = c - a + a'$ and $\phi = a - a'$, the rate of escape can be simplified to equal $k = c - \phi$. The rate of lysis of infected cells by CTLs against a particular epitope, c, can therefore be derived directly from estimates of the escape rate, k, and the fitness cost, ϕ:

$$c = k + \phi.$$

Asquith et al. found that the estimated escape rates and fitness costs are consistent across different individuals. It is therefore meaningful to use the average estimate for each of these quantities to estimate the rate of lysis of infected cells by a single CTL response. Given that the escape rate is estimated to be 0.01 day^{-1} and the fitness cost is estimated to be 0.005 day^{-1}, the rate of lysis by CTLs specific for a single epitope is estimated to be 0.02 day^{-1}. This means that an infected cell would have a lifespan of approximately 50 days if it were subject only to lysis by CTLs specific for one epitope. The estimated average lifespan of a productively infected cell is considerably shorter at approximately 1 day [35–38,40]. Taken together these figures imply that only about 2% of infected cell death is attributable to the CTL response directed against a single epitope.

The Rate of Killing of Productively Infected Cells by CTL: Estimating the Total Response Against Different Epitopes

Most infected individuals do not just elicit a CTL response against a single epitope, therefore the total degree of cell death attributable to CTL-mediated lysis within any individual is likely to be more than 2%. To extrapolate the single-epitope figure to understand the total strength of the CTL response within a typical individual, an estimate is needed for the number of epitopes targeted by CTLs within a typical host. Comprehensive epitope analysis put this average at between 14 and 19 epitopes [27,47], however, typically less than five of these epitopes will be sufficiently strong to drive CTL escape [48]. Given that the data upon which the single-epitope killing rate was based related to CTL responses strong enough to drive escape – in fact most were the patient's immunodomiant response – Asquith et al. used a range of between 5 and 14 epitopes to estimate that CTL-mediated lysis of infected cells contributes to only about 10–20% of all infected cell death.

Comparing the Strength of the CTL Response in Primary Infection and Chronic Infection

It is often assumed that the strength of CTL response in HIV infection is consistent between primary and chronic infection. For each of the 21 datasets, Asquith et al. estimated the epitope-specific rate of the killing by CTL. For this calculation a uniform fitness cost of 0.005 day^{-1} was assumed for each dataset. The results show that the rate of killing by a single CTL response is markedly slower in chronic infection compared to primary infection. This is consistent with the findings derived from analysing viral load dynamics using the target-cell limited model; namely that CTLs have some impact during acute infection, but there is no explicit evidence that they are important during chronic infection.

Comparing the Strength of the CTL Response in Humans and Macaques

Before HIV vaccine candidates are tested in humans, they undergo efficacy tests in non-human primates – typically macaques – challenged with SIV or SHIV. It is clear that, like humans, they produce a CTL response against the virus, though it is not clear how the strength of this response compares to the strength of the CTL response in humans. This is a pertinent question in view of the recent (and unequivocal) failure in humans of the Merck vaccine [11] that had previously proved to be effective in SHIV challenged macaques [12].

To compare the strength of the CTL response between humans and macaques, longitudinal CTL escape and reversion data from macaques can be analysed in exactly the same way as has been described for the human data. Asquith et al. [14] presented a summary of 35 escape events and 9 reversion events observed in macaques that were used for this comparison. In the majority (23/35) of the escape events, the outgrowth of the CTL escape variant was so rapid in comparison to the sampling interval that it was only possible to put a lower bound on the rate of escape. In the humans the same was true for 10/21 escape events. Across all of these

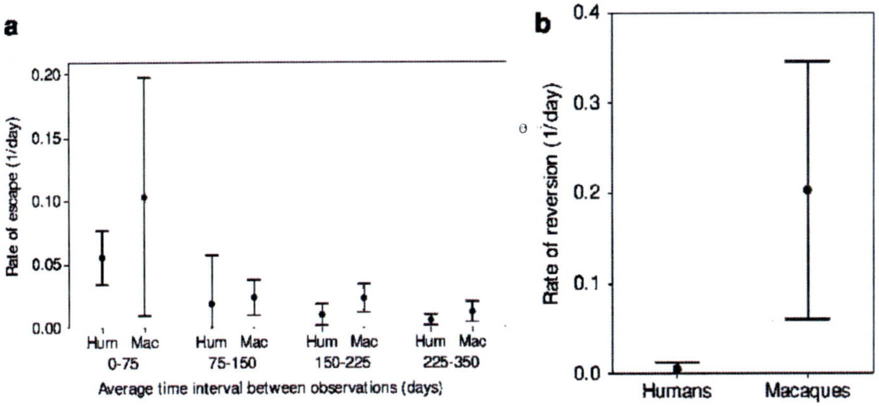

Fig. 18.7 A comparison of escape and reversion rates in humans and macaques. (**a**) To compare escape rates the data is stratified according to the average time interval between observations. For each grouping the rate of escape is faster in macaques than in humans. (**b**) A direct comparison of reversion rates reveals that reversion is faster in macaques than in humans [13]

events (amongst both human and macaque), a strong negative correlation was found between the rate of escape and the average sampling interval. To compare the rate of escape in humans with the rate of escape in macaques, Asquith et al. controlled for the length of the sampling interval by dividing the dataset into four groups with similar average sampling intervals. For each group the rate of escape in humans was compared to the corresponding rate of escape in macaques (Fig. 18.7). The rate in macaques was found to be significantly faster than the rate in humans. This result held true even when data relating to acutely infected macaques were excluded. In fact, unlike in humans, there was no statistical difference between the rate of escape in acute and chronic infection.

To determine whether the faster escape rates observed in macaques could simply be explained by lower fitness costs, fitness costs for nine reversion events were estimated in the same way as described for humans. For the majority of events (6/9) the fitness cost could be determined precisely, therefore a direct comparison was made between the macaque and human estimates (Fig. 18.7). This analysis revealed that fitness costs of escape mutants are significantly higher in macaques than in humans. The faster escape rates observed in macaques therefore cannot be explained by differences in fitness costs.

Since both the escape rate, k, and fitness cost, ϕ, are higher in macaques than in humans, the rate at which a CTL response to one epitope kills productively infected cells ($c = k + \phi$) is also higher in macaques. The mean difference in the escape rate between macaques and humans is estimated to be 0.02 day^{-1}. The mean difference in fitness cost is estimated to be 0.2 day^{-1}. The difference in the rate of CTL-mediated lysis can therefore be estimated to be 0.22 day^{-1} (0.02 + 0.2), meaning that infected cells are killed almost ten times more rapidly in macaques than in humans.

In summary, this model has been used to estimate the selection pressure exerted by CTL responses that drive the emergence of immune escape variants, thereby enabling the efficiency of HIV-1 specific CTLs to be quantified. Firstly, this analysis has revealed that only around 2% of productively infected cell death is attributable to CTLs recognising a single epitope. When this figure is scaled up to consider the total CTL response by an individual, the implication is that CTLs are not responsible for the majority of infected cell death, thus the basis upon which hopes of a CTL-based vaccine are pinned, may be flawed. Secondly, the analysis has revealed that the killing rate is significantly faster during acute infection than during chronic infection. The CTL response therefore may be functionally important during acute HIV infection, but not during chronic infection. Finally, the analysis has shown that infected cells are killed by CTLs almost ten times more rapidly in macaques than in humans, highlighting that SIV/SHIV infected macaques are a less than perfect model for testing HIV vaccines. This may explain the recent failure of the CTL-based Merck vaccine that was efficacious in macaques, but entirely ineffective in humans.

Discussion

The combination of mathematical modelling and data analysis to elucidate various aspects of the within-host dynamics of HIV infections has been one of the great success stories of mathematical biology in the last decade. A number of factors have contributed to this success. First, data on the within host dynamics of HIV is amply available. HIV circulates in the blood and there are reliable quantitative methods of detecting viral load in the blood. Thus longitudinal measures of viral abundance in the blood of infected individuals are quite common. Second, the models that have been developed to interrogate that data have remained really rather simple and transparent with many of their important parameters accessible from observed data. Despite this simplicity these models have had substantial explanatory power for the observed data. In combination these good data and good models have yielded new insights that would not have been available unless both models and data were brought to bear upon the questions. Because of this, the study of ‘viral dynamics’ in HIV infection (often meaning mathematical modelling of immune system-virus interactions) is very widely accepted in a broad church of HIV biologists.

Probably the first new insight generated by these interdisciplinary interactions was that, during HIV infection, infected cells turnover on a very rapid timescale, most of them not surviving beyond 1 day after infection. At the time this was a big surprise to many researchers in the field. With time, repeated observation of the same phenomena and further corroborating data this once novel insight is now widely accepted.

The mathematical models presented here offer three insights about the role of CTLs in immunodeficiency virus infections:

1. In chronically infected humans fewer than 20% of cells are killed by specific CTL responses.

2. In acutely infected humans CTL killing is more efficient, but not dramatically so.
3. In simian infections CTL efficacy is very dramatically higher.

The insights described in this chapter, although also based on simple models and abundant, repeatable data, are yet to gain widespread acceptance. This is despite the fact that their major prediction (that CTL vaccines might work in macaques but not in humans) was amply borne out by the STEP vaccine trial. Nevertheless we contend that this is another example of modelling elucidating previously confusing biological data that will, in time, gain wide acceptance.

References

1. UNAIDS (2008) 2008 Report on global AIDS epidemic.
2. Flynn NM, Forthal DN, Harro CD, Judson FN, Mayer KH, Para MF (2005) Placebo-controlled phase 3 trial of a recombinant glycoprotein 120 vaccine to prevent HIV-1 infection. J Infect Dis 191:654–665
3. Borrow P, Lewicki H, Hahn BH, Shaw GM, Oldstone MB (1994) Virus-specific $CD8^{+}$ cytotoxic T-lymphocyte activity associated with control of viremia in primary human immunodeficiency virus type 1 infection. J Virol 68:6103–6110
4. Koup RA, Safrit JT, Cao Y, Andrews CA, McLeod G, Borkowsky W, Farthing C, Ho DD (1994) Temporal association of cellular immune responses with the initial control of viremia in primary human immunodeficiency virus type 1 syndrome. J Virol 68:4650–4655
5. Borrow P, Lewicki H, Wei X, Horwitz MS, Peffer N, Meyers H, Nelson JA, Gairin JE, Hahn BH, Oldstone MB, Shaw GM (1997) Antiviral pressure exerted by HIV-1-specific cytotoxic T lymphocytes (CTLs) during primary infection demonstrated by rapid selection of CTL escape virus. Nature 3:205–211
6. Price DA, Goulder PJ, Klenerman P, Sewell AK, Easterbrook PJ, Troop M, Bangham CR, Phillips RE (1997) Positive selection of HIV-1 cytotoxic T lymphocyte escape variants during primary infection. Proc Natl Acad Sci USA 94:1890–1895
7. Carrington M, Nelson GW, Martin MP, Kissner T, Vlahov D, Goedert JJ, Kaslow R, Buchbinder S, Hoots K, O'Brien SJ (1999) HLA and HIV-1: heterozygote advantage and b*35-cw*04 disadvantage. Science 283:1748–1752
8. Gao X, Nelson GW, Karacki P, Martin MP, Phair J, Kaslow R, Goedert JJ, Buchbinder S, Hoots K, Vlahov D, O'Brien SJ, Carrington M (2001) Effect of a single amino acid change in MHC class I molecules on the rate of progression to AIDS. N Engl J Med 344:1668–1675
9. Carrington M, O'Brien SJ (2003) The influence of HLA genotype on AIDS. Annu Rev Med 54:535–551
10. Cohen J (2007) AIDS research. Did Merck's failed HIV vaccine cause harm? Science 318:1048–1049
11. Cohen J (2007) AIDS research. Promising AIDS vaccine's failure leaves field reeling. Science 318:28–29
12. Shiver JW, Fu TM, Chen L, Casimiro DR, Davies ME, Evans RK, Zhang ZQ, Simon AJ, Trigona WL, Dubey SA, Huang L, Harris VA, Long RS, Liang X, Handt L, Schleif WA, Zhu L, Freed DC, Persaud NV, Guan L, Punt KS, Tang A, Chen M, Wilson KA, Collins KB, Heidecker GJ, Fernandez VR, Perry HC, Joyce JG, Grimm KM, Cook JC, Keller PM, Kresock DS, Mach H, Troutman RD, Isopi LA, Williams DM, Xu Z, Bohannon KE, Volkin DB, Montefiori DC, Miura A, Krivulka GR, Lifton MA, Kuroda MJ, Schmitz JE, Letvin NL, Caulfield MJ, Bett AJ, Youil R, Kaslow DC, Emini EA (2002) Replication-incompetent adenoviral vaccine vector elicits effective anti-immunodeficiency-virus immunity. Nature 415:331–335
13. Asquith B, Edwards CT, Lipsitch M, McLean AR (2006) Inefficient cytotoxic T lymphocyte-mediated killing of HIV-1-infected cells in vivo. PLoS Biol 4:e90

14. Asquith B, McLean AR (2007) In vivo CD8$^+$ T cell control of immunodeficiency virus infection in humans and macaques. Proc Natl Acad Sci USA 104:6365–6370
15. HVTN (2007) The step study. Summary. Press briefing. November 7, 2007
16. Walker BD, Chakrabarti S, Moss B, Paradis TJ, Flynn T, Durno AG, Blumberg RS, Kaplan JC, Hirsch MS, Schooley RT (1987) HIV-specific cytotoxic T lymphocytes in seropositive individuals. Nature 328:345–348
17. Buseyne F, Fevrier M, Garcia S, Gougeon ML, Riviere Y (1996) Dual function of a human immunodeficiency virus (HIV)-specific cytotoxic T-lymphocyte clone: inhibition of HIV replication by noncytolytic mechanisms and lysis of HIV-infected CD4$^+$ cells. Virology 225:248–253
18. Yang OO, Kalams SA, Trocha A, Cao H, Luster A, Johnson RP, Walker BD (1997) Suppression of human immunodeficiency virus type 1 replication by CD8$^+$ cells: evidence for HLA class i-restricted triggering of cytolytic and noncytolytic mechanisms. J Virol 71:3120–3128
19. Van Baalen CA, Schutten M, Huisman RC, Boers PH, Gruters RA, Osterhaus AD (1998) Kinetics of antiviral activity by human immunodeficiency virus type 1-specific cytotoxic T lymphocytes (CTL) and rapid selection of CTL escape virus in vitro. J Virol 72:6851–6857
20. Matano T, Shibata R, Siemon C, Connors M, Lane HC, Martin MA (1998) Administration of an anti-CD8 monoclonal antibody interferes with the clearance of chimeric simian/human immunodeficiency virus during primary infections of rhesus macaques. J Virol 72:164–169
21. Jin X, Bauer DE, Tuttleton SE, Lewin S, Gettie A, Blanchard J, Irwin CE, Safrit JT, Mittler J, Weinberger L, Kostrikis LG, Zhang L, Perelson AS, Ho DD (1999) Dramatic rise in plasma viremia after CD8$^+$ T cell depletion in simian immunodeficiency virus-infected macaques. J Exp Med 189:991–998
22. Schmitz JE, Kuroda MJ, Santra S, Sasseville VG, Simon MA, Lifton MA, Racz P, Tenner-Racz K, Dalesandro M, Scallon BJ, Ghrayeb J, Forman MA, Montefiori DC, Rieber EP, Letvin NL, Reimann KA (1999) Control of viremia in simian immunodeficiency virus infection by CD8$^+$ lymphocytes. Science 283:857–860
23. Harrer T, Harrer E, Kalams SA, Elbeik T, Staprans SI, Feinberg MB, Cao Y, Ho DD, Yilma T, Caliendo AM, Johnson RP, Buchbinder SP, Walker BD (1996) Strong cytotoxic T cell and weak neutralizing antibody responses in a subset of persons with stable nonprogressing HIV type 1 infection. AIDS Res Hum Retroviruses 12:585–592
24. Ogg GS, Jin X, Bonhoeffer S, Dunbar PR, Nowak MA, Monard S, Segal JP, Cao Y, Rowland-Jones SL, Cerundolo V, Hurley A, Markowitz M, Ho DD, Nixon DF, McMichael AJ (1998) Quantitation of HIV-1-specific cytotoxic T lymphocytes and plasma load of viral RNA. Science 279:2103–2106
25. Edwards BH, Bansal A, Sabbaj S, Bakari J, Mulligan MJ, Goepfert PA (2002) Magnitude of functional CD8$^+$ T-cell responses to the gag protein of human immunodeficiency virus type 1 correlates inversely with viral load in plasma. J Virol 76:2298–2305
26. Oxenius A, McLean AR, Fischer M, Price DA, Dawson SJ, Hafner R, Schneider C, Joller H, Hirschel B, Phillips RE, Weber R, Gunthard HF (2002) Human immunodeficiency virus-specific CD8$^+$ t-cell responses do not predict viral growth and clearance rates during structured intermittent antiretroviral therapy. J Virol 76:10169–10176
27. Addo MM, Yu XG, Rathod A, Cohen D, Eldridge RL, Strick D, Johnston MN, Corcoran C, Wurcel AG, Fitzpatrick CA, Feeney ME, Rodriguez WR, Basgoz N, Draenert R, Stone DR, Brander C, Goulder PJ, Rosenberg ES, Altfeld M, Walker BD (2003) Comprehensive epitope analysis of human immunodeficiency virus type 1 (HIV-1)-specific T-cell responses directed against the entire expressed HIV-1 genome demonstrate broadly directed responses, but no correlation to viral load. J Virol 77:2081–2092
28. Kiepiela P, Ngumbela K, Thobakgale C, Ramduth D, Honeyborne I, Moodley E, Reddy S, de Pierres C, Mncube Z, Mkhwanazi N, Bishop K, van der Stok M, Nair K, Khan N, Crawford H, Payne R, Leslie A, Prado J, Prendergast A, Frater J, McCarthy N, Brander C, Learn GH, Nickle D, Rousseau C, Coovadia H, Mullins JI, Heckerman D, Walker BD, Goulder P (2007) CD8$^+$ t-cell responses to different HIV proteins have discordant associations with viral load. Nature 13:46–53

29. Betts MR, Nason MC, West SM, De Rosa SC, Migueles SA, Abraham J, Lederman MM, Benito JM, Goepfert PA, Connors M, Roederer M, Koup RA (2006) HIV nonprogressors preferentially maintain highly functional HIV-specific CD8$^+$ T cells. Blood 107:4781–4789
30. Daucher M, Price DA, Brenchley JM, Lamoreaux L, Metcalf JA, Rehm C, Nies-Kraske E, Urban E, Yoder C, Rock D, Gumkowski J, Betts MR, Dybul MR, Douek DC (2008) Virological outcome after structured interruption of antiretroviral therapy for human immunodeficiency virus infection is associated with the functional profile of virus-specific CD8$^+$ T cells. J Virol 82:4102–4114
31. McMichael AJ, Phillips RE (1997) Escape of human immunodeficiency virus from immune control. Annu Rev Immunol 15:271–296
32. Perelson AS (2002) Modelling viral and immune system dynamics. Nature 2:28–36
33. Phillips AN (1996) Reduction of HIV concentration during acute infection: independence from a specific immune response. Science 271:497–499
34. Stafford MA, Corey L, Cao Y, Daar ES, Ho DD, Perelson AS (2000) Modeling plasma virus concentration during primary HIV infection. J Theor Biol 203:285–301
35. Perelson AS, Neumann AU, Markowitz M, Leonard JM, Ho DD (1996) HIV-1 dynamics in vivo: virion clearance rate, infected cell life-span, and viral generation time. Science 271:1582–1586
36. Ho DD, Neumann AU, Perelson AS, Chen W, Leonard JM, Markowitz M (1995) Rapid turnover of plasma virions and CD4 lymphocytes in HIV-1 infection. Nature 373:123–126
37. Wei X, Ghosh SK, Taylor ME, Johnson VA, Emini EA, Deutsch P, Lifson JD, Bonhoeffer S, Nowak MA, Hahn BH, et al (1995) Viral dynamics in human immunodeficiency virus type 1 infection. Nature 373:117–122
38. Rodrigo AG, Shpaer EG, Delwart EL, Iversen AK, Gallo MV, Brojatsch J, Hirsch MS, Walker BD, Mullins JI (1999) Coalescent estimates of HIV-1 generation time in vivo. Proc Natl Acad Sci USA 96:2187–2191
39. Macallan DC, Asquith B, Irvine AJ, Wallace DL, Worth A, Ghattas H, Zhang Y, Griffin GE, Tough DF, Beverley PC (2003) Measurement and modeling of human T cell kinetics. Eur J Immunol 33:2316–2326
40. Markowitz M, Louie M, Hurley A, Sun E, Di Mascio M, Perelson AS, Ho DD (2003) A novel antiviral intervention results in more accurate assessment of human immunodeficiency virus type 1 replication dynamics and t-cell decay in vivo. J Virol 77:5037–5038
41. Friedrich TC, Dodds EJ, Yant LJ, Vojnov L, Rudersdorf R, Cullen C, Evans DT, Desrosiers RC, Mothe BR, Sidney J, Sette A, Kunstman K, Wolinsky S, Piatak M, Lifson J, Hughes AL, Wilson N, O'Connor DH, Watkins DI (2004) Reversion of CTL escape-variant immunodeficiency viruses in vivo. Nature 10:275–281
42. Leslie AJ, Pfafferott KJ, Chetty P, Draenert R, Addo MM, Feeney M, Tang Y, Holmes EC, Allen T, Prado JG, Altfeld M, Brander C, Dixon C, Ramduth D, Jeena P, Thomas SA, St John A, Roach TA, Kupfer B, Luzzi G, Edwards A, Taylor G, Lyall H, Tudor-Williams G, Novelli V, Martinez-Picado J, Kiepiela P, Walker BD, Goulder PJ (2004) HIV evolution: CTL escape mutation and reversion after transmission. Nature 10:282–289
43. Draenert R, Le Gall S, Pfafferott KJ, Leslie AJ, Chetty P, Brander C, Holmes EC, Chang SC, Feeney ME, Addo MM, Ruiz L, Ramduth D, Jeena P, Altfeld M, Thomas S, Tang Y, Verrill CL, Dixon C, Prado JG, Kiepiela P, Martinez-Picado J, Walker BD, Goulder PJ (2004) Immune selection for altered antigen processing leads to cytotoxic T lymphocyte escape in chronic HIV-1 infection. J Exp Med 199:905–915
44. Goulder PJ, Brander C, Tang Y, Tremblay C, Colbert RA, Addo MM, Rosenberg ES, Nguyen T, Allen R, Trocha A, Altfeld M, He S, Bunce M, Funkhouser R, Pelton SI, Burchett SK, McIntosh K, Korber BT, Walker BD (2001) Evolution and transmission of stable CTL escape mutations in HIV infection. Nature 412:334–338
45. Moore CB, John M, James IR, Christiansen FT, Witt CS, Mallal SA (2002) Evidence of HIV-1 adaptation to HLA-restricted immune responses at a population level. Science 296:1439–1443
46. Kawashima Y, Pfafferott K, Frater J, Matthews P, Payne R, Addo M, Gatanaga H, Fujiwara M, Hachiya A, Koizumi H, Kuse N, Oka S, Duda A, Prendergast A, Crawford H, Leslie A,

Brumme Z, Brumme C, Allen T, Brander C, Kaslow R, Tang J, Hunter E, Allen S, Mulenga J, Branch S, Roach T, John M, Mallal S, Ogwu A, Shapiro R, Prado JG, Fidler S, Weber J, Pybus OG, Klenerman P, Ndung'u T, Phillips R, Heckerman D, Harrigan PR, Walker BD, Takiguchi M, Goulder P (2009) Adaptation of HIV-1 to human leukocyte antigen class i. Nature 458: 641–645

47. Frahm N, Korber BT, Adams CM, Szinger JJ, Draenert R, Addo MM, Feeney ME, Yusim K, Sango K, Brown NV, SenGupta D, Piechocka-Trocha A, Simonis T, Marincola FM, Wurcel AG, Stone DR, Russell CJ, Adolf P, Cohen D, Roach T, StJohn A, Khatri A, Davis K, Mullins J, Goulder PJ, Walker BD, Brander C (2004) Consistent cytotoxic-t-lymphocyte targeting of immunodominant regions in human immunodeficiency virus across multiple ethnicities. J Virol 78:2187–2200
48. Geels MJ, Cornelissen M, Schuitemaker H, Anderson K, Kwa D, Maas J, Dekker JT, Baan E, Zorgdrager F, van den Burg R, van Beelen M, Lukashov VV, Fu TM, Paxton WA, van der Hoek L, Dubey SA, Shiver JW, Goudsmit J (2003) Identification of sequential viral escape mutants associated with altered t-cell responses in a human immunodeficiency virus type 1-infected individual. J Virol 77:12430–12440

Chapter 19
Viral Immunity and Persistence

Stephen Hickling and Rodney Phillips

Abstract Viruses survive and propagate in human populations, despite the anti-viral immune response. Viruses have developed two principal mechanisms in order to persist within populations. Some viruses, such as measles, are highly infectious and quickly spread through susceptible populations. However measles is rapidly cleared by the immune response. In contrast other less transmissible viruses such as Epstein-Barr Virus (EBV) have developed a strategy of persisting within a population by subverting or escaping from the host immune response.

The immune response has a number of defensive layers, which inhibit viral replication. The innate response is non-specific and mediated by cytokines that prevent the spread of virus through tissues and promote the initiation of highly specific adaptive immune responses. The adaptive response consists of antibodies, which bind and neutralise free virus, and $CD8^+$ T cells that identify virally infected cells and eradicate them.

Different immune mechanisms can by evaded by viruses. Cytokines can be inhibited either through preventing cytokine expression or reducing cytokine potency through the production of viroreceptors. Additionally, immune evasion can occur through antigenic variation. The adaptive immune response may cope with antigenic variation but some variability can defy the capacity of the response. Human Immunodeficiency Virus (HIV) escapes adaptive immune responses by generating antigens unrecognizable by $CD8^+$ T cell and antibody responses.

The interaction between the virus and host is a constant evolutionary struggle, which imposes forces that drive out adapted viruses. The ability of viruses to adapt allows them to persist and propagate within different populations.

Viruses are obligate intracellular parasites that hijack the infected cell machinery in order to replicate and be transmitted. The balance between the pathogenicity of a virus infection and its resolution by the immune response shapes viral survival strategy. It is not to the viruses' advantage to be highly pathogenic, as rapidly killing the

R. Phillips (✉)
Peter Medawar Building for Pathogen Research, University of Oxford, South Parks Road, Oxford OX1 3SY, UK
e-mail: rodney.phillips@ndm.ox.ac.uk

C. Molina-París and G. Lythe (eds.), *Mathematical Models and Immune Cell Biology*,
DOI 10.1007/978-1-4419-7725-0_19,

host would prevent transmission and survival within a population. However, the virus needs to be sufficiently pathogenic in order to replicate and be transmitted in a population, so a balance needs to be achieved. Consequently viruses have evolved two principal strategies for promoting their survival within a population, which can acquire immunity. Some viruses such as Morbilivirus (Measles) have a survival strategy described as hit and run [1]. They are able to infect a large number of potential hosts before being resolved by the immune response [2,3]. However, other viruses have evolved the alternative strategy of persistence. These viruses such as the Herpes viruses have low transmissibility but are generally poorly resolved by the immune response. This tactic allows them to persist within a population. Countering viral proliferation is the host immune response. The antiviral immune response consists of both innate and adaptive immune responses. In terms of antiviral immunity the innate response refers to the action of cytokines and natural killer cells (NK). The adaptive immune response to viruses includes the production of specific antiviral antibodies by B cells and the $CD8^+$ T lymphocyte response, which eliminate virally infected cells.

Immunological Mechanisms in Antiviral Responses

The innate immune response to viral infection is mediated through the action of cytokines. Cells of the innate immune response, such as macrophage and monocytes, produce a range of antiviral cytokines. The most important being Interleukin-1, -2, Tumour Necrosis Factor (TNF), Interferon (IFN)-$\alpha\beta$ and γ [4–6]. The action of these cytokines can promote an antiviral state in bystander cells, inducing the expression of proteins that destroy viral genomes and inhibit replication [6–8]. Cytokines such as IFNγ can also upregulate the surface expression of Major Histocompatibility Complex molecules (MHC) – termed Human leukocyte antigens (HLA) in humans [9].

The innate immune response also comprises NK cells. These cells are able to kill virally infected cells. NK cells act by killing targets that lack HLA molecules which act as the inhibitor self signal [10]. This is important for viral infections that attempt to subvert the $CD8^+$ T cell response by downregulation of HLA class I molecules.

Adaptive Immune Responses

The adaptive immune response, unlike the innate response, is highly specific. The production of antibody by B cells is an important effector of the adaptive immune response [11, 12]. Specific antibody can bind to viral surface glycoproteins or viral capsids [11]. Antibody binding can promote viral destruction either by antibody dependent cell mediated cytotoxicity or induction of phagocytosis [11, 12]. Additionally, antibody binding can inhibit viral entry by physically blocking viral receptors binding to complementary receptors [12].

The role of antiviral antibody is crucial for protection from a particular virus upon repeated exposure, such as measles [2, 3]. The importance of the protective effect of antibody is highlighted by the fact that all successful viral vaccinations induce protective antibody responses. In addition to antibody, $CD8^+$ T lymphocytes are crucial to the adaptive antiviral immune response. $CD8^+$ T lymphocytes are able to kill cells infected with viruses. $CD8^+$ T cells recognise virally infected cells because HLA Class I molecules display viral peptide antigens on the cell surface [13]. Class I HLA molecules are found on the surface of all nucleated cells. The $CD8^+$ T lymphocyte recognises this epitope/HLA complex through its unique T Cell Receptor (TCR) [14, 15]. This interaction can result in $CD8^+$ T cell activation. Activated $CD8^+$ T lymphocytes are able to kill the target cell (expressing the Class I HLA with viral epitope) and are termed Cytotoxic T lymphocytes (CTL) [13].

The CTL response is crucial for protection from viral infections, highlighted by individuals with Severe Combined Immunodeficiency disease and DiGeorges syndrome who lack CTL. These individuals are extremely susceptible to viral infections which are non-pathogenic in healthy people (reviewed in [16]).

HLA Class 1 Molecules and the Presentation of Viral Epitopes to the Adaptive Immune Response

Structure of the HLA Molecules

The critical role that Class I murine MHC molecules play in the presentation of endogenous, nonself peptides was first identified in 1982 [13]. At this time it became clear that Class I molecules present only short fragments of peptides, termed epitopes, to $CD8^+$ T lymphocytes. This was later confirmed in 1987 when the crystal structure of HLA Class I molecules was resolved [17].

HLA molecules are encoded within the highly polymorphic major histocompatability complex loci, located in chromosome 6p. This complex spans over 3.6 Mbp and contains approximately 118 genes. In addition to HLA Class I, the MHC also encodes Class II HLA molecules as well as a number of cytokines and elements of the innate immune system.

Since the initial discovery of HLA Class I molecules it became evident that there is huge diversity in HLA Class I molecules, with 702 alleles identified to date. These alleles can be further divided into HLA Class IA, IB and IC isotypes. Each individual has the potential to express up to two different HLA molecules of each isotype, six in total. It is thought that most CTL are reactive against peptides presented on HLA-A and B molecules.

The HLA Class I molecule is a non-covalently linked trimer (Fig. 19.1). The trimer consists of the polymorphic heavy chain (encoded in the MHC), β_2 microglobulin (β_2m) and the bound epitope [17]. The heavy chain consists of three

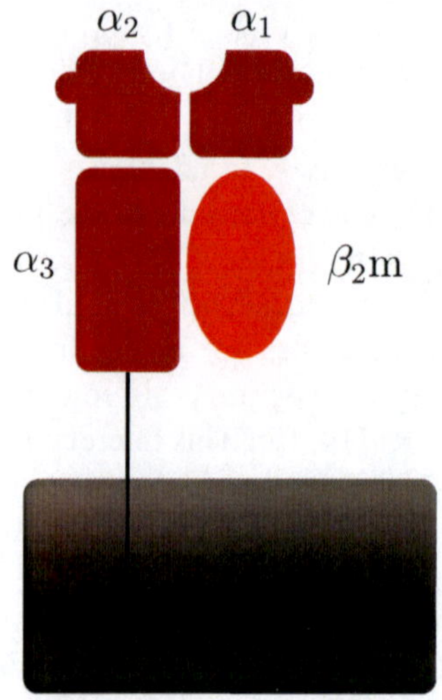

Fig. 19.1 Structure of HLA class 1 molecule. The HLA heavy chain is shown in *purple* (α_1, α_2 and α_3 domains). The Heavy chain forms a binding groove, where the specific peptide binds, between the 1 and 2 domains. The Heavy chain/peptide complex is associated with 2 microglobulin (β_2m), shown in *red*

domains α_1, α_2 and α_3 with a molecular weight of approximately 43 kDa. The α_3 domain spans the cell membrane [17]. The α_1 and α_2 domains form eight β-pleated sheets, upon which sit two anti-parallel α helices [17]. This structure forms a platform with a groove, within which the epitope sits [17, 18]. It is the α_2 and α_3 domains that form this structure which have highly polymorphic regions, allowing a greater range of peptides to bind the groove [18].

The structure of the groove places biochemical restrictions on the peptides that can stably bind. The groove can only accommodate peptide between 8 and 12 amino acids in length [18]. The peptide and HLA groove interact through the formation of hydrogen bonds at both the N- and C-terminals. The residues in the peptide that form the hydrogen bonds are termed anchor residues.

These anchor residues are determined by the amino acids found both in the β-pleated sheets and α helices of the HLA molecule. The amino acid sequence in the rest of the bound peptide interact with the groove, however there is much greater flexibility in these amino acid sequences [17, 18].

HLA Class I Antigen Presentation

Presentation of peptides on HLA Class I molecules is crucial for an effective immune response. The cellular process for the production and presentation of optimal peptides is a highly orchestrated process, which can be broadly divided into peptide cleavage and HLA loading and surface expression (Fig. 19.2).

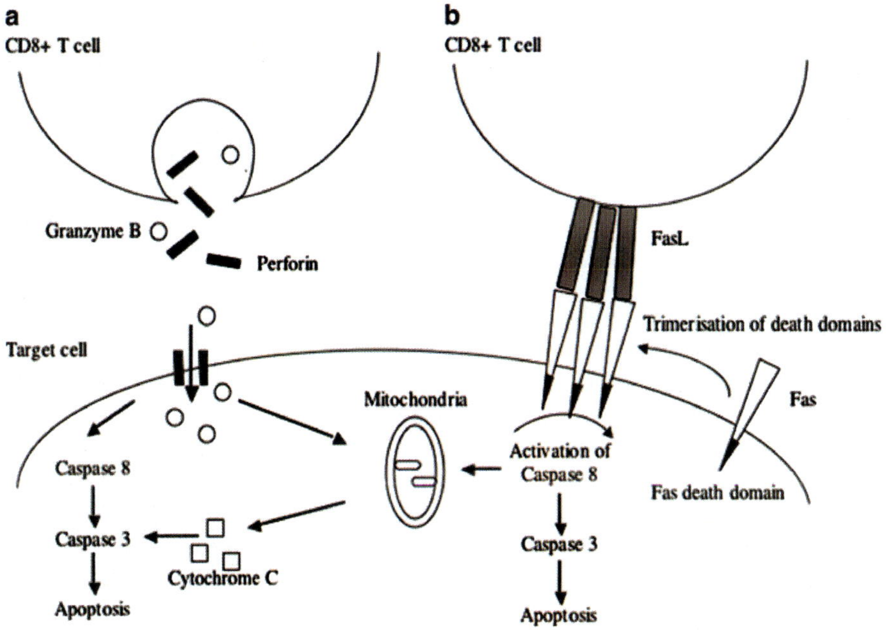

Fig. 19.2 Mechanisms of CD8$^+$ T cell killing. (**a**) Schematic showing the granzyme exocytosis model of killing. Granzyme B enters the target cell by perforin. Granzyme triggers the pro-apoptotic caspase cascade in the target cell. (**b**) Schematic depicts receptor mediated target cell death. Engagement of receptors containing death domains (such as Fas) with the appropriate ligand on the CTL causes trimerisation of death domains, which triggers the pro-apoptotic caspase cascade

The majority of peptides loaded into HLA Class I molecules are derived from the cytoplasm including self-peptides [20]. In non-infected cells, proteins are constitutively cleaved into peptide fragments by the cylindrical proteasome, also called the 20S or core particle proteasome [21]. The 20S proteasome cleaves proteins into oligo-peptides between 3 and 15 amino acids in length [22]. The proteasome's cylindrical shape is formed from two α rings, between which are two β rings [21, 23]. Each ring is composed of either seven α or seven β subunits [23]. The $\beta1$,$\beta2$ and $\beta5$ subunits of the 20S proteasome have peptidylglutamyl, trypsin and chymotrypsin activities, which primarily act at the C terminus of the peptide [24, 25]. The remaining β subunits are N-terminal acting threonine proteases [25]. The α rings are thought to control the entry of peptides into the catalytic β rings, with poly-ubiquitinated peptides being preferentially cleaved [21].

Under the influence of IFNγ, the 20S proteasome is modulated by the addition of 19S subunits, which can bind to either one, or both ends of the 20S proteasome [21]. IFNγ also upregulates the expression of other catalytic subunits, LMP2, LMP7 and MECL-1 [26, 27]. These subunits are incorporated into the β rings, replacing the $\beta1$, $\beta2$ and $\beta5$ subunits. This IFNγ induced complex forms the 26S immunoproteasomes. The modified β subunits alter the peptide cleavage specificity of the immunoproteasome [26, 27]. The trypsin and chymotrpsin activity is increased at the carboxy side of basic and hydrophobic residues while caspase activity at acidic residues is reduced. This consequently causes the immunoproteasome to produce longer peptides, which have more hydrophobic and basic residues at the Carboxy termini [26,27]. Residues with these biochemical characteristics tend to have higher binding affinity for HLA Class I molecules [17, 28]. Since basic and hydrophobic residues are more likely to form anchor residues.

Peptides produced by the immunoproteasome are transported into the Endoplasmic Reticulum (ER) [29]. The ER is where newly synthesised HLA Class I heavy chains are located. Peptides are actively transported into the ER via the Transporter Associated with Antigen presentation (TAP) [29]. TAP is a transmembrane heterodimer, encoded in the MHC [30] which preferentially transports basic and hydrophobic peptides into the ER lumen [30,31]. In the ER peptides are further optimized at the N-terminus by ER amino-peptidase (ERAAP) [32,33]. ERAAP cleaves peptides of different lengths at different rates. 10mers are cleaved at higher rate then 9mers and little activity is directed against 8mers. This hierarchy of ERAAP activity increases the production of optimal peptides for HLA Class I loading [34, 35].

TAP is also a crucial component in the peptide loading complex (PLC) [36]. The PLC allows HLA Class I heavy chains to maintain structural conformation before peptide loading [36]. The complex comprises TAP, unloaded Class I Heavy chain associated with β_2m and the molecular chaperone tapasin [37]. This complex is always found with the same stoichiometry of 1 TAP: 4 tapasin: 4 heavy chains [38]. Binding of the optimal peptides to the PLC confers conformational change, releasing the Class I trimer complex from the PLC. This release consequently allow trafficking of peptide loaded Class I HLA molecules to the cell surface, to be sampled by patrolling CD8$^+$ T lymphocytes [39].

Activation of CD8$^+$ T Lymphocytes and Killing of Virally Infected Cells

Interaction of MHC Class 1 and the TCR

Loaded HLA Class I molecules expressed on the cell surface are sampled by CD8$^+$ T lymphocytes. The CD8$^+$ T lymphocyte TCR transiently binds the surface of HLA Class I molecules, contacting both the α_2 and α_3 domains and the epitope [14, 15]. When a particular TCR encounters its complementary HLA Class I-epitope complex, CD8$^+$ T lymphocyte activation can occur (Fig. 19.3).

Despite TCR engagement, TCRs lack the ability to initiate intracellular signalling. To overcome this TCRs are non-covalently associated with the transmembrane CD3 complex [40]. The CD3 complex is composed of single γ and δ sub-units in addition to two ϵ and two ζ subunits [40, 41] (Fig. 19.4). Each CD3 associates with two TCR dimers [40]. Consequently the overall structure of the com-

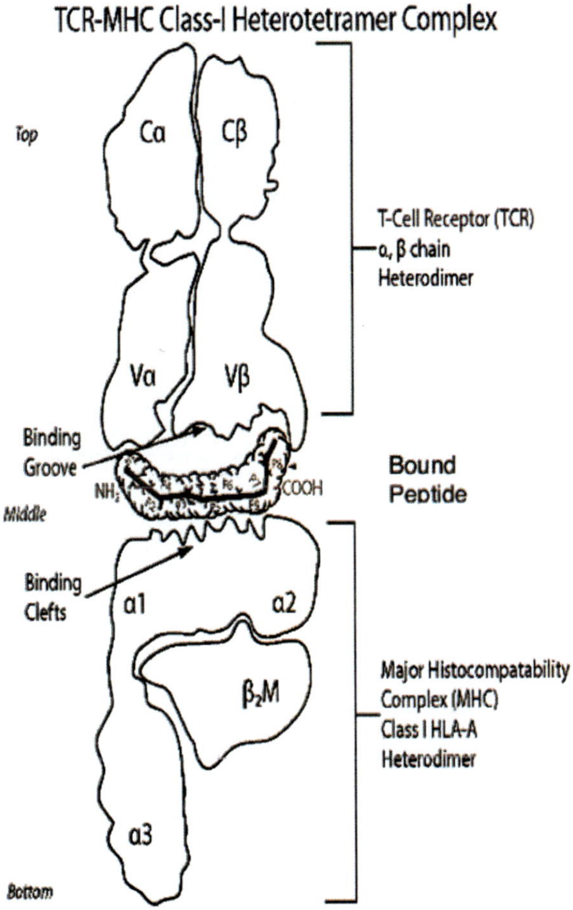

Fig. 19.3 The $CD8^+$ T cell interacts with the HLA class I molecule through its T cell receptor (TCR). The TCR binding groove, formed by the $V\alpha$ and $V\beta$ chains of the TCR, contact the surface of the HLA/peptide molecule. The TCR contacts the HLA heavy chain α_1, α_2 and the presented peptide. (Picture taken and modified from: Atlas of Genetics and cytogenetics Atlasgeneticsoncology.org/Deep/mapping Author Gerald. J. Mizejewski)

plex is $TCR_2CD3\gamma\delta\epsilon_2\zeta_2$ [42]. The cytoplasmic domains of CD3 subunits contain immuno-receptor tyrosine based motifs (ITAMS). These ITAMS can phosphorylate intracellular kinases, which facilitate intracellular signalling [43, 44].

The TCR–HLA interaction alone however is not sufficient to activate $CD8^+$ T lymphocytes. A substantial number of other receptors, termed co-receptors, are also involved; CD8 itself is involved in amplifying the signal [45]. The CD8 co-stimulatory molecule localizes to the engaged TCR and amplifies $CD8^+$ T cell activation [46]. This occurs through CD8 binding in an epitope independent manner to the α_3 domain of the HLA Class I molecule so prolonging cell engagement [45,46]. The degree that the CD8 interaction influences activation is thought to be a function of the affinity of the TCR for the HLA-epitope complex [47, 48]. TCR/HLA

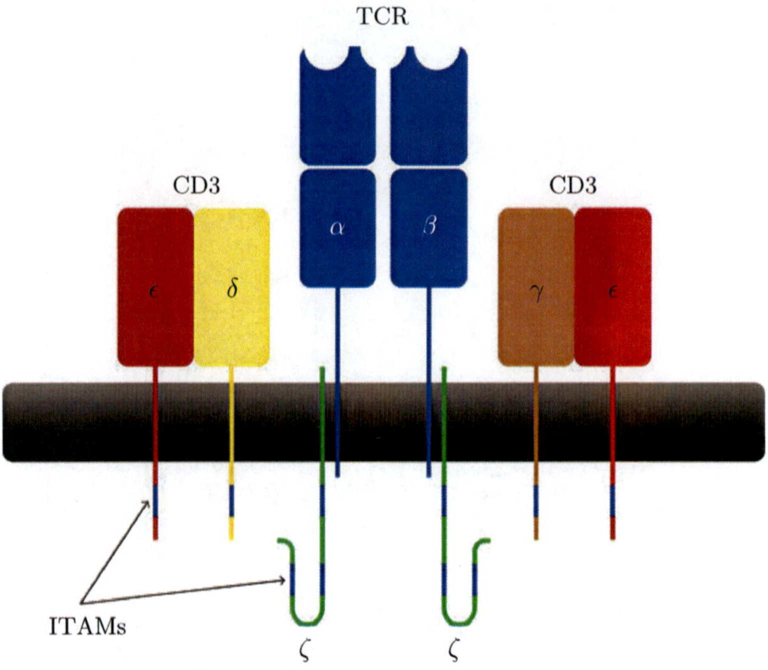

Fig. 19.4 The TCR complexes with CD3, that comprises of single and sub-units in addition to two ϵ and two ζ subunits. The overall structure of the complex is $TCR_2CD3\gamma\delta\epsilon_2\zeta_2$. The cytoplasmic domains of CD3 subunits contain immuno-receptor tyrosine based motifs (ITAMS) which facilitate intracellular signalling

and co-receptor engagement leads to surface protein rearrangement [49]. HLA molecules and co-receptors tend to be polarized to the point of initial HLA/TCR engagement [49, 50]. This leads to multiple HLA/TCR engagements. These multiple interactions form an immunological synapse, bringing the $CD8^+$ T lymphocyte and the target cell membranes into close contact [49, 50]. Multiple molecular interactions promote $CD8^+$ T cell activation [51].

Mechanisms of CTL Killing

Once activated, $CD8^+$ T lymphocytes have cytotoxic capability and are consequently termed cytotoxic T Lymphocytes (CTL). Both CTL and NK cells employ a number of different mechanisms in order to destroy virally infected cells. Cytotoxicity is mediated by direct cellular interactions such as granule exocytosis or engagement of death domains [52,53]. All these mechanisms direct target killing by inducing cellular apoptosis.

Apoptosis is an energy dependent, highly ordered process of cell death. Apoptosis causes cells to detach from the local environment, condense and lose membrane integrity. Apoptosis is orchestrated by a subset of proteases called caspases [54].

Caspases cleave substrates at C-terminal aspartate residues, and disrupt intracellular structure and DNA integrity [54]. This highly controlled process limits cell death to the target. This is in contrast to necrosis, which results in uncontrolled cellular damage resulting in inflammation (reviewed in [55]).

Granule Mediate Cytotoxicity

CTL contain granules containing effector molecules that can induce apoptosis to the target cells. The most important of these are perforin and granzyme B, members of the caspase family [56]. When CTL interact with target cells the lytic granules are polarized towards the contact point with the target [57]. This led to the granule exocytosis model of CTL-induced killing, which postulates that the CTL exocytose perforin and granzyme B into the cleft between the T cell membrane and the target cell. It was originally hypothesised that perforin monomers polymerise in the target cell membrane in the presence of Ca^{2+} [58–60]. This results in the formation of transmembrane pores in the target, allowing activated granzyme B to enter the target cell cytoplasm [61]. However, in perforin deficient mice, CTL can still kill target cells via granzyme B [62]. This, along with other observations, has led to the hypothesis that granzyme B can be internalised into target cells by receptor mediated endocytosis, with perforin playing a role in granzyme B release from vesicles [63, 64].

Granzyme B cleaves procaspase-3 and procaspase-8 [65, 66]. Both these procaspases require cleavage for activation, and form part of the caspase cascade which induce apoptosis. However, granzyme B can also induce apoptosis in a caspase independent manner [67]. Granzyme B signals through the mitochondria, disrupting energy metabolism resulting in the release of cytochrome *c*, a pro-apoptotic protein [67].

Although granzyme B is the main effector molecule for the induction of apoptosis other granzymes (A–H) and serine esterases are also released into the cell cytoplasm. These effector molecules are also members of the caspase family and help to induce apoptosis though cleavage of their molecular targets [67, 68].

Signalling Induced Apoptosis

In addition to the granule exocytosis model of cyotoxicity, it has been shown that CTL can kill targets independently of granzyme B, by receptor mediated cytotoxicity [52, 53]. This mechanism of cytotoxicity is principally used in lymphocyte selection, but has recently been associated with auto-immunity [69].

CTL kill through the CD95 (Fas) pathway [52]. Fas is a member of the tumour-necrosis factor receptor family of death receptors [70, 71]. Fas is expressed on the surface of many cell types; however, Fas ligand (FasL) is only expressed on the surface of activated CTL [52]. Ligation of Fas by FasL causes trimerization of Fas on the cell surface. Each Fas molecule has an intracellular death domain, which recruits and activates caspase 8, inducing the caspase signalling cascade [71].

Viral Persistence and Subversion of the Immune Response

Viral Subversion of the Immune System

The coexistence of the viral pathogen and the host is an extremely fine balance. The virus needs a large enough host population in order to survive and propagate, so rapidly killing the host is not a successful evolutionary strategy. Viruses with low transmission rates tend to be poorly resolved by the immune response. Thee viruses persist within hosts in order to survive. In order to do this viruses have developed strategies, which can interfere with the immune response designed to eliminate it. Interference occurs against both the innate and the adaptive responses.

Subversion of Cytokine Action

Cytokines are critical for immune signalling and promoting immunological responses. Consequently, viruses have evolved a varied number of strategies to either inhibit cytokine production or prevent their action [66]. The most common cytokines targeted are inevitably the cytokines with the most potent antiviral effects. These include Interleukin (IL)-1, -2, Tumour necrosis factor (TNF), Interferon (IFN)-α-β and γ [4,6,8].

IFNγ signalling causes the induction and expression of a number of genes, termed IFN stimulated genes (ISG). This occurs through IFN regulatory factor 3 (IRF-3), that translocates to the nucleus and binds the transcription co-activator CBP [72]. The Adenovirus E1A protein binds to and inhibits the IRF–CBP interaction preventing ISG expression [72]. Hepatitis B employs a similar strategy with the terminal protein inhibiting the IRF–CBP interaction [73].

In addition to reducing the production of cytokines, a number of other viruses prevent their action. One mechanism, employed by the Pox family of viruses, is the secretion of viroreceptors [74]. Viroreceptors can mimic the receptor for a given cytokine and have potent cytokine binding capability [74]. The myxoma virus viroreceptor, M-T2, and cowpox virus crmB both mimic the TNF receptor [75,76]. Viroreceptors compete with and sequester secreted TNF [75]. This consequently reduces the influence of TNF. Other viroreceptors, such as MT-7 produced by myxoma are thought to be promiscuous for a number of cytokines including IFNγ, IL2 and IL5 [77,78].

Herpes viruses, such as the Human Cytomegalovirus (HCMV), also express viroreceptors [79]. The HCMV viroreceptor US28 is expressed in the infected cell membrane and has cell-signalling capabilities [79]. US28 binds Macrophapge Inflammatory Protein (MIP)-1α/β and RANTES (Chemokine ligand 5) [79]. This binding capability of US28 has two principle effects. Firstly, binding of the chemokine causes US28 internalization and thus reduces cytokine concentration in the local environment. Secondly, signalling through US28 causes an intracellular calcium flux, which can inhibit cellular apoptosis and enhances viral replication [79].

Subversion of the Adaptive Immune Response

Antibody can bind to viral surface glycoproteins or viral capsids [11]. Antibody binding can opsonisize viral clearance either by antibody dependent cell mediated cytotoxicity or promotion of phagocytosis. Additionally, antibody binding can inhibit viral entry into target cells, by physically blocking the viral receptor interacting with its complementary receptor [11, 12].

Consequently, viruses have evolved to inhibit the antiviral effects of antibody. One evolutionary strategy is that closely related viruses bear antigenic diversity. Viruses with antigenic variation elicit a different antibody response defining them as different serotypes [80]. Examples of viruses with different serotypes include influenza A and Poliovirus [81]. Consequently, an individual primed with one serotype may not have cross protective antibody protection against another serotype of the same virus [80, 81].

Viral serotypes may also carry further antigenic diversity. This mainly occurs through antigenic drift, a process where viral antigens acquire mutations over time. Influenza surface variation is targeted by antibody [1, 81]. Antibody-binding site variation may inhibit antibody binding [1, 81]. HIV also displays antigenic drift to escape antibody responses [82]. However, HIV escapes antibody at a much faster pace than Influenza. Antibody taken from a HIV patient at a particular time cannot neutralize the viral isolate taken at the same time. However, that antibody can neutralize HIV isolates taken from the same patient 3–6 months earlier [83, 84]. This indicates that HIV is continually escaping the antibody response through antigenic drift and the antibody response is continually playing 'catch up'.

Antigenic drift causes small changes in antigenic variation, but major changes can occur by antigenic shift. Antigenic shift occurs when two or more viruses of different serotypes infect the same cell. This allows recombination of genetic material between viruses, creating a new serotype [85, 86]. This new serotype will have the surface antigens derived from both strains of virus. This genetically distinct virus can, consequently, escape any antibody responses present in the general population to either of the progenitor viruses. This type of antigenic shift is mostly associated with Influenza A and is the main source of new influenza pandemic strains [81, 85].

Subversion of the CD8 Response

For persistent viruses, escape from the CD8 response and its effector mechanisms is a crucial survival mechanism. Viruses have evolved diverse mechanisms that interfere with both the presentation of viral antigens and the apoptotic effects of CTL. See Table 19.1.

Inhibiting Protease Function

The proteasome is responsible for the degradation of self and virally derived peptides, and allows them to be presented on HLA Class I molecules [20, 21]. Consequently, interfering with the degradation and production of optimal peptides is an effective evasion mechanism for viruses

Epstein Barr Virus (EBV), a member of the Herpes family, expresses an early nuclear protein called EBNA 1. EBNA 1 expression dominates during latent EBV infection, but elicits a poor CTL response [93]. This is mainly as EBNA-1 has long repeats of glycine and alanine at the C-terminus. This makes the protein resistant to proteasomal cleavage [94], so EBNA-1 epitopes are rarely presented.

Table 19.1 Viral subversion targets in the innate immune response

Target Pathway	Virus	Effector Protein	Mechanism of Action	Reference
IFNγ signalling	Adenovirus	E1A	Prevents IRF-3 induced gene expression	[72]
	Hepatitis B	Terminal protein	Prevents IRF-3 induced gene expression	[73]
	Epstein-Barr Virus	EBNA-2	Downregulates transcription	[87]
	Vaccinia	B8R	Secretion of IFNγ viroreceptor	[88]
	Myxoma	MT-7	Secretion of viroreceptor	[67, 78]
	Human herpes virus	VIRF-2	Prevents expression	[89]
IFNα signalling	Vaccinia	B18R	Secretion of IFNα viroreceptor	[88]
TNF signalling	Myxoma	MT-2	Secretion of TNF viroreceptor	[90]
	Cowpox	CrmB	Secretion of viroreceptor	[48, 75, 76]
	Human cytomegalovirus	UL144	Viroreceptor, retained in infected cell	[91]
Interleukin signalling	Myxoma	MT-2	Viroreceptor binds IL-2 and 5	[90]
	Herpes simplex virus	UL37	Modulates IL-8 signalling	[92]
	Human cytomegalovirus	UL28	MIP-1α/β Viroreceptor	[79]

Inhibition of TAP Dependent Transport and the PLC

TAP is crucial for the translocation of proteasomal derived peptides into the ER [29–31, 95]. TAP also plays a crucial role in the PLC for efficient HLA Class I molecule loading [36]. Herpes Simplex virus expresses ICP-47, a protein, which is a pseudosubstate for TAP [96]. ICP-47 binds to TAP and induces conformational changes. This conformational change inhibits the ability of TAP to translocate peptides into the ER [96]. Consequently, HLA Class I molecules remain unloaded and are retained in the ER.

HCMV also targets TAP to reduce surface expression of virally loaded HLA. HCMV expresses US6, which binds to the TAP heterodimer, like ICP-47 [97]. However, US6 prevents TAP from recruiting ATP and thus inhibits active transport of peptide into the ER [97]. HCMV also expresses other proteins that interfere with the protein loading complex [98, 99]. US2 and US11 redirect HLA expressed in the ER into the cytoplasm, to the proteasome. This promotes degradation of HLA heavy chains [98]. US3 interferes with the peptide loading pathway by an unknown mechanism [99].

Modulation of HLA Class I Surface Expression

In addition to interfering with the processing of peptide and subsequent HLA Class I loading, HLA surface expression can also be inhibited. The HIV accessory protein, Nef, downregulates surface expression of HLA [100]. Nef binds to a specific motif on the cytoplasmic tail of HLA molecules. This motif is only found on the tails of HLA-A and B molecules, but not HLA-C [101]. The binding of Nef prevents trafficking of the HLA molecules to the cell surface [100]. Most CTL are reactive against peptides presented on HLA-A and B molecules so reduction of their surface expression inhibits CTL action. However, by not downregulating HLA class C, NK cells also receive the inhibitory self signal and thus do not kill the target cell. Thus, by selectively downregulating HLA-A and B, Nef inhibits CTL action and NK mediated cytotoxicity.

HCMV downregulates the surface expression of MHC by interfering with antigen processing. However, in order to avoid NK cell killing due to lack of surface HLA, HCMV expresses a HLA homolog [102]. This homolog, UL18 has 25% homology to HLA Class I molecules with a similar α_1, α_2 and α_3 structure. UL18 associates with β_2m, binds peptide and is stably expressed on the cell surface [103, 104]. It is thought that UL18 acts to inhibit NK cell function by binding inhibitory NK cell receptors [105]. The presence of this homologue may also act as competition for TCR binding of patrolling CTL from HLA Class I loaded with viral peptides [105].

Inhibition of Cellular Apoptosis

Preventing the effector mechanisms of CTL is crucial for survival; consequently viruses have evolved mechanisms to inhibit cellular apoptosis. This can either be through modulation of apoptotic protein expression or inhibition of caspases.

Adenoviruses express the protein E3. E3 upregulates the internalisation and degradation of Fas from the cell surface [106, 107]. This inhibits the Fas/FasL cytotoxic pathway [107]. Additionally, the Adenovirus E1A and E1B protein expression prevents the accumulation of p53 [108]. p53 is crucial in controlling the cell cycle and controlling apoptosis. Thus by preventing its accumulation the induction of apoptosis is inhibited [109]. Papilloma virus employs a similar strategy by targeting p53 for proteasomal degradation [110].

As caspases are the cellular mediators of apoptosis, they are common targets for inhibiting apoptosis. A number of viruses including the Herpes and Pox family of viruses encode a homologue of the mammalian caspase inhibitor Flice caspase-8 inhibitor (FLIP) [111]. The FLIP proteins contain death effector domains (DED), which bind and inhibit caspase 8. This consequently inhibits apoptosis induced through death domains associated with Fas [111]. However, FLIP proteins do not inhibit apoptosis induced by granzyme B.

The CrmA protein expressed by the Pox family of viruses is a potent inhibitor of caspase-1 [112, 113]. CrmA inhibits the action of caspase 1 by being a pseudosubstrate, which binds irreversibly to the caspase enzyme [94]. In addition to caspase 1, CrmA has also been shown to be a potent inhibitor of caspase 8, 9 and 10 [114]. Importantly the CrmA caspase inhibition prevents the induction of the apoptotic caspase cascade induced by granzyme B [114].

CTL Immune Escape Through Antigenic Variation

DNA based viruses possess genes for many of the mechanisms described to inhibit the actions of CTL. They have large genomes which can accommodate a larger number of viral genes. RNA based viruses such as Human Immunodeficiency Virus (HIV) have small genomes and cannot accommodate many of these adaptations [115]. However, these viruses have evolved strategies to escape the CTL response through antigenic variation.

In order to successfully replicate, HIV must transcribe its RNA genome into DNA, using Reverse Transcriptase (RT). RT has a high replication error rate, estimated to be 1 error every 20,000 bases [116]. This is in comparison the DNA polymerase which has an error rate of approximately 1 error every 10^6 bases [117]. The errors generated by RT introduce antigenic variation in the HIV genome by altering the amino acid sequence. These sequence variations create huge variation in the viral population within a single host [116, 118, 119].

Amino acid sequence variations can alter proteasomal cleavage [120, 121], epitope binding and the conformation of epitopes within the HLA groove [122, 123].

This can result in loss of epitope presentation, or recognition by TCRs [120, 122–124]. CTL evasion contributes to HIV persistence and is apparent as adaptability within human populations.

Alteration of amino acid sequences either side of an optimal epitope can alter a peptide's biochemistry and change the cleavage points of the immunoproteasome [120, 121]. Alteration in immunoproteasome cleavage can diminish the generation of the optimal epitopes [120, 121]. Consequently, this reduces Class I HLA that can be loaded with viral epitopes, reducing CTL antiviral activity.

Amino acid changes within an optimal peptide also alter their presentation [123, 124]. The HLA binding groove places biochemical restrictions on the peptide sequence that can be accommodated [18]. Amino acid mutations occurring at anchor residues reduce or inhibit the ability of peptides to bind stably in the HLA groove (Fig. 19.5). This consequently prevents the HLA molecules from being loaded and presented to the CTL, allowing the virus to escape the immune response [42, 119, 125].

Other amino acid polymorphisms within HIV epitopes, which do not occur at anchor residues, can also alter CTL recognition [122, 123]. Changes in the peptide sequence can change the conformation the peptide adopts within the HLA groove. As the HLA-epitope TCR interaction is highly specific any conformational changes may alter the affinity of this interaction [120, 122–124]. This can consequently impede the formation of immunological synapses or prevent the ability of the TCR to recognise its complementary ligand.

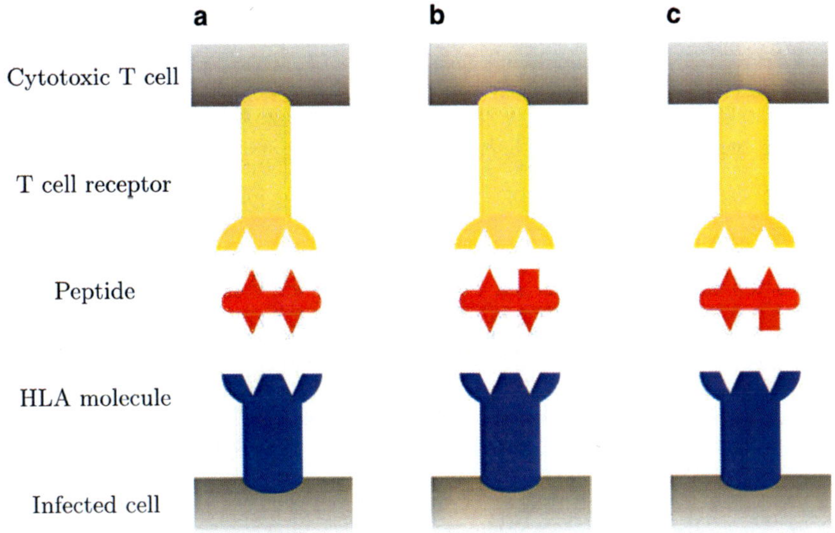

Fig. 19.5 (**a**) Activated immune response: Peptide from virus fits HLA and T cell receptor. Leading to a immune response. (**b**) Partial or no immune response: Peptide from mutated virus fits HLA but changed shape leads to a poor fit to T cell receptor. (**c**) No immune response: Peptide from mutated virus does not fit into HLA molecule

Generation of antigenic variants allows HIV to evolve continuously a in the face of immunological pressure created by by the CTL response [119]. However, alterations in the amino acid sequence can also influence the replication fitness of the virus [126, 127]. Consequently, the fixation of a particular escape mutation is a balance between the advantage conferred by escaping the CTL response and the replication fitness of the virus.

References

1. Hilleman MR (2004) Strategies and mechanisms for host and pathogen survival in acute and persistent viral infections. Proc Natl Acad Sci USA 101:14560–14566
2. Norrby E, Gollmar Y (1972) Appearance and persistence of antibodies against different virus components after regular measles infections. Infect Immun 6:240–247
3. Norrby E, Orvell C, Vandvik B, Cherry JD (1981) Antibodies against measles virus polypeptides in different disease conditions. Infect Immun 34:718–724
4. Deonarain R, Alcami A, Alexiou M, Dallman MJ, Gewert DR, Porter AC (2000) Impaired antiviral response and α/β interferon induction in mice lacking beta interferon. J Virol 74:3404–3409
5. Heron I, Berg K (1978) The immune regulatory function of interferon. Ugeskr Laeger 140:3210–3211
6. Protzer U, Nassal M, Chiang PW, Kirschfink M, Schaller H (1999) Interferon gene transfer by a hepatitis b virus vector efficiently suppresses wild-type virus infection. Proc Natl Acad Sci USA 96:10818–10823
7. Guidotti LG, Borrow P, Hobbs MV, Matzke B, Gresser I, Oldstone MB, Chisari FV (1996) Viral cross talk: intracellular inactivation of the hepatitis b virus during an unrelated viral infection of the liver. Proc Natl Acad Sci USA 93:4589–4594
8. Moskophidis D, Battegay M, Bruendler MA, Laine E, Gresser I, Zinkernagel RM (1994) Resistance of lymphocytic choriomeningitis virus to α/β interferon and to γ interferon. J Virol 68:1951–1955
9. Heron I, Hokland M, Berg K (1978) Enhanced expression of β2-microglobulin and HLA antigens on human lymphoid cells by interferon. Proc Natl Acad Sci USA 75:6215–6219
10. Biron CA (1997) Activation and function of natural killer cell responses during viral infections. Curr Opin Immunol 9:24–34
11. Cooper NR, Nemerow GR (1984) The role of antibody and complement in the control of viral infections. J Invest Dermatol 83:121s–127s
12. Levine B, Hardwick JM, Trapp BD, Crawford TO, Bollinger RC, Griffin DE (1991) Antibody-mediated clearance of αvirus infection from neurons. Science 254:856–860
13. Townsend AR, Skehel JJ (1982) Influenza a specific cytotoxic T-cell clones that do not recognize viral glycoproteins. Nature 300:655–657
14. Garboczi DN, Ghosh P, Utz U, Fan QR, Biddison WE, Wiley DC (1996) Structure of the complex between human T-cell receptor, viral peptide and HLA-A2. Nature 384:134–141
15. Garcia KC, Degano M, Stanfield RL, Brunmark A, Jackson MR, Peterson PA, Teyton L, Wilson IA (1996) An $\alpha\beta$ T cell receptor structure at 2.5 a and its orientation in the TCRMHC complex. Science 274:209–219
16. Goldmuntz E (2005) Digeorge syndrome: new insights. Clin Perinatol 32:963–978, ix–x
17. Bjorkman PJ, Saper MA, Samraoui B, Bennett WS, Strominger JL, Wiley DC (1987) Structure of the human class I histocompatibility antigen, HLA-A2. Nature 329:506–512
18. Bjorkman PJ, Saper MA, Samraoui B, Bennett WS, Strominger JL, Wiley DC (1987) The foreign antigen binding site and T cell recognition regions of class I histocompatibility antigens. Nature 329:512–518

19. Yewdell J, Reits E, Neefjes J (2003) Making sense of mass destruction: quantitating MHC class I antigen presentation. Nat Rev Immunol 3:952–961
20. Varshavsky A (2005) Regulated protein degradation. Trends Biochem Sci 30:283–286
21. Baumeister W, Walz J, Zuhl F, Seemuller E (1998) The proteasome: paradigm of a self-compartmentalizing protease. Cell 92:367–380
22. Tanaka K (2009) The proteasome: overview of structure and functions. Proc Jpn Acad Ser B Phys Biol Sciences 85:12–36
23. Yoshimura T, Kameyama K, Takagi T, Ikai A, Tokunaga F, Koide T, Tanahashi N, Tamura T, Cejka Z, Baumeister W, et al (1993) Molecular characterization of the "26s" proteasome complex from rat liver. J Struct Biol 111:200–211
24. Groll M, Heinemeyer W, Jager S, Ullrich T, Bochtler M, Wolf DH, Huber R (1999) The catalytic sites of 20s proteasomes and their role in subunit maturation: a mutational and crystallographic study. Proc Natl Acad Sci USA 96:10976–10983
25. Tanaka K, Kasahara M (1998) The MHC class I ligand-generating system: roles of immunoproteasomes and the interferon-gamma-inducible proteasome activator pa28. Immunol Rev 163:161–176
26. Driscoll J, Brown MG, Finley D, Monaco JJ (1993) MHC-linked lmp gene products specifically alter peptidase activities of the proteasome. Nature 365:262–264
27. Gaczynska M, Rock KL, Spies T, Goldberg AL (1994) Peptidase activities of proteasomes are differentially regulated by the major histocompatibility complex-encoded genes for LMP2 and LMP7. Proc Natl Acad Sci USA 91:9213–9217
28. Sibille C, Gould KG, Willard-Gallo K, Thomson S, Rivett AJ, Powis S, Butcher GW, De Baetselier P (1995) LMP2$^+$ proteasomes are required for the presentation of specific antigens to cytotoxic T lymphocytes. Curr Opin Biol 5:923–930
29. Androlewicz MJ, Anderson KS, Cresswell P (1993) Evidence that transporters associated with antigen processing translocate a major histocompatibility complex class I-binding peptide into the endoplasmic reticulum in an ATP-dependent manner. Proc Natl Acad Sci USA 90:9130–9134
30. Cerundolo V, Alexander J, Anderson K, Lamb C, Cresswell P, McMichael A, Gotch F, Townsend A (1990) Presentation of viral antigen controlled by a gene in the major histocompatibility complex. Nature 345:449–452
31. Powis SJ, Townsend AR, Deverson EV, Bastin J, Butcher GW, Howard JC (1991) Restoration of antigen presentation to the mutant cell line rma-s by an MHClinked transporter. Nature 354:528–531
32. Beninga J, Rock KL, Goldberg AL (1998) Interferon-γ can stimulate post-proteasomal trimming of the n terminus of an antigenic peptide by inducing leucine aminopeptidase. J Biol Chem 273:18734–18742
33. Stoltze L, Schirle M, Schwarz G, Schroter C, Thompson MW, Hersh LB, Kalbacher H, Stevanovic S, Rammensee HG, Schild H (2000) Two new proteases in the MHC class I processing pathway. Nat Immunol 1:413–418
34. Saric T, Chang SC, Hattori A, York IA, Markant S, Rock KL, Tsujimoto M, Goldberg AL (2002) An IFN-γ-induced aminopeptidase in the ER, ERAP1, trims precursors to MHC class I-presented peptides. Nat Immunol 3:1169–1176
35. York IA, Chang SC, Saric T, Keys JA, Favreau JM, Goldberg AL, Rock KL (2002) The ER aminopeptidase ERAP1 enhances or limits antigen presentation by trimming epitopes to 8–9 residues. Nat Immunol 3:1177–1184
36. Ortmann B, Androlewicz MJ, Cresswell P (1994) MHC class I/β 2-microglobulin complexes associate with tap transporters before peptide binding. Nature 368:864–867
37. Androlewicz MJ, Ortmann B, van Endert PM, Spies T, Cresswell P (1994) Characteristics of peptide and major histocompatibility complex class I/$\beta_2$microglobulin binding to the transporters associated with antigen processing (tap1 and tap2). Proc Natl Acad Sci USA 91:12716–12720
38. Bangia N, Cresswell P (2005) Stoichiometric tapasin interactions in the catalysis of major histocompatibility complex class I molecule assembly. Immunology 114:346–353

39. Suh WK, Cohen-Doyle MF, Fruh K, Wang K, Peterson PA, Williams DB (1994) Interaction of MHC class I molecules with the transporter associated with antigen processing. Science 264:1322–1326
40. Meuer SC, Acuto O, Hussey RE, Hodgdon JC, Fitzgerald KA, Schlossman SF, Reinherz EL (1983) Evidence for the T3-associated 90k heterodimer as the T-cell antigen receptor. Nature 303:808–810
41. Borst J, van de Griend RJ, van Oostveen JW, Ang SL, Melief CJ, Seidman JG, Bolhuis RL (1987) A T-cell receptor γ/CD3 complex found on cloned functional lymphocytes. Nature 325:683–688
42. Blumberg RS, Alarcon B, Sancho J, McDermott FV, Lopez P, Breitmeyer J, Terhorst C (1990) Assembly and function of the T cell antigen receptor. requirement of either the lysine or arginine residues in the transmembrane region of the α chain. J Biol Chem 265:14036–14043
43. Cantrell D, Davies AA, Londei M, Feldman M, Crumpton MJ (1987) Association of phosphorylation of the T3 antigen with immune activation of T lymphocytes. Nature 325:540–542
44. Patel MD, Samelson LE, Klausner RD (1987) Multiple kinases and signal transduction. phosphorylation of the T cell antigen receptor complex. J Biol Chem 262:5831–5838
45. Salter RD, Benjamin RJ, Wesley PK, Buxton SE, Garrett TP, Clayberger C, Krensky AM, Norment AM, Littman DR, Parham P (1990) A binding site for the T-cell co-receptor CD8 on the α_3 domain of HLA-A2. Nature 345:41–46
46. Connolly JM, Hansen TH, Ingold AL, Potter TA (1990) Recognition by CD8 on cytotoxic T lymphocytes is ablated by several substitutions in the class I α_3 domain: CD8 and the T-cell receptor recognize the same class I molecule. Proc Natl Acad Sci USA 87:2137–2141
47. al Ramadi BK, Jelonek MT, Boyd LF, Margulies DH, Bothwell AL (1995) Lack of strict correlation of functional sensitization with the apparent affinity of MHC/peptide complexes for the TCR. J Immunol 155:662–673
48. Holler PD, Kranz DM (2003) Quantitative analysis of the contribution of TCR/pepMHC affinity and CD8 to T cell activation. Immunity 18:255–264
49. Geiger B, Rosen D, Berke G (1982) Spatial relationships of microtubule-organizing centers and the contact area of cytotoxic T lymphocytes and target cells. J Cell Biol 95:137–143
50. Monks CR, Freiberg BA, Kupfer H, Sciaky N, Kupfer A (1998) Three-dimensional segregation of supramolecular activation clusters in T cells. Nature 395:82–86
51. Grakoui A, Bromley SK, Sumen C, Davis MM, Shaw AS, Allen PM, Dustin ML (1999) The immunological synapse: a molecular machine controlling T cell activation. Science 285: 221–227
52. Itoh N, Yonehara S, Ishii A, Yonehara M, Mizushima S, Sameshima M, Hase A, Seto Y, Nagata S (1991) The polypeptide encoded by the cdna for human cell surface antigen fas can mediate apoptosis. Cell 66:233–243
53. Smith CA, Davis T, Anderson D, Solam L, Beckmann MP, Jerzy R, Dower SK, Cosman D, Goodwin RG (1990) A receptor for tumor necrosis factor defines an unusual family of cellular and viral proteins. Science 248:1019–1023
54. Thornberry NA, Lazebnik Y (1998) Caspases: enemies within. Science 281:1312–1316
55. Halestrap A (2005) Biochemistry: a pore way to die. Nature 434:578–579
56. Darmon AJ, Nicholson DW, Bleackley RC (1995) Activation of the apoptotic protease CPP32 by cytotoxic T-cell-derived granzyme B. Nature 377:446–448
57. Geiger B, Avnur Z, Schlessinger J (1982) Restricted mobility of membrane constituents in cell-substrate focal contacts of chicken fibroblasts. J Cell Biol 93:495–500
58. Dennert G, Podack ER (1983) Cytolysis by H-2-specific T killer cells. Assembly of tubular complexes on target membranes. J Exp Med 157:1483–1495
59. Dourmashkin RR, Deteix P, Simone CB, Henkart P (1980) Electron microscopic demonstration of lesions in target cell membranes associated with antibody-dependent cellular cytotoxicity. Clin Exp Immunol 42:554–560
60. Podack ER, Dennert G (1983) Assembly of two types of tubules with putative cytolytic function by cloned natural killer cells. Nature 302:442–445
61. Stinchcombe JC, Bossi G, Booth S, Griffiths GM (2001) The immunological synapse of CTL contains a secretory domain and membrane bridges. Immunity 15:751–761

62. Kagi D, Ledermann B, Burki K, Seiler P, Odermatt B, Olsen KJ, Podack ER, Zinkernagel RM, Hengartner H (1994) Cytotoxicity mediated by T cells and natural killer cells is greatly impaired in perforin-deficient mice. Nature 369:31–37
63. Motyka B, Korbutt G, Pinkoski MJ, Heibein JA, Caputo A, Hobman M, Barry M, Shostak I, Sawchuk T, Holmes CF, Gauldie J, Bleackley RC (2000) Mannose 6-phosphate/insulin-like growth factor II receptor is a death receptor for granzyme b during cytotoxic T cell-induced apoptosis. Cell 103:491–500
64. Shi L, Mai S, Israels S, Browne K, Trapani JA, Greenberg AH (1997) Granzyme B (GraB) autonomously crosses the cell membrane and perforin initiates apoptosis and GraB nuclear localization. J Exp Med 185:855–866
65. Atkinson EA, Barry M, Darmon AJ, Shostak I, Turner PC, Moyer RW, Bleackley RC (1998) Cytotoxic T lymphocyte-assisted suicide. caspase 3 activation is primarily the result of the direct action of granzyme b. J Biol Chem 273:21261–21266
66. Medema JP, Toes RE, Scaffidi C, Zheng TS, Flavell RA, Melief CJ, Peter ME, Offringa R, Krammer PH (1997) Cleavage of flice (caspase-8) by granzyme b during cytotoxic T lymphocyte-induced apoptosis. Eur J Immunol 27:3492–3498
67. Kroemer G, Reed JC (2000) Mitochondrial control of cell death. Nat Med 6:513–519
68. Pereira RA, Simon MM, Simmons A (2000) Granzyme a, a noncytolytic component of $CD8^+$ cell granules, restricts the spread of herpes simplex virus in the peripheral nervous systems of experimentally infected mice. J Virol 74:1029–1032
69. Santamaria P (2001) Effector lymphocytes in autoimmunity. Curr Opin Immunol 13:663–669
70. Ostergaard HL, Kane KP, Mescher MF, Clark WR (1987) Cytotoxic T lymphocyte mediated lysis without release of serine esterase. Nature 330:71–72
71. Rouvier E, Luciani MF, Golstein P (1993) Fas involvement in Ca^{2+}-independent T cell-mediated cytotoxicity. J Exp Med 177:195–200
72. Lin R, Heylbroeck C, Pitha PM, Hiscott J (1998) Virus-dependent phosphorylation of the irf-3 transcription factor regulates nuclear translocation, transactivation potential, and proteasome-mediated degradation. Mol Cell Biol 18:2986–2996
73. Foster GR, Ackrill AM, Goldin RD, Kerr IM, Thomas HC, Stark GR (1991) Expression of the terminal protein region of hepatitis b virus inhibits cellular responses to interferons α and γ and double-stranded RNA. Proc Natl Acad Sci USA 88:2888–2892
74. Novick D, Engelmann H, Wallach D, Rubinstein M (1989) Soluble cytokine receptors are present in normal human urine. J Exp Med 170:1409–1414
75. Hu FQ, Smith CA, Pickup DJ (1994) Cowpox virus contains two copies of an early gene encoding a soluble secreted form of the type II TNF receptor. Virology 204:343–356
76. Loparev VN, Parsons JM, Knight JC, Panus JF, Ray CA, Buller RM, Pickup DJ, Esposito JJ (1998) A third distinct tumor necrosis factor receptor of orthopoxviruses. Proc Natl Acad Sci USA 95:3786–3791
77. Lalani AS, Graham K, Mossman K, Rajarathnam K, Clark-Lewis I, Kelvin D, McFadden G (1997) The purified myxoma virus γ interferon receptor homolog m-t7 interacts with the heparin-binding domains of chemokines. J Virol 71:4356–4363
78. Mossman K, Upton C, McFadden G (1995) The myxoma virus-soluble interferon-γ receptor homolog, M-T7, inhibits interferon-γ in a species-specific manner. J Biol Chem 270: 3031–3038
79. Billstrom MA, Johnson GL, Avdi NJ, Worthen GS (1998) Intracellular signaling by the chemokine receptor us28 during human cytomegalovirus infection. J Virol 72:5535–5544
80. Charan S, Hengartner H, Zinkernagel RM (1987) Antibodies against the two serotypes of vesicular stomatitis virus measured by enzyme-linked immunosorbent assay: immunodominance of serotype-specific determinants and induction of asymmetrically cross-reactive antibodies. J Virol 61:2509–2514
81. Hilleman MR (2002) Realities and enigmas of human viral influenza: pathogenesis, epidemiology and control. Vaccine 20:3068–87
82. Pellegrin I, Legrand E, Neau D, Bonot P, Masquelier B, Pellegrin JL, Ragnaud JM, Bernard N, Fleury HJ (1996) Kinetics of appearance of neutralizing antibodies in 12 patients with primary

or recent HIV-1 infection and relationship with plasma and cellular viral loads. J Acquir Immune Defic Syndr Hum Retrovirol 11:438–447
83. Richman DD, Wrin T, Little SJ, Petropoulos CJ (2003) Rapid evolution of the neutralizing antibody response to HIV type 1 infection. Proc Natl Acad Sci USA 100:4144–4149
84. Wei X, Decker JM, Wang S, Hui H, Kappes JC, Wu X, Salazar-Gonzalez JF, Salazar MG, Kilby JM, Saag MS, Komarova NL, Nowak MA, Hahn BH, Kwong PD, Shaw GM (2003) Antibody neutralization and escape by HIV-1. Nature 422:307–312
85. Kawaoka Y, Krauss S, Webster RG (1989) Avian-to-human transmission of the pb1 gene of influenza a viruses in the 1957 and 1968 pandemics. J Virol 63:4603–4608
86. Schafer JR, Kawaoka Y, Bean WJ, Suss J, Senne D, Webster RG (1993) Origin of the pandemic 1957 H2 influenza A virus and the persistence of its possible progenitors in the avian reservoir. Virology 194:781–788
87. Kanda K, Decker T, Aman P, Wahlstrom M, von Gabain A, Kallin B (1992) The ebna2-related resistance towards α interferon (IFN-α) in burkitt's lymphoma cells effects induction of IFN-induced genes but not the activation of transcription factor ISGF-3. Mol Cell Biol 12:4930–4936
88. Symons JA, Alcami A, Smith GL (1995) Vaccinia virus encodes a soluble type I interferon receptor of novel structure and broad species specificity. Cell 81:551–560
89. Burysek L, Yeow WS, Lubyova B, Kellum M, Schafer SL, Huang YQ, Pitha PM (1999) Functional analysis of human herpesvirus 8-encoded viral interferon regulatory factor 1 and its association with cellular interferon regulatory factors and p300. J Virol 73:7334–7342
90. Upton C, Macen JL, Schreiber M, McFadden G (1991) Myxoma virus expresses a secreted protein with homology to the tumor necrosis factor receptor gene family that contributes to viral virulence. Virology 184:370–382
91. Benedict CA, Butrovich KD, Lurain NS, Corbeil J, Rooney I, Schneider P, Tschopp J, Ware CF (1999) Cutting edge: a novel viral tnf receptor superfamily member in virulent strains of human cytomegalovirus. J Immunol 162:6967–6970
92. Liu X, Fitzgerald K, Kurt-Jones E, Finberg R, Knipe DM (2008) Herpesvirus tegument protein activates NF-κB signaling through the traf6 adaptor protein. Proc Natl Acad Sci USA 105:11335–11339
93. Khanna R, Burrows SR, Kurilla MG, Jacob CA, Misko IS, Sculley TB, Kieff E, Moss DJ (1992) Localization of Epstein-Barr virus cytotoxic T cell epitopes using recombinant vaccinia: implications for vaccine development. J Exp Med 176:169–176
94. Levitskaya J, Coram M, Levitsky V, Imreh S, Steigerwald-Mullen PM, Klein G, Kurilla MG, Masucci MG (1995) Inhibition of antigen processing by the internal repeat region of the Epstein-Barr virus nuclear antigen-1. Nature 375:685–688
95. Anderson K, Cresswell P, Gammon M, Hermes J, Williamson A, Zweerink H (1991) Endogenously synthesized peptide with an endoplasmic reticulum signal sequence sensitizes antigen processing mutant cells to class I-restricted cell-mediated lysis. J Exp Med 174:489–492
96. Fruh K, Ahn K, Djaballah H, Sempe P, van Endert PM, Tampe R, Peterson PA, Yang Y (1995) A viral inhibitor of peptide transporters for antigen presentation. Nature 375:415–418
97. Ahn K, Gruhler A, Galocha B, Jones TR, Wiertz EJ, Ploegh HL, Peterson PA, Yang Y, Fruh K (1997) The er-luminal domain of the hcmv glycoprotein us6 inhibits peptide translocation by tap. Immunity 6:613–621
98. Ahn K, Angulo A, Ghazal P, Peterson PA, Yang Y, Fruh K (1996) Human cytomegalovirus inhibits antigen presentation by a sequential multistep process. Proc Natl Acad Sci USA 93:10990–10995
99. Jones TR, Wiertz EJ, Sun L, Fish KN, Nelson JA, Ploegh HL (1996) Human cytomegalovirus us3 impairs transport and maturation of major histocompatibility complex class I heavy chains. Proc Natl Acad Sci USA 93:11327–11333
100. Collins KL, Chen BK, Kalams SA, Walker BD, Baltimore D (1998) HIV-1 nef protein protects infected primary cells against killing by cytotoxic T lymphocytes. Nature 391:397–401
101. Williams M, Roeth JF, Kasper MR, Fleis RI, Przybycin CG, Collins KL (2002) Direct binding of human immunodeficiency virus type 1 nef to the major histocompatibility complex class I (MHC-I) cytoplasmic tail disrupts MHC-I trafficking. J Virol 76:12173–12184

102. Beck S, Barrell BG (1988) Human cytomegalovirus encodes a glycoprotein homologous to MHC class-i antigens. Nature 331:269–272
103. Browne H, Smith G, Beck S, Minson T (1990) A complex between the MHC class I homologue encoded by human cytomegalovirus and β 2 microglobulin. Nature 347:770–772
104. Fahnestock ML, Johnson JL, Feldman RM, Neveu JM, Lane WS, Bjorkman PJ (1995) The MHC class I homolog encoded by human cytomegalovirus binds endogenous peptides. Immunity 3:583–590
105. Reyburn HT, Mandelboim O, Vales-Gomez M, Davis DM, Pazmany L, Strominger JL (1997) The class I MHC homologue of human cytomegalovirus inhibits attack by natural killer cells. Nature 386:514–517
106. Shisler J, Yang C, Walter B, Ware CF, Gooding LR (1997) The adenovirus E3-10.4K/14.5K complex mediates loss of cell surface fas (CD95) and resistance to fas-induced apoptosis. J Virol 71:8299–8306
107. Tollefson AE, Hermiston TW, Lichtenstein DL, Colle CF, Tripp RA, Dimitrov T, Toth K, Wells CE, Doherty PC, Wold WS (1998) Forced degradation of fas inhibits apoptosis in adenovirus-infected cells. Nature 392:726–730
108. Barker DD, Berk AJ (1987) Adenovirus proteins from both e1b reading frames are required for transformation of rodent cells by viral infection and dna transfection. Virology 156: 107–121
109. Kao CC, Yew PR, Berk AJ (1990) Domains required for in vitro association between the cellular p53 and the adenovirus 2 e1b 55k proteins. Virology 179:806–814
110. Dyson N, Howley PM, Munger K, Harlow E (1989) The human papilloma virus-16 E7 oncoprotein is able to bind to the retinoblastoma gene product. Science 243:934–937
111. Irmler M, Thome M, Hahne M, Schneider P, Hofmann K, Steiner V, Bodmer JL, Schroter M, Burns K, Mattmann C, Rimoldi D, French LE, Tschopp J (1997) Inhibition of death receptor signals by cellular flip. Nature 388:190–195
112. Komiyama T, Quan LT, Salvesen GS (1996) Inhibition of cysteine and serine proteinases by the cowpox virus serpin CRMA. Adv Exp Med Biol 389:173–176
113. Tewari M, Telford WG, Miller RA, Dixit VM (1995) Crma, a poxvirus-encoded serpin, inhibits cytotoxic T-lymphocyte-mediated apoptosis. J Biol Chem 270:22705–22708
114. Zhou Q, Salvesen GS (2000) Viral caspase inhibitors CrmA and p35. Methods Enzymol 322:143–154
115. Elena SF, Sanjuan R (2005) Adaptive value of high mutation rates of RNA viruses: separating causes from consequences. J Virol 79:11555–11558
116. Jenkins GM, Rambaut A, Pybus OG, Holmes EC (2002) Rates of molecular evolution in rna viruses: a quantitative phylogenetic analysis. J Mol Evol 54:156–165
117. Drake JW, Charlesworth B, Charlesworth D, Crow JF (1998) Rates of spontaneous mutation. Genetics 148:1667–1686
118. Borrow P, Lewicki H, Wei X, Horwitz MS, Peffer N, Meyers H, Nelson JA, Gairin JE, Hahn BH, Oldstone MB, Shaw GM (1997) Antiviral pressure exerted by HIV-1-specific cytotoxic T lymphocytes (CTLs) during primary infection demonstrated by rapid selection of CTL escape virus. Nat Med 3:205–211
119. Phillips RE, Rowland-Jones S, Nixon DF, Gotch FM, Edwards JP, Ogunlesi AO, Elvin JG, Rothbard JA, Bangham CR, Rizza CR, et al (1991) Human immunodeficiency virus genetic variation that can escape cytotoxic T cell recognition. Nature 354:453–459
120. Beekman NJ, van Veelen PA, van Hall T, Neisig A, Sijts A, Camps M, Kloetzel PM, Neefjes JJ, Melief CJ, Ossendorp F (2000) Abrogation of CTL epitope processing by single amino acid substitution flanking the c-terminal proteasome cleavage site. J Immunol 164: 1898–1905
121. Zimbwa P, Milicic A, Frater J, Scriba TJ, Willis A, Goulder PJ, Pillay T, Gunthard H, Weber JN, Zhang HT, Phillips RE (2007) Precise identification of a human immunodeficiency virus type 1 antigen processing mutant. J Virol 81:2031–2038
122. Feeney ME, Tang Y, Roosevelt KA, Leslie AJ, McIntosh K, Karthas N, Walker BD, Goulder PJ (2004) Immune escape precedes breakthrough human immunodeficiency virus

type 1 viremia and broadening of the cytotoxic T-lymphocyte response in an HLA-B27-positive long-term-nonprogressing child. J Virol 78:8927–8930
123. Kelleher AD, Long C, Holmes EC, Allen RL, Wilson J, Conlon C, Workman C, Shaunak S, Olson K, Goulder P, Brander C, Ogg G, Sullivan JS, Dyer W, Jones I, McMichael AJ, Rowland-Jones S, Phillips RE (2001) Clustered mutations in HIV-1 gag are consistently required for escape from HLA-B27-restricted cytotoxic T lymphocyte responses. J Exp Med 193:375–386
124. Price DA, Goulder PJ, Klenerman P, Sewell AK, Easterbrook PJ, Troop M, Bangham CR, Phillips RE (1997) Positive selection of HIV-1 cytotoxic T lymphocyte escape variants during primary infection. Proc Natl Acad Sci USA 94:1890–1895
125. Goulder PJ, Phillips RE, Colbert RA, McAdam S, Ogg G, Nowak MA, Giangrande P, Luzzi G, Morgan B, Edwards A, McMichael AJ, Rowland-Jones S (1997) Late escape from an immunodominant cytotoxic t-lymphocyte response associated with progression to aids. Nat Med 3:212–217
126. Asquith B, McLean AR (2007) In vivo $CD8^+$ T cell control of immunodeficiency virus infection in humans and macaques. Proc Natl Acad Sci USA 104:6365–6370
127. Martinez-Picado J, Prado JG, Fry EE, Pfafferott K, Leslie A, Chetty S, Thobakgale C, Honeyborne I, Crawford H, Matthews P, Pillay T, Rousseau C, Mullins JI, Brander C, Walker BD, Stuart DI, Kiepiela P, Goulder P (2006) Fitness cost of escape mutations in p24 gag in association with control of human immunodeficiency virus type 1. J Virol 80: 3617–3623

Index

C. Molina-París and G. Lythe (eds.), *Mathematical Models and Immune Cell Biology*,
DOI 10.1007/978-1-4419-7725-0,